Energy Transfers in Fluid Flows

Turbulence remains an unsolved problem due to the complex nonlinear interactions among a large number of multiscale structures. For hydrodynamic turbulence, Kolmogorov's theory provides quantitative measures of energy contents of the fluid structures and energy flux. However, this theory based on real space description does not quantify various scale-by-scale energy transfers. In addition, generalisations of Kolmogorov's theory to more complex system—magnetohydrodynamic and buoyancy-driven turbulence, anisotropic flows, etc.—are quite involved. Fortunately, spectral or Fourier space description, which is the theme of this monograph, overcomes some of these deficiencies.

To quantify energy transfers in turbulence, Verma and his collaborators developed a set of important spectral tools: mode-to-mode energy transfers, various energy fluxes, shell-to-shell and ring-to-ring energy transfers, variable energy flux, etc. These diagnostics are quite general, and they do not require the flows to be homogeneous or isotropic, as is assumed in Kolmogorov's theory. Researchers have used the above tools to compute important quantities for various turbulent systems. This analysis provides many valuable insights, e.g., energy transfers responsible for the magnetic energy growth in astrophysical bodies, dynamics of turbulent thermal convection.

In this monograph, Verma systematically describes various techniques of energy transfers in turbulence. These tools include mode-to-mode transfers, fluxes, shell-to-shell and ring-to-ring transfers of energy, as well as enstrophy, kinetic helicity, and magnetic helicity. After developing the framework, the author employs them to turbulence in hydrodynamics, magnetohydrodynamics, passive scalar, buoyancy-driven flows, rotating flows, active scalar and vector, compressible flows, etc. The book describes energy transfers in both real and Fourier space, but the focus is on the latter. The energy transfer diagnostics provide many valuable insights, which have been described throughout the book.

Mahendra K. Verma is a leading researcher in the field of turbulence. Presently he is a Professor at the Physics Department of Indian Institute of Technology Kanpur, India. He is a recipient of Swarnajayanti fellowship, INSA Teachers Award, and Dr APJ Abdul Kalam Cray HPC Award. In addition to this book, he has authored *Introduction to Mechanics* and *Physics of Buoyant Flows: From Instabilities to Turbulence.* His other research interests include nonlinear dynamics, high-performance computing, and non-equilibrium statistical physics.

Energy Transfers in Fluid Flows

Multiscale and Spectral Perspectives

Mahendra K. Verma

CAMBRIDGE
UNIVERSITY PRESS

CAMBRIDGE
UNIVERSITY PRESS

University Printing House, Cambridge CB2 8BS, United Kingdom

One Liberty Plaza, 20th Floor, New York, NY 10006, USA

477 Williamstown Road, Port Melbourne, VIC 3207, Australia

314 to 321, 3rd Floor, Plot No.3, Splendor Forum, Jasola District Centre, New Delhi 110025, India

79 Anson Road, #06–04/06, Singapore 079906

Cambridge University Press is part of the University of Cambridge.

It furthers the University's mission by disseminating knowledge in the pursuit of education, learning and research at the highest international levels of excellence.

www.cambridge.org
Information on this title: www.cambridge.org/9781107176195

First published 2019

Printed in India by Rajkamal Electric Press

A catalogue record for this publication is available from the British Library

ISBN 978-1-107-17619-5 Hardback

To the seekers of knowledge, who guide the world
To the workers of the world, who run the world
and
To the commoners of the world, who keep humanity alive

Contents

Preface

The Navier–Stokes equations, first formulated in 1822, still remain unsolved in the turbulent regime. Over time, scientists and engineers have been studying turbulence in more complex systems, such as thermal convection, scalar flows, flows in magnetofluids, boundary layers, flow across bluff bodies, etc. Turbulent flows have many complex issues—boundary layers, inhomogeneity and anisotropy, multiscale complex structures, spatiotemporal correlations among various fields, etc. In this book we address several fundamental issues of turbulence such as energy transfers in hydrodynamic, magnetohydrodynamic, scalar, compressible, anisotropic, and other forms of turbulence; variable energy flux; effects of enstrophy and kinetic helicity on turbulence.

Kolmogorov (1941) formulated a theory of hydrodynamic turbulence according to which the energy flux that flows from large scales to small scale is related to the third order structure function. This real-space formulation is applicable to isotropic and homogeneous turbulence, and its extension to anisotropic turbulence is difficult. On the other hand, such computations are relatively easier in spectral space. In 1959, Kraichnan derived a formula for the combined energy transfer for a wavenumber triad of hydrodynamic turbulence. This formalism is useful, but its scope is limited. For example, the combined energy transfer formula does not yield the energy transfers that are responsible for the generation of the large-scale magnetic field in dynamos. Curiously, most research works on turbulence report energy spectrum, but energy transfers such as energy flux have not been discussed often.

We (Gaurav Dar, Vinayak Eswaran, and I) started to work on quantifying energy transfers in magnetohydrodynamic (MHD) turbulence way back in 1999–2000. During the investigation, we discovered a very nice formalism called *mode-to-mode energy transfer* that helped us quantify all the energy transfers, energy fluxes, and shell-to-shell transfers in hydrodynamic and MHD turbulence, as well as in dynamos. After publication of this work (with some difficulties) in 2001, we extended this formalism to the following systems and obtained many interesting insights (also see Acknowledgments):

- Passive scalar turbulence
- Helical turbulence
- Buoyancy-driven turbulence—stably stratified turbulence and turbulent thermal convection
- Large-scale and small-scale dynamos
- Anisotropic turbulence
- Rotating turbulence
- Quasi-static MHD turbulence
- Field-theoretic computation of energy transfers, in particular, energy fluxes
- Role of energy transfers in pattern formation

In addition, Daniele Carati and Olivier Debliquy performed energy transfer computations of MHD turbulence using large-resolution data; Bogdan Teaca, Franck Jenko, and coworkers extended this formalism to gyrokinetic plasma turbulence; and Rodion Stepanov and Franck Plunian employed similar techniques to shell models of turbulence.

Several research groups have computed the energy fluxes in various turbulent systems using the numerical procedure of Dar et al. and Frisch. Yet, the underlying formalism of mode-to-mode energy transfer has remained somewhat unnoticed. This book is an attempt to present this powerful formalism and its applications to a variety of turbulent flows in a coherent and general framework so as to reach the turbulence and fluid community. In the monograph I present the fluxes, shell-to-shell and ring-to-ring transfers of kinetic energy, kinematic helicity, enstrophy, etc., for hydrodynamic, MHD, and scalar turbulence. I also describe popular turbulence phenomenologies, and their verification using numerical simulations and experiments. The earlier derivation of mode-to-mode formalism by Dar et al. had an uncomfortable issue of *circulating energy transfer*. In the present book, using tensor analysis and the structure of the nonlinear terms, I show that the circulating energy transfer is zero, thus resolving the ambiguity of the earlier derivation. The present monograph illustrates the usefulness of the energy transfers for understanding turbulence.

In addition to the earlier works (listed above) on the energy transfers, the book contains several new works on energy transfers, which are as follows:

- Derivation of mode-to-mode transfer formulas for kinetic helicity, magnetic helicity, and enstrophy
- Energy transfers in compressible turbulence
- Energy transfers in electron MHD
- Energy transfers in Craya–Herring and helical basis
- Energy transfers in tensor flows, and in spherical flows
- Field-theoretic computation of energy transfers in Craya–Herring basis

The presentations on compressible turbulence, electron MHD, and flows on a sphere are quite elementary and nascent. I believe that a lot more can be done in these areas.

For a thematic presentation, the book is divided into four parts. The first part deals with hydrodynamic flows—two-dimensional and three-dimensional turbulence, helical turbulence, enstrophy transfers, Craya–Herring and helical basis, and Kolmogorov theory of turbulence. In the second part, flows with a scalar that includes passive scalar, stably stratified flows, thermal convection, and binary fluid are covered. The third part is dedicated to flows with vectors, namely MHD and electron MHD. In the last part, compressible and Burgers turbulence, shell model, flows in spherical geometry, etc., have been discussed. The monograph includes works of many researchers, including those of my collaborators, who are listed in Acknowledgments. Yet, many important works could not be included due to lack of space.

Though the book focuses on energy transfers in fluid flows, one can observe that energy transfers would be useful for studying the nature of interactions in other nonequilibrium systems. For example, we expect the detailed balance to be broken in a generic nonequilibrium system, which would lead to directional energy transfers in time, space, and across scales. These transfers also yield direction to time in terms of evolution of the system. We believe that such analysis would be very useful for studying many nonequilibrium systems—quantum turbulence, financial market, coarsening in material science, etc. I do hope that the ideas of energy transfers would be employed to these systems.

Lastly, I hope that the book will be useful to students and researchers. I would greatly welcome comments, criticisms, and ideas on the contents of the book at my email mkv@iitk.ac.in.

Mahendra Verma
IIT Kanpur

Acknowledgments

The present monograph includes research works of many of my collaborators. It is great pleasure to acknowledge their contributions.

The *mode-to-mode energy transfer* formalism was discovered in collaboration with Gaurav Dar and Vinayak Eswaran around 1999. This work became part of Garav's PhD thesis in which he analyzed energy transfers in 2D MHD turbulence. I am very grateful to Gaurav and Eswaran for this collaboration.

Daniele Carati and Olivier Debliquy performed numerical computation of energy fluxes for 3D MHD turbulence. Later, Bogdan Teaca, Bernard Knaepen, and Thomas Lessinnes joined this collaboration, and we worked out energy transfers in anisotropic MHD turbulence, shell model, dynamo, etc. We also developed ring spectrum and ring-to-ring energy transfers for anisotropic turbulence. Arvind Ayyer, Amar Chandra, and V. Avinash collaborated with me to employ field-theoretic techniques to analyze helical turbulence and locality in turbulence.

Abhishek Kumar and I worked out energy transfers in stably stratified turbulence and turbulent thermal convection. Using variable energy flux we could show that stably stratified turbulence with moderate stratification shows Bolgiano–Obukhov scaling, while thermal convection is closer to Kolmogorov's scaling for hydrodynamic turbulence. I thank Abhishek Kumar for this collaboration. For thermal convection, I also collaborated with Ambrish Pandey, Pankaj Mishra, Mani Chandra, Anando Chatterjee, Dinesh Nath, Shashwat Bhattacharjee, Sumit Vashishtha, Krishna Kumar, Supriyo Paul, Pinaki Pal, and Roshan Samuel. I also benefited by my collaboration with Jai Sukhatme, Anirban Guha, and Shadab Alam on stably stratified turbulence.

The energy transfer studies on quasi-static MHD turbulence was performed in collaboration with Sandeep Reddy and Raghwendra Kumar, while those on dynamo and dynamo transition was with Rohit Kumar, Ravi Samtaney, and Rakesh Yadav. In recent times, I collaborated with Manohar Sharma, Sagar Chakraborty, and Abhishek Kumar on rotating turbulence, and with Rohith Jayaraman and Akanksha Gupta on two-dimensional turbulence. In addition, I worked with Alex Alexakis and Sita Sundar on anisotropic MHD turbulence.

Recently, I collaborated with Rodion Stepanov, Peter Frick, Valeriy Titov, Andrei Teimurazov, Andrei Sukhanovskii, Franck Plunian, Shubhadeep Sadhukhan, Abhishek Kumar, Satyajit Barman, and Ravi Samtaney under an Indo-Russian project. During this project we developed the framework of kinetic helicity, magnetic helicity, cross helicity, and enstrophy transfers, and performed numerical computations of these quantities in hydrodynamic, MHD, and convective turbulence. We also performed a very high resolution hydrodynamic turbulence simulation, and analyzed energy spectra and fluxes in the inertial–dissipation range.

I started my research career under the mentorship of Melvyn Goldstein and Aaron Roberts. Many thanks to them. I am grateful to Daniele Carati, Ravi Samtaney, Franck Plunian, and Rodion Stepanov for long-term and fruitful collaborations, and camaraderie. I also thank Stephan Fauve, Katepalli Sreenivasan (Sreeni), and Jayant Bhattacharya for many interesting suggestions, encouragement, and constant help throughout my career.

I am very grateful to my other friends and collaborators—William Matthaeus, Jörg Schumacher, G. Ravindra Kumar, Mustansir Barma, Arnab Rai Choudhuri, Arul Lakshminarayan, Edgar Knobloch, Peter Davidson, Detlef Lohse, K.-Q. Xia, Maurice Rossi, Jaywant Arakari, Supratik Banerjee, Avinash Khare, Amita Das, and Shayak Bhattacharya—for many useful ideas and discussions. In addition, I thank my family members—my parents, sisters, and brothers-in-law—and friends—Anurag Gupta, Reema Mittal, Vishal Garg, Vivek Garg—for encouragement and kind comments on the manuscript. A large body of the monograph was written in CCD at IIT Kanpur, and I owe a great deal to the CCD staff for kind hospitality and an excellent environment. I also thank all the members of our turbulence group for proof reading, and Manmohan Dewbanshi for assistance on printing, etc. I thank Abhishek Kumar, Mohammad Anas, Manohar Sharma, Shubhadeep Sadhukhan, Shashwat Bhattacharya, Shadab Alam, Akanksha Gupta, and Roshan Bhaskaran for preparing several figures of the book. I also thank the editors of Cambridge University Press for constant support in editing and various suggestions.

All the energy transfer algorithms are implemented in our spectral software TARANG. Anando Chatterjee and other group members of our turbulence group contributed to the development and testing of this software, for which I am greatly indebted to them. I am thankful to King Abdullah University of Science and Technology (KAUST) for providing computational access on SHAHEEN I and SHAHEEN II through project K1052. Some of the numerical simulations were also performed on HPC2010, HPC2013, CHAOS clusters of IIT Kanpur, and PARAM YUVA of CDAC.

I gratefully acknowledge the financial and computational support from research projects—Swaranajayanti fellowship from Department of Science and Technology; PLANEX/PHY/2015239 from ISRO; SERB/F/3279/2013-14 from Science and Engineering Research Board; Indo-Russian project (DST-RSF) INT/RUS/RSF/P-03 and RSF-16-41-02012; and research grants from DoRD and DoRA, IIT Kanpur. I benefited from the participation in the scientific meetings: *Summer school and discussion meeting on Buoyancy-driven Flows* and *Turbulence from Angstorms to light* held at ICTS Bengaluru in 2017, and *Turbulence Mixing and Beyond 2017* held at ICTP Trieste; I thank ICTS and ICTP for the same.

Mahendra Verma
IIT Kanpur

Part I

FORMALISM OF ENERGY TRANSFERS

Chapter 1

Introduction

Fluid flows exhibit rich behavior—regular flow in viscous regime; patterns and chaos in weakly nonlinear regime; and turbulence in strongly nonlinear regime. Such complex behavior arises due to nonlinearity. A good understanding of some of these features, specially turbulence, is still lacking even after sustained efforts extending over two centuries.

Hydrodynamics broadly deals with the analysis of the Navier–Stokes equations and their generalization to magnetohydrodynamics, convection, passive scalars, rotating flows, etc. The large-scale structures in such flows contribute to various instabilities, patterns, and chaos, but turbulence is governed by structures at all scales. Researchers have studied these phenomena using the energy exchanges among various Fourier modes, a topic which has not been studied in great detail. In this book we attempt to fill this gap by focusing on energy transfers in fluid flows, and their role in turbulence. Note, however, that energy transfer formalism developed in the book are general, and they could be applied to study pattern formation and chaos as well.

The energy transfers in fluid flows are generic, and they arise in linear as well as in nonlinear systems. A pendulum is an example of a linear system that exhibits a periodic exchange of kinetic energy and potential energy during its oscillations. Similar transfers occur in surface gravity waves and in internal gravity waves. Also note that in the unstable configuration of the pendulum (vertically standing up), it is the potential energy that drives the kinetic energy and makes the pendulum unstable. The buoyancy-driven instabilities—thermal instability, Rayleigh–Taylor instability—have energy transfers similar to that in the unstable configuration of the pendulum.

In nonlinear hydrodynamics, the energy transfers among the interacting modes arise due to quadratic nonlinearities, for example, $\mathbf{u}\cdot\nabla\mathbf{u}$ in fluids, $\mathbf{u}\cdot\nabla T$ in thermal convection, $\nabla\times(\mathbf{u}\times\mathbf{b})$ and $\mathbf{b}\cdot\nabla\mathbf{b}$ in magnetohydrodynamics, where $\mathbf{u},\mathbf{b}$, and T are respectively the velocity, magnetic, and temperature fields. Therefore, we develop a general formulation to compute the energy transfers that help us formulate interactions among the participating modes, energy flux, shell-to-shell and ring-to-ring energy transfers, etc. The energy transfers among the large-scale modes help us understand patterns and chaos. On the other hand, the energy flux, and the shell-to-shell and ring-to-ring energy transfers provide useful information about turbulence in hydrodynamics, magnetohydrodynamics, and buoyancy-driven flows. In this book, we present these topics thematically, duly emphasizing the common features among them. Also, the present book focusses on the nonlinear energy transfers.

In the next section we introduce a generic nonlinear equation that contains basic features of nonlinear energy transfers.

1.1 A Generic Nonlinear Equation

In a nonlinear system, nonlinearity induces interactions among the Fourier modes. These interactions are illustrated using the following equation:

$$\frac{\partial}{\partial t}f(x,t) = \frac{\partial^2}{\partial x^2}f(x,t) + a[f(x,t)]^2, \tag{1.1}$$

where a is a constant. In this equation, f^2 is the nonlinear term. We assume the field f to be contained in a periodic box of length L. We decompose the field $f(x)$ into Fourier modes $f(k)$:

$$f(x,t) = \sum_k f(k,t)\exp(ikx), \tag{1.2}$$

where $k = 2n\pi/L$ with n as an integer. The corresponding inverse transform is given by

$$f(k,t) = \frac{1}{L}\int_0^L dx f(x,t)\exp(-ikx). \tag{1.3}$$

Using this definition, we derive an equation for $f(k,t)$ in Fourier space. We start with

$$\frac{\partial}{\partial t}\sum f(k)\exp(ikx) = \frac{\partial^2}{\partial x^2}\sum f(k)\exp(ikx) + a\sum f(p)\exp(ipx)\sum f(q)\exp(iqx). \tag{1.4}$$

Using the orthogonality relation,

$$\frac{1}{L}\int_0^L dx\exp(i(p+q-k)x) = \delta_{p+q,k}, \tag{1.5}$$

we obtain the requisite equation:

$$\frac{d}{dt}f(k) = -k^2 f(k) + a\sum_p f(p)f(k-p). \tag{1.6}$$

Note that the nonlinear term has become a convolution in Fourier space.

In the absence of the nonlinear term, Eq. (1.6) reduces to diffusion equation:

$$\frac{d}{dt}f(k) = -k^2 f(k) \tag{1.7}$$

that has a very simple solution—each Fourier mode decays exponentially as

$$f(k,t) = f(0)\exp(-k^2 t). \tag{1.8}$$

However, the nonlinear term, $a\sum_p f(p)f(k-p)$, of Eq. (1.6), couples the Fourier modes in an intricate manner. In Eq. (1.6), nonlinear interactions involve three Fourier modes, for example, $f(k)$, $f(p)$, and $f(k-p)$. In general, there is no analytical solution for Eq. (1.6) because of the nonlinearity. Note however that analytical solution may be possible for some specific cases.

The modal energy $|f(k)|^2/2$ is the energy of the Fourier mode $f(k)$. By multiplying Eq. (1.6) with $f^*(k)$ and summing the resulting equation with its complex conjugate, we obtain an equation for the time evolution of $|f(k)|^2$ as

$$\frac{d}{dt}\frac{1}{2}|f(k)|^2 = -2k^2\frac{1}{2}|f(k)|^2 + a\Re\left[\sum_p f(k-p)f(p)f^*(k)\right], \tag{1.9}$$

where $\Re(.)$ stands for the real part of the argument. The last term of Eq. (1.9) represents the nonlinear energy transfer. The modal energy decays exponentially in the absence of such transfer. Note, however, that the nonlinear term could stabilize the modal energy or make it time dependent. We remark that on many occasions, energy is more convenient to analyze than the Fourier modes. For example, $|f(k)|^2$ is real in contrast to complex $f(k)$.

In the aforementioned equation, the decaying term $-k^2|f(k)|^2$ provides energy transfer from *coherent energy* to *heat* or *dissipation*. The nonlinear term $\Re(f(k-p)f(p)f^*(k))$ represents the energy transfers from the modes $f(p)$ and $f(k-p)$ to $f(k)$. Note that these energy transfers occur among various length scales, and it differs from the energy transport in real space, for example, $\int f^2 d\mathbf{S}$ where $d\mathbf{S}$ is an elemental area.

The aforementioned behavior is a generic feature of nonlinear partial differential equations including the Navier–Stokes equation. In the present book, we will present energy transfer formalism for hydrodynamics, magnetohydrodynamics, scalar flows, buoyancy-driven flows, etc. Most of our discussion will be focused on incompressible flows.

Before we start our detailed discussion on energy transfers, we outline the contents of the book.

1.2 Outline of the Book

The book is divided into four parts. Part I contains discussion on various aspects of hydrodynamics, while Parts II–IV cover more complex applications. Though the book focuses on energy transfers, we also cover physics of the flows under consideration—hydrodynamics, scalar flows, magnetohydrodynamics, thermal convection, etc.

In Part I, Chapters 2 and 3 describe the governing equations of hydrodynamics in real and Fourier spaces respectively. The formalism of energy transfers in hydrodynamics is developed in Chapter 4. Chapters 5 and 7 describe phenomenologies of three-dimensional and two-dimensional hydrodynamic turbulence respectively. Here we describe energy fluxes and other related diagnostics. Chapter 6 contains a discussion on enstrophy transfers in turbulent flows.

In Chapter 8, we describe helical turbulence. Chapter 9 introduces Craya–Herring and helical basis that are very useful for describing flow properties. In Chapter 10 we briefly describe field-theoretic treatment of energy transfers. These computations provide many useful insights into turbulence dynamics. Chapter 11 contains formulation of energy transfer in anisotropic turbulence.

Energy transfers in real space—energy flow from one scale to another—can be described using structure function. This scheme was first derived by Kolmogorov (1941a,b,c). In Chapter 12, we describe this formalism, and relate it to the energy transfers in Fourier space. Note, however, that the present book focuses on energy transfers in Fourier space.

In nature and in engineering, we encounter flows that advect scalar, vector, or tensor fields along with it. There are many examples of such flows, as listed in Table 1.1. In subsequent chapters, we will cover properties, especially related to the energy transfers, of such flows.

Table 1.1 Examples of scalars, vectors, and tensors in flows.

Field	Examples
Scalar	Densities of dust particles and pollution, fluid density, temperature, binary fluid
Vector	Magnetic field, flock velocity, dipoles
Tensor	Polymer, elastic fluid

In Part II (Chapters 13 to 17), we systematically cover flows with scalars. We describe properties of passive scalar flows, stably stratified flows, and thermal convection. We also cover binary fluid mixture in this part.

In Part III (Chapters 18 to 25), we describe flows with vectors that include passive vector flows, magnetohydrodynamics (MHD), and electron MHD. The topics covered are energy transfers in MHD, turbulence phenomenologies, dynamo, etc.

In Part IV (Chapters 26 and 31), we include miscellaneous topics that go beyond scalar and vector flows. These topics include tensor flows, rotating turbulence, shell model of turbulence, Burgers turbulence, and compressible turbulence. The last two topics in this part are on compressible hydrodynamics—an exception to the theme of the book. We conclude in Chapter 32.

There are a large number of excellent textbooks and reviews on various kinds of turbulent flows (hydrodynamic, magnetohydrodynamic, buoyancy-driven). In Table 1.2 we list some of the important references, especially those connected to our discussion. The present book attempts to highlight common features in the aforementioned flows using energy transfers, energy fluxes, etc., and show that energy transfer formalism provides important insights into turbulent flows.

Table 1.2 Key references.

Hydrodynamic turbulence	Leslie (1973); McComb (1990, 2014); Frisch (1995); Mathieu and Scott (2000); Pope (2000); Davidson (2004); Lesieur (2008); Kraichnan (1959); Alexakis and Biferale (2018)
Craya–Herring basis	Sagaut and Cambon (2008); Lesieur (2008)
Buoyancy-driven turbulence	Manneville (2014); Verma (2018); Lohse and Xia (2010); Verma et al. (2017)
MHD turbulence	Biskamp (2003); Davidson (2017); Verma (2004)
Shell model	Ditlevsen (2010); Plunian et al. (2012); Verma and Kumar (2016)
Anisotropy	Davidson (2013); Verma (2017, 2018)

Chapter 2

Basics of Hydrodynamics

In this chapter we describe the governing equations of incompressible hydrodynamics in real space. We will also discuss the conserved quantities of hydrodynamics.

2.1 Governing Equations of Incompressible Flows

Under continuum[1] and incompressibility[2] approximations, the fluid flow is described by the incompressible Navier–Stokes (NS) equations:

$$\frac{\partial \mathbf{u}}{\partial t} + \mathbf{u} \cdot \nabla \mathbf{u} = -\frac{1}{\rho}\nabla p + \frac{1}{\rho}\mathbf{F}_u + \nu \nabla^2 \mathbf{u}, \tag{2.1a}$$

$$\nabla \cdot \mathbf{u} = 0, \tag{2.1b}$$

where $\mathbf{u}(\mathbf{r}, t)$ is the velocity field, $p(\mathbf{r}, t)$ is the pressure field, $\mathbf{F}_u(\mathbf{r}, t)$ is the external force field, ν is the kinematic viscosity, and ρ is the density of the fluid. In the incompressible limit, ρ can be treated as a constant.

[1] In continuum approximation, the mean free path length between two collisions experienced by a microscopic particle is much smaller than the system size.

[2] In an incompressible flow, in the comoving frame with a fluid parcel, the density ρ is constant, i.e., $d\rho/dt = 0$. From the continuity equation,

$$\frac{\partial \rho}{\partial t} + \nabla \cdot (\rho \mathbf{u}) = 0,$$

we deduce that

$$\frac{d\rho}{dt} + \rho \nabla \cdot \mathbf{u} = 0.$$

Hence, the incompressibility approximation yields $\nabla \cdot \mathbf{u} = 0$.

It is customary to nondimensionalize the Navier–Stokes equations. We use the large-scale velocity U_0 as the velocity scale, the system size L as the length scale, and L/U_0 as the time scale. Consequently, the nondimensional variables are

$$x' = x/L; \quad \mathbf{u}' = \mathbf{u}/U_0; \quad t' = t/(L/U_0); \tag{2.2}$$

$$\nabla' = L\nabla; \quad p' = p/(\rho U_0^2); \quad \mathbf{F}'_u = L\mathbf{F}_u/(\rho U_0^2). \tag{2.3}$$

In terms of these nondimensional variables, the Navier–Stokes equations are

$$\frac{\partial \mathbf{u}'}{\partial t'} + \mathbf{u}' \cdot \nabla' \mathbf{u}' = -\nabla' p + \mathbf{F}'_u + \frac{1}{\mathrm{Re}} \nabla'^2 \mathbf{u}', \tag{2.4a}$$

$$\nabla' \cdot \mathbf{u}' = 0, \tag{2.4b}$$

where

$$\mathrm{Re} = \frac{U_0 L}{\nu} \tag{2.5}$$

is a nondimensional number called the *Reynolds number*. It is a measure of the ratio of the nonlinear term $(\mathbf{u} \cdot \nabla \mathbf{u})$ and the viscous term $(\nu \nabla^2 \mathbf{u})$. Turbulent flows have $\mathrm{Re} \gg 1$, while viscous flows have $\mathrm{Re} \lesssim 1$. For convenience, in the above equations we drop the primes and set $\mathrm{Re} = 1/\nu$. Hence, the new equations are

$$\frac{\partial \mathbf{u}}{\partial t} + \mathbf{u} \cdot \nabla \mathbf{u} = -\nabla p + \mathbf{F}_u + \nu \nabla^2 \mathbf{u}, \tag{2.6a}$$

$$\nabla \cdot \mathbf{u} = 0. \tag{2.6b}$$

The aforementioned equations in tensorial form are

$$\frac{\partial u_i}{\partial t} + \partial_j (u_j u_i) = -\partial_i p + F_{u,i} + \nu \partial_{jj} u_i, \tag{2.7a}$$

$$\partial_i u_i = 0. \tag{2.7b}$$

Here, the indices i, j take values $\{1, 2, 3\}$ or $\{x, y, z\}$. In Eqs. (2.7) we follow Einstein convention according to which the repeated indices are summed. We will study the properties of Navier–Stokes equations in subsequent discussions.

For incompressible flows, the pressure field is a dependent variable; it is a function of the velocity field. By taking a divergence of Eq. (2.6a), and by assuming that the force is divergence free, i.e., $\nabla \cdot \mathbf{F}_u = 0$, we obtain the following equation for the pressure:

$$-\nabla^2 p = \nabla \cdot [(\mathbf{u} \cdot \nabla)\mathbf{u}], \tag{2.8}$$

which is Poisson's equation. Thus, the pressure field is uniquely determined given $\mathbf{u}$ and the boundary condition.

A useful identity for fluid flows is

$$\frac{Df}{Dt} = \frac{\partial f}{\partial t} + (\mathbf{u} \cdot \nabla) f, \tag{2.9}$$

where $f(\mathbf{r}, t)$ could be a scalar or a vector field. Here, Df/Dt is the material derivative or the total derivative of f, as it measures the time derivative of f while moving with the flow.

2.2 Vorticity and its Equation

The vorticity field $\boldsymbol{\omega}$ is defined as the curl of the velocity field, i.e.,

$$\boldsymbol{\omega} = \nabla \times \mathbf{u}. \tag{2.10}$$

Physically, vorticity is related to the circulation or curliness of the velocity field. In a rotating flow, a cyclone has a finite vorticity.

To derive a dynamical equation for the vorticity field, we employ a vector identity

$$(\mathbf{u} \cdot \nabla)\mathbf{u} = -\mathbf{u} \times \boldsymbol{\omega} + \nabla \frac{u^2}{2} \tag{2.11}$$

and substitute it in Eq. (2.6a) that yields

$$\frac{\partial \mathbf{u}}{\partial t} = -\nabla \left(p + \frac{u^2}{2} \right) + \mathbf{u} \times \boldsymbol{\omega} + \mathbf{F}_u + \nu \nabla^2 \mathbf{u}. \tag{2.12}$$

Taking a curl of the aforementioned equation yields the following dynamical equation for $\boldsymbol{\omega}$:

$$\frac{\partial \boldsymbol{\omega}}{\partial t} = \nabla \times (\mathbf{u} \times \boldsymbol{\omega}) + \mathbf{F}_\omega + \nu \nabla^2 \boldsymbol{\omega}, \tag{2.13}$$

where $\mathbf{F}_\omega = \nabla \times \mathbf{F}_u$. The vector identities

$$\nabla \times (\mathbf{A} \times \mathbf{B}) = \mathbf{A} \nabla \cdot \mathbf{B} - \mathbf{B} \nabla \cdot \mathbf{A} + (\mathbf{B} \cdot \nabla)\mathbf{A} - (\mathbf{A} \cdot \nabla)\mathbf{B}, \tag{2.14}$$

$$\nabla \cdot \boldsymbol{\omega} = \nabla \cdot (\nabla \times \mathbf{u}) = 0, \tag{2.15}$$

and the incompressibility condition $\nabla \cdot \mathbf{u} = 0$ yields

$$\frac{\partial \boldsymbol{\omega}}{\partial t} + (\mathbf{u} \cdot \nabla)\boldsymbol{\omega} = \boldsymbol{\omega} \cdot \nabla \mathbf{u} + \mathbf{F}_\omega + \nu \nabla^2 \boldsymbol{\omega}, \tag{2.16}$$

or

$$\frac{D\boldsymbol{\omega}}{Dt} = \boldsymbol{\omega} \cdot \nabla \mathbf{u} + \mathbf{F}_\omega + \nu \nabla^2 \boldsymbol{\omega}. \tag{2.17}$$

This is another equation for $\boldsymbol{\omega}$.

For a two-dimensional flow,

$$\mathbf{u} = u_x(x,y)\hat{x} + u_y(x,y)\hat{y}. \tag{2.18}$$

Hence,

$$\boldsymbol{\omega} = \omega \hat{z} = (\partial_x u_y - \partial_y u_x)\hat{z} \tag{2.19}$$

is perpendicular to the plane of the velocity field, and

$$\frac{D\omega}{Dt} = \frac{\partial \omega}{\partial t} + (\mathbf{u} \cdot \nabla)\omega = F_\omega + \nu \nabla^2 \omega, \tag{2.20}$$

where

$$F_\omega = (\partial_x F_y - \partial_y F_x). \tag{2.21}$$

For a force-free and inviscid ($\nu = 0$) flow,

$$\frac{D\omega}{Dt} = 0. \tag{2.22}$$

This result is related to Kelvin's circulation theorem, which will be discussed in Section 2.4.

2.3 Quadratic Quantities in Hydrodynamics

For fluid flows, we construct several scalars using the velocity and vorticity fields:

$$\text{Kinetic energy density } E_u(\mathbf{r}) = \frac{1}{2}u^2, \tag{2.23a}$$

$$\text{Kinetic helicity density } H_K(\mathbf{r}) = \frac{1}{2}\mathbf{u} \cdot \boldsymbol{\omega}, \tag{2.23b}$$

$$\text{Enstrophy density } E_\omega(\mathbf{r}) = \frac{1}{2}\omega^2. \tag{2.23c}$$

We sum these quantities over the volume and obtain:

$$\text{Total kinetic energy } E_u = \frac{1}{2}\int d\mathbf{r}u^2, \tag{2.24a}$$

$$\text{Total kinetic helicity } H_K = \frac{1}{2}\int d\mathbf{r}(\mathbf{u}\cdot\boldsymbol{\omega}), \tag{2.24b}$$

$$\text{Total enstrophy } E_\omega = \frac{1}{2}\int d\mathbf{r}\omega^2. \tag{2.24c}$$

The total kinetic energy (KE in short), the total kinetic helicity, and the total enstrophy are called *quadratic quantities* since they are products of two field variables. Note that the kinetic energy density of a fluid flow is typically defined as $\rho u^2/2$. Since ρ is a constant, we set it to unity.

In the following discussion, we derive equations for the aforementioned quadratic quantities. By taking a scalar product of Eq. (2.6a) with $\mathbf{u}$, we obtain the following dynamical equation for kinetic energy:

$$\frac{\partial}{\partial t}\frac{u^2}{2} + \nabla\cdot\left(\frac{1}{2}u^2\mathbf{u}\right) = -\nabla\cdot(p\mathbf{u}) + \mathbf{F}_u\cdot\mathbf{u} + \nu\mathbf{u}\cdot\nabla^2\mathbf{u}. \tag{2.25}$$

Using the following identity (Spiegel, 2010)[3]

$$\mathbf{u}\cdot\nabla^2\mathbf{u} = -\omega^2 + \nabla\cdot[\mathbf{u}\times\boldsymbol{\omega}], \tag{2.26}$$

we rewrite the equation for kinetic energy as

$$\frac{\partial}{\partial t}\frac{u^2}{2} + \nabla\cdot\left[\frac{u^2}{2}\mathbf{u}\right] = -\nabla\cdot(p\mathbf{u} - \nu\mathbf{u}\times\boldsymbol{\omega}) + \mathbf{F}_u\cdot\mathbf{u} - \nu\omega^2. \tag{2.27}$$

Using a similar procedure, the Navier–Stokes equations and Eq. (2.16) yield the following dynamical equations for enstrophy and kinetic helicity (with $\mathbf{F}_u = 0$):

[3]**Proof:**

$$\begin{aligned}
-\omega^2 + \nabla\cdot[\mathbf{u}\times\boldsymbol{\omega}] &= -\epsilon_{kij}\epsilon_{klm}(\partial_i u_j)(\partial_l u_m) + \epsilon_{ijk}\partial_i[u_j\omega_k] \\
&= -\epsilon_{kij}\epsilon_{klm}(\partial_i u_j)(\partial_l u_m) + \epsilon_{kij}\epsilon_{klm}\partial_i[u_j(\partial_l u_m)] \\
&= -(\partial_i u_j)(\partial_i u_j) + (\partial_i u_j)(\partial_j u_i) + \partial_i(u_j\partial_i u_j) - \partial_i(u_j\partial_j u_i) \\
&= u_j\partial_i^2 u_j - u_j\partial_j\partial_i u_i \\
&= \mathbf{u}\cdot\nabla^2\mathbf{u} - \mathbf{u}\cdot\nabla(\nabla\cdot\mathbf{u}).
\end{aligned}$$

For incompressible flows, $\nabla\cdot\mathbf{u} = 0$; hence

$-\omega^2 + \nabla\cdot[\mathbf{u}\times\boldsymbol{\omega}] = \mathbf{u}\cdot\nabla^2\mathbf{u}.$

$$\frac{\partial}{\partial t}\frac{\omega^2}{2} + \nabla \cdot \left[\frac{\omega^2}{2}\mathbf{u}\right] = \boldsymbol{\omega} \cdot [\boldsymbol{\omega} \cdot \nabla \mathbf{u}] + \nu \boldsymbol{\omega} \cdot (\nabla^2 \boldsymbol{\omega})$$

and

$$\begin{aligned} \frac{D}{Dt}(\mathbf{u} \cdot \boldsymbol{\omega}) &= \boldsymbol{\omega} \cdot \frac{D\mathbf{u}}{Dt} + \mathbf{u} \cdot \frac{D\boldsymbol{\omega}}{Dt} \\ &= -\nabla \cdot [p\boldsymbol{\omega}] + \mathbf{u} \cdot (\boldsymbol{\omega} \cdot \nabla)\mathbf{u} + \nu \left[\boldsymbol{\omega} \cdot (\nabla^2 \mathbf{u}) + \mathbf{u} \cdot (\nabla^2 \boldsymbol{\omega})\right]. \end{aligned} \quad (2.28)$$

Now, let us write the KE equation in an integral form. We perform volume integrals of each term of Eq. (2.25) over an arbitrary volume that yields

$$\frac{d}{dt}\int \frac{u^2}{2} d\mathbf{r} = -\oint \left(\frac{1}{2}u^2\mathbf{u}\right) \cdot d\mathbf{S} - \oint (p\mathbf{u}) \cdot d\mathbf{S} + \int d\mathbf{r}(\mathbf{F}_u \cdot \mathbf{u}) + \int d\mathbf{r}\nu\mathbf{u} \cdot \nabla^2 \mathbf{u}, \quad (2.29)$$

where

$$\text{Kinetic energy flux (real space)} = \frac{1}{2}u^2\mathbf{u}, \quad (2.30a)$$

$$\text{Pressure energy flux (real space)} = p\mathbf{u}, \quad (2.30b)$$

$$\text{Viscous dissipation rate (real space)} = \nu\mathbf{u} \cdot \nabla^2 \mathbf{u}, \quad (2.30c)$$

$$\text{Energy feed rate by forcing (real space)} = \mathbf{F}_u \cdot \mathbf{u}. \quad (2.30d)$$

A physical interpretation of Eq. (2.29) is that the rate of change of kinetic energy in a volume equals the sum of the four terms of Eq. (2.29). Viscous dissipation always depletes KE. However, the other terms either enhance or deplete KE depending on their sign. The term $\mathbf{F}_u \cdot \mathbf{u}$ can take both positive or negative values. The KE and pressure energy fluxes redistribute kinetic and pressure energies from one region to another, while viscous dissipation occurs locally.

The energetics arguments discussed up to now are for real space. In Section 4.4, we will provide complimentary energy balance arguments for Fourier space. The energetics in Fourier space provides scale-by-scale energy transfers and energy flux in hydrodynamics. In Chapter 12, we will describe how structure function also captures energy transfers.

Enstrophy has a similar integral form:

$$\frac{d}{dt}\int \frac{\omega^2}{2} d\mathbf{r} = -\oint \left(\frac{1}{2}\omega^2\mathbf{u}\right) \cdot d\mathbf{S} + \int \boldsymbol{\omega} \cdot [\boldsymbol{\omega} \cdot \nabla \mathbf{u}] d\mathbf{r} + \int d\mathbf{r}\nu\boldsymbol{\omega} \cdot \nabla^2 \boldsymbol{\omega}. \quad (2.31)$$

The enstrophy flux $\int (\omega^2\mathbf{u}/2) \cdot d\mathbf{S}$ transfers enstrophy from one region to another. However, the second term of the right-hand-side (RHS) of Eq. (2.31) represents the enstrophy production due to vortex stretching by the velocity field. The integral form of kinetic helicity can be derived in a similar manner.

Before closing this section, we prove a vector identity. For a scalar field f, in a fixed volume,

$$\frac{d}{dt}\int f d\mathbf{r} = \lim_{\Delta t \to 0} \frac{1}{\Delta t}\int [f(\mathbf{r}, t+\Delta t) - f(\mathbf{r}, t)] d\mathbf{r} = \int \frac{\partial f}{\partial t} d\mathbf{r}, \tag{2.32}$$

where $d\mathbf{r}$ represents a volume element that is assumed to be fixed in the aforementioned process. If the velocity field or f satisfies vanishing or periodic boundary condition, then

$$\int (\mathbf{u}\cdot\nabla) f d\mathbf{r} = \int \nabla\cdot(\mathbf{u}f) d\mathbf{r} = \int f\mathbf{u}\cdot d\mathbf{S} = 0. \tag{2.33}$$

Therefore, for $\mathbf{u}$ or f satisfying vanishing or periodic boundary condition,

$$\frac{d}{dt}\int f d\mathbf{r} = \int \left[\frac{\partial f}{\partial t} + (\mathbf{u}\cdot\nabla) f\right] d\mathbf{r} = \int \frac{Df}{Dt} d\mathbf{r}. \tag{2.34}$$

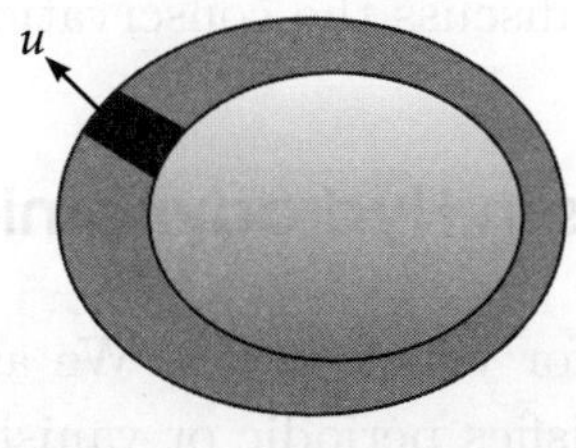

Figure 2.1 Initial volume (light grey region) expands to a larger volume (dark grey region) due to the velocity field. A volume element dr is represented by a dark patch.

We prove another useful theorem:

$$\frac{D}{Dt} d\mathbf{r} = (\nabla\cdot\mathbf{u}) d\mathbf{r}. \tag{2.35}$$

Proof: Consider an arbitrary volume within a flow. See Fig. 2.1 for an illustration. In a small time dt, the volume changes due to the velocity field at the surface:

$$V(t+dt) - V(t) = \oint (\mathbf{u}dt)\cdot d\mathbf{S}. \tag{2.36}$$

Hence,

$$\frac{D}{Dt}\int d\mathbf{r} = \oint \mathbf{u}\cdot d\mathbf{S}. \tag{2.37}$$

An application of Gauss theorem yields

$$\frac{D}{Dt}\int d\mathbf{r} = \int (\nabla\cdot\mathbf{u})d\mathbf{r}. \tag{2.38}$$

This equation is evaluated for an arbitrary volume. Hence, we take a small volume element $d\mathbf{r}$ for which $(\nabla\cdot\mathbf{u})$ is constant. Therefore,

$$\frac{D}{Dt}d\mathbf{r} = (\nabla\cdot\mathbf{u})d\mathbf{r}.$$

Thus, we prove the above theorem.

We can also prove this theorem by taking $d\mathbf{r} = dxdydz$ and then employing $Ddx/Dt = dx\partial u_x/\partial x$, etc. For incompressible flows, $\nabla\cdot\mathbf{u} = 0$. Hence, according to Eq. (2.35), volume of an incompressible fluid element does not change with time, as expected.

In the next section we will discuss the conservation laws in hydrodynamics.

2.4 Conservation Laws in Hydrodynamics

We consider an inviscid flow for which $\nu = 0$. We assume that the external force, $\mathbf{F}_u = 0$, and that the flow satisfies periodic or vanishing boundary conditions. For such flows, in three dimensions (3D), the total kinetic energy and the total kinetic helicity are conserved. However, in two dimensions (2D), the total kinetic energy and the total enstrophy are conserved. In this section, we discuss these conservation laws along with several others.

We start with Eq. (2.29) and employ the integral over a volume with periodic or vanishing boundary conditions, and take $\nu = 0$ and $\mathbf{F}_u = 0$. Under these approximations, the surface integrals and the viscous dissipation term vanish, hence we obtain

$$\frac{d}{dt}\int \frac{u^2}{2}d\mathbf{r} = 0, \tag{2.39}$$

which is the statement of *conservation of total kinetic energy* for inviscid flows. Note that this proof is valid for both 2D and 3D flows.

Now we prove the conservation of kinetic helicity in 3D. For the same, we employ Eqs. (2.28, 2.34) with $\nu = 0$, and obtain

$$\begin{aligned}
\frac{d}{dt}\int \frac{1}{2}\mathbf{u}\cdot\boldsymbol{\omega} d\mathbf{r} &= \frac{1}{2}\int \frac{D}{Dt}(\mathbf{u}\cdot\boldsymbol{\omega}) d\mathbf{r} \\
&= \frac{1}{2}\int \left(\boldsymbol{\omega}\cdot\frac{D\mathbf{u}}{Dt} + \mathbf{u}\cdot\frac{D\boldsymbol{\omega}}{Dt}\right) d\mathbf{r} \\
&= \frac{1}{2}\int \left[\boldsymbol{\omega}\cdot(-\nabla p) + \mathbf{u}\cdot(\boldsymbol{\omega}\cdot\nabla\mathbf{u})\right] d\mathbf{r} \\
&= \frac{1}{2}\int \left[\left(\frac{u^2}{2} - p\right)\boldsymbol{\omega}\right]\cdot d\mathbf{S} \\
&= 0
\end{aligned}$$

under the periodic or vanishing boundary condition. Thus, we prove the *conservation of kinetic helicity* for inviscid flows. In 2D, kinetic helicity is trivially zero since the vorticity and velocity fields are perpendicular to each other. Hence, it is trivially conserved.

Now we show why enstrophy is not conserved in 3D, but is conserved in 2D. The equation for total enstrophy with $\nu = 0$ is

$$\frac{d}{dt}\int \frac{\omega^2}{2} d\mathbf{r} = -\oint \left(\frac{1}{2}\omega^2\mathbf{u}\right)\cdot d\mathbf{S} + \int d\mathbf{r}\boldsymbol{\omega}\cdot\left[(\boldsymbol{\omega}\cdot\nabla)\mathbf{u}\right]. \tag{2.40}$$

The first term of the RHS is zero for the periodic or vanishing boundary condition. However, the second term is nonzero, and it corresponds to the enstrophy production due to vortex stretching. Hence, enstrophy is not conserved in 3D hydrodynamics. Note however that in 2D hydrodynamics, $(\boldsymbol{\omega}\cdot\nabla)\mathbf{u} = 0$, and hence enstrophy is conserved.

So far we dealt with conservation of quadratic quantities. Now, we will discuss the following non-quadratic conserved quantity called *circulation:*

$$\Gamma(t) = \oint \mathbf{u}\cdot d\mathbf{l}. \tag{2.41}$$

The aforementioned integral is defined for a closed and comoving contour; here $d\mathbf{l}$ is the line element. An application of Stokes' theorem yields

$$\Gamma(t) = \int \boldsymbol{\omega}\cdot d\mathbf{S}, \tag{2.42}$$

where the surface integral is performed over any surface whose edge is the aforementioned contour. In the inviscid limit, the circulation defined here is conserved. This is *Kelvin's circulation theorem.*

Proof: The time derivative of the circulation is[4]

$$\begin{aligned} \frac{d}{dt}\oint \mathbf{u}\cdot d\mathbf{l} &= \oint \frac{D\mathbf{u}}{Dt}\cdot d\mathbf{l} + \mathbf{u}\cdot\frac{D d\mathbf{l}}{Dt} \\ &= \oint (-\nabla p)\cdot d\mathbf{l} + \mathbf{u}\cdot[d\mathbf{l}\cdot\nabla\mathbf{u}] \\ &= \oint -dp + du^2/2 = 0. \end{aligned} \tag{2.43}$$

Hence, circulation is conserved. Q.E.D.

Example 2.1: Prove that for an inviscid and incompressible two-dimensional flow, the integral $\int \omega^n dA$ over a closed and comoving contour is conserved. Here, the exponent n is a constant, and dA is the area element.

Solution: The time derivative of $\int \omega^n dA$ is

$$\begin{aligned} \frac{d}{dt}\int \omega^n dA &= \int \frac{D\omega^n}{Dt} dA + \int \omega^n \frac{DdA}{Dt} \\ &= \int n\omega^{n-1}\frac{D\omega}{Dt} dA + \int \omega^n \nabla\cdot\mathbf{u} dA \\ &= \int n\omega^{n-1}(\nu\nabla^2\omega) dA + 0 \\ &= 0 \end{aligned}$$

because $\nabla\cdot\mathbf{u} = 0$ and $\nu = 0$.

In the aforementioned derivation, we employ $DdA/Dt = \nabla\cdot\mathbf{u}dA$ that follows from Eq. (2.35). For this case, we consider a cylindrical volume element $d\mathbf{r} = dAdh$, where height dh is perpendicular to the plane of the velocity field.

---

[4]In the derivation of Eq. (2.43), we employ $Dd\mathbf{l}/Dt = d\mathbf{u}$, which is proven as follows. Let us denote $d\mathbf{l}$ as separation between two points A and B whose position vectors are $\mathbf{r}$ and $\mathbf{r} + d\mathbf{l}$. Therefore,

$$\frac{Dd\mathbf{l}}{Dt} = \mathbf{u}_B - \mathbf{u}_A = d\mathbf{u} = d\mathbf{l}\cdot\nabla\mathbf{u}.$$

Example 2.2: Consider the following two-dimensional fluid flow in a two-dimensional box of size $\pi \times \pi$:

$$\mathbf{u} = 4A(\hat{x} \sin x \cos y - \hat{y} \cos x \sin y).$$

Show that the fluid is incompressible. Compute the vorticity, the average kinetic energy, and the average enstrophy for the flow. Plot the velocity field.

Solution: For the flow,

$$\nabla \cdot \mathbf{u} = 4A(\cos x \cos y - \cos x \cos y) = 0.$$

Hence, the fluid is incompressible. The vorticity of the fluid is $\boldsymbol{\omega} = \omega \hat{z}$, where

$$\omega = \partial_x u_y - \partial_y u_x = 8A \sin x \sin y.$$

The average kinetic energy of the flow is

$$\begin{aligned} E_u &= \frac{1}{2}\langle u^2 \rangle = \frac{1}{\pi^2} \int_0^\pi dx dy \frac{1}{2}(u_x^2 + u_y^2) \\ &= 8A^2(\langle \sin^2 x \cos^2 y \rangle + \langle \cos^2 x \sin^2 y \rangle) \\ &= 4A^2. \end{aligned}$$

The average enstrophy of the flow is

$$E_\omega = \frac{1}{2}\langle \omega^2 \rangle = 8A^2.$$

The vector plot of the velocity field is shown in Fig. 2.2.

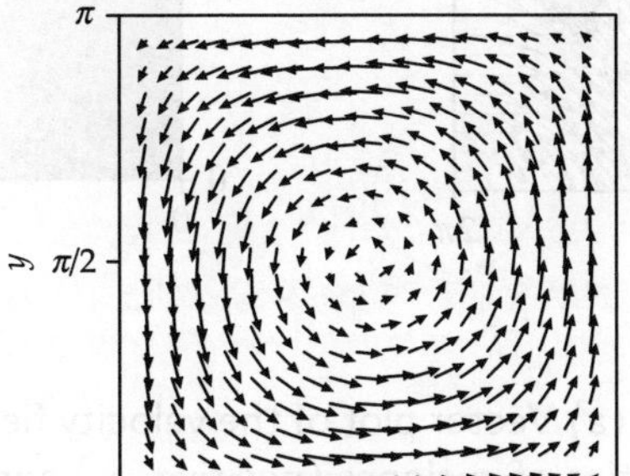

Figure 2.2 Example 2.2: Vector plot of the velocity field.

Example 2.3: In a periodic box of size $(2\pi)^2$, the velocity field of a 2D flow is $2(-\sin y, \sin x, 0)$. Compute the vorticity field, the stream function, the average kinetic energy and enstrophy of the flow. Plot the stream function and the velocity field.

Solution: Using the definition of the stream function ψ,

$$u_x = \frac{\partial \psi}{\partial y}; \quad u_y = -\frac{\partial \psi}{\partial x},$$

we derive

$$\psi = 2(\cos x + \cos y).$$

The vorticity of the field is

$$\boldsymbol{\omega} = \nabla \times \mathbf{u} = 2(\cos x + \cos y)\hat{z}.$$

The vector plot of the velocity field and the density plot of the stream function are shown in Fig. 2.3. Note that for this example, $\omega_z = \psi$. In the figures we observe a periodic lattice of cyclones (positive ω_z) and anticyclones (negative ω_z).

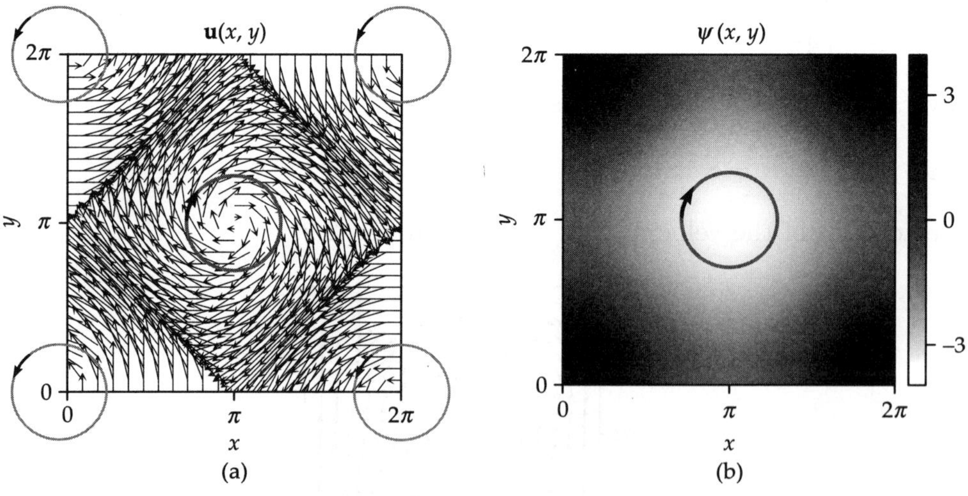

Figure 2.3 Example 2.3: (a) Vector plot of the velocity field. The figure illustrates a periodic lattice of cyclones (positive ω_z) and anticyclones (negative ω_z). (b) Density plot of the stream function, which is same as ω_z.

The average kinetic energy and enstrophy of the field are

$$E_u = \frac{1}{2}\langle u^2 \rangle = 2\langle \sin^2 y \rangle + 2\langle \cos^2 y \rangle = 2,$$

$$E_\omega = \frac{1}{2}\langle \omega^2 \rangle = 2\langle (\cos x + \cos y)^2 \rangle = 2.$$

The kinetic helicity of the flow is zero because $\mathbf{u} \perp \boldsymbol{\omega}$.

Example 2.4: How does the results of Example 2.3 change if the velocity field is $2(-\sin y, -\sin x, 0)$?

Solution: Following the same procedure as in Example 2.3, we can derive the stream function of the flow as

$$\psi = 2(-\cos x + \cos y).$$

Clearly, the velocity field is shifted horizontally (along $\hat{x}$) by π, while the properties of the flow remains the same.

Example 2.5: Consider a 2D flow with three components of the velocity field in a periodic box of size $(2\pi)^2$. The velocity field is $2(-\sin y, \sin x, \cos x + \cos y)$. Make a vector plot of $\mathbf{u}_\perp = (u_x, u_y)$, and density plot of u_z. What are the vorticity field and kinetic helicity of the flow?

Solution: The vorticity field $\nabla \times \mathbf{u}$ is

$$\boldsymbol{\omega} = (-2\sin y, 2\sin x, 2(\cos x + \cos y)).$$

Hence, the average kinetic helicity of the field is

$$H_K = \frac{1}{2}\langle \mathbf{u} \cdot \boldsymbol{\omega} \rangle = 4.$$

Note that $\boldsymbol{\omega} = \mathbf{u}$; hence, the flow field is maximally helical.

The vector plot of the velocity fields $\mathbf{u}_\perp = (u_x, u_y)$, and the density plot of u_z (which is same as ω_z) is the same as that plotted in Fig. 2.3. There is an anticyclone at the centre (π, π), and cyclones at the corners. In addition, in the centre, both ω_z and u_z are negative.

Example 2.6: If the velocity field of Example 2.5 was $2(-\sin y, \sin x, -\cos x - \cos y)$, how does the result change?

Solution: For the velocity field $2(-\sin y, \sin x, -\cos x - \cos y)$, the vorticity field $\nabla \times \mathbf{u}$ is

$$\boldsymbol{\omega} = (2\sin y, -2\sin x, 2(\cos x + \cos y)).$$

Clearly, $\boldsymbol{\omega} = -\mathbf{u}$; hence, the flow field is maximally helical with negative kinetic helicity. The average kinetic helicity of the field is

$$H_K = \frac{1}{2}\langle \mathbf{u} \cdot \boldsymbol{\omega} \rangle = -4.$$

The vector plot of the velocity field $\mathbf{u}_\perp = (u_x, u_y)$ is same as that of Fig. 2.3(a), while u_z is negative of that shown in Fig. 2.3(b). Thus, in the centre of the figure, $\omega_z > 0$, but $u_z < 0$. This corresponds to maximal helicity with a negative sign.

Example 2.7: Consider the following fluid flow in a 3D box of size $\pi \times \pi \times \pi$:

$$\begin{aligned} \mathbf{u} &= 4C(\hat{x}\sin x \cos z - \hat{z}\cos x \sin z) + 4B(\hat{y}\sin y \cos z - \hat{z}\cos y \sin z) \\ &\quad + 8A(-\hat{x}\sin x \cos y \cos 2z - \hat{y}\cos x \sin y \cos 2z + \hat{z}\cos x \cos y \sin 2z). \end{aligned}$$

Show that the fluid is incompressible. Also, compute the average kinetic energy of the flow.

Solution: We find that $\nabla \cdot \mathbf{u} = 0$, hence the flow is incompressible. The average kinetic energy of the field is

$$E_u = \frac{1}{2}\langle u^2 \rangle = 4C^2 + 4B^2 + 12A^2.$$

In the next chapter we will describe the fluid equations in Fourier space.

Further Reading

There are many excellent textbooks on fluid dynamics. See for example, Kundu et al. (2015), Choudhuri (1998), Tritton (1988), and Landau and Lifshitz (1987).

Exercises

1. Consider a 2D flow with $\mathbf{u}(\rho, \phi) = (1/\rho)\hat{\phi}$. Sketch the velocity field, and compute its curl and divergence. Is the flow incompressible?
2. Show that the Navier–Stokes equation is Galilean invariant.
3. Consider the following flow field in a 2D box of size $[\pi, \pi]$:

$$\mathbf{u} = 4B(\hat{y}\sin 2x \cos 2y - \hat{y}\cos 2x \sin 2y) + 4A(\hat{x}\sin x \cos y - \hat{y}\cos x \sin y).$$

Plot the vector field for $A = 1, B = 1/2$, and for $A = -1, B = 1/2$. Compute the vorticity, total kinetic energy, total enstrophy, and total kinetic helicity of the flow.

Chapter 3

Fourier Space Description of Hydrodynamics

In fluid flows, energy transfer across multiple scales is nicely described using Fourier modes. In the present chapter we will describe flow equations and associated quantities (e.g., kinetic energy) in Fourier space. In the linear limit, the Fourier modes evolve independently. Nonlinear interactions introduce coupling among the Fourier modes that facilitates energy transfers among them. In the next chapter, we will describe these energy transfers among Fourier modes.

We begin the present chapter with the definition and properties of Fourier transform.

3.1 Fourier Transform and its Properties

Fourier series is used to represent a velocity field in a periodic box.[1] Such representation is very useful for studying turbulence properties away from the walls.

We consider a periodic box of size $L_x \times L_y \times L_z$. In Fourier representation, the velocity field is expanded using the Fourier modes as follows:

$$\mathbf{u}(\mathbf{r},t) = \sum_{\mathbf{k}} \mathbf{u}(\mathbf{k},t) \exp(i\mathbf{k}\cdot\mathbf{r}), \tag{3.1}$$

[1]In this book, we consider Fourier transform of space coordinate $\mathbf{r}$ only, but not of time.

where $\mathbf{r}$ is the real space coordinate, $\mathbf{k} = (k_x, k_y, k_z)$ is the wavenumber with

$$k_x = \frac{2l\pi}{L_x}; \quad k_y = \frac{2m\pi}{L_y}; \quad k_z = \frac{2n\pi}{L_z}. \tag{3.2}$$

Here l, m, n are integers (from $-\infty$ to ∞). The variables $\mathbf{u}(\mathbf{k}, t)$ is Fourier amplitude of the velocity field. The associated inverse transform is

$$\mathbf{u}(\mathbf{k}, t) = \frac{1}{L_x L_y L_z} \int d\mathbf{r}[\mathbf{u}(\mathbf{r}, t) \exp(-i\mathbf{k} \cdot \mathbf{r})], \tag{3.3}$$

where the aforementioned integrals are performed over the whole box. Note that the velocity field is real, hence

$$\mathbf{u}(-\mathbf{k}, t) = \frac{1}{L_x L_y L_z} \int d\mathbf{r}[\mathbf{u}(\mathbf{r}, t) \exp(-i\mathbf{k} \cdot \mathbf{r})] = \mathbf{u}^*(\mathbf{k}, t). \tag{3.4}$$

Similarly, for pressure and other scalar fields,

$$p(-\mathbf{k}, t) = p^*(\mathbf{k}, t). \tag{3.5}$$

As a result, in a computer simulation, we need to store only half the Fourier modes that reduces computational and storage requirements.

The Fourier transforms have very interesting properties. They are listed as follows:

1. Fourier transform of constant function, C, is $C\delta_{\mathbf{k},0}$, that is,

$$\frac{1}{L_x L_y L_z} \int d\mathbf{r} C \; \exp(-i\mathbf{k} \cdot \mathbf{r}) = C\delta_{\mathbf{k},0}. \tag{3.6}$$

2. The Fourier transform of a product of two real functions $f(\mathbf{r})$ and $g(\mathbf{r})$ is a convolution, that is,

$$(fg)(\mathbf{k}) = \sum_{\mathbf{p}} f(\mathbf{k} - \mathbf{p}) g(\mathbf{p}), \tag{3.7}$$

Proof:

$$\begin{aligned} (fg)(\mathbf{k}) &= \frac{1}{L_x L_y L_z} \int d\mathbf{r} f(\mathbf{r}) g(\mathbf{r}) \exp(-i\mathbf{k} \cdot \mathbf{r}) \\ &= \frac{1}{L_x L_y L_z} \sum_{\mathbf{p}} \sum_{\mathbf{q}} f(\mathbf{q}) g(\mathbf{p}) \int d\mathbf{r} \exp(i(\mathbf{p} + \mathbf{q} - \mathbf{k}) \cdot \mathbf{r}) \end{aligned}$$

$$= \frac{1}{L_x L_y L_z} \sum_{\mathbf{p}} \sum_{\mathbf{q}} f(\mathbf{q}) g(\mathbf{p}) (L_x L_y L_z) \delta_{\mathbf{p}+\mathbf{q}-\mathbf{k}}$$

$$= \sum_{\mathbf{p}} f(\mathbf{k}-\mathbf{p}) g(\mathbf{p}),$$

3. The Fourier transform of the derivative of a real function:

$$(\partial f/\partial x_j)(\mathbf{k}) = i k_j f(\mathbf{k}). \tag{3.8}$$

Proof:

$$\begin{aligned} (\partial f/\partial x_j)(\mathbf{k}) &= \frac{1}{L_x L_y L_z} \int d\mathbf{r} \frac{\partial f(\mathbf{r})}{\partial x_j} \exp(-i\mathbf{k}\cdot\mathbf{r}) \\ &= \frac{1}{L_x L_y L_z} \left\{ [\exp(-i\mathbf{k}\cdot\mathbf{r}) f(\mathbf{r})]_{\text{surface}} + i k_j \int d\mathbf{r} \exp(-i\mathbf{k}\cdot\mathbf{r}) f(\mathbf{r}) \right\} \\ &= i k_j f(\mathbf{k}). \end{aligned}$$

The first term in the second step vanishes due to the periodic boundary condition.

4. Parseval's theorem:

$$\langle f(\mathbf{r}) g(\mathbf{r}) \rangle = \sum_{\mathbf{p}} \Re[f^*(\mathbf{p}) g(\mathbf{p})], \tag{3.9}$$

where $\langle . \rangle$ stands for the spatial average of the quantity at hand.

Proof:

$$\begin{aligned} \langle f(\mathbf{r}) g(\mathbf{r}) \rangle &= \frac{1}{L_x L_y L_z} \int d\mathbf{r} f(\mathbf{r}) g(\mathbf{r}) \\ &= \frac{1}{L_x L_y L_z} \sum_{\mathbf{p}} \sum_{\mathbf{q}} f(\mathbf{q}) g(\mathbf{p}) \int d\mathbf{r} \exp(i(\mathbf{p}+\mathbf{q})\cdot\mathbf{r}) \\ &= \frac{1}{L_x L_y L_z} \sum_{\mathbf{p}} \sum_{\mathbf{q}} f(\mathbf{q}) g(\mathbf{p}) (L_x L_y L_z) \delta_{\mathbf{p}+\mathbf{q}} \\ &= \sum_{\mathbf{p}} f(-\mathbf{p}) g(\mathbf{p}) = \frac{1}{2} \left[\sum_{\mathbf{p}} f^*(\mathbf{p}) g(\mathbf{p}) + \sum_{-\mathbf{p}} f^*(\mathbf{p}) g(\mathbf{p}) \right] \\ &= \sum_{\mathbf{p}} \Re[f^*(\mathbf{p}) g(\mathbf{p})], \end{aligned}$$

where $\Re[.]$ stands for the real part of the argument. When $f = g$, we obtain

$$\langle [f(\mathbf{r})]^2 \rangle = \sum_{\mathbf{p}} \Re[f^*(\mathbf{p}) f(-\mathbf{p})] = \sum_{\mathbf{p}} |f(\mathbf{p})|^2. \tag{3.10}$$

5. Fourier transform of the correlation function is the power spectrum, that is,

$$\frac{1}{L_x L_y L_z} \int d\mathbf{l} \langle f(\mathbf{r}) g(\mathbf{r}+\mathbf{l}) \rangle \exp(-i(\mathbf{k}\cdot\mathbf{l})) = f(-\mathbf{k}) g(\mathbf{k}) \tag{3.11}$$

Proof:

$$\begin{aligned}
&\frac{1}{L_x L_y L_z} \int d\mathbf{l} \langle f(\mathbf{r}) g(\mathbf{r}+\mathbf{l}) \rangle \exp(-i(\mathbf{k}\cdot\mathbf{l})) \\
&= \frac{1}{(L_x L_y L_z)^2} \int\int d\mathbf{l} d\mathbf{r} f(\mathbf{r}) g(\mathbf{r}+\mathbf{l}) \exp(-i(\mathbf{k}\cdot\mathbf{l})) \\
&= \frac{1}{(L_x L_y L_z)^2} \int\int d\mathbf{l} d\mathbf{r} \sum_{\mathbf{p}} \sum_{\mathbf{q}} f(\mathbf{p}) g(\mathbf{q}) \exp(i\mathbf{p}\cdot\mathbf{r}) \exp(i\mathbf{q}\cdot(\mathbf{r}+\mathbf{l})) \\
&\qquad \times \exp(-i(\mathbf{k}\cdot\mathbf{l})) \\
&= \sum_{\mathbf{p}} \sum_{\mathbf{q}} \delta_{\mathbf{q},\mathbf{k}} \delta_{\mathbf{p},-\mathbf{q}} f(\mathbf{p}) g(\mathbf{q}) \\
&= f(-\mathbf{k}) g(\mathbf{k}).
\end{aligned}$$

When $f = g$, we obtain

$$\frac{1}{L_x L_y L_z} \int d\mathbf{l} \langle f(\mathbf{r}) f(\mathbf{r}+\mathbf{l}) \rangle \exp(-i(\mathbf{k}\cdot\mathbf{l})) = |f(\mathbf{k})|^2. \tag{3.12}$$

The energy of a velocity Fourier mode, called *modal kinetic energy,* is given by

$$E_u(\mathbf{k}) = \frac{1}{2} |\mathbf{u}(\mathbf{k})|^2. \tag{3.13}$$

Using Parseval's theorem, we obtain

$$E_u = \frac{1}{2} \langle u^2 \rangle = \sum_{\mathbf{k}} \frac{1}{2} |\mathbf{u}(\mathbf{k})|^2. \tag{3.14}$$

E_u of Eq. (3.14) is the *total kinetic energy* , and is a sum over all the Fourier modes. In the next section, we present Navier–Stokes equations in Fourier space.

3.2 Flow Equations in Fourier Space

Using the identities stated in the previous sections, we represent the incompressible Navier–Stokes equations in Fourier space. That is, we rewrite Eq. (2.6) in Fourier space as

$$\frac{d}{dt}\mathbf{u}(\mathbf{k}) + \mathbf{N}_u(\mathbf{k}) = -i\mathbf{k}p(\mathbf{k}) + \mathbf{F}_u(\mathbf{k}) - \nu k^2\mathbf{u}(\mathbf{k}), \tag{3.15a}$$

$$\mathbf{k}\cdot\mathbf{u}(\mathbf{k}) = 0, \tag{3.15b}$$

where the nonlinear term is

$$\mathbf{N}_u(\mathbf{k}) = i\sum_{\mathbf{p}}\{\mathbf{k}\cdot\mathbf{u}(\mathbf{q})\}\mathbf{u}(\mathbf{p}) \tag{3.16}$$

with $\mathbf{q} = \mathbf{k} - \mathbf{p}$. Equations (3.15, 3.16) are the equations of motion for mode $\mathbf{u}(\mathbf{k})$. Equation (3.15) can be written in tensorial form as

$$\frac{d}{dt}u_i(\mathbf{k}) = -ik_i p(\mathbf{k}) - ik_j\sum_{\mathbf{p}} u_j(\mathbf{q})u_i(\mathbf{p}) + F_{u,i}(\mathbf{k}) - \nu k^2 u_i(\mathbf{k}), \tag{3.17a}$$

$$k_i u_i(\mathbf{k}) = 0. \tag{3.17b}$$

In Eq. (3.17), i represents $\sqrt{-1}$ in front of the $k_i p(\mathbf{k})$ term, and x, y, z components in u_i. Note that the pressure $p(\mathbf{k})$ is derived by taking dot product of Eq. (3.15a) with $i\mathbf{k}$ and by employing $\mathbf{k}\cdot\mathbf{u}(\mathbf{k}) = 0$:

$$p(\mathbf{k}) = \frac{i}{k^2}\mathbf{k}\cdot\{\mathbf{N}_u(\mathbf{k}) - \mathbf{F}_u(\mathbf{k})\}. \tag{3.18}$$

To derive a dynamical equation for the modal KE, $|\mathbf{u}(\mathbf{k})|^2/2$, we perform a dot product of Eq. (3.15a) with $\mathbf{u}^*(\mathbf{k})$, and add the resultant equation with its complex conjugate. These operations yield

$$\frac{d}{dt}E_u(\mathbf{k}) = \sum_{\mathbf{p}}\Im\left[\{\mathbf{k}\cdot\mathbf{u}(\mathbf{q})\}\{\mathbf{u}(\mathbf{p})\cdot\mathbf{u}^*(\mathbf{k})\}\right] + \Re[\mathbf{F}_u(\mathbf{k})\cdot\mathbf{u}^*(\mathbf{k})] - 2\nu k^2 E_u(\mathbf{k}), \tag{3.19}$$

where $\mathbf{q} = \mathbf{k} - \mathbf{p}$, and $\Re[.]$, $\Im[.]$ stand respectively for the real and imaginary parts of the argument. The modal energy $E_u(\mathbf{k})$ changes with time due to the three terms in the right-hand side of Eq. (3.19). These terms have the following physical interpretations:

1. Term 1: The nonlinear energy transfers from all the Fourier modes to $\mathbf{u}(\mathbf{k})$. It is usually denoted by $T_u(\mathbf{k})$.
2. Term 2: The energy supply rate by the external force

$$\mathcal{F}_u(\mathbf{k}) = \Re[\mathbf{F}_u(\mathbf{k})\cdot\mathbf{u}^*(\mathbf{k})]. \tag{3.20}$$

3. Term 3: The viscous dissipation rate

$$D_u(\mathbf{k}) = -2\nu k^2 E_u(\mathbf{k}). \tag{3.21}$$

From the above, we deduce that the total viscous dissipation rate in the flow is

$$\epsilon_u = \sum_{\mathbf{k}} D_u(\mathbf{k}), \tag{3.22}$$

and the total energy supply rate to the flow by the external force is

$$\mathcal{F}_u = \sum_{\mathbf{k}} \mathcal{F}_u(\mathbf{k}). \tag{3.23}$$

For a homogeneous and isotropic turbulence, the modal kinetic energies of all the modes in a thin wavenumber shell are statistically equal. The averaging process could be either ensemble or temporal. Hence, it is customary to define the following one-dimensional KE spectrum, $E_u(k)$, which is a sum of all the modes of a shell of unit width around wavenumber k:

$$E_u(k) = \sum_{k-1<k'\leq k} \frac{1}{2}|\mathbf{u}(\mathbf{k}')|^2. \tag{3.24}$$

The dynamical equations for $E_u(k)$ is obtained by summing Eq. (3.19) over all the Fourier modes of the shell:

$$\begin{aligned}\frac{d}{dt}E_u(k) &= \sum_{k-1<k''\leq k} \sum_{\mathbf{p}} \Im\left[\{\mathbf{k}''\cdot\mathbf{u}(\mathbf{q})\}\{\mathbf{u}(\mathbf{p})\cdot\mathbf{u}^*(\mathbf{k}'')\}\right] - \sum_{k-1<k''\leq k} 2\nu k''^2 E_u(\mathbf{k}'') \\ &\quad + \sum_{k-1<k''\leq k} \Re[\mathbf{F}_u(\mathbf{k}'')\cdot\mathbf{u}^*(\mathbf{k}'')]\end{aligned} \tag{3.25}$$

with $\mathbf{q} = \mathbf{k}'' - \mathbf{p}$.

For homogeneous and isotropic turbulence, the energy contents of the three components of $\mathbf{u}(\mathbf{k})$ are equal (on an average), that is,

$$\langle|u_x(\mathbf{k})|^2\rangle = \langle|u_y(\mathbf{k})|^2\rangle = \langle|u_z(\mathbf{k})|^2\rangle. \tag{3.26}$$

However, when the flow is under an influence of an external field, it becomes anisotropic. We assume the direction of the external field or that of anisotropy to be along $\hat{z}$. Hence, the flow has azimuthal symmetry, and

$$\langle |u_x(\mathbf{k})|^2 \rangle = \langle |u_y(\mathbf{k})|^2 \rangle \neq \langle |u_z(\mathbf{k})|^2 \rangle. \tag{3.27}$$

For a more general notation, we denote $u_z = u_{\|}$, and

$$\mathbf{u}_{\perp} = u_x \hat{x} + u_y \hat{y}. \tag{3.28}$$

The modal KE of these fields are defined as follows:

$$E_{u,\|}(\mathbf{k}) = \frac{1}{2} |u_{\|}(\mathbf{k})|^2, \tag{3.29a}$$

$$E_{u,\perp}(\mathbf{k}) = \frac{1}{2} |\mathbf{u}_{\perp}(\mathbf{k})|^2, \tag{3.29b}$$

and their corresponding one-dimensional KE spectra are as follows:

$$E_{u,\|}((k) = \sum_{k-1<k' \leq k} \frac{1}{2} |u_{\|}(\mathbf{k}')|^2, \tag{3.30a}$$

$$E_{u,\perp}(k) = \sum_{k-1<k' \leq k} \frac{1}{2} |\mathbf{u}_{\perp}(\mathbf{k}')|^2. \tag{3.30b}$$

In Chapter 11, we also employ the following anisotropic parameter $A(k)$ to quantify anisotropy in turbulence:

$$A(k) = \frac{E_{u,\perp}(k)}{2E_{u,\|}(k)}. \tag{3.31}$$

Chapter 11 contains a detailed discussion on quantification of anisotropic turbulence.

3.3 Vorticity, Kinetic Helicity, and Enstrophy

In Fourier space, the vorticity field $\boldsymbol{\omega} = \nabla \times \mathbf{u}$ gets transformed to

$$\boldsymbol{\omega}(\mathbf{k}) = i\mathbf{k} \times \mathbf{u}(\mathbf{k}). \tag{3.32}$$

Using $\mathbf{k} \cdot \mathbf{u}(\mathbf{k}) = 0$, we invert Eq. (3.32) as follows:

$$\mathbf{u}(\mathbf{k}) = \frac{i}{k^2} \mathbf{k} \times \boldsymbol{\omega}(\mathbf{k}). \tag{3.33}$$

Note that for a 2D velocity field, $\boldsymbol{\omega}(\mathbf{k})$ is perpendicular to the plane of the velocity field.

Using Eqs. (2.13), we derive the dynamical equation for the vorticity in Fourier space as

$$\frac{d}{dt}\boldsymbol{\omega}(\mathbf{k}) + \mathbf{N}_\omega(\mathbf{k}) = \mathbf{F}_\omega(\mathbf{k}) - \nu k^2 \boldsymbol{\omega}(\mathbf{k}), \tag{3.34}$$

where

$$\mathbf{N}_\omega(\mathbf{k}) = -i\mathbf{k} \times \sum_{\mathbf{p}} \{\mathbf{u}(\mathbf{q}) \times \boldsymbol{\omega}(\mathbf{p})\} \tag{3.35}$$

or

$$\mathbf{N}_\omega(\mathbf{k}) = i \sum_{\mathbf{p}} \{\mathbf{k} \cdot \mathbf{u}(\mathbf{q})\}\boldsymbol{\omega}(\mathbf{p}) - \{\mathbf{k} \cdot \boldsymbol{\omega}(\mathbf{q})\}\mathbf{u}(\mathbf{p}), \tag{3.36}$$

and

$$\mathbf{F}_\omega(\mathbf{k}) = i\mathbf{k} \times \mathbf{F}_u(\mathbf{k}). \tag{3.37}$$

There are two important quadratic quantities related to the vorticity—kinetic helicity and enstrophy. The modal kinetic helicity is defined as follows:

$$H_K(\mathbf{k}) = \frac{1}{2}\Re[\mathbf{u}(\mathbf{k}) \cdot \boldsymbol{\omega}^*(\mathbf{k})]. \tag{3.38}$$

Using Eq. (3.33), we obtain

$$H_K(\mathbf{k}) = \frac{1}{2k^2}\Re[i\mathbf{k} \cdot \{\boldsymbol{\omega}(\mathbf{k}) \times \boldsymbol{\omega}^*(\mathbf{k})\}]. \tag{3.39}$$

The modal enstrophy is defined as

$$E_\omega(\mathbf{k}) = \frac{1}{2}|\boldsymbol{\omega}(\mathbf{k})|^2. \tag{3.40}$$

Using Parseval's theorem, we deduce that the total kinetic helicity and total enstrophy are

$$H_K = \frac{1}{2}\langle \mathbf{u} \cdot \boldsymbol{\omega} \rangle = \sum_{\mathbf{k}} \frac{1}{2}\Re[\mathbf{u}(\mathbf{k}) \cdot \boldsymbol{\omega}^*(\mathbf{k})] \tag{3.41}$$

and

$$E_\omega = \frac{1}{2}\langle |\boldsymbol{\omega}|^2 \rangle = \sum_{\mathbf{k}} \frac{1}{2}|\boldsymbol{\omega}(\mathbf{k})|^2. \tag{3.42}$$

The evolution equations for the kinetic helicity can be derived by taking time derivative of Eq. (3.39):

$$\frac{d}{dt}H_K(\mathbf{k}) = \frac{1}{2k^2}\Re[i\mathbf{k}\cdot\{\dot{\boldsymbol{\omega}}(\mathbf{k})\times\boldsymbol{\omega}^*(\mathbf{k})+\boldsymbol{\omega}(\mathbf{k})\times\dot{\boldsymbol{\omega}}^*(\mathbf{k})\}]. \tag{3.43}$$

Since

$$i\mathbf{k}\cdot\{\boldsymbol{\omega}(\mathbf{k})\times\dot{\boldsymbol{\omega}}^*(\mathbf{k})\} = -i\mathbf{k}\cdot\{\dot{\boldsymbol{\omega}}^*(\mathbf{k})\times\boldsymbol{\omega}(\mathbf{k})\} = [i\mathbf{k}\cdot\{\dot{\boldsymbol{\omega}}(\mathbf{k})\times\boldsymbol{\omega}^*(\mathbf{k})\}]^*, \tag{3.44}$$

Equation (3.43) is simplified to

$$\frac{d}{dt}H_K(\mathbf{k}) = \frac{1}{k^2}\Re[i\mathbf{k}\cdot\{\dot{\boldsymbol{\omega}}(\mathbf{k})\times\boldsymbol{\omega}^*(\mathbf{k})\}]. \tag{3.45}$$

Note that for $\nu = 0$ and $\mathbf{F}_u = 0$,

$$\begin{aligned} i\mathbf{k}\cdot\{\dot{\boldsymbol{\omega}}(\mathbf{k})\times\boldsymbol{\omega}^*(\mathbf{k})\} &= i\mathbf{k}\cdot\sum_{\mathbf{p}}[i\{\mathbf{k}\times(\mathbf{u}(\mathbf{q})\times\boldsymbol{\omega}(\mathbf{p}))\}\times\boldsymbol{\omega}^*(\mathbf{k})] \\ &= \mathbf{k}\cdot\sum_{\mathbf{p}}[\mathbf{k}\{\mathbf{u}(\mathbf{q})\times\boldsymbol{\omega}(\mathbf{p})\}\cdot\boldsymbol{\omega}^*(\mathbf{k})] \\ &= \sum_{\mathbf{p}} k^2\mathbf{u}(\mathbf{q})\cdot\{\boldsymbol{\omega}(\mathbf{p})\times\boldsymbol{\omega}^*(\mathbf{k})\}, \end{aligned} \tag{3.46}$$

where $\mathbf{q} = \mathbf{k} - \mathbf{p}$. Therefore, with ν and $\mathbf{F}_u$, Eq. (3.45) yields

$$\frac{d}{dt}H_K(\mathbf{k}) = \sum_{\mathbf{p}}\Re[\mathbf{u}(\mathbf{q})\cdot\{\boldsymbol{\omega}(\mathbf{p})\times\boldsymbol{\omega}^*(\mathbf{k})\}] + \mathcal{F}_{H_K}(\mathbf{k}) - \nu k^2 H_K(\mathbf{k}). \tag{3.47}$$

where

$$\mathcal{F}_{H_K}(\mathbf{k}) = \frac{1}{2}\Re[\dot{\mathbf{u}}(\mathbf{k})\cdot\boldsymbol{\omega}^*(\mathbf{k})] + \frac{1}{2}\Re[\mathbf{u}(\mathbf{k})\cdot\dot{\boldsymbol{\omega}}^*(\mathbf{k})] = \Re[\boldsymbol{\omega}^*(\mathbf{k})\cdot\mathbf{F}_u(\mathbf{k})] \tag{3.48}$$

is the kinetic helicity supply rate by the external force.

Following a similar procedure as earlier, we deduce the following dynamical equation for the enstrophy:

$$\begin{aligned} \frac{d}{dt}E_\omega(\mathbf{k}) &= \Re[\dot{\boldsymbol{\omega}}(\mathbf{k})\cdot\boldsymbol{\omega}^*(\mathbf{k})] \\ &= \sum_{\mathbf{p}}\Im[\{\mathbf{k}\cdot\mathbf{u}(\mathbf{q})\}\{\boldsymbol{\omega}(\mathbf{p})\cdot\boldsymbol{\omega}^*(\mathbf{k})\}] - \Im[\{\mathbf{k}\cdot\boldsymbol{\omega}(\mathbf{q})\}\{\mathbf{u}(\mathbf{p})\cdot\boldsymbol{\omega}^*(\mathbf{k})\}] \\ &\quad + \mathcal{F}_\omega(\mathbf{k}) - 2\nu k^2 E_\omega(\mathbf{k}) \end{aligned} \tag{3.49}$$

where

$$\mathcal{F}_\omega(\mathbf{k}) = \Re[i\mathbf{k}\times\mathbf{F}_u(\mathbf{k})\cdot\boldsymbol{\omega}^*(\mathbf{k})] = k^2\Re[\mathbf{u}^*(\mathbf{k})\cdot\mathbf{F}_u(\mathbf{k})] \tag{3.50}$$

is the enstrophy supply rate by the external force. Equation (3.49) indicates that the enstrophy is enhanced by nonlinearity. Note however that a single mode with $\mathbf{u}(\mathbf{k}) \parallel \boldsymbol{\omega}(\mathbf{k})$ does not lead to stretching of a vortex. That is, vortex stretching is a nonlinear phenomena.

In a 2D inviscid flow, $\boldsymbol{\omega}(\mathbf{k}) \perp \mathbf{u}(\mathbf{k})$, hence $\mathbf{u}(\mathbf{p}) \cdot \boldsymbol{\omega}^*(\mathbf{k})) = 0$. Therefore,

$$\frac{d}{dt}E_\omega(\mathbf{k}) = \sum_{\mathbf{p}} \Im\left[\{\mathbf{k}\cdot\mathbf{u}(\mathbf{q})\}\{\omega(\mathbf{p})\omega^*(\mathbf{k})\}\right] + \mathcal{F}_\omega(\mathbf{k}) - 2\nu k^2 E_\omega(\mathbf{k}). \tag{3.51}$$

We will revisit these equations when we study kinetic energy, kinetic helicity, and enstrophy transfers among the Fourier modes.

Example 3.1: Consider the following two-dimensional fluid flow in a periodic box of size $2\pi \times 2\pi$:

$$\mathbf{u} = 4A(\hat{x}\sin x\cos y - \hat{y}\cos x\sin y).$$

Compute the amplitudes of $\mathbf{u}(\mathbf{k})$ and that of the vorticity field. What are the total kinetic energy and enstrophy of the flow?

Solution: We expand the components of the velocity field as

$$u_x = \frac{A}{i}\left[\exp(ix) - \exp(-ix)\right]\left[\exp(iy) + \exp(-iy)\right],$$

$$u_y = -\frac{A}{i}\left[\exp(ix) + \exp(-ix)\right]\left[\exp(iy) - \exp(-iy)\right],$$

from which we can extract Fourier amplitudes $\mathbf{u}(\pm 1, \pm 1)$. These amplitudes are listed in Table 3.1. We can compute the corresponding amplitudes for the vorticity field using

$$\boldsymbol{\omega}(\mathbf{k}) = i\mathbf{k} \times \mathbf{u}(\mathbf{k}).$$

The modal energy and modal enstrophy are computed using

$$E_u(\mathbf{k}) = \frac{1}{2}|\mathbf{u}(\mathbf{k})|^2; \quad E_\omega(\mathbf{k}) = \frac{1}{2}|\boldsymbol{\omega}(\mathbf{k})|^2. \tag{3.52}$$

These values are also listed in Table 3.1. The total kinetic energy and enstrophy of the flow are $4A^2$ and $8A^2$ respectively. One can easily verify that these results are consistent with those presented in Example 2.2 for real space.

Table 3.1 Example 3.1: Amplitudes of the velocity and vorticity Fourier modes. The modal kinetic energy and modal enstrophy are also listed.

mode	$u_x(\mathbf{k})$	$u_y(\mathbf{k})$	$\omega_z(\mathbf{k})$	$E_u(\mathbf{k})$	$E_\omega(\mathbf{k})$
(1,1)	A/i	$-A/i$	$-2A$	A^2	$2A^2$
$(1,-1)$	A/i	A/i	$2A$	A^2	$2A^2$
$(-1,1)$	$-A/i$	$-A/i$	$2A$	A^2	$2A^2$
$(-1,-1)$	$-A/i$	A/i	$-2A$	A^2	$2A^2$

Example 3.2: Consider the two-dimensional fluid flows in a periodic box of size $2\pi \times 2\pi$:

$$\mathbf{u} = 2(-\sin y, \sin x, 0).$$

Compute the amplitudes of $\mathbf{u}(\mathbf{k})$ and that of the vorticity field. What are the total kinetic energy and enstrophy of the flow?

Solution: We compute the desired quantities following a similar procedure as in the previous example. The results are listed in Table 3.2. Note that the total kinetic energy and total enstrophy of the flow are 2 units each, which are the same as those computed in Example 2.3 for real space.

Table 3.2 Example 3.2: Amplitudes of the velocity and vorticity Fourier modes. The modal kinetic energy and modal enstrophy are also listed.

Mode	$u_x(\mathbf{k})$	$u_y(\mathbf{k})$	$\omega_z(\mathbf{k})$	$E_u(\mathbf{k})$	$E_\omega(\mathbf{k})$
(1,0)	0	$1/i$	1	1/2	1/2
$(-1,0)$	0	$-1/i$	1	1/2	1/2
(0,1)	$-1/i$	0	1	1/2	1/2
$(0,-1)$	$1/i$	0	1	1/2	1/2

Example 3.3: Consider the following two-dimensional fluid flows with three components:

$$(a): \quad \mathbf{u} = 2(-\sin y, \sin x, \cos x + \cos y).$$

$$(b): \quad \mathbf{u} = 2(-\sin y, \sin x, -\cos x - \cos y).$$

Compute the Fourier amplitudes of the velocity and vorticity fields, as well as the total kinetic energy, enstrophy, and kinetic helicity.

Solution: Following a similar procedure as in earlier exercises, we compute the desired quantities, which are listed in Table 3.3. For the velocity field (a), $\mathbf{u}(\mathbf{k}) = \boldsymbol{\omega}(\mathbf{k})$, and the kinetic helicity is maximally positive. The total kinetic energy, enstrophy, and kinetic helicity are 4 each. For the velocity field (b), $\mathbf{u}(\mathbf{k}) = -\boldsymbol{\omega}(\mathbf{k})$, and the kinetic helicity is maximally negative. Here, the total kinetic energy and enstrophy are 4 each, while the total kinetic helicity is -4. These results are consistent with those presented in Examples 2.5 and 2.6 of Chapter 2.

Table 3.3 Example 3.3: Amplitudes of the velocity and vorticity Fourier modes for the given velocity fields. The modal kinetic energy, modal enstrophy, and modal kinetic helicity are also listed.

Flow	Mode	$\mathbf{u}(\mathbf{k})$	$\boldsymbol{\omega}(\mathbf{k})$	$E_u(\mathbf{k})$	$E_\omega(\mathbf{k})$	$H_K(\mathbf{k})$
(a)	(1,0)	$(0,-i,1)$	$(0,-i,1)$	1	1	1
	$(-1,0)$	$(0,i,1)$	$(0,i,1)$	1	1	1
	(0,1)	$(i,0,1)$	$(i,0,1)$	1	1	1
	$(0,-1)$	$(-i,0,1)$	$(-i,0,1)$	1	1	1
(b)	(1,0)	$(0,-i,-1)$	$(0,i,1)$	1	1	-1
	$(-1,0)$	$(0,i,-1)$	$(0,-i,1)$	1	1	-1
	(0,1)	$(i,0,-1)$	$(-i,0,1)$	1	1	-1
	$(0,-1)$	$(-i,0,-1)$	$(i,0,1)$	1	1	-1

Example 3.4: Consider the following flow field:

$$\mathbf{u} = \hat{x}2B\cos y + \hat{y}2C\cos x + (\hat{x} - \hat{y})2A\sin(x + y).$$

List the active Fourier modes. Compute the total kinetic energy of the flow. Assuming that $\mathbf{F}_u = 0$, derive equations of motion for A, B, and C.

Solution: In Fourier basis, the associated wavenumbers of the velocity field are $(\pm 1, 0)$, $(0, \pm 1)$, $(1,1)$, and $(-1,-1)$. The Fourier amplitudes of the Fourier modes are listed in Table 3.4.

The total kinetic energy of the flow is

$$E_u = \sum_{\mathbf{k}} \frac{1}{2}|\mathbf{u}(\mathbf{k})|^2 = C^2 + B^2 + 2A^2.$$

Table 3.4 Example 3.4: Amplitudes of the velocity Fourier modes.

Mode	u_x	u_y
(1,0)	0	C
$(-1,0)$	0	C
(0,1)	B	0
$(0,-1)$	B	0
(1,1)	$\frac{A}{i}$	$-\frac{A}{i}$
$(-1,-1)$	$-\frac{A}{i}$	$\frac{A}{i}$

In Figure 3.1, we illustrate the interacting Fourier wavenumbers. For the derivation of the equations of motion for the modes, we focus on a single triad: $\mathbf{q} = (1,0)$, $\mathbf{p} = (0,1)$, and $\mathbf{k} = (1,1)$. Note that $\mathbf{k} = \mathbf{p} + \mathbf{q}$. Using Eq. (3.16), we compute the nonlinear term $\mathbf{N}_u$ as

$$\begin{aligned}
\mathbf{N}_u(\mathbf{k}) &= i\mathbf{k} \cdot \mathbf{u}(\mathbf{q})\mathbf{u}(\mathbf{p}) + i\mathbf{k} \cdot \mathbf{u}(\mathbf{p})\mathbf{u}(\mathbf{q}) \\
&= iu_y(\mathbf{q})u_x(\mathbf{p})\hat{x} + iu_x(\mathbf{p})u_y(\mathbf{q})\hat{y} = iBC(\hat{x} + \hat{y}), \\
\mathbf{N}_u(\mathbf{q}) &= i\mathbf{q} \cdot \mathbf{u}(\mathbf{k})\mathbf{u}(-\mathbf{p}) + i\mathbf{q} \cdot \mathbf{u}(-\mathbf{p})\mathbf{u}(\mathbf{k}) \\
&= iu_x(\mathbf{k})u_x^*(\mathbf{p})\hat{x} + iu_x^*(\mathbf{p})\mathbf{u}(\mathbf{k}) = AB(2\hat{x} - \hat{y}), \\
\mathbf{N}_u(\mathbf{p}) &= AC(\hat{x} - 2\hat{y}).
\end{aligned}$$

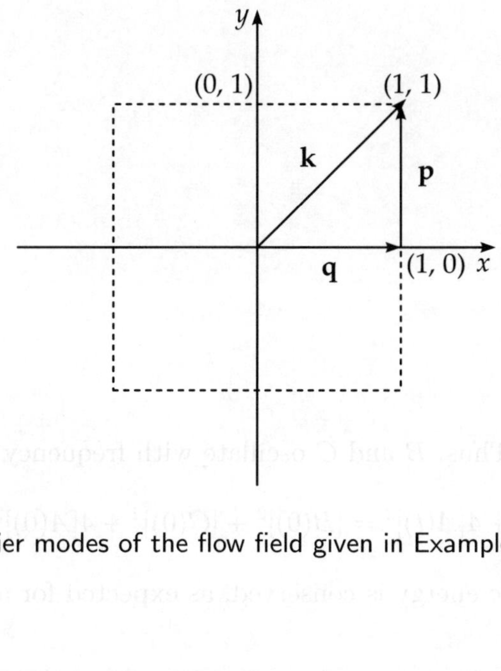

Figure 3.1 Fourier modes of the flow field given in Example 3.4

Using Eq. (3.18) we compute the pressure Fourier modes as

$$\begin{aligned}
p(\mathbf{k}) &= -BC, \\
p(\mathbf{p}) &= -2iAC, \\
p(\mathbf{q}) &= 2iAB.
\end{aligned}$$

Now we substitute $\mathbf{N}_u(\mathbf{k})$ and $p(\mathbf{k})$ in Eq. (3.15a) and set $\mathbf{F}_u(\mathbf{k}) = 0$ that yields

$$\begin{aligned}
\frac{d}{dt}\mathbf{u}(\mathbf{k}) &= -\mathbf{N}_u(\mathbf{k}) - i\mathbf{k}p(k) - \nu k^2 \mathbf{u}(\mathbf{k}) \\
&= -iBC(\hat{x}+\hat{y}) - i(\hat{x}+\hat{y})(-BC) - \nu 2\mathbf{u}(\mathbf{k}) \\
&= -2\nu\mathbf{u}(\mathbf{k}).
\end{aligned}$$

Similar analysis yields

$$\begin{aligned}
\frac{d}{dt}\mathbf{u}(\mathbf{q}) &= AB\hat{y} - \nu\mathbf{u}(\mathbf{q}), \\
\frac{d}{dt}\mathbf{u}(\mathbf{p}) &= -AC\hat{x} - \nu\mathbf{u}(\mathbf{p}).
\end{aligned}$$

Let us consider the inviscid limit for which $\nu = 0$. Since $\mathbf{u}(\mathbf{k}) = (A/i)(\hat{x} - \hat{y})$, $\mathbf{u}(\mathbf{q}) = C\hat{y}$, and $\mathbf{u}(\mathbf{p}) = B\hat{x}$, the equations of motion are

$$\begin{aligned}
\dot{A} &= 0, \\
\dot{B} &= -AC, \\
\dot{C} &= AB,
\end{aligned}$$

whose solution are

$$\begin{aligned}
A &= \text{constant}, \\
B &= c\cos(At), \\
C &= c\sin(At)
\end{aligned}$$

with c as a constant. Thus, B and C oscillate with frequency A. Note that

$$[B(t)]^2 + [C(t)]^2 + 4[A(t)]^2 = [B(0)]^2 + [C(0)]^2 + 4[A(0)]^2 = \text{const.}$$

Hence, the total kinetic energy is conserved, as expected for an inviscid flow.

The dynamical equations are more easily derived when we employ dot products of equations for $\dot{\mathbf{u}}(\mathbf{k})$, $\dot{\mathbf{u}}(\mathbf{p})$, $\dot{\mathbf{u}}(\mathbf{q})$ with $\hat{x} - \hat{y}$, $\hat{x}$, and $\hat{y}$ respectively. This process eliminates the pressure term automatically and yields the following dynamical equations:

$$
\begin{aligned}
\frac{2\dot{A}}{i} &= -\mathbf{N}_u(\mathbf{k}) \cdot (\hat{x} - \hat{y}) = 0, \\
\dot{B} &= -\mathbf{N}_u(\mathbf{p}) \cdot \hat{x} = -AC, \\
\dot{C} &= -\mathbf{N}_u(\mathbf{q}) \cdot \hat{y} = AB,
\end{aligned}
$$

which are same as those obtained earlier.

We will obtain the same set of equations for A, B, and C if we use the other triads, for example, $\{(-1,0),(0,-1),(-1,-1)\}$. This is because the amplitudes of all the modes are related to each other as shown in Table 3.4.

Example 3.5: In the light of the previous example, what is the equation of motion of the following flow field:

$$
\mathbf{u} = \hat{x}2B\cos y + \hat{y}2C\cos x + 2A\cos(x+y)(\hat{x} - \hat{y}).
$$

Solution: We follow the same procedure as in the previous example with $\mathbf{q} = (1,0)$, $\mathbf{p} = (0,1)$, and $\mathbf{k} = (1,1)$. For $\nu = 0$, we obtain

$$
\begin{aligned}
\frac{d}{dt}\mathbf{u}(\mathbf{k}) &= 0 \\
\frac{d}{dt}\mathbf{u}(\mathbf{q}) &= iAB\hat{y} \\
\frac{d}{dt}\mathbf{u}(\mathbf{p}) &= -iAC\hat{x}
\end{aligned}
$$

that leads to

$$
\begin{aligned}
\dot{A} &= 0, \\
\dot{B} &= -iAC, \\
\dot{C} &= iAB,
\end{aligned}
$$

Since A, B, C are real, we obtain

$$
\dot{B} = 0; \quad \dot{C} = 0; \quad AC = 0; \quad AB = 0,
$$

whose solution is

$$A = 0, \quad B = \text{const}, \quad C = \text{const}.$$

$A = 0$ implies that the nonlinear interactions among the modes are turned off.

Example 3.6: Consider the velocity field

$$\begin{aligned}\mathbf{u} &= 4C(\hat{x}\sin x\cos z - \hat{z}\cos x\sin z) + 4B(\hat{y}\sin y\cos z - \hat{z}\cos y\sin z) \\ &\quad +8A(-\hat{x}\sin x\cos y\cos 2z - \hat{y}\cos x\sin y\cos 2z + \hat{z}\cos x\cos y\sin 2z).\end{aligned}$$

Compute the vorticity, the total kinetic energy, and the total kinetic helicity of the flow. Derive the equations of motion for A, B, and C.

Solution: The wavenumbers associated with the flow field are $(\pm 1, 0, \pm 1)$, $(0, \pm 1, \pm 1)$, and $(\pm 1, \pm 1, \pm 2)$. In Table 3.5 we present the Fourier amplitudes for the wavenumbers (1,0,1), (0,1,1), and (1,1,2). The amplitudes of other Fourier modes can be written easily using the properties of sin and cos functions. Table 3.5 also contains the amplitudes of the vorticity Fourier modes, as well as the modal kinetic energy and modal kinetic helicity.

Table 3.5 Example 3.6: Fourier amplitudes of the velocity and vorticity field, as well as the modal kinetic energy and kinetic helicity.

Mode	$\mathbf{u}(\mathbf{k})$	$\boldsymbol{\omega}(\mathbf{k})$	$E_u(\mathbf{k})$	$H_K(\mathbf{k})$
(1,1,2)	$(-\frac{A}{i}, -\frac{A}{i}, \frac{A}{i})$	$(3A, -3A, 0)$	$3A^2/2$	0
(1,0,1)	$(\frac{C}{i}, 0, -\frac{C}{i})$	$(0, 2C, 0)$	C^2	0
(0,1,1)	$(0, \frac{B}{i}, -\frac{B}{i})$	$(-2B, 0, 0)$	B^2	0

The total kinetic energy of the flow is

$$E_u = \sum_{\mathbf{k}} \frac{1}{2}|\mathbf{u}(\mathbf{k})|^2 = 4 \times C^2 + 4 \times B^2 + 8 \times \frac{3}{2}A^2 = 4(C^2 + B^2 + 3A^2).$$

Here the prefactors are the number of Fourier modes in all the quadrants in Fourier space. The vorticity field is computed using

$$\boldsymbol{\omega}(\mathbf{k}) = i\mathbf{k} \times \mathbf{u}(\mathbf{k}),$$

and they are listed in Table 3.5. The total kinetic helicity of the flow is

$$H_K = \sum_{\mathbf{k}} \frac{1}{2} \Re(\mathbf{u}(\mathbf{k}) \cdot \boldsymbol{\omega}^*(\mathbf{k})) = 0.$$

Hence, the flow has zero kinetic helicity. Note that the kinetic helicity for each mode is zero.

We focus on the following interacting triads:

$$(1,1,2) \bigoplus (-1,0,-1) \bigoplus (0,-1,-1) = 0,$$

$$(-1,1,2) \bigoplus (1,0,-1) \bigoplus (0,-1,-1) = 0,$$

$$(1,-1,2) \bigoplus (-1,0,-1) \bigoplus (0,1,-1) = 0,$$

which are depicted in Fig. 3.2. Here the symbol $\bigoplus$ represent nonlinear interactions among the modes. The nonlinear interactions involving the aforementioned modes can occur among the following triads only:[2]

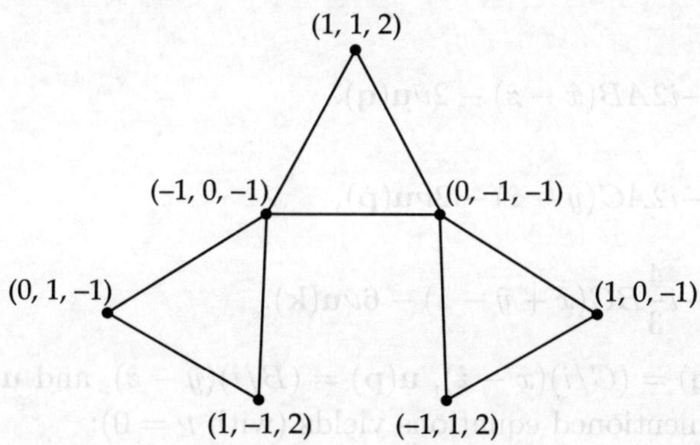

Figure 3.2 Example 3.6. The interacting triads. The wavenumbers in a triad add up to zero.

---

[2]**Proof:** If $k_x, k_y = \pm 1$ and $k_z = \pm 2$, then the nonlinear interactions occur only through the following triads:

$$\begin{aligned}
(k_x, k_y, k_z) &= (k_x, 0, k_z') \bigoplus (0, k_y, k_z''), \\
(k_x, 0, k_z') &= [(k_x, k_y, k_z) \bigoplus (0, -k_y, -k_z'')] + [(k_x, -k_y, k_z) \bigoplus (0, k_y, -k_z'')] \\
(0, k_y, k_z') &= [(k_x, k_y, k_z) \bigoplus (-k_x, 0, -k_z'')] + [(-k_x, k_y, k_z) \bigoplus (k_x, 0, -k_z'')].
\end{aligned}$$

In the aforementioned expressions, when $k_z = 2 \implies k_z' = 1, k_z'' = 1$, and when $k_z = -2 \implies k_z' = -1, k_z'' = -1$.

$$\mathbf{k} = (1,1,2) = (1,0,1) \bigoplus (0,1,1),$$

$$\mathbf{q} = (1,0,1) = [(0,-1,-1) \bigoplus (1,1,2)] + [(0,1,-1) \bigoplus (1,-1,2)],$$

$$\mathbf{p} = (0,1,1) = [(-1,0,-1) \bigoplus (1,1,2)] + [(1,0,-1) \bigoplus (-1,1,2)].$$

Using Eqs. (3.16, 3.18), we compute the nonlinear terms and pressure as

$$\begin{aligned}
\mathbf{N}_u(\mathbf{k}) &= iBC(\hat{x} + \hat{y} - 2\hat{z}),\\
\mathbf{N}_u(\mathbf{q}) &= i2AB(\hat{x} - \hat{z}),\\
\mathbf{N}_u(\mathbf{p}) &= i2AC(\hat{y} - \hat{z}),\\
p(\mathbf{q}) &= p(\mathbf{p}) = 0,\\
p(\mathbf{k}) &= \frac{1}{3}BC.
\end{aligned}$$

Substitution of $\mathbf{N}_u(\mathbf{k})$ and $p(\mathbf{k})$ in Eq. (3.15a), and an assumption that $\mathbf{F}_u(\mathbf{k}) = 0$ yields

$$\begin{aligned}
\frac{d}{dt}\mathbf{u}(\mathbf{q}) &= -i2AB(\hat{x} - \hat{z}) - 2\nu\mathbf{u}(\mathbf{q}),\\
\frac{d}{dt}\mathbf{u}(\mathbf{p}) &= -i2AC(\hat{y} - \hat{z}) - 2\nu\mathbf{u}(\mathbf{p}),\\
\frac{d}{dt}\mathbf{u}(\mathbf{k}) &= -i\frac{4}{3}BC(\hat{x} + \hat{y} - \hat{z}) - 6\nu\mathbf{u}(\mathbf{k}).
\end{aligned}$$

Substitution of $\mathbf{u}(\mathbf{q}) = (C/i)(\hat{x} - \hat{z})$, $\mathbf{u}(\mathbf{p}) = (B/i)(\hat{y} - \hat{z})$, and $\mathbf{u}(\mathbf{k}) = (-A/i)(\hat{x} + \hat{y} - \hat{z})$ in the aforementioned equations yields (with $\nu = 0$):

$$\begin{aligned}
\dot{A} &= -\frac{4}{3}BC,\\
\dot{B} &= 2AC,\\
\dot{C} &= 2AB.
\end{aligned}$$

In Fig. 3.3, we plot the time series of A, B, C for an initial condition $B = 1, C = 0.2, A = 0.2$. Here, B oscillates around a mean value, and A, C oscillate around zero. The role of B and C could have been flipped. Note that if A or C are zero, then the modes will not couple nonlinearly, and B will remain a constant. We also remark that the aforementioned equations have the same form as those of an asymmetric top (Landau and Lifshitz, 1976; Verma, 2016).

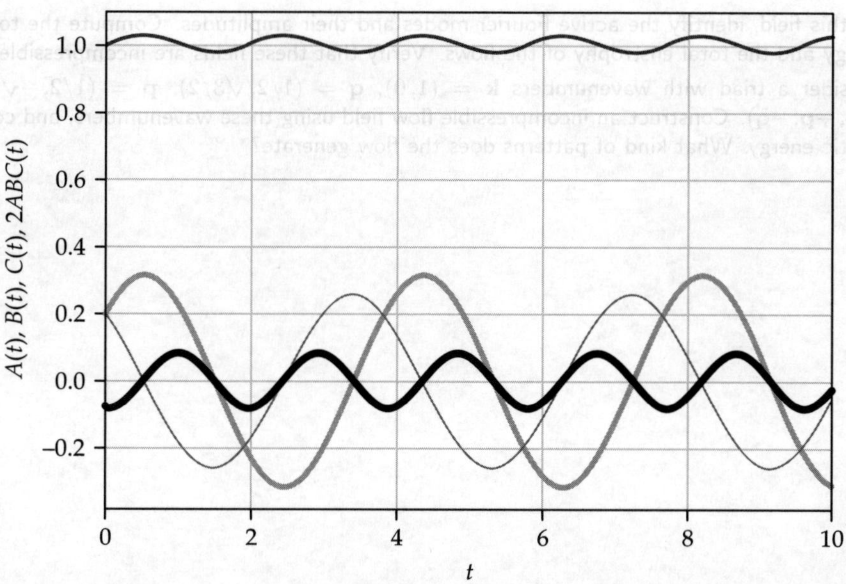

Figure 3.3 Example 3.6: Time series of A, B, C and $-2ABC$ shown respectively as curves with increasing thickness. Note that $-2ABC$, the energy transfer from $\mathbf{u}(\mathbf{p})$ to $\mathbf{u}(\mathbf{k}')$ (to be discussed in Example 4.4), oscillates around zero.

The next chapter contains discussion on the energy transfers among Fourier modes.

Further Reading

For Fourier transform and its application to partial differential equations, refer to texts on mathematical physics, such as Kreyszig et al. (2011). For applications of Fourier transforms to fluid dynamics, refer to textbooks by Leslie (1973), Lesieur (2008), McComb (1990), Davidson (2004), and Pope (2000).

Exercises

1. Consider an incompressible velocity field in the xy plane whose $u_x = 4A \sin ax \cos by$ with $a, b > 0$. Construct the full velocity field.
2. Consider the following flow field in a two-dimensional box of size $[\pi, \pi]$:

$$\begin{aligned}\mathbf{u} &= 4C(\hat{x} \sin 3x \cos y - \hat{y} 3 \cos 3x \sin y) + 4B(\hat{y} \sin 2x \cos 2y - \hat{y} \cos 2x \sin 2y) \\ &\quad +4A(\hat{x} \sin x \cos y - \hat{y} \cos x \sin y).\end{aligned}$$

For this field, identify the active Fourier modes and their amplitudes. Compute the total kinetic energy and the total enstrophy of the flows. Verify that these fields are incompressible.

3. Consider a triad with wavenumbers $\mathbf{k} = (1,0)$, $\mathbf{q} = (1/2, \sqrt{3}/2)$, $\mathbf{p} = (1/2, -\sqrt{3}/2)$, and $(-\mathbf{k}, -\mathbf{p}, -\mathbf{q})$. Construct an incompressible flow field using these wavenumbers, and compute its kinetic energy. What kind of patterns does the flow generate?

Chapter 4

Energy Transfers in Hydrodynamic Flows

In the last chapter we introduced the dynamical equations for fluid flows in Fourier space. According to the energy equation, a Fourier mode gains or loses energy to the other Fourier modes via nonlinear interactions. In this chapter we will quantify energy transfers among the Fourier modes, as well as discuss related quantities such as energy flux.

Kraichnan (1959) was the first to derive a formula for the energy transfers in hydrodynamics. Later, Dar et al. (2001) and Verma (2004) derived a more general formula called *mode-to-mode energy transfer*, which will be described in the present chapter. These transfers are related to the energy flux of Kolmogorov's theory (Kolmogorov, 1941a,c), but they are not the same. We remark that the formulas derived in this chapter are applicable only to incompressible flows. Compressible flows have additional complexities that will be briefly described in Chapter 30.

It is important to note that the energy transfers discussed here refer to energy exchanges across length scales. It differs from the usual definition of real-space energy flux, which is the rate of energy crossing a unit area. Also, in this chapter we use the phrase *energy* for $u^2/2$, which is actually the kinetic energy. Here, this notation is unambiguous because hydrodynamics has only one field, $\mathbf{u}$. In later chapters, we differentiate kinetic energy from other quadratic quantities like magnetic energy, kinetic helicity, etc.

We also remark that linear systems too can exhibit energy exchange. For example, in a simple oscillator, the kinetic and potential energies are exchanged via the spring force. However, the energy exchanges to be discussed in this chapter are very different; they occur via nonlinear interactions.

We start this chapter with the derivation of the mode-to-mode energy transfers in hydrodynamics.

4.1 Mode-to-mode Energy Transfers in Hydrodynamics

We start with Eq. (3.19) for the evolution of modal kinetic energy $|\mathbf{u}(\mathbf{k})|^2/2$:

$$\frac{d}{dt}E_u(\mathbf{k}) = \sum_{\mathbf{p}} \Im\left[\{\mathbf{k}\cdot\mathbf{u}(\mathbf{q})\}\{\mathbf{u}(\mathbf{p})\cdot\mathbf{u}^*(\mathbf{k})\}\right] + \Re[\mathbf{F}_u(\mathbf{k})\cdot\mathbf{u}^*(\mathbf{k})] - 2\nu k^2 E_u(\mathbf{k}). \quad (4.1)$$

Clearly, the Fourier mode $\mathbf{u}(\mathbf{k})$ receives energy from other Fourier modes via nonlinear interactions. The basic unit of interaction is a triad formed by three Fourier modes with wavenumbers $(\mathbf{k},\mathbf{p},\mathbf{q})$ satisfying a condition $\mathbf{k}=\mathbf{p}+\mathbf{q}$. To derive the formula for the mode-to-mode energy transfer, we consider a flow formed by a pair of triads—$(\mathbf{k},\mathbf{p},\mathbf{q})$ and $(-\mathbf{k},-\mathbf{p},-\mathbf{q})$. See Fig. 4.1 for an illustration. The velocity Fourier modes at these wavenumbers are $\mathbf{u}(\pm\mathbf{k})$, $\mathbf{u}(\pm\mathbf{p})$, and $\mathbf{u}(\pm\mathbf{q})$. Note that $\mathbf{u}(-\mathbf{k}) = \mathbf{u}^*(\mathbf{k})$ due to the reality condition. We also simplify the system by making it dissipationless ($\nu = 0$) and force-free ($\mathbf{F}_u = 0$). Note that the total energy is conserved in this limit.

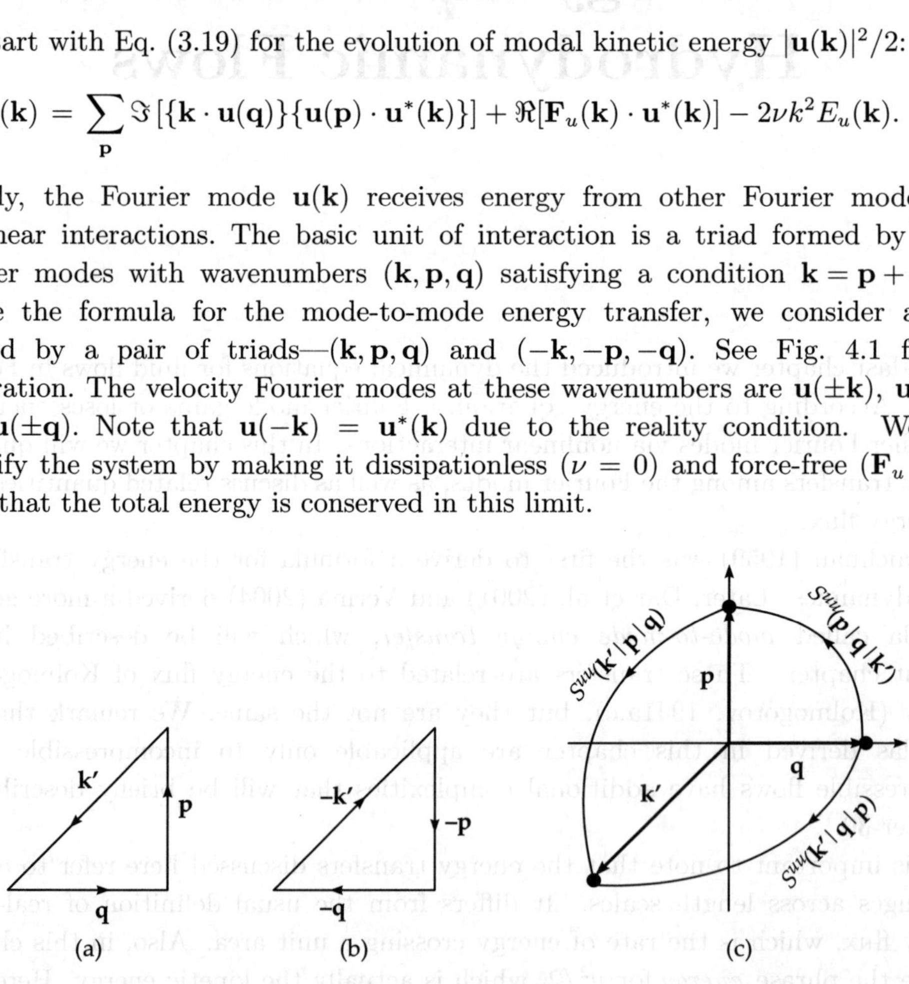

Figure 4.1 (a,b) Two interacting wavenumber triads $(\mathbf{k}',\mathbf{p},\mathbf{q})$ and $(-\mathbf{k}',-\mathbf{p},-\mathbf{q})$. (c) The energy transfer $S^{uu}(\mathbf{X}|\mathbf{Y}|\mathbf{Z})$: from $\mathbf{u}(\mathbf{Y})$ to $\mathbf{u}(\mathbf{X})$ with $\mathbf{u}(\mathbf{Z})$ acting as a mediator.

When we work with triad wavenumbers $(\mathbf{k}', \mathbf{p}, \mathbf{q})$ with $\mathbf{k}' + \mathbf{p} + \mathbf{q} = 0$, the equations become symmetric in $\mathbf{k}', \mathbf{p}, \mathbf{q}$ providing a major simplification. Note that $\mathbf{k}' = -\mathbf{k}$. Using Eq. (4.1) we derive an equation for the modal energy $E_u(\mathbf{k}) = |\mathbf{u}(\mathbf{k})|^2/2$ in the aforementioned triad as

$$\begin{aligned}\frac{d}{dt}E_u(\mathbf{k}', t) &= -\Im\left[\{\mathbf{k}' \cdot \mathbf{u}(\mathbf{q})\}\{\mathbf{u}(\mathbf{p}) \cdot \mathbf{u}(\mathbf{k}')\} + \{\mathbf{k}' \cdot \mathbf{u}(\mathbf{p})\}\{\mathbf{u}(\mathbf{q}) \cdot \mathbf{u}(\mathbf{k}')\}\right], \\ &= S^{uu}(\mathbf{k}'|\mathbf{p}, \mathbf{q}), \end{aligned} \tag{4.2}$$

where $\Im[.]$ denotes the imaginary part of the argument. Kraichnan (1959) was the first to derive Eq. (4.2). $S^{uu}(\mathbf{k}'|\mathbf{p}, \mathbf{q})$[1] is termed as the *combined energy transfer* to $\mathbf{u}(\mathbf{k})$ from $\mathbf{u}(\mathbf{p})$ and $\mathbf{u}(\mathbf{q})$. The evolution equations for $E_u(\mathbf{p})$ and $E_u(\mathbf{q})$ can be derived in a similar fashion:

$$\begin{aligned}\frac{d}{dt}E_u(\mathbf{p}, t) &= -\Im\left[\{\mathbf{p} \cdot \mathbf{u}(\mathbf{q})\}\{\mathbf{u}(\mathbf{p}) \cdot \mathbf{u}(\mathbf{k}')\} + \{\mathbf{p} \cdot \mathbf{u}(\mathbf{k}')\}\{\mathbf{u}(\mathbf{p}) \cdot \mathbf{u}(\mathbf{q})\}\right] \\ &= S^{uu}(\mathbf{p}|\mathbf{q}, \mathbf{k}), \end{aligned} \tag{4.3}$$

$$\begin{aligned}\frac{d}{dt}E_u(\mathbf{q}, t) &= -\Im\left[\{\mathbf{q} \cdot \mathbf{u}(\mathbf{p})\}\{\mathbf{u}(\mathbf{q}) \cdot \mathbf{u}(\mathbf{k}')\} + \{\mathbf{q} \cdot \mathbf{u}(\mathbf{k}')\}\{\mathbf{u}(\mathbf{q}) \cdot \mathbf{u}(\mathbf{p})\}\right] \\ &= S^{uu}(\mathbf{q}|\mathbf{p}, \mathbf{k}). \end{aligned} \tag{4.4}$$

By adding the above three equations, we obtain

$$\frac{d}{dt}\left[E_u(\mathbf{k}', t) + E_u(\mathbf{p}, t) + E_u(\mathbf{q}, t)\right] = 0 \tag{4.5}$$

because $\mathbf{k}' \cdot \mathbf{u}(\mathbf{k}') = \mathbf{p} \cdot \mathbf{u}(\mathbf{p}) = \mathbf{p} \cdot \mathbf{u}(\mathbf{p}) = 0$. Thus, for an interacting triad,

$$E_u(\mathbf{k}', t) + E_u(\mathbf{p}, t) + E_u(\mathbf{q}, t) = \text{constant} \tag{4.6}$$

in the inviscid limit. This is the *law of detailed energy conservation* (Kraichnan, 1959; Lesieur, 2008). Note that the incompressibility condition is critical for this conservation law. In compressible flows, there is an exchange of kinetic and the internal energies (to be discussed in Chapter 30).

An important question is whether we can derive an expression for mode-to-mode energy transfer rates from mode $\mathbf{u}(\mathbf{p})$ to mode $\mathbf{u}(\mathbf{k}')$, and from mode $\mathbf{u}(\mathbf{q})$ to mode $\mathbf{u}(\mathbf{k}')$ separately. These individual energy transfers are not provided by Kraichnan (1959). Dar et al. (2001) and Verma (2004) derived formulas for these transfer rates. We denote these desired quantities by $S^{uu}(\mathbf{k}'|\mathbf{p}|\mathbf{q})$ and $S^{uu}(\mathbf{k}'|\mathbf{q}|\mathbf{p})$ respectively, as shown in Fig. 4.1(c). Here the first

[1]The superscript uu represents u to u energy transfers. In magnetohydrodynamics, we will also consider S^{bb} and S^{ub} that represent the magnetic to magnetic, and magnetic to kinetic energy transfers respectively.

argument is the *receiver* wavenumber, the second argument is the *giver* or *donor* wavenumber, and the third argument is the *mediator* wavenumber. For $dE_u(\mathbf{p})/dt$, the corresponding energy transfer functions are $S^{uu}(\mathbf{p}|\mathbf{k}'|\mathbf{q})$ and $S^{uu}(\mathbf{p}|\mathbf{q}|\mathbf{k}')$, and for $dE_u(\mathbf{q})/dt$, they are $S^{uu}(\mathbf{q}|\mathbf{k}'|\mathbf{p})$ and $S^{uu}(\mathbf{q}|\mathbf{p}|\mathbf{k}')$. Note that the mode-to-mode energy transfer $S^{uu}(\mathbf{k}'|\mathbf{p}|\mathbf{q})$ is very different from the combined energy transfer $S(\mathbf{k}'|\mathbf{p},\mathbf{q})$.

According to Dar et al. (2001) and Verma (2004), *the rate of mode-to-mode energy transfer* from mode $\mathbf{u}(\mathbf{p})$ to mode $\mathbf{u}(\mathbf{k}')$ with $\mathbf{u}(\mathbf{q})$ acting as a mediator is

$$S^{uu}(\mathbf{k}'|\mathbf{p}|\mathbf{q}) = -\Im\left[\{\mathbf{k}'\cdot\mathbf{u}(\mathbf{q})\}\{\mathbf{u}(\mathbf{p})\cdot\mathbf{u}(\mathbf{k}')\}\right]. \tag{4.7}$$

For brevity, we drop the term "rate of" and refer to S^{uu} of Eq. (4.7) as *mode-to-mode energy transfer.* In the aforementioned formula, the Fourier amplitudes of the receiver and giver modes are dotted together, while the receiver wavenumber is dotted with the amplitude of the mediator mode. Using this recipe, the formula for other energy transfers in the triad are as follows:

$$S^{uu}(\mathbf{k}'|\mathbf{q}|\mathbf{p}) = -\Im\left[\{\mathbf{k}'\cdot\mathbf{u}(\mathbf{p})\}\{\mathbf{u}(\mathbf{q})\cdot\mathbf{u}(\mathbf{k}')\}\right], \tag{4.8a}$$

$$S^{uu}(\mathbf{p}|\mathbf{k}'|\mathbf{q}) = -\Im\left[\{\mathbf{p}\cdot\mathbf{u}(\mathbf{q})\}\{\mathbf{u}(\mathbf{p})\cdot\mathbf{u}(\mathbf{k}')\}\right], \tag{4.8b}$$

$$S^{uu}(\mathbf{p}|\mathbf{q}|\mathbf{k}') = -\Im\left[\{\mathbf{p}\cdot\mathbf{u}(\mathbf{k}')\}\{\mathbf{u}(\mathbf{p})\cdot\mathbf{u}(\mathbf{q})\}\right], \tag{4.8c}$$

$$S^{uu}(\mathbf{q}|\mathbf{k}'|\mathbf{p}) = -\Im\left[\{\mathbf{q}\cdot\mathbf{u}(\mathbf{p})\}\{\mathbf{u}(\mathbf{q})\cdot\mathbf{u}(\mathbf{k}')\}\right], \tag{4.8d}$$

$$S^{uu}(\mathbf{q}|\mathbf{p}|\mathbf{k}') = -\Im\left[\{\mathbf{q}\cdot\mathbf{u}(\mathbf{k}')\}\{\mathbf{u}(\mathbf{p})\cdot\mathbf{u}(\mathbf{q})\}\right]. \tag{4.8e}$$

Since $\mathbf{k}' = -\mathbf{k}$, the formula $S^{uu}(\mathbf{k}'|\mathbf{p}|\mathbf{q})$ is also written as

$$S^{uu}(\mathbf{k}|\mathbf{p}|\mathbf{q}) = \Im\left[\{\mathbf{k}\cdot\mathbf{u}(\mathbf{q})\}\{\mathbf{u}(\mathbf{p})\cdot\mathbf{u}^*(\mathbf{k})\}\right]. \tag{4.9}$$

In the following, we sketch the proof for the aforementioned formulas.

The functions S^{uu}'s satisfy the following properties. For convenience, we label the wavenumbers as $\mathbf{X},\mathbf{Y},\mathbf{Z}$ with $\mathbf{X}+\mathbf{Y}+\mathbf{Z}=0$.

1. The sum of $S^{uu}(\mathbf{X}|\mathbf{Y}|\mathbf{Z})$ and $S^{uu}(\mathbf{X}|\mathbf{Z}|\mathbf{Y})$ is the combined energy transfer to mode $\mathbf{u}(\mathbf{X})$ from modes $\mathbf{u}(\mathbf{Y})$ and $\mathbf{u}(\mathbf{Z})$ (see Eq. (4.2)). Therefore,

$$S^{uu}(\mathbf{k}'|\mathbf{p}|\mathbf{q}) + S^{uu}(\mathbf{k}'|\mathbf{q}|\mathbf{p}) = S^{uu}(\mathbf{k}'|\mathbf{p},\mathbf{q}), \tag{4.10a}$$

$$S^{uu}(\mathbf{p}|\mathbf{k}'|\mathbf{q}) + S^{uu}(\mathbf{p}|\mathbf{q}|\mathbf{k}') = S^{uu}(\mathbf{p}|\mathbf{k}',\mathbf{q}), \tag{4.10b}$$

$$S^{uu}(\mathbf{q}|\mathbf{k}'|\mathbf{p}) + S^{uu}(\mathbf{q}|\mathbf{p}|\mathbf{k}') = S^{uu}(\mathbf{q}|\mathbf{k}',\mathbf{p}). \tag{4.10c}$$

2. By definition, the energy transfer from mode $\mathbf{u}(\mathbf{Y})$ to the mode $\mathbf{u}(\mathbf{X})$, $S^{uu}(\mathbf{X}|\mathbf{Y}|\mathbf{Z})$, is equal and opposite to the energy transfer from mode $\mathbf{u}(\mathbf{X})$ to mode $\mathbf{u}(\mathbf{Y})$, $S^{uu}(\mathbf{Y}|\mathbf{X}|\mathbf{Z})$. Therefore,
$$S^{uu}(\mathbf{k}'|\mathbf{p}|\mathbf{q}) + S^{uu}(\mathbf{p}|\mathbf{k}'|\mathbf{q}) = 0, \tag{4.11a}$$
$$S^{uu}(\mathbf{k}'|\mathbf{q}|\mathbf{p}) + S^{uu}(\mathbf{q}|\mathbf{k}'|\mathbf{p}) = 0, \tag{4.11b}$$
$$S^{uu}(\mathbf{p}|\mathbf{q}|\mathbf{k}') + S^{uu}(\mathbf{q}|\mathbf{p}|\mathbf{k}') = 0. \tag{4.11c}$$
Note that the terms in the derivation of Eq. (4.5) are cancelled due to the identities of Eqs. (4.11).

The aforementioned conditions arise due to the principles of transactions. Similar conditions would apply to financial transactions as well. Note that the mediator mode does not receive any energy during a transaction, but its presence is required. There is no transaction without the mediator mode.

The aforementioned six equations have six unknowns. It is easy to verify that the mode-to-mode energy transfers of Eqs. (4.7, 4.8) satisfy Eqs. (4.10, 4.11). Equations (4.10) are satisfied by definition, while Eqs. (4.11) are satisfied due to the incompressibility condition $\mathbf{q}\cdot\mathbf{u}(\mathbf{q}) = \mathbf{p}\cdot\mathbf{u}(\mathbf{p}) = \mathbf{k}'\cdot\mathbf{u}(\mathbf{k}') = 0$. For example,

$$\begin{aligned}
& S^{uu}(\mathbf{k}'|\mathbf{p}|\mathbf{q}) + S^{uu}(\mathbf{p}|\mathbf{k}'|\mathbf{q}) \\
={}& -\Im\left[\{\mathbf{k}'\cdot\mathbf{u}(\mathbf{q})\}\{\mathbf{u}(\mathbf{p})\cdot\mathbf{u}(\mathbf{k}')\}\right] - \Im\left[\{\mathbf{p}\cdot\mathbf{u}(\mathbf{q})\}\{\mathbf{u}(\mathbf{p})\cdot\mathbf{u}(\mathbf{k}')\}\right] \\
={}& -\Im\left[\{(\mathbf{k}'+\mathbf{p})\cdot\mathbf{u}(\mathbf{q})\}\{\mathbf{u}(\mathbf{p})\cdot\mathbf{u}(\mathbf{k}')\}\right] \\
={}& \Im\left[\{\mathbf{q}\cdot\mathbf{u}(\mathbf{q})\}\{\mathbf{u}(\mathbf{p})\cdot\mathbf{u}(\mathbf{k}')\}\right] = 0.
\end{aligned}$$

Hence, the function of Eqs. (4.7, 4.8) form a solution for the energy transfers in a triad.

We can rewrite Eqs. (4.10, 4.11) in the following matrix form:

$$\begin{pmatrix} 1 & 1 & 0 & 0 & 0 & 0 \\ 0 & 0 & 1 & 1 & 0 & 0 \\ 0 & 0 & 0 & 0 & 1 & 1 \\ 1 & 0 & 1 & 0 & 0 & 0 \\ 0 & 1 & 0 & 0 & 1 & 0 \\ 0 & 0 & 0 & 1 & 0 & 1 \end{pmatrix} \begin{pmatrix} S^{uu}(\mathbf{k}'|\mathbf{p}|\mathbf{q}) \\ S^{uu}(\mathbf{k}'|\mathbf{q}|\mathbf{p}) \\ S^{uu}(\mathbf{p}|\mathbf{k}'|\mathbf{q}) \\ S^{uu}(\mathbf{p}|\mathbf{q}|\mathbf{k}') \\ S^{uu}(\mathbf{q}|\mathbf{k}'|\mathbf{p}) \\ S^{uu}(\mathbf{q}|\mathbf{p}|\mathbf{k}') \end{pmatrix} = \begin{pmatrix} S^{uu}(\mathbf{k}'|\mathbf{p},\mathbf{q}) \\ S^{uu}(\mathbf{p}|\mathbf{k}',\mathbf{q}) \\ S^{uu}(\mathbf{q}|\mathbf{k}',\mathbf{p}) \\ 0 \\ 0 \\ 0 \end{pmatrix} \tag{4.12}$$

The determinant of the matrix in the above equation is zero. Hence, the set S^{uu}'s of Eqs. (4.7, 4.8) is not a unique solution of Eqs. (4.10, 4.11). It is straightforward to show that an arbitrary *circulating transfer* Δ as shown in Fig. 4.2(b) can be added to S^{uu}s. The new solution still satisfy Eqs. (4.10, 4.11). In the present book, we however argue that the above circulating transfer is zero.

Note that the mode-to-mode energy transfers S^{uu}'s cannot be measured using numerical simulation or experimental data. We can make an analogy with electrodynamics: S^{uu}'s are similar to the scalar and vector potentials of electrodynamics, while Δ is like a gauge. Dar et al. (2001) and Verma (2004) argued that the trivial circulating transfer Δ flows via $\mathbf{p} \to \mathbf{k}' \to \mathbf{q} \to \mathbf{p}$, and hence it does not alter physical observables such as $dE_u(\mathbf{k})/dt$. For each wavenumber, it gets cancelled by the two S^{uu}'s. Most importantly, in this book, we prove that $\Delta = 0$.

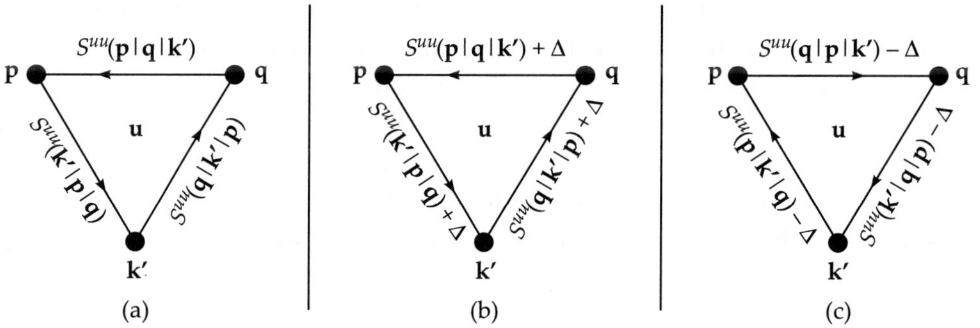

Figure 4.2 (a) A set of solution to Eqs. (4.10, 4.11). They are S^{uu}'s of Eqs. (4.7, 4.8). (b) We can add Δ to the solution of (a); they still satisfy Eqs. (4.7, 4.8). (c) Energy transfers opposite to those of (b).

In the following two subsections, we demonstrate uniqueness of the solution of Eqs. (4.7, 4.8) using physical and mathematical arguments. The mathematical arguments involve additional constraints imposed on S^{uu}'s.

4.1.1 A physical argument

The mode-to-mode energy transfer formula $S^{uu}(\mathbf{k}'|\mathbf{p}|\mathbf{q})$ of Eq. (4.7) can be interpreted in the following manner. The Navier–Stokes equations for a flow with a mean velocity $\mathbf{U}_0$ are

$$\frac{\partial \mathbf{u}'}{\partial t} + \mathbf{U}_0 \cdot \nabla \mathbf{u}' + \mathbf{u}' \cdot \nabla \mathbf{u}' = -\frac{1}{\rho}\nabla p + \frac{1}{\rho}\mathbf{F}_u + \nu \nabla^2 \mathbf{u}'. \tag{4.13}$$

Here, the net velocity $\mathbf{u} = \mathbf{U}_0 + \mathbf{u}'$. In the above equations, the mean flow $\mathbf{U}_0$ advects the fluctuations $\mathbf{u}'$. In similar lines, in the nonlinear term $(\mathbf{u}' \cdot \nabla)\mathbf{u}'$, $\mathbf{u}'$ in left of the ∇ operator is expected to advect the $\mathbf{u}'$ in the right.[2] This phenomena is more transparent in the following equation for a scalar θ:

$$\frac{\partial \theta}{\partial t} + (\mathbf{u} \cdot \nabla)\theta = \kappa \nabla^2 \theta, \tag{4.14}$$

where $\mathbf{u}$ in $(\mathbf{u} \cdot \nabla)\theta$ advects the scalar θ. Here, κ is the scalar diffusion coefficient.

Hence, in the nonlinear term of the energy equation, $[(\mathbf{u} \cdot \nabla)\mathbf{u}] \cdot \mathbf{u}$, $\mathbf{u}$ in left of the ∇ operator advects the field in the right of the ∇ operator, and it only mediates the energy transfers among the $\mathbf{u}$ fields in the right of the ∇ operator.[3] Hence, the Fourier mode $\mathbf{u}(\mathbf{q})$ that appears in the scalar product with $\mathbf{k}$ must be the mediator in the energy transfer. Therefore, $S^{uu}(\mathbf{k}'|\mathbf{p}|\mathbf{q})$ of Eq. (4.7) provides the energy transfer from $\mathbf{u}(\mathbf{p})$ to $\mathbf{u}(\mathbf{k}')$. See Fig. 4.3 for an illustration.

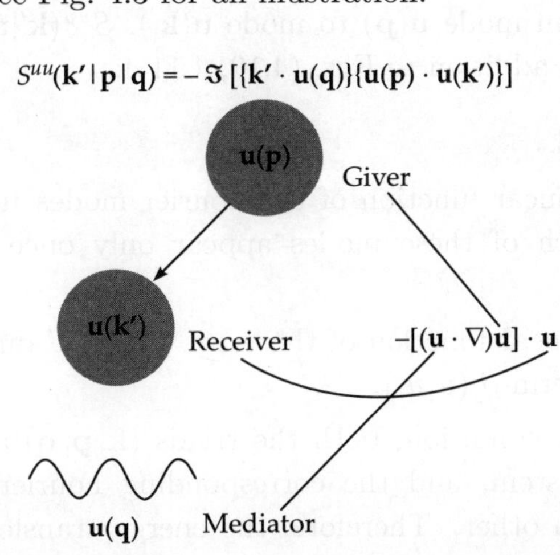

Figure 4.3 A schematic diagram exhibiting the mode-to-mode energy transfer from mode $\mathbf{u}(\mathbf{p})$ to mode $\mathbf{u}(\mathbf{k}')$ with mode $\mathbf{u}(\mathbf{q})$ acting as a mediator. During the energy transfer, the mode $\mathbf{u}(\mathbf{q})$ advects the modes $\mathbf{u}(\mathbf{p})$ and $\mathbf{u}(\mathbf{k}')$ who exchange energy among themselves. The mode $\mathbf{u}(\mathbf{q})$ does not receive any energy during the transaction.

A corollary to the above argument is that the mean flow $\mathbf{U}_0$ does not facilitate any energy transfer. The mean flow corresponds to $\mathbf{q} = 0$; hence, $\mathbf{p} = -\mathbf{k}'$. Therefore,

[2]Kraichnan (1964) argued that the large-scale velocity field *sweeps* the velocity fluctuations. The aforementioned arguments of fluctuations advecting fluctuations could be connected to the sweeping effect.

[3]This interpretation also works for the K41 formula (Kolmogorov, 1941a):

$$\epsilon_u = -T_u(l) = -\frac{1}{4}\nabla_l \cdot \langle \delta\mathbf{u}[\delta\mathbf{u} \cdot \delta\mathbf{u}] \rangle. \tag{4.15}$$

Here, the first argument of the ∇_l operator is the mediator field, while the other two $\delta\mathbf{u}$'s are the giver and receiver fields. See Section 12.3 for more details.

$$S^{uu}(\mathbf{k}'|\mathbf{p}|\mathbf{q}) = -\Im[\{\mathbf{k}'.\mathbf{U}_0\}\{\mathbf{u}(-\mathbf{k}')\cdot\mathbf{u}(\mathbf{k})\}] = -\Im[\{\mathbf{k}'.\mathbf{U}_0\}|\mathbf{u}(\mathbf{k})|^2] = 0 \quad (4.16)$$

because $\mathbf{U}_0$ is real. Thus, the energy transfer formula captures the effects of the mean velocity quite correctly.

In the next subsection we will provide a mathematical argument to claim that $S^{uu}(\mathbf{k}'|\mathbf{p}|\mathbf{q})$ is the correct energy transfer formula.

4.1.2 A mathematical argument based on tensor analysis

The following mathematical argument to prove uniqueness of the solution given by Eqs. (4.7, 4.8) is based on tensor analysis and the energy equations. Here we focus on the mode-to-mode transfer $S^{uu}(\mathbf{k}'|\mathbf{p}|\mathbf{q})$. From the energy equation, Eq. (4.2), the energy transfer from mode $\mathbf{u}(\mathbf{p})$ to mode $\mathbf{u}(\mathbf{k}')$, $S^{uu}(\mathbf{k}'|\mathbf{p}|\mathbf{q})$, must satisfy the following properties (in addition to Eqs. (4.10, 4.11)):

1. $S^{uu}(\mathbf{k}'|\mathbf{p}|\mathbf{q})$ is real.
2. $S^{uu}(\mathbf{k}'|\mathbf{p}|\mathbf{q})$ is a linear function of the Fourier modes $\mathbf{u}(\mathbf{k}')$, $\mathbf{u}(\mathbf{p})$, and $\mathbf{u}(\mathbf{q})$. Consequently, each of these modes appear only once in the expression of $S^{uu}(\mathbf{k}'|\mathbf{p}|\mathbf{q})$.
3. $S^{uu}(\mathbf{k}'|\mathbf{p}|\mathbf{q})$ is a linear function of the wave vector $\mathbf{k}'$ due to the derivative ∂_j in the nonlinear term $\partial_j(u_j u_i)$.
4. Due to the reality condition, both the triads $(\mathbf{k},\mathbf{p},\mathbf{q})$ and $(-\mathbf{k},-\mathbf{p},-\mathbf{q})$ are present in the system, and the corresponding Fourier modes are complex conjugates of each other. Therefore, the energy transfer from $\mathbf{u}(\mathbf{p})$ to $\mathbf{u}(\mathbf{k}')$ should be the same as $\mathbf{u}(-\mathbf{p})$ to $\mathbf{u}(-\mathbf{k}')$, that is,

$$S^{uu}(-\mathbf{k}'|-\mathbf{p}|-\mathbf{q}) = S^{uu}(\mathbf{k}'|\mathbf{p}|\mathbf{q}). \quad (4.17)$$

Given the above properties, the tensor analysis yields the following form for $S^{uu}(\mathbf{k}'|\mathbf{p}|\mathbf{q})$:

$$\begin{aligned} S^{uu}(\mathbf{k}'|\mathbf{p}|\mathbf{q}) &= c_1\Re\left[\{\mathbf{k}'\cdot\mathbf{u}(\mathbf{q})\}\{\mathbf{u}(\mathbf{p})\cdot\mathbf{u}(\mathbf{k}')\}\right] \\ &+ c_2\Re\left[\{\mathbf{k}'\cdot\mathbf{u}(\mathbf{p})\}\{\mathbf{u}(\mathbf{q})\cdot\mathbf{u}(\mathbf{k}')\}\right] \\ &+ c_3\Re\left[\{\mathbf{k}'\cdot\mathbf{u}(\mathbf{k}')\}\{\mathbf{u}(\mathbf{p})\cdot\mathbf{u}(\mathbf{q})\}\right] \\ &+ c_4\Im\left[\{\mathbf{k}'\cdot\mathbf{u}(\mathbf{q})\}\{\mathbf{u}(\mathbf{p})\cdot\mathbf{u}(\mathbf{k}')\}\right] \\ &+ c_5\Im\left[\{\mathbf{k}'\cdot\mathbf{u}(\mathbf{p})\}\{\mathbf{u}(\mathbf{q})\cdot\mathbf{u}(\mathbf{k}')\}\right] \\ &+ c_6\Im\left[\{\mathbf{k}'\cdot\mathbf{u}(\mathbf{k}')\}\{\mathbf{u}(\mathbf{p})\cdot\mathbf{u}(\mathbf{q})\}\right], \end{aligned} \quad (4.18)$$

where c_i's are constants. The fourth condition, Eq. (4.17), makes $c_1 = c_2 = c_3 = 0$ because the corresponding formulas change sign under the $(\mathbf{k}', \mathbf{p}, \mathbf{q}) \to (-\mathbf{k}', -\mathbf{p}, -\mathbf{q})$ transformation. For example,

$$\Re\left[\{\mathbf{k}' \cdot \mathbf{u}(\mathbf{q})\}\{\mathbf{u}(\mathbf{p}) \cdot \mathbf{u}(\mathbf{k}')\}\right] \to \Re\left[\{-\mathbf{k}' \cdot \mathbf{u}(-\mathbf{q})\}\{\mathbf{u}(-\mathbf{p}) \cdot \mathbf{u}(-\mathbf{k}')\}\right]$$

$$= -\Re\left[\{\mathbf{k}' \cdot \mathbf{u}^*(\mathbf{q})\}\{\mathbf{u}^*(\mathbf{p}) \cdot \mathbf{u}^*(\mathbf{k}')\}\right] = -\Re\left[\{\mathbf{k}' \cdot \mathbf{u}(\mathbf{q})\}\{\mathbf{u}(\mathbf{p}) \cdot \mathbf{u}(\mathbf{k}')\}\right].$$

In addition, the sixth term of Eq. (4.18) is zero because $\mathbf{k}' \cdot \mathbf{u}(\mathbf{k}') = 0$. Now we are left with

$$\begin{aligned} S^{uu}(\mathbf{k}'|\mathbf{p}|\mathbf{q}) &= c_4 \Im\left[\{\mathbf{k}' \cdot \mathbf{u}(\mathbf{q})\}\{\mathbf{u}(\mathbf{p}) \cdot \mathbf{u}(\mathbf{k}')\}\right] \\ &+ c_5 \Im\left[\{\mathbf{k}' \cdot \mathbf{u}(\mathbf{p})\}\{\mathbf{u}(\mathbf{q}) \cdot \mathbf{u}(\mathbf{k}')\}\right]. \end{aligned} \tag{4.19}$$

When we substitute Eq. (4.19) in Eq. (4.11a), we obtain

$$c_5 \Im\left[\{\mathbf{k}' \cdot \mathbf{u}(\mathbf{p})\}\{\mathbf{u}(\mathbf{q}) \cdot \mathbf{u}(\mathbf{k}')\}\right] + c_5 \Im\left[\{\mathbf{p} \cdot \mathbf{u}(\mathbf{k}')\}\{\mathbf{u}(\mathbf{q}) \cdot \mathbf{u}(\mathbf{p})\}\right] = 0. \tag{4.20}$$

Hence, $c_5 = 0$, and

$$S^{uu}(\mathbf{k}'|\mathbf{p}|\mathbf{q}) = c_4 \Im\left[\{\mathbf{k}' \cdot \mathbf{u}(\mathbf{q})\}\{\mathbf{u}(\mathbf{p}) \cdot \mathbf{u}(\mathbf{k}')\}\right]. \tag{4.21}$$

Using similar arguments, we can show that the other energy transfers between modes in a triad are given by Eqs. (4.8) with a factor of c_4 as in Eq. (4.21). When we substitute them in Eq. (4.10a), we obtain

$$\begin{aligned} c_4 \left[\Im\left[\{\mathbf{k}' \cdot \mathbf{u}(\mathbf{q})\}\{\mathbf{u}(\mathbf{p}) \cdot \mathbf{u}(\mathbf{k}')\}\right]\right] &+ c_4 \left[\Im\left[\{\mathbf{k}' \cdot \mathbf{u}(\mathbf{p})\}\{\mathbf{u}(\mathbf{q}) \cdot \mathbf{u}(\mathbf{k}')\}\right]\right] \\ &= S^{uu}(\mathbf{k}'|\mathbf{p}, \mathbf{q}). \end{aligned} \tag{4.22}$$

Comparison of this equation with Eq. (4.2) yields $c_4 = -1$. Thus, we show that Eqs. (4.7, 4.8) provide unique formulas for the mode-to-mode energy transfers in a triad. Q.E.D.

Both the triads $(\mathbf{k}, \mathbf{p}, \mathbf{q})$ and $(-\mathbf{k}, -\mathbf{p}, -\mathbf{q})$ exist in the system. Using $S^{uu}(\mathbf{k}| - \mathbf{p}| - \mathbf{q}) = S^{uu}(\mathbf{k}'|\mathbf{p}|\mathbf{q})$ we obtain

$$\frac{d}{dt} E_u(\mathbf{k}', t) = S^{uu}(\mathbf{k}'|\mathbf{p}|\mathbf{q}) + S^{uu}(\mathbf{k}'|\mathbf{q}|\mathbf{p}) \tag{4.23a}$$

$$\frac{d}{dt} E_u(\mathbf{k}, t) = S^{uu}(\mathbf{k}| - \mathbf{p}| - \mathbf{q}) + S^{uu}(\mathbf{k}| - \mathbf{q}| - \mathbf{p}). \tag{4.23b}$$

Therefore,

$$\frac{d}{dt} E_u(\mathbf{k}', t) = \frac{d}{dt} E_u(\mathbf{k}, t), \tag{4.24}$$

or the net energy transfers to the modes $\mathbf{u}(\mathbf{k})$ and $\mathbf{u}(-\mathbf{k})$ via the corresponding nonlinear triads are identical. Due to these reasons, $S(\mathbf{k}'|\mathbf{p}|\mathbf{q})$ is also written as $S(\mathbf{k}|\mathbf{p}|\mathbf{q})$:

$$\begin{aligned} S^{uu}(\mathbf{k}|\mathbf{p}|\mathbf{q}) &= -\Im\left[\{\mathbf{k}'\cdot\mathbf{u}(\mathbf{q})\}\{\mathbf{u}(\mathbf{p})\cdot\mathbf{u}(\mathbf{k}')\}\right] \\ &= \Im\left[\{\mathbf{k}\cdot\mathbf{u}(\mathbf{q})\}\{\mathbf{u}(\mathbf{p})\cdot\mathbf{u}^*(\mathbf{k})\}\right]. \end{aligned} \tag{4.25}$$

Example 4.1: Using Eq. (4.7) explicitly prove that $S^{uu}(-\mathbf{k}'|-\mathbf{p}|-\mathbf{q}) = S^{uu}(\mathbf{k}'|\mathbf{p}|\mathbf{q})$.

Solution: We apply Eq. (4.7) to the triad $(-\mathbf{k}', -\mathbf{p}, -\mathbf{q})$ and derive

$$\begin{aligned} S^{uu}(-\mathbf{k}'|-\mathbf{p}|-\mathbf{q}) &= -\Im\left[\{-\mathbf{k}'\cdot\mathbf{u}(-\mathbf{q})\}\{\mathbf{u}(-\mathbf{p})\cdot\mathbf{u}(-\mathbf{k}')\}\right] \\ &= \Im\left[\{\mathbf{k}'\cdot\mathbf{u}^*(\mathbf{q})\}\{\mathbf{u}^*(\mathbf{p})\cdot\mathbf{u}^*(\mathbf{k}')\}\right] \\ &= -\Im\left[\{\mathbf{k}'\cdot\mathbf{u}(\mathbf{q})\}\{\mathbf{u}(\mathbf{p})\cdot\mathbf{u}(\mathbf{k}')\}\right] \\ &= S^{uu}(\mathbf{k}'|\mathbf{p}|\mathbf{q}) \end{aligned}$$

since $\Im(f^*) = -\Im(f)$.

Example 4.2: Show that the energy transfers for a triad ($\mathbf{k} = \mathbf{a}$, $\mathbf{p} = 3\mathbf{a}/4$, $\mathbf{q} = \mathbf{a}/4$) are zero. Here $\mathbf{a}$ is a constant vector.

Solution: The vectors $\mathbf{k}, \mathbf{p}, \mathbf{q}$ are collinear along $\mathbf{a}$. Therefore, all the Fourier modes $\mathbf{u}(\mathbf{k}), \mathbf{u}(\mathbf{k}), \mathbf{u}(\mathbf{q})$ would be perpendicular to $\mathbf{a}$. Hence, for the energy transfers, the receiver wavenumber is always perpendicular to the vector direction of the mediator mode. For example, $\mathbf{k}\cdot\mathbf{u}(\mathbf{q}) = 0$. Therefore, using Eq. (4.9) we deduce that the energy transfers between any two modes is zero.

A typical flow involves a large number of interacting modes. In the following section, we describe these collective interactions.

4.2 Energy Transfers in the Presence of Many Triads

Now we write down an equation for the energy in the presence of a large number of triads. Following Eq. (4.1) we obtain

$$\frac{d}{dt}E_u(\mathbf{k}) = \sum_{\mathbf{p}} \Im\left[\{\mathbf{k}\cdot\mathbf{u}(\mathbf{q})\}\{\mathbf{u}(\mathbf{p})\cdot\mathbf{u}^*(\mathbf{k})\}\right] + \Re[\mathbf{F}_u(\mathbf{k})\cdot\mathbf{u}^*(\mathbf{k})] - 2\nu k^2 E_u(\mathbf{k})$$

$$= \sum_{\mathbf{p}} S^{uu}(\mathbf{k}|\mathbf{p}|\mathbf{q}) + \mathcal{F}_u(\mathbf{k}) - D_u(\mathbf{k})$$
$$= T_u(\mathbf{k}) + \mathcal{F}_u(\mathbf{k}) - D_u(\mathbf{k}), \tag{4.26}$$

where $\mathbf{q} = \mathbf{k} - \mathbf{p}$, $T_u(\mathbf{k})$ is the total energy transfer to $\mathbf{u}(\mathbf{k})$ via nonlinear interactions,

$$\mathcal{F}_u(\mathbf{k}) = \Re[\mathbf{F}_u(\mathbf{k}) \cdot \mathbf{u}^*(\mathbf{k})] \tag{4.27}$$

is the energy supply by the external force, and

$$D_u(\mathbf{k}) = -2\nu k^2 E_u(\mathbf{k}) \tag{4.28}$$

is the viscous dissipation rate of $E_u(\mathbf{k})$. The sum $\sum_{\mathbf{p}}$ is performed over all the modes except $\mathbf{k}$. Thus, the sum includes the energy transfer from wavenumber $\mathbf{q} = \mathbf{k} - \mathbf{p}$ to $\mathbf{k}$. This is the reason why we do not write $S^{uu}(\mathbf{k}|\mathbf{q}|\mathbf{p})$ in the aforementioned sum.

For the total energy $E_u = \sum_{\mathbf{k}} E_u(\mathbf{k})$, in the absence of viscous dissipation and eternal force,

$$\frac{d}{dt} E_u = \sum_{\mathbf{k}} \sum_{\mathbf{p}} S(\mathbf{k}|\mathbf{p}|\mathbf{q}) = 0 \tag{4.29}$$

because in the above equation, all the energy transfers appear in pairs, and they cancel each other since

$$S(\mathbf{k}'|\mathbf{p}|\mathbf{q}) + S(\mathbf{p}|\mathbf{k}'|\mathbf{q}) = 0. \tag{4.30}$$

Thus, we show that the total energy in the Fourier space is conserved. The nonlinear interactions only exchange energy among the modes while keeping the total energy constant. Using similar arguments as above, we can show that

$$\sum_{\mathbf{k}' \in A} \sum_{\mathbf{p} \in A} S^{uu}(\mathbf{k}'|\mathbf{p}|\mathbf{q}) = 0, \tag{4.31}$$

where A represents a wavenumber region.

Example 4.3: Consider the Fourier modes of Example 3.4. Compute the energy transfers among the modes.

Solution: The interacting triad formed by the wavenumber is shown in Fig. 3.1. The energy transfers among the modes of the triad are

$$\begin{aligned}
S^{uu}\{(-1,-1)|(0,1)|(1,0)\} &= -\Im([(-\hat{x}-\hat{y})\cdot(C\hat{y})][(B\hat{x})\cdot\frac{A}{i}(-\hat{x}+\hat{y})]) \\
&= ABC = -S^{uu}\{(0,1)|(-1,-1)|(1,0)\}, \\
S^{uu}\{(-1,-1)|(1,0)|(0,1)\} &= -\Im([(-\hat{x}-\hat{y})\cdot(B\hat{x})][(C\hat{y})\cdot\frac{A}{i}(-\hat{x}+\hat{y})]) \\
&= -ABC = -S^{uu}\{(1,0)|(-1,-1)|(0,1)\}, \\
S^{uu}\{(1,0)|(0,1)|(-1,-1))\} &= -\Im([\hat{x}\cdot\frac{A}{i}(-\hat{x}+\hat{y})][C\hat{y}\cdot B\hat{x}) \\
&= 0 = S^{uu}\{(0,1)|(1,0)|(-1,-1))\}.
\end{aligned}$$

Hence, the energy ABC flows from $\mathbf{u}(0,1)$ to $\mathbf{u}(-1,-1)$, which is then transferred to $\mathbf{u}(1,0)$. Therefore, the net energy gain by the mode $\mathbf{u}(-1,-1)$ is zero, and $\dot{A}=0$. We show in Example 3.4 that A = constant, $B = c\cos(At)$, and $C = c\sin(At)$. Therefore, ABC changes sign with a frequency of $2A$. We depict the aforementioned energy transfers in Fig. 4.4(a). Note that there is no energy transfer between the modes $\mathbf{u}(0,1)$ and $\mathbf{u}(1,0)$.

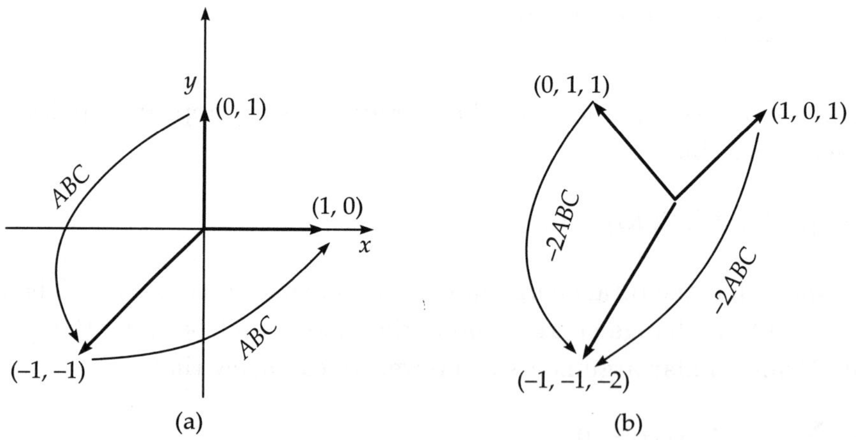

Figure 4.4 Energy transfers among the modes in the triads of (a) Example 4.3 and (b) Example 4.5.

The energy transfers in the second triad $\{(1,1),(0,-1),(-1,0)\}$ can be computed by following a similar procedure. Using the properties of the Fourier modes, we can show that the energy transfers among the modes $\mathbf{u}(0,-1)$, $\mathbf{u}(-1,0)$, and $\mathbf{u}(1,1)$ are identical to the corresponding transfers among $\mathbf{u}(0,1)$, $\mathbf{u}(1,0)$, and $\mathbf{u}(-1,-1)$.

Example 4.4: Compute the energy transfers among the modes of Example 3.5.

Solution: The flow of Example 3.5 is very similar to that of Example 3.4, except that the amplitudes of the modes are all real. Note that the mode-to-mode energy transfer is proportional to the imaginary part of the products of the Fourier modes. Therefore, the energy transfers vanish for this example. As a result, the mode amplitudes do not change with time, which is consistent with the results of Example 3.5.

Example 4.5: Compute the energy transfers among the modes of Example 3.6.

Solution: The Fourier wavenumbers that constitute the flow are $(\pm1, 0, \pm1)$, $(0, \pm1, \pm1)$, and $(\pm1, \pm1, \pm2)$, and they form a number of triads. Here we focus on energy transfers in one of the triads whose wavenumbers are $\mathbf{q} = (1, 0, 1)$, $\mathbf{p} = (0, 1, 1)$, and $\mathbf{k}' = (-1, -1, -2)$. In this triad, the energy transfers are

$$\begin{aligned} S^{uu}(\mathbf{k}'|\mathbf{p}|\mathbf{q}) &= -\Im\left[\{\mathbf{k}' \cdot \mathbf{u}(\mathbf{q})\}\{\mathbf{u}(\mathbf{p}) \cdot \mathbf{u}(\mathbf{k}')\}\right] = -2ABC \\ S^{uu}(\mathbf{k}'|\mathbf{q}|\mathbf{p}) &= -\Im\left[\{\mathbf{k}' \cdot \mathbf{u}(\mathbf{p})\}\{\mathbf{u}(\mathbf{q}) \cdot \mathbf{u}(\mathbf{k}')\}\right] = -2ABC \\ S^{uu}(\mathbf{q}|\mathbf{p}|\mathbf{k}') &= -\Im\left[\{\mathbf{q} \cdot \mathbf{u}(\mathbf{k}')\}\{\mathbf{u}(\mathbf{p}) \cdot \mathbf{u}(\mathbf{q})\}\right] = 0 \end{aligned}$$

We depict the energy transfers in Fig. 4.4(b).

The energy flows from the modes $\mathbf{u}(\mathbf{p})$ and $\mathbf{u}(\mathbf{q})$ to the mode $\mathbf{u}(\mathbf{k})$ by an amount $-2ABC$ each. Therefore, net rate of energy transfer to $\mathbf{u}(\mathbf{k}')$ is $-4ABC$. As discussed in Example 3.6, two quantities among A, B, C oscillates around zero, while the third one oscillates around a mean value. Hence, $S^{uu}(\mathbf{k}'|\mathbf{p}|\mathbf{q}) = -2ABC$ oscillates around zero as shown in Fig. 3.3.

The energy transfers in other triads can be computed easily. We observe that all the modes $\mathbf{u}(\pm1, \pm1, \pm2)$ gain energy by an amount $-4ABC$.

The energy transfer formula of Eq. (4.7) is general, and it is applicable to both 2D and 3D fluid flows. The formulas in 2D however get simplified, as we will illustrate in the next section.

4.3 Energy Transfers and Equations of Motion for a Two-dimensional Flow

In the present section, we consider a two-dimensional flow in the xy plane. We derive the energy transfer formulas for this flow.

In a 2D incompressible flow, the amplitude of a Fourier mode can be expressed using a single complex number due to the constraint $\mathbf{k} \cdot \mathbf{u}(\mathbf{k}) = 0$. The velocity vector $\mathbf{u}(\mathbf{k})$ is in the direction of the unit vector $\hat{k} \times \hat{z}$, where $\hat{k}$ is the unit vector along $\mathbf{k}$, and $\hat{z}$ is perpendicular to the plane containing the flow.[4] Hence, we can express $\mathbf{u}(\mathbf{k})$ as

$$\mathbf{u}(\mathbf{k}) = u(\mathbf{k})(\hat{k} \times \hat{z}), \tag{4.32}$$

where $u(\mathbf{k})$ is the amplitude of the velocity Fourier mode. See Fig. 4.5(a) for an illustration. This decomposition is related to the *Craya–Herring decomposition*, and will be discussed in more detail in Chapter 9. For convenience, and to relate the present discussion to the Craya–Herring decomposition, we denote

$$e_1(\mathbf{k}) = \hat{k} \times \hat{z}. \tag{4.33}$$

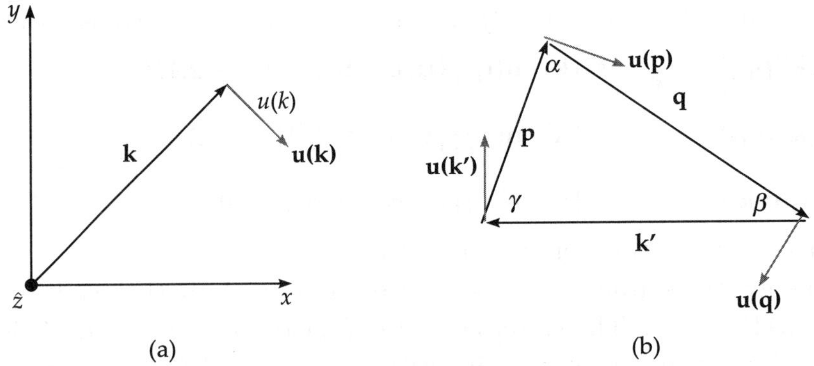

Figure 4.5 For 2D hydrodynamics: (a) Representation of a velocity Fourier mode $\mathbf{u}(\mathbf{k})$; (b) An interacting wavenumber triad. Here $\mathbf{k}' + \mathbf{p} + \mathbf{q} = 0$. The figure illustrates the Fourier modes as well as the internal angles.

Now let us rephrase the formulas for the energy transfers in a triad for a 2D flow. These formulas become somewhat simpler:

$$\begin{aligned} S^{uu}(\mathbf{k}'|\mathbf{p}|\mathbf{q}) &= -\Im\left[\{\mathbf{k}' \cdot \mathbf{u}(\mathbf{q})\}\{\mathbf{u}(\mathbf{p}) \cdot \mathbf{u}(\mathbf{k}')\}\right] \\ &= -[\mathbf{k}' \cdot (\hat{q} \times \hat{n})][(\hat{p} \times \hat{n}) \cdot (\hat{k}' \times \hat{n})]\Im\{u(\mathbf{q})u(\mathbf{p})u(\mathbf{k}')\} \\ &= -k'[(\hat{k}' \times \hat{q}) \cdot \hat{n}](\hat{k}' \cdot \hat{p})\Im\{u(\mathbf{q})u(\mathbf{p})u(\mathbf{k}')\} \\ &= k' \sin\beta \cos\gamma \Im\{u(\mathbf{q})u(\mathbf{p})u(\mathbf{k}')\}, \end{aligned} \tag{4.34}$$

where the angles α, β, γ are shown in Fig. 4.5(b). Note that $\hat{k}' \cdot \hat{p} = -\cos\gamma$. Similar

[4] $\hat{k} \times \hat{z}$ is along $\hat{e}_1$ of the Craya–Herring basis that will be discussed in Chapter 9.

computations yield

$$S^{uu}(\mathbf{p}|\mathbf{q}|\mathbf{k}') = -p[(\hat{p}\times\hat{k}')\cdot\hat{n}](\hat{p}\cdot\hat{q})\Im\{u(\mathbf{q})u(\mathbf{p})u(\mathbf{k}')\}$$
$$= p\sin\gamma\cos\alpha\Im\{u(\mathbf{q})u(\mathbf{p})u(\mathbf{k}')\}. \tag{4.35a}$$
$$S^{uu}(\mathbf{q}|\mathbf{k}'|\mathbf{p}) = -q[(\hat{q}\times\hat{p})\cdot\hat{n}](\hat{q}\cdot\hat{k}')\Im\{u(\mathbf{q})u(\mathbf{p})u(\mathbf{k}')\}$$
$$= q\sin\gamma\cos\beta\Im\{u(\mathbf{q})u(\mathbf{p})u(\mathbf{k}')\}, \tag{4.35b}$$

A quick way to construct the above formulas is as follows. When we traverse the wavenumbers $(\mathbf{a},\mathbf{b},\mathbf{c})$ in a clockwise direction as in Fig. 4.5(b), the energy transfer $S^{uu}(\mathbf{a}|\mathbf{b}|\mathbf{c})$ is a product of four terms: (a) the magnitude of the receiver wavenumber, (b) $\Im\{u(\mathbf{a})u(\mathbf{b})u(\mathbf{c})\}$, (c) $\sin\zeta$, (b) $\cos\psi$, where ζ,ψ are respectively the angles of the triangle at the tail and head of the receiver wavenumber. The aforementioned formulas shows that the energy transfer is zero if the giver and the receiver Fourier modes are perpendicular to each other.

We can also derive equations of motion for $u(\mathbf{k})$ and other two modes. Equation (4.32) is a convenient representation because of the automatic elimination of the pressure term. Moreover, we have only one equation for every wavenumber. If we focus on a single triad, then the equations of motion for the modes are derived as follows.

$$\frac{d}{dt}\mathbf{u}(\mathbf{k}') = -i[\mathbf{k}'\cdot\mathbf{u}(-\mathbf{q})]\mathbf{u}(-\mathbf{p})-i[\mathbf{k}'\cdot\mathbf{u}(-\mathbf{p})]\mathbf{u}(-\mathbf{q})-i\mathbf{k}p(\mathbf{k})-\nu k^2\mathbf{u}(\mathbf{k}') \tag{4.36}$$

since $\mathbf{k}'=-\mathbf{p}-\mathbf{q}$. By taking a dot product of Eq. (4.36) with $\hat{e}_1(\mathbf{k}')$, we obtain

$$\dot{u}(\mathbf{k}') = [-ik'\sin\beta\hat{e}_1(\mathbf{p})\cdot\hat{e}_1(\mathbf{k}')+ik'\sin\gamma\hat{e}_1(\mathbf{q})\cdot\hat{e}_1(\mathbf{k}')]u^*(\mathbf{q})u^*(\mathbf{p})-\nu k^2u(\mathbf{k}')$$
$$= ik'\sin(\beta-\gamma)u^*(\mathbf{p})u^*(\mathbf{q})-\nu k^2u(\mathbf{k}'). \tag{4.37}$$

We derive the equations for the other two modes in a similar fashion:

$$\dot{u}(\mathbf{p}) = ip\sin(\gamma-\alpha)u^*(\mathbf{k}')u^*(\mathbf{q})-\nu p^2u(\mathbf{p}), \tag{4.38a}$$
$$\dot{u}(\mathbf{q}) = iq\sin(\alpha-\beta)u^*(\mathbf{k}')u^*(\mathbf{p})-\nu q^2u(\mathbf{q}). \tag{4.38b}$$

Example 4.6: Rework the energy transfer rates for the triad $\mathbf{q}=(1,0)$, $\mathbf{p}=(0,1)$, $\mathbf{k}'=(-1,-1)$ of Example 3.4 using the formula derived in Section 4.3.

Solution: The set of wavenumbers $(\mathbf{k}',\mathbf{p},\mathbf{q})$ form a triad. Following the convention described in the present section, we deduce that the velocity Fourier modes are aligned as shown in Fig. 4.6 with the amplitudes of the modes being

$$u(\mathbf{q}) = u(1,0) = -C; \quad u(\mathbf{p}) = u(0,1) = B; \quad u(\mathbf{k}') = u(-1,-1) = A\sqrt{2}/i.$$

The internal angles of the triangle are

$$\alpha = 90°; \quad \beta = 45°; \quad \gamma = 45°.$$

Substitution of these values in Eqs. (4.34) yields

$$\begin{aligned} S^{uu}(\mathbf{k}'|\mathbf{p}|\mathbf{q}) &= k' \sin\beta \cos\gamma \Im\{u(\mathbf{q})u(\mathbf{p})u(\mathbf{k}')\} \\ &= \sqrt{2} \times \frac{1}{2} \Im\left[B(-C)(\frac{A}{i}\sqrt{2})\right] = ABC. \end{aligned}$$

Following a similar procedure, we derive that

$$\begin{aligned} S^{uu}(\mathbf{p}|\mathbf{q}|\mathbf{k}') &= 0, \\ S^{uu}(\mathbf{q}|\mathbf{k}'|\mathbf{p}) &= ABC. \end{aligned}$$

These results are the same as those derived in Example 4.3.

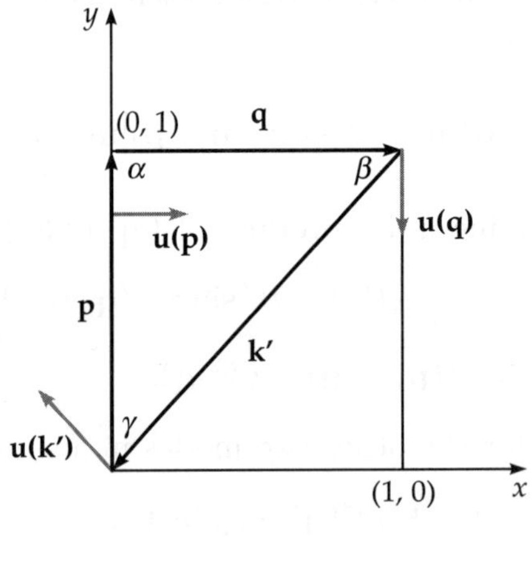

Figure 4.6 Example 4.6.

The aforementioned formalism of mode-to-mode energy transfers is very useful for analyzing pattern formation, energy flux, shell-to-shell energy transfers, etc. In this book we will frequently use this formalism to derive expressions for the energy flux and shell-to-shell energy transfers in turbulence.

4.4 Spectral Energy Flux

The flow configuration of a fully developed turbulence is an organized state.[5] The phases of the Fourier modes are related to each other in an intricate manner so as to transfer kinetic energy from large scales to small scales. Note that random phases for the modes will yield zero energy flux. It is the complex correlations among the modes that yield finite energy flux in turbulence. Thus, a turbulent flow is not random as in white noise.

In the subsequent part of this chapter we will derive collective measures of energy transfers—energy flux and shell-to-shell energy transfers. In this section we start with energy flux. In a generic three-dimensional flow, the energy supplied at the large scales cascades to the intermediate range (called inertial range), and then to the dissipation range. We will define these wavenumber ranges more precisely in the next chapter. Quantitatively, energy flux $\Pi(k_0)$ of a sphere of radius k_0 is defined as the cumulative energy transfer rate from the modes inside the sphere to the modes outside the sphere. See Fig. 4.7 for an illustration of the energy flux.

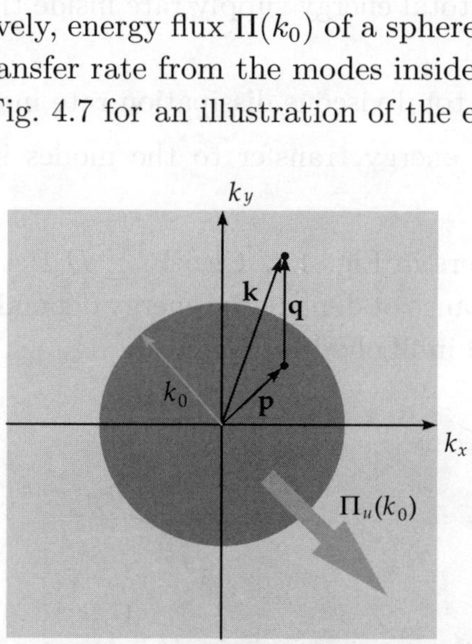

Figure 4.7 Energy flux from a wavenumber sphere of radius k_0. It is a sum of the energy transfers from all the modes inside the sphere to all the modes outside the sphere. One such transfer is from mode $\mathbf{u}(\mathbf{p})$ residing inside the sphere to mode $\mathbf{u}(\mathbf{k})$ that is outside the sphere.

This formula for the flux can be easily derived using the mode-to-mode energy transfers. We need to sum over all the energy transfers for which the giver mode is inside the sphere and the receiver mode is outside the sphere. Thus,

[5]Some researchers claim that fully-developed turbulence is a self-organized critical state. This is an interesting field of research, but is beyond the scope of this book.

$$\Pi_u(k_0) = \sum_{|\mathbf{p}|\le k_0} \sum_{|\mathbf{k}|>k_0} S^{uu}(\mathbf{k}|\mathbf{p}|\mathbf{q}) \tag{4.39}$$

with $\mathbf{k} = \mathbf{p} + \mathbf{q}$.

We can derive the formula for the energy flux in several other ways. Using Eq. (4.26), we derive

$$\frac{d}{dt} \sum_{|\mathbf{k}|\le k_0} E_u(\mathbf{k}) = \sum_{|\mathbf{k}|\le k_0} T_u(\mathbf{k}) + \sum_{|\mathbf{k}|\le k_0} \mathcal{F}_u(\mathbf{k}) - \sum_{|\mathbf{k}|\le k_0} D_u(\mathbf{k}). \tag{4.40}$$

Physical interpretations of the three terms in the right-hand side of Eq. (4.40) are as follows:

1. $\sum_{|\mathbf{k}|\le k_0} \mathcal{F}_u(\mathbf{k})$ is the total energy supply rate inside the sphere by the external force.
2. $\sum_{|\mathbf{k}|\le k_0} D_u(\mathbf{k})$ is the total viscous dissipation rate inside the sphere.
3. $\sum_{|\mathbf{k}|\le k_0} T_u(\mathbf{k})$ is the energy transfer to the modes inside the sphere due to nonlinearity.

We illustrate these transfers in Fig. 4.8. Clearly $\sum D_u(\mathbf{k})$ depletes the energy, but $\sum \mathcal{F}_u(\mathbf{k})$ and $\sum T_u(\mathbf{k})$ enhance or deplete the energy depending on their signs. Under steady state, $\sum T_u(\mathbf{k}) < 0$ in 3D hydrodynamics.

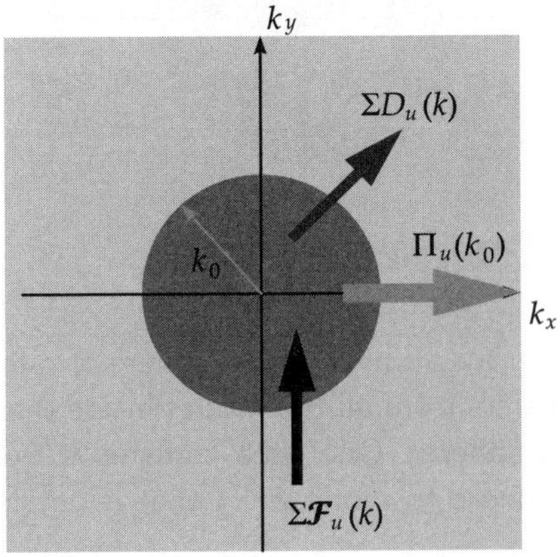

Figure 4.8 The energy of the sphere is affected due to the energy supply rate by the forcing $\sum \mathcal{F}_u(\mathbf{k})$, the energy flux $\Pi_u(k_0)$, and the viscous dissipation rate $\sum D_u(\mathbf{k})$.

It is important to compare this energy balance in Fourier space with the corresponding energy balance in real space (see Eq. (2.29)). The Fourier space counterpart provides scale-by-scale energy transfer, which is not possible in real space. Kolmogorov's theory based on third order structure function however provide a measure of the inertial range flux; this topic will be discussed in Chapter 12.

Since the total kinetic energy is conserved for nonlinear interactions, the energy lost by the sphere equals the gain of energy in the outer region. This energy transfer is the kinetic energy flux. Hence,

$$\Pi_u(k_0) = -\sum_{|\mathbf{k}|\le k_0} T_u(\mathbf{k}). \tag{4.41}$$

We can show equivalence between Eq. (4.39) and Eq. (4.41) using the following arguments:

$$\begin{aligned}
\Pi_u(k_0) &= -\sum_{|\mathbf{k}'|\le k_0} T_u(\mathbf{k}) \\
&= -\sum_{|\mathbf{k}'|\le k_0} \sum_{\mathbf{p}} S^{uu}(\mathbf{k}'|\mathbf{p}|\mathbf{q}) \\
&= -\sum_{|\mathbf{k}'|\le k_0} \sum_{|\mathbf{p}|\le k_0} S^{uu}(\mathbf{k}'|\mathbf{p}|\mathbf{q}) - \sum_{|\mathbf{k}|\le k_0} \sum_{|\mathbf{p}|> k_0} S^{uu}(\mathbf{k}'|\mathbf{p}|\mathbf{q}) \\
&= 0 + \sum_{|\mathbf{p}|> k_0} \sum_{|\mathbf{k}'|\le k_0} S^{uu}(\mathbf{p}|\mathbf{k}'|\mathbf{q}) \\
&= \sum_{|\mathbf{k}'|> k_0} \sum_{|\mathbf{p}|\le k} S^{uu}(\mathbf{k}'|\mathbf{p}|\mathbf{q}),
\end{aligned} \tag{4.42}$$

which is the same as Eq. (4.39).

Similar interpretations apply to the region outside the sphere for which the energy equation is

$$\frac{d}{dt} \sum_{|\mathbf{k}|>k_0} E_u(\mathbf{k}) = \sum_{|\mathbf{k}|>k_0} T_u(\mathbf{k}) + \sum_{|\mathbf{k}|>k_0} \mathcal{F}_u(\mathbf{k}) - \sum_{|\mathbf{k}|>k_0} D_u(\mathbf{k}). \tag{4.43}$$

The modes outside the sphere gain energy via $\Pi(k_0)$, which is

$$\begin{aligned}
\Pi_u(k_0) &= \sum_{|\mathbf{k}|>k_0} T_u(\mathbf{k}) \\
&= \sum_{|\mathbf{k}'|>k_0} \sum_{\mathbf{p}} S^{uu}(\mathbf{k}'|\mathbf{p}|\mathbf{q})
\end{aligned}$$

$$= \sum_{|\mathbf{k}'|>k_0} \sum_{|\mathbf{p}|\le k_0} S^{uu}(\mathbf{k}'|\mathbf{p}|\mathbf{q}) + \sum_{|\mathbf{k}'|>k_0} \sum_{|\mathbf{p}|>k_0} S^{uu}(\mathbf{k}'|\mathbf{p}|\mathbf{q})$$

$$= \sum_{|\mathbf{k}'|>k_0} \sum_{|\mathbf{p}|\le k_0} S^{uu}(\mathbf{k}'|\mathbf{p}|\mathbf{q}). \qquad (4.44)$$

This is same as Eq. (4.39).

We rewrite Eq. (4.40) for the spheres of radii k and $k+dk$ and take a difference between the two equations that yields

$$\frac{d}{dt} \sum_{k<k'\le k+dk} E_u(k') = \sum_{k<k'\le k+dk} T_u(\mathbf{k}') + \sum_{k<k'\le k+dk} \mathcal{F}_u(\mathbf{k}') - \sum_{k<k'\le k+dk} D(\mathbf{k}')$$

$$= [-\Pi_u(k+dk) + \Pi_u(k)] + \sum_{k<k'\le k+dk} \mathcal{F}_u(\mathbf{k}') - \sum_{k<k'\le k+dk} D(\mathbf{k}'), \qquad (4.45)$$

where $dk \to 0$. We illustrate the aforementioned energetics in Fig. 4.9. The rate of change of energy in a wavenumber shell is determined by the flux difference $\Pi_u(k+dk) - \Pi_u(k)$, the energy supply rate by the external force, and the viscous dissipation rate in the shell.

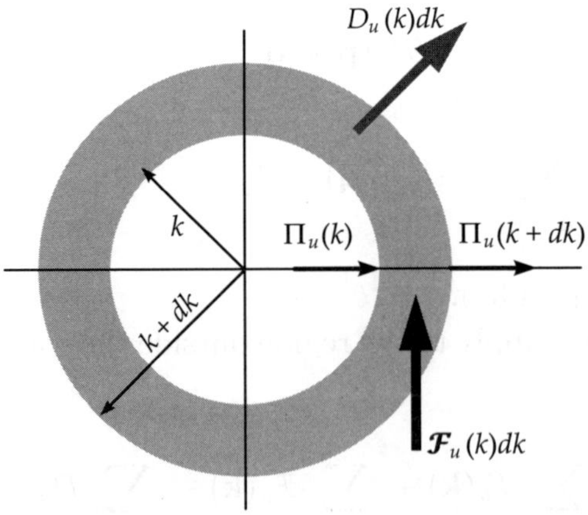

Figure 4.9 The rate of change of energy in a shell is given by the energy flux difference $\Pi_u(k+dk) - \Pi_u(k)$, the energy supply rate by the external force $\mathcal{F}_u(k)dk$, and the viscous dissipation rate $D_u(k)dk$.

Now taking the limit $dk \to 0$ yields

$$\frac{\partial}{\partial t} E_u(k,t) = -\frac{\partial}{\partial k} \Pi_u(k,t) + \mathcal{F}_u(k,t) - D_u(k,t), \qquad (4.46)$$

where

$$\mathcal{F}_u(k)dk = \sum_{k<k'\leq k+dk} \Re[\mathbf{F}_u(\mathbf{k}') \cdot \mathbf{u}^*(\mathbf{k}')], \tag{4.47}$$

$$D_u(k)dk = 2\nu \sum_{k<k'\leq k+dk} k'^2 E_u(\mathbf{k}'). \tag{4.48}$$

Equation (4.46) is quite important in turbulence research, and it helps us understand the flow dynamics. Note that in numerical simulations, the above sums are performed over the modes in the shell $(k-1, k)$. Refer to Eq. (3.25).

Before closing this section, we remark that the energy flux is not related to the conservation of energy, rather it is the outcome of the nonlinear interactions that facilitates energy transfers. Note that an energy flux is defined even when the energy conservation breaks down, for example, in the presence of an external force like buoyancy. Another important point to keep in mind is that the spectral energy flux describes energy transfers across multiple scales. It is very different from energy flow in real space.

4.5 Variable Energy Flux

Equation (4.46) is derived based on energetics considerations and it provides useful insights into the physics of turbulence. Let us analyze a steady state of a turbulent flow. For such flows, $\partial E_u(k,t)/\partial t = 0$. Consequently,

$$\frac{d}{dk}\Pi_u(k) = \mathcal{F}_u(k) - D_u(k) \tag{4.49}$$

for a wavenumber shell. Hence, the energy flux $\Pi_u(k)$ varies with k depending on $\mathcal{F}_u(k)$ and $D_u(k)$. The dissipation rate $D_u(k)$ is significant in the dissipative range, which is $k \gtrsim k_{\mathrm{DI}}$ in Fig. 4.10(b).

For the time being, we focus on the inertial range where $D_u(k)$ is negligible. We assume an energy injection rate of $\mathcal{F}_{\mathrm{in}}(k)$ in the inertial range, and $\mathcal{F}_{\mathrm{LS}}(k)$ at large scales. Here, the subscripts "in" and "LS" stand for the inertial range and large scales, respectively. Hence, in the inertial range,

$$\frac{d}{dk}\Pi_u(k) = \mathcal{F}_{\mathrm{in}}(k). \tag{4.50}$$

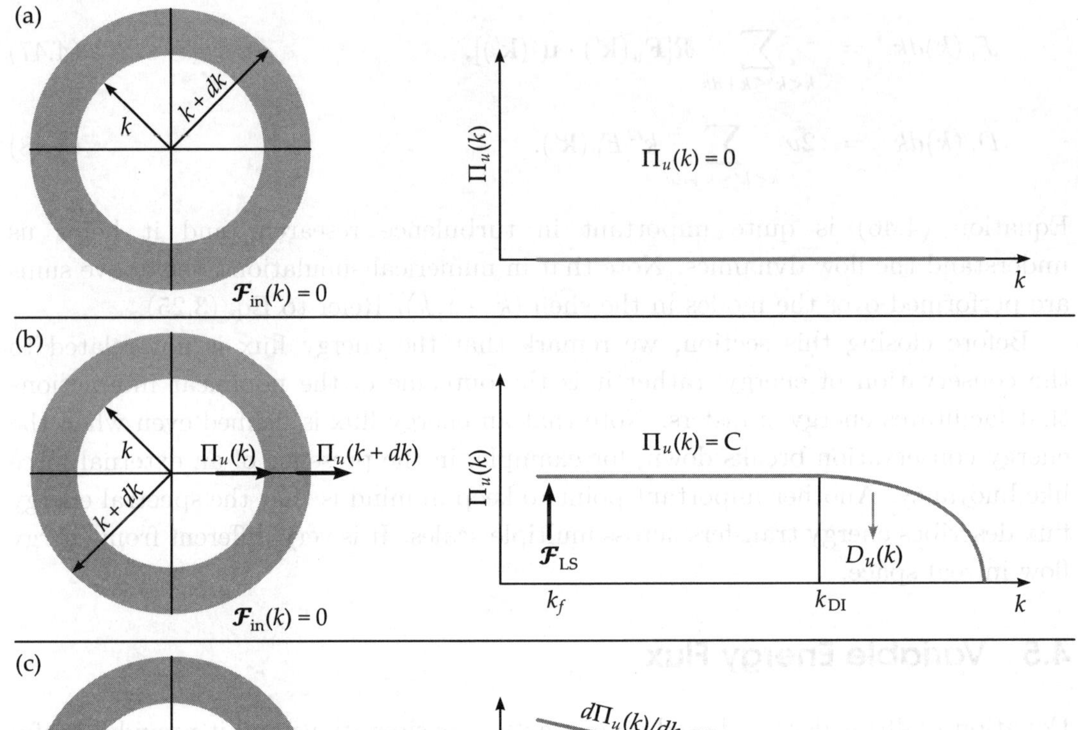

Figure 4.10 Schematic diagrams illustrating variations of energy flux $\Pi_u(k)$ for various energy injection rates $\mathcal{F}_{\rm in}(k)$: (a) $\mathcal{F}_{\rm in}(k) = 0$ and $\Pi_u(k) = 0$; (b) $\mathcal{F}_{\rm in}(k) = 0$ and $\Pi_u(k) = C > 0$; (c) $\mathcal{F}_{\rm in}(k) < 0$ and $d\Pi_u(k)/dk < 0$; (d) $\mathcal{F}_{\rm in}(k) > 0$ and $d\Pi_u(k)/dk > 0$.

Based on the nature of $\mathcal{F}_{\text{in}}(k)$, we can classify the turbulent systems into four categories:

1. $\mathcal{F}_{\text{in}}(k) = 0$ and $\Pi_u(k) = 0$: Substitution of $\mathcal{F}_{\text{in}}(k) = 0$ in Eq. (4.50) yields $\Pi_u(k) = C = \text{const.}$ For an equilibrium configuration, $C = 0$ or $\Pi_u(k) = 0$; thus there is no energy flux in the inertial range. See Fig. 4.10(a) for an illustration. For such cases, we expect the energy transfers among the velocity modes to vanish on an average, i.e., $\langle S(\mathbf{k}'|\mathbf{p}|\mathbf{q})\rangle = 0$. Note however that $S(\mathbf{k}'|\mathbf{p}|\mathbf{q})$, the energy transfer from $\mathbf{u}(\mathbf{p})$ to $\mathbf{u}(\mathbf{k}')$, could fluctuate around zero. $\langle S(\mathbf{k}'|\mathbf{p}|\mathbf{q})\rangle = 0$ implies a detailed balance of energy transfers among any pair of modes. Such a scenario is analogous to a thermodynamic equilibrium in which there is no energy transfer from one region to another, both in real space and in Fourier space.

   In this case, there is no energy supply at large scales, and there is no viscous dissipation at small scale. Also, for $\langle S(\mathbf{k}'|\mathbf{p}|\mathbf{q})\rangle = 0$ to hold, we expect the phases of the modes $\mathbf{u}(\mathbf{k}')$, $\mathbf{u}(\mathbf{p})$, and $\mathbf{u}(\mathbf{q})$ to be random. Such assumptions are typically made for describing equilibrium phenomena.

2. $\mathcal{F}_{\text{in}}(k) = 0$ and $\Pi_u(k) > 0$: As in case (1), substitution of $\mathcal{F}_{\text{in}}(k) = 0$ in Eq. (4.50) yields $\Pi_u(k) = C = \text{const.}$ However, when the external force at large scale feeds kinetic energy and the viscous dissipation at small scales kills this kinetic energy, a constant energy flux is maintained in the inertial range, that is, $\Pi_u(k) = C > 0$. See Fig. 4.10(b) for an illustration. The energy flux is destroyed in the dissipation range by viscosity.

   For this case,

$$
\begin{aligned}
\langle S(\mathbf{k}|\mathbf{p}|\mathbf{q})\rangle &> 0 \quad \text{for } k > p, \\
\langle S(\mathbf{k}|\mathbf{p}|\mathbf{q})\rangle &< 0 \quad \text{for } k < p.
\end{aligned}
$$

   Thus, there is no detailed balance of energy transfers among the modes. This is a nonequilibrium scenario, and it differs from the equilibrium case discussed in (1).[6] We also remark that the nonzero $\langle S(\mathbf{k}'|\mathbf{p}|\mathbf{q})\rangle$ requires certain dynamical phase relations among the Fourier modes $\mathbf{u}(\mathbf{k}')$, $\mathbf{u}(\mathbf{p})$, and $\mathbf{u}(\mathbf{q})$; these phases are not random as in case (1). Also, on some occasions, $\Pi_u(k) < 0$, which leads to a growth of energy at large scales; we will describe this feature in 2D

[6] In some literature, it is stated that the large-scale eddies provide energy to the small-scale eddies, which are in statistical equilibrium among themselves. Strictly speaking, the phrase "equilibrium" here is meant to be "steady state". There is a steady transfer of energy from one scale to another; hence, the system is far from being in equilibrium.

hydrodynamic turbulence (see Chapter 7). In Section 4.8 we relate the steady energy flux and lack of detailed balance to the arrow of time.

3. $\mathcal{F}_{\text{in}}(k) < 0$ and $d\Pi_u(k)/dk < 0$: From Eq. (4.50), $\mathcal{F}_{\text{in}}(k) < 0$ leads to $d\Pi_u(k)/dk < 0$. Here, a fraction of the energy flux is transferred elsewhere, while the remaining flux goes to the dissipation range where it is dissipated by viscosity. See Fig. 4.10(c) for an illustration. We illustrate this case using the following examples:
   - Ekman turbulence: In quasi two-dimensional flows, Ekman friction is typically modeled as $-\alpha\mathbf{u}$, where α is a constant. This force plays an important role in atmospheric flows. Clearly, $\mathcal{F}_{\text{in}}(k) = -\alpha|\mathbf{u}(\mathbf{k})|^2 < 0$. Hence, the energy flux will decrease with k that leads to a steepening of the energy spectrum. In this case, the Ekman friction depletes the energy flux in the inertial range. This is in addition to the viscous damping that acts at the dissipation range. See Verma (2012) for details.
   - Quasi-static magnetohydrodynamic turbulence: In turbulent flows of liquid metals under the influence of an external magnetic field $\mathbf{B}_0 = B_0\hat{z}$, the electromagnetic force is proportional to $-(B_0\cos\theta)^2\mathbf{u}(\mathbf{k})$, where θ is the angle between $\mathbf{k}$ and $\hat{z}$ (Verma, 2017). Here too, $\mathcal{F}_u(k) = -(B_0\cos\theta)^2|\mathbf{u}(\mathbf{k})|^2 < 0$ that leads to a decrease of $\Pi_u(k)$ with k. In this case, the Joule dissipation depletes the energy flux in the inertial range. See Verma (2017) and Chapter 24 for details.
   - Stably-stratified turbulence: In stably-stratified turbulence, the kinetic energy is converted to potential energy, which leads to a depletion of $\Pi_u(k)$ in the inertial range. We will describe this system in Chapter 15. Also see Verma (2018).

4. $\mathcal{F}_{\text{in}}(k) > 0$ and $d\Pi_u(k)/dk > 0$: When we substitute a positive $\mathcal{F}_{\text{in}}(k)$ in Eq. (4.50), we obtain $d\Pi_u(k)/dk > 0$. See Fig. 4.10(d) for an illustration. Turbulent convection and bubbly turbulence come in this category of flows. Here, the kinetic energy feed by the external force enhances $\Pi_u(k)$ in the inertial range. For more details, refer to Chapter 16 and Verma (2018).

Thus, the formalism of variable energy flux provides valuable insights into the dynamics of turbulent flows. This formalism can be extended to the fluxes of scalars and vectors, for example, for the temperature field in turbulent convection, and the magnetic field in magnetohydrodynamics. We will describe these fluxes when we cover these systems in Parts II and III of this book. Interestingly, we can extend the ideas of multiscale flux to wealth and money transfers in an economy. We sketch these ideas in Appendix B. Using the formalism of variable energy flux, we can also model the energy flux and energy spectrum in the dissipation range of hydrodynamic turbulence. We describe these models in Chapters 5 and 7.

Researchers have derived various formulas for the energy flux. In the next section, we will show equivalence between these formulas.

4.6 Equivalence between Various Formulas of Energy Flux

In this section, we briefly describe some of the earlier derivations of energy flux. For hydrodynamics, Kraichnan (1959) formulated the combined energy transfer from two modes to the third mode in an interacting triad (see Eq. (4.2)). Using this transfer Kraichnan (1959) derived a formula for the energy flux as

$$\Pi(k_0) = T_C - T_D \tag{4.51}$$

where

$$T_C = \frac{1}{2} \sum_{k>k_0} \sum_{p,q}^{\Delta} S(\mathbf{k}|\mathbf{p},\mathbf{q}), \tag{4.52a}$$

$$T_D = \frac{1}{2} \sum_{k\le k_0} \sum_{p,q}^{\Delta} S(\mathbf{k}|\mathbf{p},\mathbf{q}). \tag{4.52b}$$

Here, Δ represents the condition that $\mathbf{k} = \mathbf{p} + \mathbf{q}$. When we compare these expressions with Eqs. (4.41, 4.44), we observe that

$$T_C = \frac{1}{2} \sum_{k>k_0} T_u(\mathbf{k}) = \frac{1}{2}\Pi(k_0), \tag{4.53a}$$

$$T_D = \frac{1}{2} \sum_{k\le k_0} T_u(\mathbf{k}) = -\frac{1}{2}\Pi(k_0). \tag{4.53b}$$

Hence, $T_C - T_D = \Pi(k_0)$. Thus, we show consistency between Kraichnan's formula and Eq. (4.39) derived using the mode-to-mode formalism.

Frisch (1995) derived the following formula for energy flux:

$$\Pi(k_0) = \left\langle \mathbf{u}^<_{k_0} \cdot \{\mathbf{u}^<_{k_0} \cdot \nabla \mathbf{u}^>_{k_0}\} \right\rangle + \left\langle \mathbf{u}^<_{k_0} \cdot \{\mathbf{u}^>_{k_0} \cdot \nabla \mathbf{u}^>_{k_0}\} \right\rangle \tag{4.54}$$

When we sum the aforementioned two terms, we obtain

$$\Pi(k_0) = \left\langle \mathbf{u}^<_{k_0} \cdot \{\mathbf{u} \cdot \nabla \mathbf{u}^>_{k_0}\} \right\rangle. \tag{4.55}$$

Here the mediator field $\mathbf{u}$ is the full velocity field. However, the giver field $\mathbf{u}^<_{k_0}$ is a sum of modes inside the sphere, while the receiver field $\mathbf{u}^>_{k_0}$ is the sum of modes outside the sphere. Thus, Frisch's formula is same as that of Eq. (4.39).

Example 4.7: Consider the flow field of Example 3.4. Compute the energy flux for any wavenumber sphere.

Solution: The wavenumbers associated with the flow are $(\pm 1, 0)$, $(0, \pm 1)$, $(1, 1)$, and $(-1, -1)$. The wavenumbers $(\pm 1, 0)$ and $(0, \pm 1)$ belong to the wavenumber sphere of radius 1, while $(\pm 1, \pm 1)$ belong to that of radius $\sqrt{2}$. In Example 4.3 we computed the energy transfers among the modes. We observed that $\mathbf{u}(0, 1)$ transfers energy to $\mathbf{u}(1, 1)$, who in turn transfers all of it to $\mathbf{u}(1, 0)$. Thus, neither of the spheres gain any net energy. Therefore, energy flux of any wavenumber sphere is zero.

Example 4.8: Consider the flow field of Example 3.6. Compute the energy flux for any wavenumber sphere.

Solution: The wavenumbers of the active Fourier modes are $(\pm 1, 0, \pm 1)$, $(0, \pm 1, \pm 1)$, and $(\pm 1, \pm 1, \pm 2)$. The wavenumbers $(\pm 1, 0, \pm 1)$, $(0, \pm 1, \pm 1)$ belong to the sphere of radius $\sqrt{2}$, while the wavenumbers $(\pm 1, \pm 1, \pm 2)$ belong to the sphere of radius $\sqrt{6}$. As shown in Example 4.5, each of wavenumbers $(\pm 1, \pm 1, \pm 2)$ receive kinetic energy of amplitude $-4ABC$ from the modes $(\pm 1, 0, \pm 1)$, $(0, \pm 1, \pm 1)$. Hence, the sphere of radius $\sqrt{6}$ receives energy at the rate of $-32ABC$. Therefore,

$$\Pi_u(k_0) = \begin{cases} -32ABC & \text{if } \sqrt{2} < k_0 < \sqrt{6} \\ 0 & \text{otherwise.} \end{cases}$$

In the next section, we will derive a formula for shell-to-shell energy transfers in hydrodynamic turbulence.

4.7 Shell-to-shell Energy Transfers

The other quantity of interest in turbulence research is the shell-to-shell energy transfer between two wavenumber shells (see Fig. 4.11). Using $S^{uu}(\mathbf{k}|\mathbf{p}|\mathbf{q})$, the energy transfer from shell m to shell n can be defined as

$$T^{u,m}_{u,n} = \sum_{\mathbf{p} \in m} \sum_{\mathbf{k} \in n} S^{uu}(\mathbf{k}|\mathbf{p}|\mathbf{q}), \tag{4.56}$$

where the $\mathbf{k}$-sum is over the shell n, $\mathbf{p}$-sum is over the shell m, and $\mathbf{k} = \mathbf{p} + \mathbf{q}$. As shown in Fig. 4.11, the aforementioned formula yields correct contributions for both the triads—type A, and type B—even though the wavenumber $\mathbf{q}$ of triad B is outside the shell m. Note that the shell-to-shell energy transfers provide a more detailed picture than the energy flux, and they are useful for testing locality (whether the energy transfers are dominant for neighboring shells or distant shells) and direction of energy transfers.

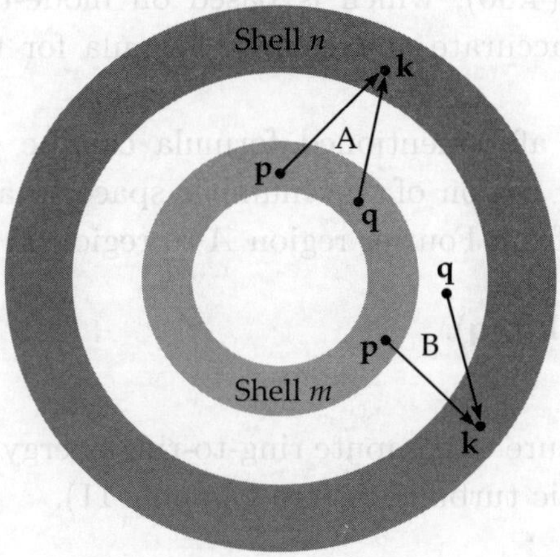

Figure 4.11 Shell-to-shell energy transfers from the wavenumber-shell m to wavenumber-shell n. The participating triads are of two types: Type A, where both $\mathbf{p}$ and $\mathbf{q}$ are inside the shell m, and Type B, where only $\mathbf{p}$ is inside shell m.

Note that we cannot compute the shell-to-shell energy transfers accurately using the combined energy transfer $S^{uu}(\mathbf{k}|\mathbf{p}, \mathbf{q})$ of Kraichnan (see Eq. (4.2)). If we write the shell-to-shell energy transfer as

$$T'^{u,m}_{u,n} = \frac{1}{2} \sum_{\mathbf{p},\mathbf{q} \in m} \sum_{\mathbf{k} \in n} S^{uu}(\mathbf{k}|\mathbf{p}, \mathbf{q}) \tag{4.57}$$

as done by Domaradzki and Rogallo (1990), Domaradzki (1992), Waleffe (1992), Zhou (1993), and Domaradzki et al. (2009), then we encounter the following difficulty. Rewriting Eq. (4.57) in terms of $S^{uu}(\mathbf{k}|\mathbf{p}|\mathbf{q})$ yields

$$\begin{aligned} T'^{u,m}_{u,n} &= \frac{1}{2} \sum_{\mathbf{p},\mathbf{q} \in m} \sum_{\mathbf{k} \in n} [S^{uu}(\mathbf{k}|\mathbf{p}|\mathbf{q}) + S^{uu}(\mathbf{k}|\mathbf{q}|\mathbf{p})] \\ &= \left[T^{u,m}_{u,n}\right]_{\text{triadA}}. \end{aligned} \tag{4.58}$$

Thus, $T'^{u,m}_{u,n}$ includes triads of Type A, but not of type B. Another alternative,

$$T''^{u,m}_{u,n} = \frac{1}{2} \sum_{\mathbf{p} \in m, \mathbf{q} \notin m} \sum_{\mathbf{k} \in n} S^{uu}(\mathbf{k}|\mathbf{p}, \mathbf{q}), \tag{4.59}$$

is also inadequate because $S^{uu}(\mathbf{k}|\mathbf{p}, \mathbf{q})$ is a symmetric function of $\mathbf{p}$ and $\mathbf{q}$. Thus, neither $T'^{u,m}_{u,n}$ nor $T''^{u,m}_{u,n}$ provide correct values for the shell-to-shell energy transfer. The researchers referred to earlier have alluded to the aforementioned difficulties. Note however that Eq. (4.56), which is based on mode-to-mode energy transfer formalism, provides an accurate and concise formula for the shell-to-shell energy transfer.

We remark that the aforementioned formula can be generalized to compute energy transfer from any region of wavenumber space to any other. For example, the energy transfer rate from Fourier region A to region B is

$$T^{u,A}_{u,B} = \sum_{\mathbf{k} \in B} \sum_{\mathbf{p} \in A} S^{uu}(\mathbf{k}|\mathbf{p}|\mathbf{q}). \tag{4.60}$$

We will employ this measure to compute ring-to-ring energy transfer, which is useful for quantifying anisotropic turbulence (see Chapter 11).

Example 4.9: Consider the flow field of Example 4.5. Compute the shell-to-shell energy transfers among wavenumber shells whose inner and outer radii are (1,2), (2,3), (3,4).

Solution: We denote the shells by $S_1 = (1,2)$, $S_2 = (2,3)$, and $S_3 = (3,4)$. The wavenumbers of the active Fourier modes are $(\pm1, 0, \pm1)$, $(0, \pm1, \pm1)$, and $(\pm1, \pm1, \pm2)$. Hence, $(\pm1, 0, \pm1) \in S_1$, $(0, \pm1, \pm1) \in S_1$, and $(\pm1, \pm1, \pm2) \in S_2$. As shown in Example 4.5, each of wavenumbers $(\pm1, \pm1, \pm2)$ receive kinetic energy of amplitude $-4ABC$ from the modes $(\pm1, 0, \pm1)$, $(0, \pm1, \pm1)$. Hence, the shell S_2 receives energy at the rate of $-32ABC$ from shell S_1. Therefore,

$$T^{u,1}_{u,2} = -32ABC,$$

and $T^{u,1}_{u,3} = T^{u,2}_{u,3} = 0$.

In the next section, we digress to relate turbulent energy flux to arrow of time.

4.8 Turbulent Energy Flux and Arrow of Time

In turbulent systems, the properties of energy transfer can be used to set the direction of time. As mentioned earlier, kinetic energy cascades from large scales to intermediate scales, and then to small scales, where it gets dissipated. Imagine that we record a video of a flow along with its energy transfers. If the video is run backward, the kinetic energy will flow from small scales to large scale, which is distinguishable from the original flow behavior. It is important to note that we can start a hydrodynamic turbulent system from an arbitrary initial condition, but it always tends to move toward a state with a constant energy flux; it never returns to its original configuration. Thus, the direction of energy transfer sets the direction of temporal evolution.

Hence, in a turbulent system, the factors that determine time's arrow are—hierarchy of structures, direction of energy transfers, and viscous dissipation at small scales. Many *nonequilibrium dissipative systems* (e.g., earthquakes, financial system) exhibit similar behavior; here too, the direction of corresponding flux would determine the arrow of time.

Contrast this phenomena with those in equilibrium systems that lacks a direction of energy transfer (see 1st category of turbulent systems in Section 4.5). Equilibrium systems behave as if they are frozen in time, apart from small fluctuations in the energy exchange among the modes. Some equilibrium systems may involve dissipation (e.g., see Langevin's equation), but it is the lack of direction of energy transfer that prevents us from determining time's arrow in these systems.

Hamiltonian and conservative nonequilibrium systems exhibit a somewhat different behavior. Such systems have hierarchical structures with energy transfers across multiple scales, and they are typically far from equilibrium. However, due to lack of dissipation, the system is able to come to its original configuration, or close to it. This process however may take a very long time. This is an intuitive statement of *Poincare's recurrence theorem* (Arnold, 1989). Such systems evolve so as to preserve the phase space volume.[7] This is called *Liouville's theorem* (Arnold, 1989). These assumptions form a basis for the *ergodic hypothesis* and *principle of equal a priori probabilities*, which are fundamental postulates of statistical physics (Ma, 1985). In summary, such systems may go back and forth from one state to another (possibly covering the available phase space with equal probability); yet they lack a definite direction of energy transfer from one scale to another. As a result, evolution of such systems do not have a definite pattern to

[7]The space formed by Fourier modes $\{\mathbf{u}(\mathbf{k})\}$ can be treated as phase space.

define the temporal direction that distinguishes forward and backward motion of the system.

Some researchers attribute the arrow of time to sensitivity to initial conditions. That is, in a nonlinear system, an arbitrary initial condition is likely to take the system to a chaotic state. Only a small set of initial conditions take a chaotic system to an ordered state. Therefore, an arbitrary initial condition is likely to take the system to a disordered state; such direction of evolution can define the arrow of time. Note that this definition of time's arrow differs significantly from that arising due to turbulent energy flux from large scales to small scales.

A key ingredient for the turbulent energy flux is viscous dissipation. It may be argued that such dissipation cannot be derived from first principles (involving only fundamental forces). It is a serious objection, but may be circumvented by saying that "organized structures" may be diffused by many-body interactions, thus providing a semblance of dissipation. Note that some recents works on superfluid turbulence (a quantum system) indicate presences of dissipation, possibly via phonon coupling (Nemirovskii, 2012). These are difficult issues requiring careful investigation.

An exception to the aforementioned energy transfers from large scales to small scales is 2D hydrodynamic turbulence in which the energy transfer is in the opposite direction. Hence, time's arrow in 2D hydrodynamic turbulence will be one in which the energy transfer is from small scales to large scales that leads to an increase in order. These issues are quite tricky and beyond the scope of present discussion.

4.9 Spectral Decomposition, Energy Transfers, and Amplitude Equations

As discussed in the present chapter, the dynamics of large-scale modes are captured quite nicely using spectral decomposition (also called *Galerkin truncation*). A set of equations thus constructed are called low-dimensional models. The energy transfers among the Fourier modes shed important light into the dynamics of the flow, as illustrated in Examples 4.3 to 4.8, as well as in the following example. Galerkin truncation is especially useful for studying patterns and chaos near the onset of instability (Verma, 2018). For example, we can study square patterns in thermal convection by choosing Fourier modes $(1,0,1)$, $(0,1,1)$, and $(1,1,2)$ as primary interacting modes. Similarly, hexagonal patterns are formed by a wavenumber triad $\{(-1,0),(1/2,\sqrt{3}/2),(1/2,-\sqrt{3}/2)\}$ and several other modes.

An alternative approach to address pattern formation is via *amplitude equation.* Here we employ scale separation between the fast and slow time scales, and then derive appropriate equations for the slow variables. The choice of interacting modes depend on the symmetries of the system. This is an extensive research field with applications in many areas of physics, but it is beyond the scope of this book. We refer readers to Manneville (2014), Cross and Greenside (2009), Hoyle (2006), Knobloch (1992), and Cross and Hohenberg (1993).

Let us contrast the essential features of amplitude equations and those of low-dimensional models constructed using spectral decomposition. An amplitude equation is typically constructed based on symmetry principles and the nature of nonlinear interactions (Hoyle, 2006). Such equations provide useful insights into pattern formation. Construction of low-dimensional models using Galerkin truncation however requires more work, as can be seen in this chapter. However, the low-dimensional models capture the onset of instability and patterns more accurately than amplitude equations. Energy transfers among modes provide further insights. These topics, though very interesting, will not be covered in this book. The interested reader is referred to the aforementioned books and Verma (2018)

In the next section, we briefly describe numerical methods based on spectral approach.

4.10 Numerical Simulations Using Spectral Method

Numerical simulations are akin to experiments. They help construct models and test various conjectures. Numerical simulations also complement experiments. For example, measurements of a high-resolution velocity field in the interiors of a liquid metal flow is still a major experimental challenge. This computation however is very easy in a numerical simulation. On the contrary, direct simulations of $\mathrm{Re} = 10^8$ is impossible even on the best supercomputers of present times, but many experimental flows have Re far beyond 10^8.

For turbulence simulations, researchers employ various numerical techniques, namely, finite difference, finite volume, finite element, pseudo-spectral, vortex methods, etc. (Ferziger and Peric, 2001). In this book, we will discuss the pseudo-spectral method (Boyd, 2003; Canuto et al., 1988) because it captures multiscale features of turbulent flows in a natural way. Moreover, spectral method is one of the most accurate numerical schemes for solving partial differential equations. In the following, we will briefly describe the pseudo-spectral method.

In Fourier space, the incompressible hydrodynamics is described by Eqs. (3.17). In spectral method, we rewrite these equations as

$$\left\{\frac{d}{dt} + \nu k^2\right\} u_i(\mathbf{k},t) = -ik_i p(\mathbf{k},t) - N_{u,i}(\mathbf{k}) + F_{u,i}(\mathbf{k}) - \nu k^2 u_i(\mathbf{k}), \quad (4.61a)$$

$$k_i u_i(\mathbf{k},t) = 0, \quad (4.61b)$$

where $N_{u,i}$ and $F_{u,i}$ are the ith components of the nonlinear term and the external force respectively. The pressure p is determined using

$$p(\mathbf{k}) = \sqrt{-1}\frac{1}{k^2} k_j \{N_{u,j}(\mathbf{k}) - F_{u,j}(\mathbf{k})\}. \quad (4.62)$$

It is convenient to assume that the real space domain of the flow is $L_x \times L_y \times L_z = (2\pi) \times (2\pi) \times (2\pi)$ so that the wavenumber components $k_i = 2\pi n_i/L_i = n_i$ are integers. We divide the domain into a $M = N_x \times N_y \times N_z$ grid. Therefore, the wavenumbers of the system are $\mathbf{k} = (k_x, k_y, k_z)$ with $k_x = (-N_x/2 : N_x/2), k_y = (-N_y/2 : N_y/2), k_z = (-N_z/2 : N_z/2)$. However, because of the reality condition, $\mathbf{u}(-k) = \mathbf{u}^*(k)$, we need to time advance only half of these wavenumbers, which are $k_x = (-N_x/2 : N_x/2), k_y = (-N_y/2 : N_y/2), k_z = (0 : N_z/2)$. The evolution of each of these Fourier modes is described by Eq. (4.61a). The objective of the computation is to solve for the field variables at a later time given the initial condition $\{\mathbf{u}(\mathbf{k}, t = 0)\}$.

The computation of the nonlinear term is the most expensive part of a spectral simulation. A naive calculation involving convolution will require $O(M^2)$ floating point operations, which is prohibitively expensive. Hence, the nonlinear terms are alternatively computed using fast Fourier transform (FFT), as suggested first by Orszag (Boyd, 2003; Canuto et al., 1988). The method is as follows:

1. Compute $\mathbf{u}(\mathbf{r})$ from $\mathbf{u}(\mathbf{k})$ using inverse FFT.
2. Compute $u_i(\mathbf{r})u_j(\mathbf{r})$ in real space by multiplying the fields at each space point.
3. Compute Fourier transform of $u_i(\mathbf{r})u_j(\mathbf{r})$ using forward FFT that yields $(u_i u_j)(\mathbf{k})$.
4. Compute $ik_j(u_i u_j)(\mathbf{k})$ by multiplying the result of Step 3 by ik_j, and then summing over all j. The resulting vector is the desired $N_{u,i}(\mathbf{k})$.

We illustrate these steps in Fig. 4.12. Since a FFT takes $O(M \log_2 M)$ floating point operations, the aforementioned method is quite efficient. The multiplication $u_i(\mathbf{r})u_j(\mathbf{r})$ is performed in real space; therefore, this procedure is called *pseudo-spectral method* instead of *spectral method*.

We need to take some care while choosing the grid size, the time-stepping scheme, and the time step. Theory of turbulence, to be discussed in Chapter 5, tells us that $M \gtrsim \mathrm{Re}^{9/4}$. For time stepping, Runge–Kutta third- or fourth-order schemes, or multistep schemes (e.g. Adam–Bashforth method) are recommended. The equations are stiff for large Re; hence, the exponential trick or Crank–Nicolson schemes are suggested for the dissipation term (Verma et al., 2013a; Chatterjee et al., 2018). The time step Δt is estimated using Courant–Friedrichs–Lewy (CFL) condition. Refer to Boyd (2003) and Canuto et al. (1988) for details of spectral methods.

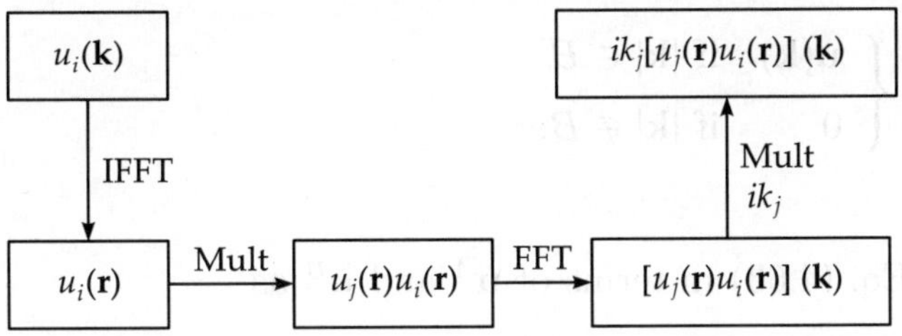

Figure 4.12 A schematic diagram depicting the computation of the nonlinear term $ik_j(u_iu_j)(\mathbf{k})$ in a pseudo-spectral method.

Spectral methods are employed to study the following properties of a turbulent flow:

1. Energy and dissipation spectra.
2. Energy flux.
3. Shell-to-shell energy transfer rates.
4. Evolution of global quantities like total energy, total enstrophy, and total kinetic helicity.
5. Time series of the amplitudes of Fourier modes.

In the next section, we describe how to compute the energy transfers using numerical and experimental data.

4.11 Computation of Energy Transfers Using Data

We can compute energy transfers at any time using an instantaneous flow profile, which may have been obtained from a numerical simulation or from an experiment. As in Eq. (4.60), the kinetic energy transfer rate from region A to region B is

$$T^{u,A}_{u,B} = \sum_{\mathbf{k}\in B}\sum_{\mathbf{p}\in A} \Im[\{\mathbf{k}\cdot\mathbf{u}(\mathbf{q})\}\{\mathbf{u}(\mathbf{p})\cdot\mathbf{u}^*(\mathbf{k})\}]. \tag{4.63}$$

Here the donor wavenumber $\mathbf{p}$ belongs to region A, the receiver wavenumber $\mathbf{k}$ to region B, and $\mathbf{q} = \mathbf{k} - \mathbf{p}$.

To perform the computation in Eq. (4.63), we define two *truncated* variables $\mathbf{u}^A$ and $\mathbf{u}^B$ as

$$\mathbf{u}^A(\mathbf{k}) = \begin{cases} \mathbf{u}(\mathbf{k}) & \text{if } |\mathbf{k}| \in A \\ 0 & \text{if } |\mathbf{k}| \notin A \end{cases} \tag{4.64a}$$

$$\mathbf{u}^B(\mathbf{k}) = \begin{cases} \mathbf{u}(\mathbf{k}) & \text{if } |\mathbf{k}| \in B \\ 0 & \text{if } |\mathbf{k}| \notin B. \end{cases} \tag{4.64b}$$

Now we rewrite Eq. (4.63) in terms of $\mathbf{u}^A$ and $\mathbf{u}^B$ as

$$T^{u,A}_{u,B} = \Im \left[\sum_{\mathbf{k}} [u_i^B(\mathbf{k})]^* \left\{ k_j \sum_{\mathbf{p}} u_j(\mathbf{k}-\mathbf{p}) u_i^A(\mathbf{p}) \right\} \right]. \tag{4.65}$$

The $\mathbf{p}$ summation in this equation is a convolution, which is computed using FFT. In fact, the term in the curly bracket is $N_{u,i}(\mathbf{k})/i$ (nonlinear term) with the truncated $\mathbf{u}^A(\mathbf{p})$, and it can be computed easily using the pseudo-spectral method described earlier. After the computation of nonlinear terms, we perform the $\mathbf{k}$ sum over region B of the wavenumber space. Thus, we compute the energy transfer from region A to region B.

For the flux computations, region A is the inner volume of the wavenumber sphere, while region B is the outer volume of the sphere. For the shell-to-shell energy transfer from shell m to shell n, regions A and B correspond to the volumes of shells m and n respectively. In Chapters 5 and 7, we will report the numerical results of energy flux and shell-to-shell energy transfers in 3D and 2D hydrodynamic turbulence. In Parts II and III, we will cover these transfers for other systems such as thermal convection and magnetohydrodynamics.

The computation of the aforementioned energy transfers requires a full dataset, which is relatively easier in numerical simulations but at a moderate Reynolds number ($\lessapprox 10^5$). High-resolution spectral simulations provide $\mathbf{u}(\mathbf{k})$ using which we can compute energy flux and shell-to-shell energy transfers. For other numerical schemes like finite difference, finite volume, and finite elements, we can interpolate the data to a uniform mesh and then employ FFT to compute $\mathbf{u}(\mathbf{k})$ (Chandra and Verma, 2011, 2013).

Equations (4.40, 4.41) provide another scheme to compute the energy flux:

$$\Pi_u(k_0) = -\frac{d}{dt} \sum_{|\mathbf{k}|\le k_0} E_u(\mathbf{k}) + \sum_{|\mathbf{k}|\le k_0} \mathcal{F}_u(\mathbf{k}) - \sum_{|\mathbf{k}|\le k_0} D_u(\mathbf{k}). \tag{4.66}$$

We measure

$$\frac{d}{dt}\sum_{|\mathbf{k}|\le k_0} E_u(\mathbf{k}) = \frac{\sum_{|\mathbf{k}|\le k_0} E_u(\mathbf{k}, t+dt) - \sum_{|\mathbf{k}|\le k_0} E_u(\mathbf{k}, t)}{dt} \tag{4.67}$$

by computing the total kinetic energy inside the sphere at two nearby times (t and $t{+}dt$). The energy supply rate by external force and the dissipation rate in the sphere can be computed using the velocity and force fields. Verma et al. (1996) adopted this strategy to compute the energy fluxes in magnetohydrodynamics (MHD) turbulence.

Experimental measurement of 3D datasets requires 3D particle image velocimetry (PIV), which is very expensive. These measurements typically provide low-resolution data that makes flux computations in the inertial range quite difficult. Note, however, that for 2D hydrodynamics, high-resolution 2D PIV provides data with sufficient resolution for the energy flux computation. Under an assumption of isotropy and homogeneity, the data obtained using 2D PIV could also be used for computing the energy flux of 3D turbulence. Researchers also employ structure functions to compute the energy flux; this topic will be discussed in Chapter 12. These discussions show that the computation of energy transfers are relatively easier using numerical simulation than using experiments.

The properties of energy transfers have also been studied analytically using field-theoretic methods (Kraichnan, 1959; Leslie, 1973; McComb, 1990; Verma, 2004; Verma et al., 2005; Yakhot and Orszag, 1986). This analytic technique, though quite involved, is very useful for computing energy flux and shell-to-shell energy transfers. We briefly describe these computations in Chapter 10.

Here we close our discussion on energy transfers in hydrodynamic turbulence. In the next chapter, we will describe some of the important properties (including energy transfers) of 3D hydrodynamic turbulence.

Further Reading

The mode-to-mode energy transfer described in this chapter was first formulated by Dar et al. (2001), and later expanded by Verma (2004). Their focus however was on magnetohydrodynamic turbulence, which is a superset of hydrodynamic turbulence. Earlier, Kraichnan (1959) derived the *combined energy transfer* of Eq. (4.2) and associated energy flux.

Domaradzki and Rogallo (1990), Domaradzki (1992), and Waleffe (1992) derived formulas for the shell-to-shell energy transfer using combined energy transfers; however, these formulas have deficiencies. Mode-to-mode energy transfers provide accurate and concise formulas for the energy flux and shell-to-shell energy transfers.

Exercises

1. Consider the following 2D fluid flow in a box of size $[\pi, \pi]$:
$$\begin{aligned}\mathbf{u} &= 4C(\hat{x}\sin 3x\cos y - \hat{y}3\cos 3x\sin y) + 4B(\hat{y}\sin 2x\cos 2y - \hat{y}\cos 2x\sin 2y) \\ &\quad +4A(\hat{x}\sin x\cos y - \hat{y}\cos x\sin y).\end{aligned}$$
Compute the Fourier amplitudes of the modes, and the energy transfers among them.
2. Consider the following 2D fluid flow in a box of size $[\pi, \pi]$:
$$\begin{aligned}\mathbf{u} &= 4D(\hat{x}\sin x\cos 3y - \hat{y}\frac{1}{3}\cos x\sin 3y) + 4C(\hat{x}\sin 3x\cos y - \hat{y}3\cos 3x\sin y) \\ &\quad +4B(\hat{y}\sin 2x\cos 2y - \hat{y}\cos 2x\sin 2y) + 4A(\hat{x}\sin x\cos y - \hat{y}\cos x\sin y).\end{aligned}$$
Compute the Fourier amplitudes of the modes, and the energy transfers among them. It is convenient to use the Craya–Herring basis (see Section 4.3). Write down the equation of motion for A, B, C, and D.
3. Consider a 2D flow field constructed using wavenumbers $\mathbf{k}' = (-1, 0)$, $\mathbf{q} = (1/2, \sqrt{3}/2)$, and $\mathbf{p} = (1/2, -\sqrt{3}/2)$. Denote the amplitudes of modes in the Craya–Herring basis as A, B, and C. Write down equations of motion for the modes in the presence of viscosity. Solve for A, B, and C. Compute the energy transfers among the aforementioned modes.
4. Consider the two-dimensional flow fields of Exercises 4.1 and 4.2. Compute the energy fluxes for these fields.

Chapter 5

Energy Spectrum and Flux of 3D Hydrodynamics

Turbulence is a complex problem with a very few analytical and exact results. In this chapter we will present Kolmogorov's theory of hydrodynamic turbulence (spectral version) that yields energy spectrum and flux for the flow. Later in the chapter we will also discuss energy spectrum and flux for the dissipative regime and for laminar flows.

Kolmogorov's theory of turbulence is a starting point for many applications such as turbulence in passive and active scalars, magnetohydrodynamic turbulence, thermal convection, etc. These systems will be discussed in Parts II and III of this book. Hydrodynamic turbulence in two dimensions (2D) and in three dimensions (3D) are quite different. In the present chapter, we cover 3D hydrodynamic turbulence. 2D hydrodynamic turbulence will be discussed in Chapter 7.

5.1 Kolmogorov's Theory for 3D Hydrodynamic Turbulence in Spectral Space

Turbulent flows are typically quite complex. However, notable simplification is achieved when we focus on the flow at intermediate and small scales away from the walls. For such flows, Kolmogorov (1941a) and Kolmogorov (1941c) provided a very important and powerful theory of turbulence. Kolmogorov's theory of turbulence has been presented in several ways. In the following discussion, we present this theory in spectral space using the constancy of energy flux and dimensional arguments. In Chapter 12 we describe Kolmogorov's original arguments that are for real space.

In Section 4.4 we showed that the time evolution of the kinetic energy (KE) spectrum $E_u(k)$ is described by the following equation:

$$\frac{\partial}{\partial t} E_u(k,t) = -\frac{\partial}{\partial k}\Pi_u(k,t) + \mathcal{F}_u(k,t) - D_u(k,t), \tag{5.1}$$

where $\Pi_u(k)$ is the energy flux, and

$$\mathcal{F}_u(k)dk = \sum_{k<k'\leq k+dk} \Re[\mathbf{F}_u(\mathbf{k}')\cdot\mathbf{u}^*(\mathbf{k}')], \tag{5.2}$$

$$D_u(k)dk = 2\nu \sum_{k<k'\leq k+dk} k'^2 E_u(\mathbf{k}') \tag{5.3}$$

are respectively the energy supply rate by the external force $\mathbf{F}_u$ and the viscous dissipation rate in the wavenumber shell k.

The key assumptions of Kolmogorov's theory are as follows:

1. The external force $\mathbf{F}_u$ is active at large length scales, which are of the order of the system size. Thus, the external force peaks at wavenumber $k_f \sim k_L \approx 1/L$. The KE feed by the external force is $\mathcal{F}_{\mathrm{LS}}(k)$. The wavenumber band near $k = k_f$ is called the *forcing range*. See Fig. 5.1 for an illustration.
2. The kinematic viscosity $\nu \to 0$, or Re $= UL/\nu \to \infty$. Hence, $D_u(k)$ is significant only in the *dissipation range*, which is at large k, due to the k^2

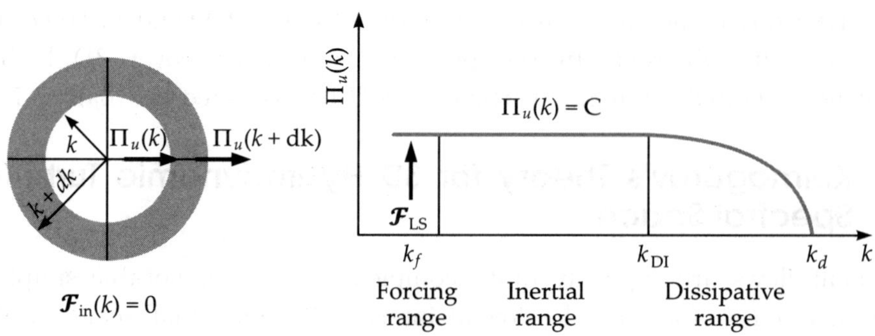

Figure 5.1 Schematic diagrams of the kinetic energy flux $\Pi_u(k)$ in hydrodynamic turbulence. The energy supplied at small k cascades to intermediate wavenumbers, and then to the dissipative wavenumbers. In the intermediate or inertial range, $\Pi_u(k) = \mathrm{const}$.

factor in $D_u(k)$. In Fig. 5.1, this range is illustrated as $k > k_{\mathrm{DI}}$, where k_{DI} is the transition wavenumber between the intermediate range and the dissipation range.

3. The energy supplied by $\mathbf{F}_u$ cascades to the intermediate scales and then to the small scales where the energy is dissipated. The intermediate wavenumber band between the forcing range and dissipation range is called *inertial range* where the nonlinear term and the pressure gradient play a dominant role (see Fig. 5.1).
4. Kolmogorov assumed that the physics of inertial range is independent of forcing and dissipative mechanisms, and that it depends on local properties—local wavenumber, and the energy flux that traverses through the local scale.
5. The flow is statistically homogeneous and isotropic at the intermediate and small scales.
6. In the early stages of evolution, typically, the velocity fluctuations are present at large scales or small k. Hence, $E_u(k) \approx 0$ at the intermediate and small scales. Nonlinear interactions quickly transfer energy from small k's to intermediate k's and then to small k's. This energy transfer is fast and local, that is, from a wavenumber shell to its neighboring shell. The flow typically reaches a quasi-steady state in an eddy turnover time, L/U, where U is the large scale velocity. In a steady state, the net energy supply rate by the external force is balanced by the total dissipation rate ϵ_u, that is,

$$\int_0^\infty dk' \mathcal{F}_u(k') = \int_0^\infty dk' D_u(k') = \epsilon_u. \tag{5.4}$$

Under a steady state, we have $\partial E_u(k)/\partial t \approx 0$. In the inertial–dissipation range, $\mathcal{F}_u(k) = 0$, hence,

$$\frac{d}{dk}\Pi_u(k,t) = -D_u(k,t). \tag{5.5}$$

In addition, $D_u(k) \to 0$ in the inertial range. Therefore, using Eq. (5.5), we deduce that

$$\Pi_u(k) = \Pi_u = \text{const}, \tag{5.6}$$

as illustrated in Fig. 5.1.

If we chose k just beyond k_f, then an integration of Eq. (5.5) yields

$$\Pi_u(k) = \Pi_u = \int_k^\infty D_u(k')dk'. \tag{5.7}$$

Assuming that the dissipation in the band $(0, k_f)$ is negligible, we can deduce that

$$\Pi_u \approx \int_0^\infty D_u(k')dk' = \epsilon_u. \tag{5.8}$$

This relation indicates that $\Pi_u \lessapprox \epsilon_u$.

Now, we derive one-dimensional KE spectrum of the flow using dimensional analysis. Since the physics of turbulence in the inertial range is independent of forcing and dissipative mechanisms, the energy spectrum $E_u(k)$ will depend only on the local wavenumber, k, and energy flux, Π_u, which is visible to all the scales. Kolmogorov's picture of energy cascade implicitly assumes local energy transfer in the inertial range. Moreover, since $E_u(k)$ does not depend on any external length scale, it will be a power law[1] in k. Therefore, we deduce the following functional form for $E_u(k)$:

$$E_u(k) = \Pi_u^\alpha k^\beta. \tag{5.9}$$

Now using the fact that

$$[E_u(k)] = [E_u/k] = [L^3/T^2]; \quad [\Pi_u] = [E_u/T] = [L^2/T^3]; \quad [k] = [L]^{-1}, \tag{5.10}$$

we derive the following formula for the one-dimensional kinetic energy spectrum:

$$E_u(k) = K_{\mathrm{Ko}} \Pi_u^{2/3} k^{-5/3}, \tag{5.11}$$

where the proportionality constant K_{Ko} is called *Kolmogorov's constant.* Numerical simulations, experiments, and analytical computations report that $K_{\mathrm{Ko}} \approx 1.6$. The aforementioned energy spectrum and energy flux have been observed in experiments and numerical simulations.

Note that Eqs. (5.6, 5.11) provide a *universal* description for hydrodynamic turbulence, and they are applicable irrespective of the boundary condition, initial condition, dissipative forces, and forcing mechanism. Due to its universal nature, Kolmogorov's theory is widely used for modeling many natural and engineering systems, e.g., turbulent flows in planets, stars, galaxies, turbines, engines, etc.

Kolmogorov's theory provides many useful insights into the physics of turbulence, some of which will be discussed in the next section.

[1]Note that functions sin, exp take nondimensional numbers as arguments that can be constructed by multiplying k with an appropriate length scale. Since the physics of turbulence in the inertial range is independent of forcing or dissipative mechanisms, such length scale is unavailable.

5.2 Insights from Kolmogorov's Theory of Turbulence

From Kolmogorov's theory of turbulence, we can make the following deductions:

1. We can estimate the large-scale or rms velocity U of a turbulent flow as follows. The total kinetic energy of the flow is

$$E_u = \int_0^\infty E_u(k)dk \approx \int_{k_L}^\infty E_u(k)dk \approx \epsilon^{2/3} L^{2/3}. \tag{5.12}$$

It is assumed that the forcing and dissipation range do not contribute significantly to this integral, and that $E_u(k)$ in the integrand is given by Eq. (5.11). Note that energy is concentrated at large scales because $E_u(k) \sim k^{-5/3}$ drops rather sharply with k. Using Eq. (5.12) and $E_u = U^2/2$, we immediately deduce that

$$U \approx \epsilon^{1/3} L^{1/3}. \tag{5.13}$$

2. In the inertial range, we can estimate the energy content at the length scale l or at the wavenumber $k = 1/l$ using

$$\frac{u_l^2}{2} = \int_k^{k+\Delta k} E_u(k)dk. \tag{5.14}$$

Due to the power law physics of turbulence (see Appendix A), it is appropriate to use $\Delta k \sim k$. Hence,

$$\frac{u_l^2}{2} = \int_k^{2k} dk[\epsilon^{2/3} k^{-5/3}] \sim \epsilon^{2/3} k^{-2/3} \tag{5.15}$$

that yields the magnitude of the velocity fluctuation and effective interaction time scale as

$$u_k \approx \epsilon_u^{1/3} k^{-1/3}, \tag{5.16a}$$

$$u_l \approx \epsilon_u^{1/3} l^{1/3}, \tag{5.16b}$$

$$\tau_l \approx \frac{l}{u_l} \approx \epsilon_u^{-1/3} l^{2/3}. \tag{5.16c}$$

3. In Kolmogorov's theory, the assumption that $\nu \to 0$ is crucial. The nonzero but small viscosity causes viscous dissipation at $k \approx \infty$, and it induces energy cascade from large scales to intermediate scales and then to small scales. In the limit of $\nu \to 0$, the dissipation in the inertial range is negligible, thus making

$\Pi_u(k)$ a constant in the inertial range. Moreover, the inertial range is infinite in this limit. Note that energy cascade from large scales to small scales makes turbulence a nonequilibrium phenomena.

4. A real fluid however has a finite viscosity. Hence, the viscous dissipation in such flows occur at a finite length scale l_d, which is estimated as follows. When we extend Kolmogorov's scaling to the forcing and dissipative scales, Eq. (5.16b) yields

$$\epsilon_u \approx \frac{U^3}{L} \approx \frac{u_{l_d}^3}{l_d}, \tag{5.17}$$

where $l_d \approx 1/k_d$ is referred to as *Kolmogorov's length.* Using $\nu \approx u_{l_d} l_d$, we obtain

$$l_d \approx \left(\frac{\nu^3}{\epsilon_u}\right)^{1/4}. \tag{5.18}$$

Note that the dissipation scale l_d is a function of ν and, surprisingly, of the energy supply rate ϵ_u. We can also derive l_d by multiplying Eq. (5.16b) with l_d and by setting $l = l_d$. The corresponding wavenumber,

$$k_d \approx \left(\frac{\epsilon_u}{\nu^3}\right)^{1/4} \tag{5.19}$$

is called *Kolmogorov wavenumber*. We remark that $k_d \gg k_{\rm DI}$ (see Fig. 5.1). For moderately large Re, Pope (2000) shows that k_d is an order of magnitude larger than $k_{\rm DI}$. A more rigorous derivation of k_d is arrived at in Section 5.5.

From the kinetic theory of dilute gas, $\nu = c_s\lambda$, where c_s is the sound speed, and λ is the mean free path length. Note that $u_{l_d} \neq c_s$ and $l_d \neq \lambda$. In fact, $u_{l_d} \ll c_s$ and $l_d \gg \lambda$. So, l_d is very different from the mean free path length. In fact, l_d is the length scale at which the velocity field is completely diffused; thus, $u_l \approx 0$ for $l < l_d$.

It is important to note that the relations $\nu \approx u_{l_d} l_d$ and $\nu = c_s\lambda$ connect the nonequilibrium processes of hydrodynamic turbulence to the equilibrium processes at kinetic scales. The detailed physical mechanism between the length scale l_d and λ that comes in the purview of kinetic theory is quite complex. There are many unresolved issues in this field.

5. We can relate the Reynolds number to L/l_d using

$$\mathrm{Re} = \frac{UL}{\nu} \approx \frac{UL}{u_{l_d} l_d} \approx \left(\frac{L}{l_d}\right)^{4/3}, \tag{5.20}$$

or

$$\frac{L}{l_d} \approx \mathrm{Re}^{3/4}. \tag{5.21}$$

The aforementioned relation provides an estimate for the range of length scales present in a turbulent flow. Clearly, a realistic numerical simulation of a turbulent flow requires minimum grid resolution of L/l_d along each direction. For example, a numerical simulation of a turbulent flow with $\mathrm{Re} = 10^4$ requires

$$\frac{L}{l_d} \approx \mathrm{Re}^{3/4} \approx 10^3. \tag{5.22}$$

Hence, we need $(10^3)^3 = 10^9$ grid points to simulate a 3D turbulent flow with $\mathrm{Re} = 10^4$. Similarly, a numerical simulation of $\mathrm{Re} = 10^8$ requires 10^{18} grid points, which is beyond the memory capacity of the best supercomputer available at present. Turbulence simulations are hence very challenging.

6. Physics of turbulence involves power laws, as illustrated by the functional form of $E_u(k)$. As shown in item (2), the velocity magnitude at wavenumber k is $u_k \sim \epsilon_u^{1/3} k^{-1/3}$. Kolmogorov's theory of turbulence makes an implicit assumption that the dominant interactions in turbulence are local (among neighboring wavenumber shells) and forward (from a lower wavenumber shell to larger wavenumber shell). We verify this assumption by estimating the shell-to-shell energy transfer from shell n to shell $(n+1)$:

$$T^{u,n}_{u,n+1} = \sum_{\mathbf{k}\in n+1} \sum_{\mathbf{p}\in n} \Im[\{\mathbf{k}\cdot\mathbf{u}(\mathbf{q})\}\{\mathbf{u}(\mathbf{p})\cdot\mathbf{u}^*(\mathbf{k})\}]. \tag{5.23}$$

Since the above shells are neighbor of each other, it is reasonable to assume, $p \sim k \sim q$. From Eq. (5.16b), we estimate $u_k \sim u_p \sim u_q \sim \epsilon_u^{1/3} k^{-1/3}$. Therefore,

$$T^{u,n}_{u,n+1} \sim k u_k^3 \sim \epsilon_u \sim \Pi_u. \tag{5.24}$$

Thus, maximal energy indeed is transferred locally, which is consistent with the assumptions of Kolmogorov's theory.

7. Taylor (1954) showed that turbulence enhances diffusion. Using Kolmogorov's theory we can derive how the distance between two particles diverges in a turbulent medium. We assume that the dispersion for inertial scales do not depend on the viscosity and large-scale forcing, but it depends only on the local length scale and the energy flux. Then, dimensional analysis yields

$$\Delta r \sim \epsilon_u^{1/2} t^{3/2}. \tag{5.25}$$

Note that the aforementioned formula, which is applicable when Δr lies in the inertial range, indicates superdiffusion. This diffusion is faster than even a ballistic process ($\sim t$) because large eddies advect the particles with speeds faster than that of the particles. Also note that the diffusion coefficient is scale dependent:

$$D_l \sim u_l l \sim \epsilon^{1/3} l^{4/3} \tag{5.26}$$

substitution of which in $\Delta r = \sqrt{D_l t}$ yields Taylor's formula.

In the next section, we will illustrate these results using numerical simulations.

5.3 Numerical Verification of Kolmogorov's Theory

For 3D hydrodynamic turbulence, the energy spectrum $E_u(k)$ of Eq. (5.11) has been verified by many experiments and numerical simulations. In this book we do not discuss those results in detail. Rather, we just illustrate the energy spectrum, energy flux, and shell-to-shell energy transfers using a high-resolution (on 4096^3 grid) simulation (Verma et al., 2018). In this simulation, $\nu = 8 \times 10^{-5}$, $\epsilon_u = 0.1$, $\mathrm{Re} = 6.8 \times 10^4$, and Kolmogorov's wavenumber $k_d \approx 660$. Hence,

$$l_d = \frac{\pi}{k_d} \approx \frac{\pi}{660}, \tag{5.27}$$

and the grid resolution is

$$\Delta = \frac{2\pi}{N} = \frac{\pi}{2048}. \tag{5.28}$$

Thus, $l_d/\Delta \approx 2048/660 \approx 3$. Therefore, the smallest velocity structure in the flow is resolved in a 3^3 simulation grid. Therefore, we conclude that the flow is well resolved[2] in our simulation. Verma et al. (2018) also observe that

$$\frac{\epsilon_u}{U^3/L} \approx 1, \tag{5.29}$$

which is consistent with earlier results.

[2]It is customary to denote the condition for resolution in direct numerical simulation as $k_{\max} l_d > 1$.

In Figs. 5.2 and 5.3 we illustrate $E_u(k)$ and $\Pi_u(k)$ computed using the simulation data. The gray region in the plots is the forcing band (2,4). The plots exhibit a wide inertial range ($4 \lessapprox k \lessapprox 100$) where

$$E_u(k) \sim k^{-5/3}; \quad \Pi_u(k) \sim \text{const.} \tag{5.30}$$

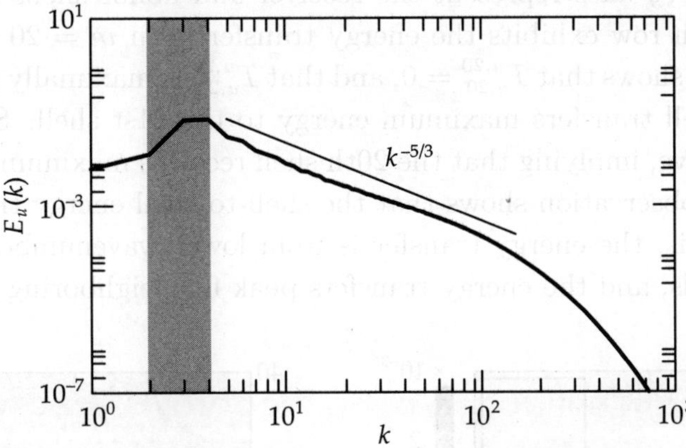

Figure 5.2 The energy spectrum $E_u(k)$ computed using spectral simulation. In the inertial range, $4 \lessapprox k \lessapprox 100$, $E_u(k) \sim k^{-5/3}$. Adopted from a figure of Verma et al. (2018).

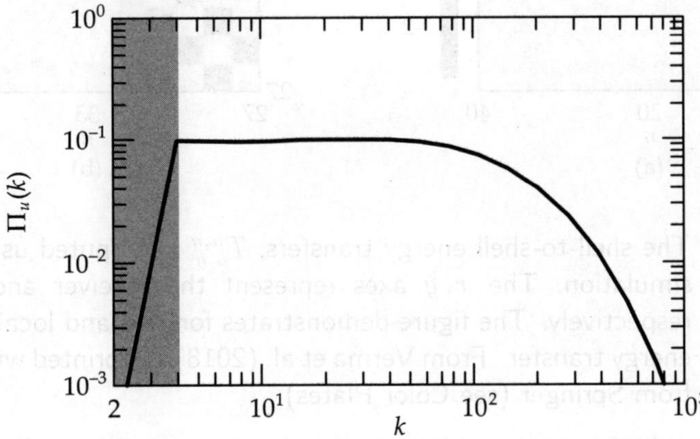

Figure 5.3 The energy flux $\Pi_u(k)$ computed using spectral simulation. $\Pi_u(k) \approx$ constant in the inertial band $4 \lessapprox k \lessapprox 100$. Adopted from a figure of Verma et al. (2018).

Using the numerical data, we compute

$$K_{\text{Ko}} = \frac{E_u(k)k^{5/3}}{\epsilon_u^{2/3}} \approx 1.75 \pm 0.05, \tag{5.31}$$

which is close to the expected value of 1.6. The viscous dissipation starts to dominate for k beyond 200. Note that $E_u(k)$ is already negligible for $k > k_d \approx 660$.

Figure 5.4(a) exhibits the density plot of the shell-to-shell energy transfer from shell m to shell n, $T^{u,m}_{u,n}$, for the same run. See Eq. (4.56) for the definition. The shell radii in the inertial range are chosen as $k_n = 2^{n/5}$ to mimic the power law physics. In the plot, the x, y axes represent the receiver and donor shells respectively. For example, the 20th row exhibits the energy transfer from $m = 20$ shell to all other shells. The figure shows that $T^{u,20}_{u,20} = 0$, and that $T^{u,20}_{u,21}$ is maximally positive implying that the 20th shell transfers maximum energy to the 21st shell. Similarly, $T^{u,20}_{u,19}$ is maximally negative, implying that the 20th shell receives maximum energy from the 19th shell. This observation shows that the shell-to-shell energy transfer is forward and local. That is, the energy transfer is from lower wavenumber shells to larger wavenumber shells, and the energy transfers peak for neighboring shells.

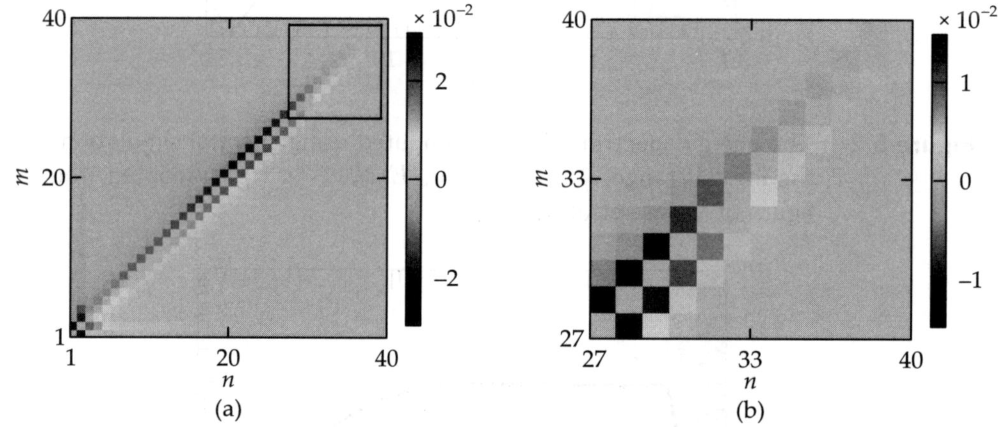

Figure 5.4 The shell-to-shell energy transfers, $T^{u,m}_{u,n}$, computed using a spectral simulation. The x, y axes represent the receiver and giver shells respectively. The figure demonstrates forward and local shell-to-shell energy transfer. From Verma et al. (2018). Reprinted with permission from Springer (see Color Plates).

The 27th shell whose wavenumber range is [194, 223] falls in the transition region between the inertial range and the dissipation range. Interestingly, the shell-to-shell energy transfers in the dissipation range (shells 27 to 40) too are local and forward, but they are relatively weaker than those in the inertial range (see Fig. 5.4(b)).

In Fig. 5.5 we plot $T^{u,m}_{u,n}$ as a function of $n - m$ for a different simulation. The figure clearly illustrates a forward cascade of energy. The transfers peak for the nearest neighbors, and then decline sharply for larger $n - m$, thus indicating locality of the energy transfer in the wavenumber space. In addition, in the inertial range,

$T^{u,m}_{u,n}$ curves for various ms almost collapse into a single curve. This feature indicates a scale-independent energy transfer in the inertial range. Thus, the shell-to-shell energy transfer in three-dimensional fluid turbulence is forward, local, and scale invariant in the inertial range. This is consistent with the power law physics of hydrodynamic turbulence in the inertial range. Also see Verma et al. (2005).

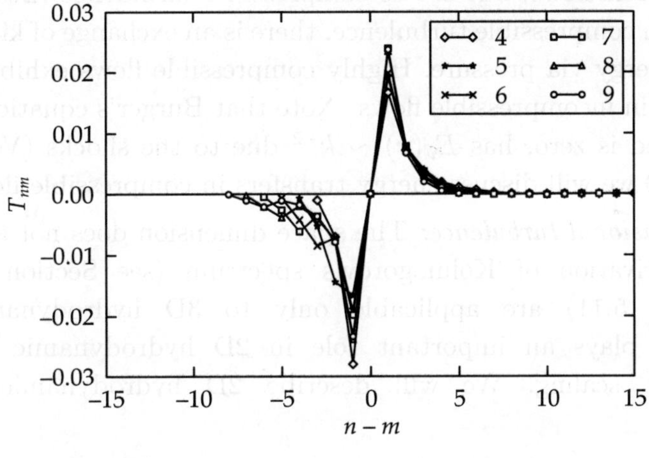

Figure 5.5 Plot of $T^{u,m}_{u,n}$ as a function of $n - m$ for 3D fluid turbulence with $\mathrm{Re} = 1100$.

Kolmogorov's theory has been quite successful in explaining many important features of incompressible hydrodynamic turbulence. Note however that it has certain limitations that will be listed in the next section.

5.4 Limitations of Kolmogorov's Theory of Turbulence

The shortcomings of Kolmogorov's theory are listed as follows:

1. *Anisotropic turbulence:* Kolmogorov's theory of turbulence assumes that the flow is statistically isotropic. Consequently, the modal energy of all the modes in a thin shell of radius k are equal, that is, $C(\mathbf{k}) = C(k)$. This assumption holds good in the absence of an external force field, and away from the boundaries. Note that external sources like rotation, external magnetic field, and buoyancy induce anisotropy in the flow. In addition, channel flow is also anisotropic. Surprisingly, in the absence of external sources, the flows far away from walls tend to be isotropic at small scales, thus making Kolmogorov's theory applicable to such flows.

   Note, however, that the energy equation [Eq. (5.1)] and the energy flux scenario illustrated in Fig. 5.1 are applicable to anisotropic flows as well. Here, the energy transfer will not be isotropic, but Eq. (5.1) is valid for a wavenumber shell of radius k. For deciphering the angular dependence of turbulence, Teaca et al. (2009) and Nath et al. (2016) proposed ring spectrum and ring-to-ring energy transfers, which will be discussed in Chapter 11.

2. *Compressible turbulence:* Kolmogorov's theory for turbulence is applicable to incompressible flows, but not to compressible turbulence (Lesieur, 2008). For example, in compressible turbulence, there is an exchange of kinetic energy and internal energy via pressure. Highly compressible flows exhibit shocks, which are absent in incompressible flows. Note that Burger's equation, for which the sound speed is zero, has $E_u(k) \sim k^{-2}$ due to the shocks (Verma, 2000). In Chapter 30 we will discuss energy transfers in compressible flows.

3. *Two-dimensional turbulence:* The space dimension does not appear explicitly in the derivation of Kolmogorov's spectrum (see Section 5.1). However, Eqs. (5.6, 5.11) are applicable only to 3D hydrodynamic turbulence. Enstrophy plays an important role in 2D hydrodynamic turbulence and affects the scaling. We will describe 2D hydrodynamic turbulence in Chapter 7.

4. *Euler turbulence:* As described in Section 5.2, finite but small viscosity is mandatory for energy cascade. It makes turbulence a nonequilibrium phenomena. On the contrary, in an ideal fluid with $\nu = 0$, the energy dissipation is zero. Such a flow may show randomness, which is referred to as *Euler turbulence.* Force-free Euler's equation conserves energy and has a Hamiltonian structure. Consequently, the phase space volume is conserved during the temporal evolution, which is the statement of *Liouville's theorem* (Arnold, 1989).

   It is assumed that the Euler turbulence is ergodic as in equilibrium statistical mechanics (Ma, 1985), that is, it scans all of the available phase space (region of constant total energy) with equal probability. Therefore, in the asymptotic limit ($t \to \infty$), all the Fourier modes have equal energy, or $E_u(\mathbf{k}) = C$. Consequently, 1D energy spectrum is $E_u(k) = 4\pi C k^2$ in 3D (Lesieur, 2008); such spectrum has been observed in numerical simulations, see for example, Krstulovic et al. (2009). This phenomena is related to the equilibrium behavior for which $\Pi_u(k) \approx 0$ [see item (1) of Section 4.5]. Using field-theoretic arguments, in Section 10.2.4 we show that $S^{uu}(\mathbf{k}|\mathbf{p}|\mathbf{q}) = 0$ and hence $\Pi_u(k) \approx 0$ for Euler turbulence [also see Verma (2004)]. Thus, Euler turbulence is not described by Kolmogorov theory.

   Dallas et al. (2015) studied the energy spectra of turbulent flows at $k < k_f$. They argued that flows at these scale are in equilibrium, and hence exhibit zero energy flux. In Fig. 5.6 we exhibit the energy flux and spectrum of one such flow that is forced at $k = 10$. For $k < 10, \Pi_u(k) = 0$ and $E_u(k) \sim k^2$, consistent with the conjecture that the modes in the band $k < k_f$ are thermalized.

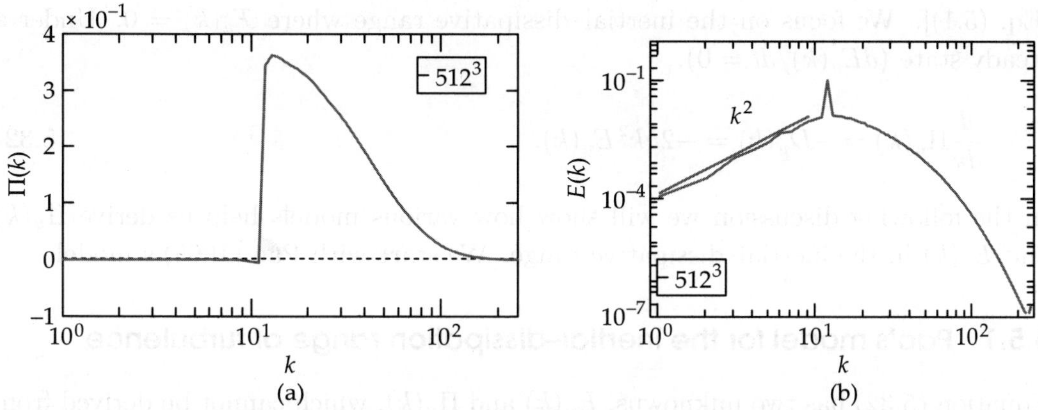

Figure 5.6 For hydrodynamic flow that is forced at $k \approx 10$: (a) The energy flux $\Pi_u(k)$, (b) $E_u(k)$. Note that for $k < 10$, $\Pi_u(k) = 0$ and $E_u(k) \sim k^2$ indicating thermalization of these modes.

5. *Fluctuations in the energy flux:* Kolmogorov's theory assumes constant energy flux, which is not strictly correct. Turbulent flows exhibit fluctuations in energy flux $\Pi_u(k)$, and $\Pi_u = \langle \Pi_u(k) \rangle$ is constant only on an average. Kolmogorov (1962) argued that $\Pi_u(k)$ follows log-normal distribution, while She and Leveque (1994) proposed $\Pi_u(k)$ to be log-Poisson. Note that such fluctuations are related to higher-order correlations or structure functions. These issues remain unresolved till date; they are discussed briefly in Chapter 12.
6. Kolmogorov's $k^{-5/3}$ spectrum is valid for the inertial range, that is, for $k_L \ll k \ll k_{\rm DI}$. In Section 5.5, we present its generalization to the inertial–dissipation range.

In the next section, we will generalize Kolmogorov's energy spectrum and flux to the dissipative range.

5.5 Energy Spectrum of Turbulent Flow in the Dissipative Regime

Kolmogorov's theory works quite well for the inertial range. In the limit $\nu \to 0$, the inertial range extends to $k = \infty$. However, real flows have finite viscosity, hence we need to generalize the energy spectrum to the dissipative regime. For turbulent flows, Pao (1965) and Pope (2000) generalized Kolmogorov's energy spectrum to the dissipative range. We will discuss these topics in this section.

For the following models, we start with the equation for the energy flux [Eq. (5.1)]. We focus on the inertial–dissipative range where $\mathcal{F}_u(k) = 0$. Under a steady state $(dE_u(k)/dt = 0)$,

$$\frac{d}{dk}\Pi_u(k) = -D_u(k) = -2\nu k^2 E_u(k). \tag{5.32}$$

In the following discussion we will show how various models help us derive $\Pi_u(k)$ and $E_u(k)$ in the inertial–dissipative range. We start with Pao (1965)'s model.

5.5.1 Pao's model for the inertial–dissipation range of turbulence

Equation (5.32) has two unknowns, $E_u(k)$ and $\Pi_u(k)$, which cannot be derived from a single equation. To overcome this difficulty, Pao (1965) assumed that in the inertial and dissipative ranges, $\Pi_u(k)/E_u(k)$ is independent of ν, and is function only of k and ϵ_u [also see Leslie (1973)]. Under this assumption, dimensional analysis yields

$$\frac{E_u(k)}{\Pi_u(k)} = K_{\mathrm{Ko}}\epsilon_u^{-1/3}k^{-5/3}. \tag{5.33}$$

This is a universal form that works in both inertial and dissipation range. Substitution of Eq. (5.33) in Eq. (5.32) yields

$$\Pi_u(k) = \epsilon_u \exp\left(-\frac{3}{2}K_{\mathrm{Ko}}(k/k_d)^{4/3}\right), \tag{5.34a}$$

$$E_u(k) = K_{\mathrm{Ko}}\epsilon_u^{2/3}k^{-5/3}\exp\left(-\frac{3}{2}K_{\mathrm{Ko}}(k/k_d)^{4/3}\right), \tag{5.34b}$$

where k_d is Kolmogorov's wavenumber. Thus, the energy spectrum in the dissipation range has an exponential form $(\exp(-(k/k_d)^{4/3}))$.

The aforementioned derivation, which based on variable energy flux, clearly shows that the energy flux in the dissipative range is nonzero. This is contrary to the often-quoted statement that the energy flux vanishes in the dissipative range. Later in this section we will verify the above scaling using numerical simulations.

5.5.2 Pope's model for the inertial–dissipation range of turbulence

Another popular model for the turbulent flow is by Pope (2000). Based on $E_u(k)$ from experimental data, Pope proposed that

$$E_u(k) = K_{\mathrm{Ko}}\epsilon^{2/3}k^{-5/3}f_L(kL)f_\eta(k/k_d) \tag{5.35}$$

with the functions $f_L(kL)$ and $f_\eta(k/k_d)$ representing the large-scale and dissipative-scale components, respectively:

$$f_L(kL) = \left(\frac{kL}{[(kL)^2 + c_L]^{1/2}}\right)^{5/3+p_0}, \tag{5.36a}$$

$$f_\eta(\tilde{k}) = \exp\left[-\beta\left\{[\tilde{k}^4 + c_\eta^4]^{1/4} - c_\eta\right\}\right], \tag{5.36b}$$

where $\tilde{k} = k/k_d$, and c_L, c_η, p_0, β are constants. Note that $f_L(kL)$ and $f_\eta(\tilde{k})$ dominate in the forcing range and dissipative range respectively, and they are approximately unity in the inertial range. As shown by Pope (2000), the choice of constants is quite tricky. One of the choices is $\beta = 5.2$, $C_L = 6.78$, $c_\eta = 0.40$, and $p_0 = 2$. To compute $\Pi_u(k)$, we substitute $E_u(k)$ of Eq. (5.35) in Eq. (5.32), and solve the differential equation numerically.

Verma et al. (2018) performed numerical simulation of turbulent flows on 512^3, 1024^3 and 4096^3 grids for Re $= 5.7\times10^3, 1.4\times10^4$ and 6.8×10^4 respectively. Using these numerical data, they computed the following normalized energy spectra and fluxes for the aforementioned three runs:

$$\frac{\tilde{E}(k)}{K_{\mathrm{Ko}}} = E_u(k)k^{5/3}\epsilon_u^{2/3}, \tag{5.37a}$$

$$\tilde{\Pi}(k) = \frac{\Pi_u(k)}{\epsilon_u}. \tag{5.37b}$$

In Fig. 5.7, we plot the above quantities along with the aforementioned predictions of Pope and Pao. Verma et al. (2018) observed that $K_{\mathrm{Ko}} = 2.2\pm 0.2, 1.85\pm0.05, 1.75\pm 0.05$ for 512^3, 1024^3, 4096^3 grids respectively. Similar variations of K_{Ko} have been reported earlier. As shown in Fig. 5.7(a), the models of Pao and Pope capture the energy spectra quite well. However, $\Pi_u(k)$ predicted by Pope's model is significantly smaller than the numerical values in the dissipative range. Pao's model however captures $\Pi_u(k)$ quite well.

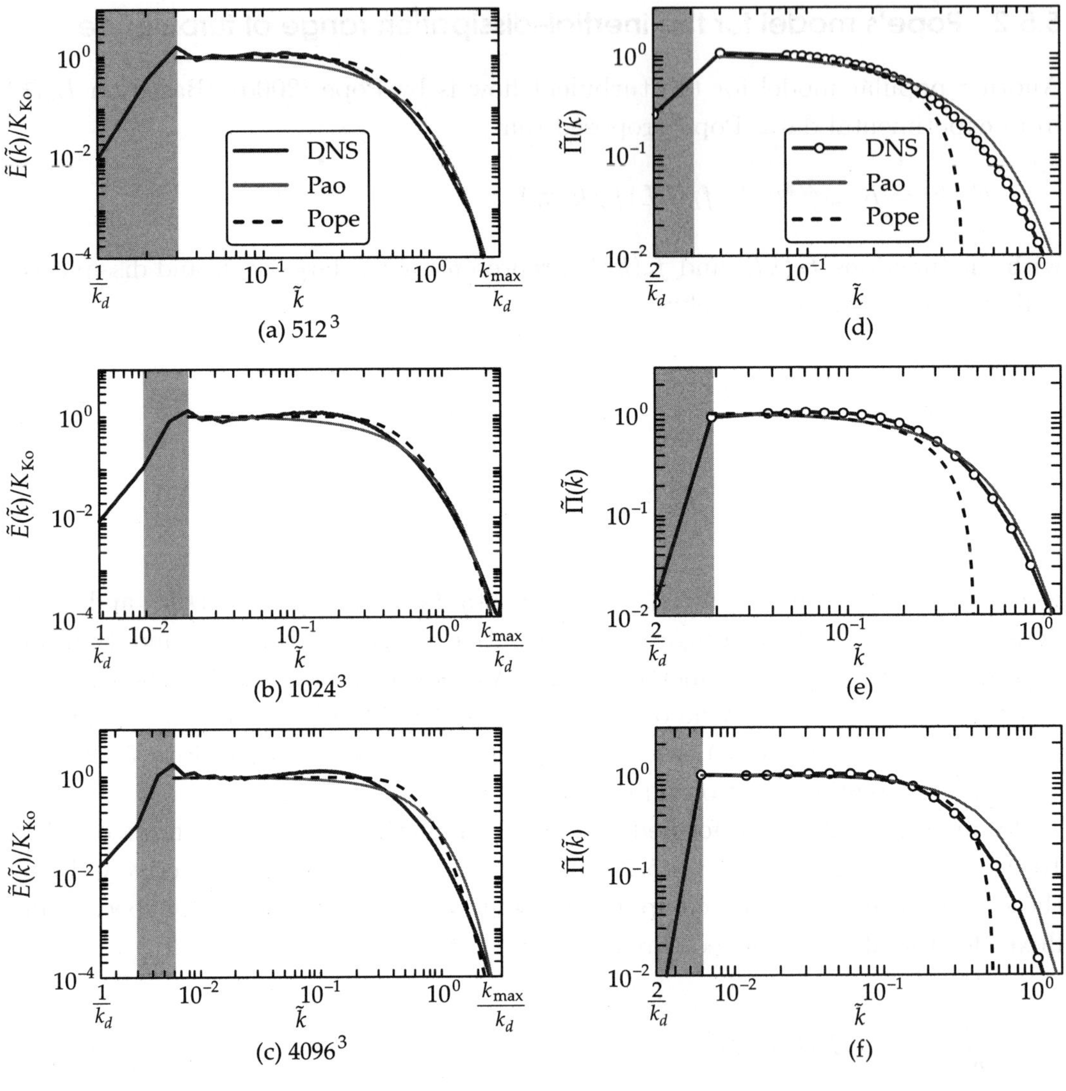

Figure 5.7 For the grid resolutions of 512^3, 1024^3, and 4096^3: (a,b,c) plots of the normalized energy spectrum $\tilde{E}(\tilde{k})/K_{\rm Ko}$ vs. $\tilde{k}$; (d,e,f) plots of normalized energy flux $\tilde{\Pi}(\tilde{k})$ vs. $k = \tilde{k}/k_d$. See Eqs. (5.37a) and Eq. (5.37b). The plots include the spectra and fluxes computed using numerical data (thick solid line), and the model prediction of Pao (thin solid line) and Pope (dashed line). From Verma et al. (2018). Reprinted with permission from Springer.

Based on these observations, we claim that Pao's model describes the energy spectrum and flux of hydrodynamic turbulence in the inertial–dissipative range quite well. A word of caution, however, is in order. Both Pao's and Pope's models do not capture the bottleneck effect of hydrodynamic turbulence (Falkovich, 1994; Verma and Donzis, 2007; Verma et al., 2018).

Before we close this section, we rederive k_d because the derivation of Eq. (5.19) appears rather ad hoc. The total dissipation rate is

$$\begin{aligned}\epsilon_u &= \int_0^\infty \nu k^2 E_u(k) dk \\ &\approx \int_0^\infty \nu K_{\mathrm{Ko}} \epsilon_u^{2/3} k^{1/3} f_\eta(k/k_d) dk \\ &\approx \nu K_{\mathrm{Ko}} \epsilon_u^{2/3} k_d^{1/3} \int_0^\infty x^{4/3} f_\eta(x) dx, \end{aligned} \tag{5.38}$$

where $x = k/k_d$. Assuming that the constants including the integral are of the order of unity, we obtain

$$k_d \approx \left(\frac{\epsilon_u}{\nu^3}\right)^{1/4}, \tag{5.39}$$

which is the same k_d of Eq. (5.19). The aforementioned argument is somewhat more rigorous.

In the next section we will discuss the energy spectrum and flux for laminar flows.

5.6 Energy Spectrum and Flux for Laminar Flows

Equations (5.34a, 5.34b) describe the inertial–dissipative range of a turbulent flow quite well, but they fail to model laminar flows with $\mathrm{Re} = O(1)$. This is expected since $E_u(k)$ of a laminar flow is not proportional to $k^{-5/3}$. It is important to note that $\Pi_u(k)$ is nonzero for such flows. In fact, $\Pi_u(k)$ would be nonzero even when Re or nonlinearity is very small.

Verma et al. (2018) proposed that for laminar flows with $\mathrm{Re} = O(1)$, the energy spectrum and flux have $\exp(-k)$ form. Under this assumption, the solution of Eq. (5.32) is

$$E_u(k) = A\frac{1}{k}\exp(-k/\bar{k}_d), \tag{5.40a}$$

$$\Pi_u(k) = A2\nu\bar{k}_d^2(1 + (k/\bar{k}_d))\exp(-k/\bar{k}_d), \tag{5.40b}$$

where $\bar{k}_d$ is the dissipation wavenumber scale for laminar flows. We estimate $\bar{k}_d$ as

$$\bar{k}_d \approx \left(\frac{\epsilon_u}{\nu^3}\right)^{1/4} \approx \left(\frac{\nu U^2}{L^2\nu^3}\right)^{1/4} \approx \frac{\sqrt{\mathrm{Re}}}{L}. \tag{5.41}$$

With $\epsilon_u \approx \nu U^2/L^2$. Numerical simulations yield the following form for the prefactor A:

$$A = \frac{U^2}{\mathrm{Re}^3}. \tag{5.42}$$

Hence,

$$E_u(k) = \frac{U^2}{\mathrm{Re}^3}\frac{1}{k}\exp(-k/\bar{k}_d), \tag{5.43a}$$

$$\Pi_u(k) = \bar{\epsilon}_u(1 + (k/\bar{k}_d))\exp(-k/\bar{k}_d), \tag{5.43b}$$

where

$$\bar{\epsilon}_u = \frac{U^2}{\mathrm{Re}^3}2\nu\bar{k}_d^2. \tag{5.44}$$

Curiously, substitution of $\bar{k}_d$ in Eq. (5.44) yields $\bar{\epsilon}_u = \nu^3/L^4$, which has the same form as Eq. (5.18) with $l_d = L$. Note that $\bar{\epsilon}_u$ is not the viscous dissipation rate; rather, it is the value of the energy flux at the wavenumber from where $\Pi_u(k)$ has functional behavior of the form of Eq. (5.43b).

In this phenomenology, the forms of A and $\bar{k}_d$ have been derived from numerical simulations. At present, we do not have a very good understanding why these quantities take such forms. We hope that more refined and detailed simulations will help us resolve this issue. In the later part of the section we will verify this phenomenology using numerical simulations.

It is important to discuss the limiting case Re $= 0$ for which the energy flux is zero due to the absence of nonlinear term. For a steady flow with Re $= 0$, the viscous term matches with the external force:

$$\nu k^2 \mathbf{u}(\mathbf{k}) = \mathbf{F_u}(\mathbf{k}). \tag{5.45}$$

Hence,

$$\mathbf{u}(\mathbf{k}) = \frac{1}{2\nu k^2}\mathbf{F_u}(\mathbf{k}) \implies E_u(\mathbf{k}) = \frac{1}{4\nu^2 k^4}|\mathbf{F_u}(\mathbf{k})|^2. \tag{5.46}$$

Thus, the energy spectrum depends on the spectrum of the external force.

For the force-free and decaying case with Re $= 0$, Eq. (5.1) yields

$$\frac{\partial}{\partial t}E_u(k,t) = -2\nu k^2 E_u(k,t), \tag{5.47}$$

whose solution is

$$E_u(k,t) = E_u(k,0)\exp(-2\nu k^2 t). \tag{5.48}$$

Thus, the kinetic energy spectrum decays exponentially in time. These results show that the behavior of energy spectrum for $\mathrm{Re} \neq 0$ (but small) and $\mathrm{Re} = 0$ are quite different. This is due to the presence of energy flux for nonzero nonlinearity. It is incorrect to assume that the energy flux is absent in laminar flows.

Verma et al. (2018) performed numerical simulations of laminar flows in a periodic box for $\mathrm{Re} = 49, 32.4, 23.1,$ 17.6. For these runs, $\bar{k}_d = 0.9, 0.7, 0.6, 0.5$ respectively. They computed $E_u(k)$ and $\Pi_u(k)$ for the steady state of these flows. For these runs, the normalized energy spectrum and flux are

$$\tilde{E}_u(k) = \frac{E_u(k)k}{A} = \exp(-k/\bar{k}_d), \tag{5.49a}$$

$$\tilde{\Pi}_u(k) = \frac{\Pi}{\bar{\epsilon}_u} = (1 + (k/\bar{k}_d))\exp(-k/\bar{k}_d), \tag{5.49b}$$

which are plotted in Fig. 5.8 respectively. For $k > k_f$, where k_f is the forcing wavenumber, numerically computed $\tilde{E}_u(\tilde{k})$ and $\tilde{\Pi}_u(\tilde{k})$ collapse nicely, and fit with the aforementioned model. Thus, we claim that the model presented here captures the energy spectrum and flux for laminar flows quite well.

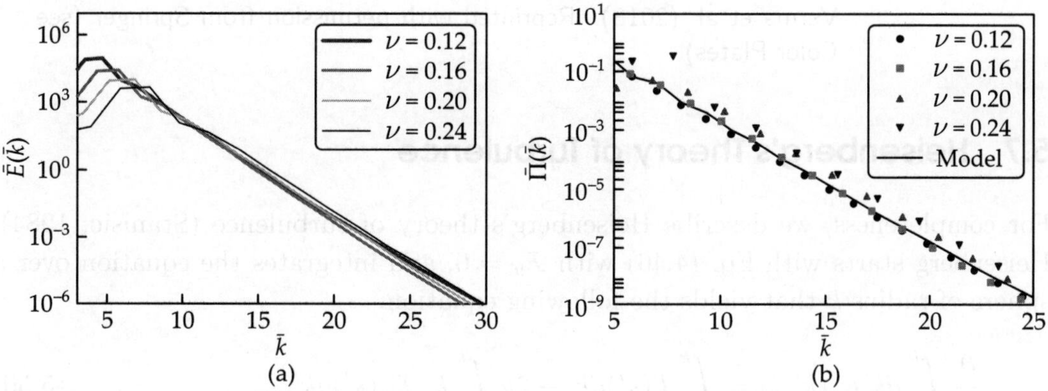

Figure 5.8 For the laminar flow simulations with $\nu = 0.12, 0.16, 0.20, 0.24$ or $\mathrm{Re} = 49, 32.4, 23.1, 17.6$, plots of (a) $\tilde{E}_u(k)$ of Eq. (5.49a), (b) $\tilde{\Pi}_u(k)$ of Eq. (5.49b). For $k > k_f$, the data collapse into single curves. From Verma et al. (2018). Reprinted with permission from Springer.

Verma et al. (2018) also computed the shell-to-shell energy transfers for the flows with $\mathrm{Re} = 49$ and 17.5. These transfers are depicted in Fig. 5.9 that exhibits nonlocal energy transfers from the shell of the forcing band to other shells. Note that the energy transfer is always from small wavenumber shells to large wavenumber shells, hence the shell-to-shell energy transfers for laminar flows are forward. Contrast these transfers with the local energy transfers in the turbulent regime (see Figs. 5.4 and 5.5).

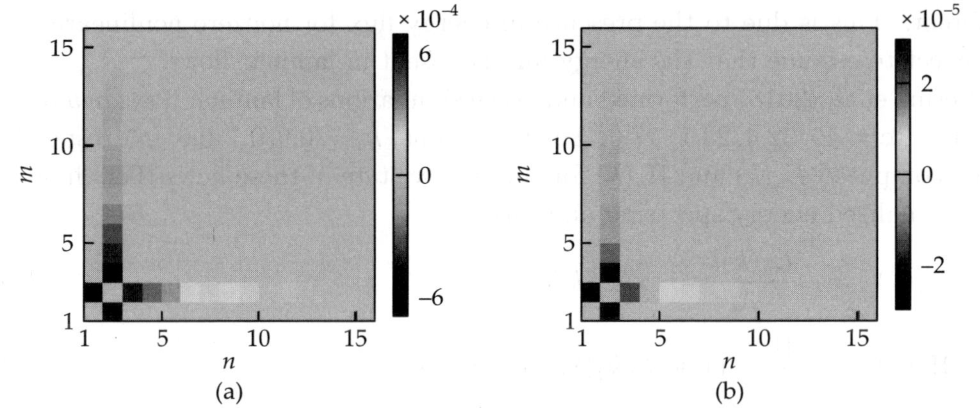

Figure 5.9 For the laminar flow simulation, the shell-to-shell energy transfer rate for (a) $\mathrm{Re} = 49$, (b) $\mathrm{Re} = 17.5$. The forcing wavenumbers (belonging to the 2nd shell) gives significant energy to the shells 3 to 10 for $\mathrm{Re} = 49$, and to the shells 3 to 7 for $\mathrm{Re} = 17.6$. Thus, the energy transfer in the laminar regime is forward and *nonlocal*. From Verma et al. (2018). Reprinted with permission from Springer (see Color Plates).

5.7 Heisenberg's Theory of Turbulence

For completeness, we describe Heisenberg's theory of turbulence (Stanisic, 1984). Heisenberg starts with Eq. (4.46) with $\mathcal{F}_u = 0$, and integrates the equation over a sphere of radius k that yields the following equation:

$$\frac{\partial}{\partial t}\int_0^k dk' E_u(k',t) = \int_0^k T(k')dk' - 2\nu \int_0^k k'^2 E_u(k')dk'. \tag{5.50}$$

He proposed that

$$\int_0^k T(k')dk' = -2\nu'(k)\int_0^k k'^2 E_u(k')dk', \tag{5.51}$$

where $\nu'(k)$ is the eddy viscosity. Using dimensional analysis, Heisenberg postulates that $\nu'(k)$ is the product of local velocity and length scales; hence,

$$\nu'(k) = u_k k^{-1} = \alpha \int_k^\infty \sqrt{\frac{E_u(k')}{k'^3}} dk', \tag{5.52}$$

where α is a constant.[3] Substitution of these in Eq. (5.50) yields

$$\frac{\partial}{\partial t} \int_0^k dk' E_u(k', t) = -2 \left(\nu + \alpha \int_k^\infty \sqrt{\frac{E_u(k')}{k'^3}} dk' \right) \int_0^k k'^2 E_u(k') dk'. \tag{5.53}$$

Heisenberg assumed that in decaying turbulence, the large-scale eddies provide energy to the smaller eddies, while the smallest eddies remain in *statistical equilibrium* among themselves. Hence,

$$\frac{\partial}{\partial t} \int_0^k dk' E_u(k', t) \approx \Pi_u. \tag{5.54}$$

Then from dimensional analysis,

$$\Pi_u \sim [E_u(k)]^{3/2} k^{5/3}, \tag{5.55}$$

that yields Kolmogorov's spectrum.

Batchelor (1953) and Chandrasekhar (Spiegel, 2010) solved Eq. (5.53) analytically and derived Kolmogorov's spectrum for the initial range, and $E_u(k) \sim k^{-7}$ for the dissipation range. In particular, Chandrasekhar (Spiegel, 2010) showed that

$$E(k) = E(k_0) \left(\frac{k_0}{k} \right)^{5/3} \frac{1}{[1 + (k/k_s)^4]^{4/3}}, \tag{5.56}$$

where $k_0 \sim 1/L$, and $k_s \sim \nu^{-3/4}$, which is related to Kolmogorov's wavenumber k_d of Eq. (5.19). Equation (5.56) yields $E(k) \sim k^{-5/3}$ for $k < k_s$, and k^{-7} for $k > k_s$.

Note however that Pao's spectrum (Eq. (5.32)) fits better with the numerical data than the above model. For the dissipative range, the aforementioned prediction for the spectrum, $E_u(k) \sim k^{-7}$, is too steep to fit with the numerical data. Hence we believe that Pao's model is more reliable than that by Heisenberg.

With this, we end our discussion on 3D hydrodynamic turbulence. The models presented here form a starting point for more complex turbulence, as in 2D hydrodynamics, scalars, magnetohydrodynamics, rotating flows, etc.

---

[3] In modern treatment, $\nu'(k) \sim \epsilon^{1/3} k^{-4/3}$. Renormalization group analysis yields the proportionality constant. See Appendix C for details.

Further Reading

The textbooks by Frisch (1995), Leslie (1973), McComb (1990), Pope (2000), Batchelor (1953), and Lesieur (2008) describe 3D hydrodynamic turbulence in great detail. The numerical simulation described in this chapter can be found in Verma et al. (2018). Heisenberg's model and its solution are discussed in Spiegel (2010) and Stanisic (1984). Also, refer to Alexakis et al. (2005a) who analyzed the shell-to-shell energy transfer and locality issues of hydrodynamic turbulence.

Exercises

1. Estimate Kolmogorov's length l_d of
   (a) A glass of turbulent water whose $\mathrm{Re} \sim 10^5$.
   (b) Turbulent atmospheric flow with $\mathrm{Re} \sim 10^6$.
   (c) In solar convection for which $L \sim 10^6$ m, $U \sim 10^2$ m/s, and $\mathrm{Re} \sim 10^{13}$.
   (d) In Earth's outer core for which $L \sim 10^6$ m, $U \sim 10^{-3}$ m/s, and $\mathrm{Re} \sim 10^8$.

   Compare these results with the correspoding mean free path lengths. Compute the grid resolutions required to simulate the aforementioned flows.
2. Argue why turbulence is a nonequilibrium process.
3. Consider two wavenumber shells $(k, 2k)$ and $(2k, 4k)$ in the inertial range of a turbulent flow. Compare the energy contents of these shells.
4. Compute the integral of Eq. (5.38) for Pao and Pope's models. Using the values of the integrals, estimate the nondimensional prefactor in the formula for k_d.

Chapter 6

Enstrophy Transfers in Hydrodynamics

Vorticity $\boldsymbol{\omega} = \nabla \times \mathbf{u}$ plays an important role in hydrodynamics. In Chapters 2, 3, and 4 we described several properties of vorticity and enstrophy ($E_\omega = \omega^2/2$). In this short chapter we will derive formulas for the enstrophy transfers in hydrodynamics.

6.1 Mode-to-mode Enstrophy Transfers in Hydrodynamics

First, we derive formulas for the mode-to-mode enstrophy transfers following similar steps as that for the mode-to-mode kinetic energy transfers (see Chapter 4). We start with Eq. (3.49), which is

$$\begin{aligned}\frac{d}{dt}E_\omega(\mathbf{k}) &= \sum_{\mathbf{p}} \Im\left[\{\mathbf{k}\cdot\mathbf{u}(\mathbf{q})\}\{\boldsymbol{\omega}(\mathbf{p})\cdot\boldsymbol{\omega}^*(\mathbf{k})\}\right] - \Im\left[\{\mathbf{k}\cdot\boldsymbol{\omega}(\mathbf{q})\}\{\mathbf{u}(\mathbf{p})\cdot\boldsymbol{\omega}^*(\mathbf{k})\}\right] \\ &\quad + \mathcal{F}_\omega(\mathbf{k}) - 2\nu k^2 E_\omega(\mathbf{k}), \end{aligned} \tag{6.1}$$

where $\mathbf{k} = \mathbf{p} + \mathbf{q}$, and $\mathcal{F}_\omega$ is the enstrophy injection rate by the external force. As in Section 4.1, we set $\nu = 0$ and $\mathcal{F}_\omega = 0$, and focus on a pair of triads, $(\mathbf{k}', \mathbf{p}, \mathbf{q})$ and $(-\mathbf{k}', -\mathbf{p}, -\mathbf{q})$, with $\mathbf{k}' = -\mathbf{k}$. See Fig. 4.1 for an illustration. Given these Fourier modes, following Eq. (6.1), the evolution equation for the modal enstrophy $E_\omega(\mathbf{k}) = |\omega(\mathbf{k})|^2/2$ is

$$\frac{d}{dt}E_\omega(\mathbf{k}') = S^{\omega\omega}(\mathbf{k}'|\mathbf{p},\mathbf{q}) + S^{\omega u}(\mathbf{k}'|\mathbf{p},\mathbf{q}), \tag{6.2}$$

where

$$S^{\omega\omega}(\mathbf{k}'|\mathbf{p},\mathbf{q}) = -\Im\left[\{\mathbf{k}'\cdot\mathbf{u}(\mathbf{q})\}\{\boldsymbol{\omega}(\mathbf{p})\cdot\boldsymbol{\omega}(\mathbf{k}')\}\right] - \Im\left[\{\mathbf{k}'\cdot\mathbf{u}(\mathbf{p})\}\{\boldsymbol{\omega}(\mathbf{q})\cdot\boldsymbol{\omega}(\mathbf{k}')\}\right], \quad (6.3a)$$

$$S^{\omega u}(\mathbf{k}'|\mathbf{p},\mathbf{q}) = \Im\left[\{\mathbf{k}'\cdot\boldsymbol{\omega}(\mathbf{q})\}\{\mathbf{u}(\mathbf{p})\cdot\boldsymbol{\omega}(\mathbf{k}')\}\right] + \Im\left[\{\mathbf{k}'\cdot\boldsymbol{\omega}(\mathbf{p})\}\{\mathbf{u}(\mathbf{q})\cdot\boldsymbol{\omega}(\mathbf{k}')\}\right]. \quad (6.3b)$$

We can derive similar equations for $E_\omega(\mathbf{p})$ and $E_\omega(\mathbf{q})$.

Using the incompressibility condition $\mathbf{k}\cdot\mathbf{u}(\mathbf{k}) = 0$, we can show that

$$S^{\omega\omega}(\mathbf{k}'|\mathbf{p},\mathbf{q}) + S^{\omega\omega}(\mathbf{p}|\mathbf{k}',\mathbf{q}) + S^{\omega\omega}(\mathbf{q}|\mathbf{k}',\mathbf{p}) = 0. \quad (6.4)$$

A physical interpretation of this equation is that the enstrophy in a triad is conserved for the nonlinear interactions $S^{\omega\omega}(\mathbf{k}'|\mathbf{p},\mathbf{q})$. These interactions correspond to advection of vortices by the velocity modes, and they involve enstrophy exchange among vorticity modes, thus conserving total enstrophy in a triad.

Note however that

$$S^{\omega u}(\mathbf{k}'|\mathbf{p},\mathbf{q}) + S^{\omega u}(\mathbf{p}|\mathbf{k}',\mathbf{q}) + S^{\omega u}(\mathbf{q}|\mathbf{k}',\mathbf{p}) \neq 0. \quad (6.5)$$

This is because $S^{\omega u}()$ represents stretching of vortices that leads to either enhancement or depletion of E_ω. This is the reason why the total enstrophy is not conserved in 3D hydrodynamics.

Now we are ready to derive expressions for the mode-to-mode enstrophy transfers—$S^{\omega\omega}(\mathbf{k}'|\mathbf{p}|\mathbf{q})$ and $S^{\omega u}(\mathbf{k}'|\mathbf{p}|\mathbf{q})$. This is the topic of the next two subsections.

6.1.1 Derivation of mode-to-mode enstrophy transfer $S^{\omega\omega}(\mathbf{k}'|\mathbf{p}|\mathbf{q})$

In this subsection we derive an expression for $S^{\omega\omega}(\mathbf{k}'|\mathbf{p}|\mathbf{q})$, which is the enstrophy transfer from $\boldsymbol{\omega}(\mathbf{p})$ to $\boldsymbol{\omega}(\mathbf{k}')$ with $\mathbf{u}(\mathbf{q})$ acting as a mediator. For convenience, we denote the wavenumbers in the triad as $\mathbf{X}, \mathbf{Y}, \mathbf{Z}$ with $\mathbf{X}+\mathbf{Y}+\mathbf{Z} = 0$. Following the derivation of Section 4.1, we deduce the following properties for $S^{\omega\omega}(\mathbf{X}|\mathbf{Y}|\mathbf{Z})$s:

1. The sum of $S^{\omega\omega}(\mathbf{X}|\mathbf{Y}|\mathbf{Z})$ and $S^{\omega\omega}(\mathbf{X}|\mathbf{Z}|\mathbf{Y})$ is the combined enstrophy transfer $S^{\omega\omega}(\mathbf{X}|\mathbf{Y},\mathbf{Z})$ defined in Eq. (6.3a). Therefore,

$$S^{\omega\omega}(\mathbf{k}'|\mathbf{p}|\mathbf{q}) + S^{\omega\omega}(\mathbf{k}'|\mathbf{q}|\mathbf{p}) = S^{\omega\omega}(\mathbf{k}'|\mathbf{p},\mathbf{q}), \quad (6.6a)$$

$$S^{\omega\omega}(\mathbf{p}|\mathbf{k}'|\mathbf{q}) + S^{\omega\omega}(\mathbf{p}|\mathbf{q}|\mathbf{k}') = S^{\omega\omega}(\mathbf{p}|\mathbf{k}',\mathbf{q}), \quad (6.6b)$$

$$S^{\omega\omega}(\mathbf{q}|\mathbf{k}'|\mathbf{p}) + S^{\omega\omega}(\mathbf{q}|\mathbf{p}|\mathbf{k}') = S^{\omega\omega}(\mathbf{q}|\mathbf{k}',\mathbf{p}). \quad (6.6c)$$

2. The enstrophy transfer from $\boldsymbol{\omega}(\mathbf{Y})$ to $\boldsymbol{\omega}(\mathbf{X})$, $S^{\omega\omega}(\mathbf{X}|\mathbf{Y}|\mathbf{Z})$, is equal and opposite to the enstrophy transfer from $\boldsymbol{\omega}(\mathbf{X})$ to $\boldsymbol{\omega}(\mathbf{Y})$, $S^{\omega\omega}(\mathbf{Y}|\mathbf{X}|\mathbf{Z})$. Therefore,

$$S^{\omega\omega}(\mathbf{k}'|\mathbf{p}|\mathbf{q}) + S^{\omega\omega}(\mathbf{p}|\mathbf{k}'|\mathbf{q}) = 0, \tag{6.7a}$$

$$S^{\omega\omega}(\mathbf{k}'|\mathbf{q}|\mathbf{p}) + S^{\omega\omega}(\mathbf{q}|\mathbf{k}'|\mathbf{p}) = 0, \tag{6.7b}$$

$$S^{\omega\omega}(\mathbf{p}|\mathbf{q}|\mathbf{k}') + S^{\omega\omega}(\mathbf{q}|\mathbf{p}|\mathbf{k}') = 0. \tag{6.7c}$$

We find that the set

$$S^{\omega\omega}(\mathbf{X}|\mathbf{Y}|\mathbf{Z}) = -\Im\left[\{\mathbf{X}\cdot\mathbf{u}(\mathbf{Z})\}\{\boldsymbol{\omega}(\mathbf{Y})\cdot\boldsymbol{\omega}(\mathbf{X})\}\right] \tag{6.8}$$

satisfies Eqs. (6.6, 6.7). Hence, it is a solution of Eqs. (6.6, 6.7). Note that for $\mathbf{X}=\mathbf{k}', \mathbf{Y}=\mathbf{p}, \mathbf{Z}=\mathbf{q}$, the formula is

$$S^{\omega\omega}(\mathbf{k}'|\mathbf{p}|\mathbf{q}) = -\Im\left[\{\mathbf{k}'\cdot\mathbf{u}(\mathbf{q})\}\{\boldsymbol{\omega}(\mathbf{p})\cdot\boldsymbol{\omega}(\mathbf{k}')\}\right]. \tag{6.9}$$

Unfortunately, this set is not a unique solution because the determinant of the matrix formed by these linear equations is zero. We show in the following that $S^{\omega\omega}(\mathbf{X}|\mathbf{Y}|\mathbf{Z})$ of Eq. (6.8) indeed yields the mode-to-mode enstrophy transfers in the triad.

The physical argument in support of the aforementioned statement is similar to those in Section 4.1.1. In real space, the nonlinear term corresponding to $S^{\omega\omega}(\mathbf{k}'|\mathbf{p}|\mathbf{q})$ is $[(\mathbf{u}\cdot\nabla)\boldsymbol{\omega}]\cdot\boldsymbol{\omega}$. Here, the velocity field $\mathbf{u}$ advects the $\boldsymbol{\omega}$ field, hence $\mathbf{u}$ does not participate directly in the enstrophy exchange. The $\boldsymbol{\omega}$ fields that appear in the right of the ∇ operator interact among themselves and exchange enstrophy. Therefore, in $S^{\omega\omega}(\mathbf{k}'|\mathbf{p}|\mathbf{q})$, $\mathbf{u}(\mathbf{q})$ must be the mediator mode, and $\boldsymbol{\omega}(\mathbf{p})$ and $\boldsymbol{\omega}(\mathbf{k}')$ must be giver and receiver vorticity modes respectively. See Fig. 6.1 for an illustration. Thus, we show that Eq. (6.9) is indeed the mode-to-mode enstrophy transfer from $\boldsymbol{\omega}(\mathbf{p})$ to $\boldsymbol{\omega}(\mathbf{k}')$ with the mediation of $\mathbf{u}(\mathbf{q})$.

Now we present mathematical arguments which are similar to those in Section 4.1. Using the structure of Eq. (6.3a), we demand that $S^{\omega\omega}(\mathbf{k}|\mathbf{p}|\mathbf{q})$ satisfies the following properties:

1. $S^{\omega\omega}(\mathbf{k}'|\mathbf{p}|\mathbf{q})$ is real.
2. $S^{\omega\omega}(\mathbf{k}'|\mathbf{p}|\mathbf{q})$ is a linear function of the wave vector $\mathbf{k}'$.
3. $S^{\omega\omega}(\mathbf{k}'|\mathbf{p}|\mathbf{q})$ is a linear function of one of the Fourier modes $\mathbf{u}(\mathbf{k}')$, $\mathbf{u}(\mathbf{p})$, and $\mathbf{u}(\mathbf{q})$. That is, the expressions of $S^{\omega\omega}(\mathbf{k}'|\mathbf{p}|\mathbf{q})$ includes one of the three velocity modes.

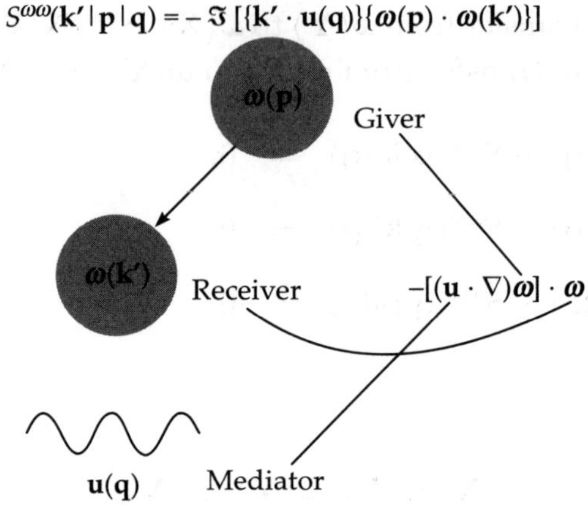

Figure 6.1 A schematic diagram of $S^{\omega\omega}(\mathbf{k}'|\mathbf{p}|\mathbf{q})$, the mode-to-mode enstrophy transfer from mode $\boldsymbol{\omega}(\mathbf{p})$ to mode $\boldsymbol{\omega}(\mathbf{k}')$ with the mediation of mode $\mathbf{u}(\mathbf{q})$. The modes $\mathbf{u}(\mathbf{q})$ advects the vorticity modes who exchange enstrophy among themselves.

4. $S^{\omega\omega}(\mathbf{k}'|\mathbf{p}|\mathbf{q})$ is a linear function of two of the three vorticity modes $\boldsymbol{\omega}(\mathbf{k}')$, $\boldsymbol{\omega}(\mathbf{p})$, and $\boldsymbol{\omega}(\mathbf{q})$. The arguments of $\boldsymbol{\omega}$'s differ from that of $\mathbf{u}$.
5. Due to the equivalence of the triads $(\mathbf{k},\mathbf{p},\mathbf{q})$ and $(-\mathbf{k},-\mathbf{p},-\mathbf{q})$,

$$S^{\omega\omega}(-\mathbf{k}'|-\mathbf{p}|-\mathbf{q}) = S^{\omega\omega}(\mathbf{k}'|\mathbf{p}|\mathbf{q}). \tag{6.10}$$

Given the aforementioned properties, the tensor analysis yields the following form for $S^{uu}(\mathbf{k}'|\mathbf{p}|\mathbf{q})$:

$$\begin{aligned} S^{\omega\omega}(\mathbf{k}'|\mathbf{p}|\mathbf{q}) &= c_1\Im\left[\{\mathbf{k}'\cdot\mathbf{u}(\mathbf{q})\}\{\boldsymbol{\omega}(\mathbf{p})\cdot\boldsymbol{\omega}(\mathbf{k}')\}\right] \\ &+ c_2\Im\left[\{\mathbf{k}'\cdot\mathbf{u}(\mathbf{p})\}\{\boldsymbol{\omega}(\mathbf{q})\cdot\boldsymbol{\omega}(\mathbf{k}')\}\right] \\ &+ c_3\Im\left[\{\mathbf{k}'\cdot\boldsymbol{\omega}(\mathbf{q})\}\{\boldsymbol{\omega}(\mathbf{p})\cdot\mathbf{u}(\mathbf{k}')\}\right] \\ &+ c_4\Im\left[\{\mathbf{k}'\cdot\boldsymbol{\omega}(\mathbf{q})\}\{\boldsymbol{\omega}(\mathbf{k}')\cdot\mathbf{u}(\mathbf{p})\}\right] \\ &+ c_5\Im\left[\{\mathbf{k}'\cdot\boldsymbol{\omega}(\mathbf{p})\}\{\boldsymbol{\omega}(\mathbf{q})\cdot\mathbf{u}(\mathbf{k}')\}\right] \\ &+ c_6\Im\left[\{\mathbf{k}'\cdot\boldsymbol{\omega}(\mathbf{p})\}\{\boldsymbol{\omega}(\mathbf{k}')\cdot\mathbf{u}(\mathbf{q})\}\right]. \end{aligned} \tag{6.11}$$

Here we have eliminated some terms by exploiting the conditions—$\mathbf{k}'\cdot\mathbf{u}(\mathbf{k}') = 0$ and $\mathbf{k}'\cdot\boldsymbol{\omega}(\mathbf{k}') = 0$. From the structure of $S^{\omega\omega}(\mathbf{k}'|\mathbf{p},\mathbf{q})$ in Eq. (6.3a), we deduce that the last four terms do not belong here; they are part of the solution for $S^{\omega u}(\mathbf{k}'|\mathbf{p}|\mathbf{q})$ that will be discussed in the next subsection. Note that $\Re[.]$ does not appear in Eq. (6.11) due to Eq. (6.10) (see Section 4.1). Therefore,

$$\begin{aligned} S^{\omega\omega}(\mathbf{k}'|\mathbf{p}|\mathbf{q}) &= c_1\Im\left[\{\mathbf{k}'\cdot\mathbf{u}(\mathbf{q})\}\{\boldsymbol{\omega}(\mathbf{p})\cdot\boldsymbol{\omega}(\mathbf{k}')\}\right] \\ &+ c_2\Im\left[\{\mathbf{k}'\cdot\mathbf{u}(\mathbf{p})\}\{\boldsymbol{\omega}(\mathbf{q})\cdot\boldsymbol{\omega}(\mathbf{k}')\}\right]. \end{aligned} \tag{6.12}$$

Now, an application of Eq. (6.7a) leads to $c_2 = 0$, while that of Eq. (6.6a) yields $c_1 = -1$. Thus, we deduce that

$$S^{\omega\omega}(\mathbf{k}|\mathbf{p}|\mathbf{q}) = -\Im\left[\{\mathbf{k}'\cdot\mathbf{u}(\mathbf{q})\}\{\boldsymbol{\omega}(\mathbf{p})\cdot\boldsymbol{\omega}(\mathbf{k}')\}\right] \tag{6.13}$$

is the formula for the enstrophy transfer from $\boldsymbol{\omega}(\mathbf{p})$ to $\boldsymbol{\omega}(\mathbf{k}')$ with the mediation of $\mathbf{u}(\mathbf{q})$.

It is important to note that for the wavenumber triad,

$$\sum_{\mathbf{k}}\sum_{\mathbf{p}} S^{\omega\omega}(\mathbf{k}|\mathbf{p}|\mathbf{q}) = 0. \tag{6.14}$$

This is the law of enstrophy conservation (via $S^{\omega\omega}$ interactions) for any wavenumber region.

In the next subsection, we derive a formula for $S^{\omega u}(\mathbf{k}'|\mathbf{p}|\mathbf{q})$.

6.1.2 Derivation of mode-to-mode enstrophy transfer $S^{\omega u}(\mathbf{k}'|\mathbf{p}|\mathbf{q})$

In this subsection we derive a formula for $S^{\omega u}(\mathbf{k}'|\mathbf{p}|\mathbf{q})$ that represents the mode-to-mode enstrophy transfer from $\mathbf{u}(\mathbf{p})$ to $\boldsymbol{\omega}(\mathbf{k}')$ with the mediation of $\boldsymbol{\omega}(\mathbf{q})$. The derivation is very similar to that for $S^{\omega\omega}(\mathbf{k}'|\mathbf{p},\mathbf{q})$, but with several critical differences.

The enstrophy transfers $S^{\omega u}(\mathbf{X}|\mathbf{Y}|\mathbf{Z})$ satisfy the following property. The sum of $S^{\omega u}(\mathbf{X}|\mathbf{Y}|\mathbf{Z})$ and $S^{\omega u}(\mathbf{X}|\mathbf{Z}|\mathbf{Y})$ is the combined enstrophy transfer $S^{\omega u}(\mathbf{X}|\mathbf{Y},\mathbf{Z})$ defined in Eq. (6.3b). Therefore,

$$S^{\omega u}(\mathbf{k}'|\mathbf{p}|\mathbf{q}) + S^{\omega u}(\mathbf{k}'|\mathbf{q}|\mathbf{p}) = S^{\omega u}(\mathbf{k}'|\mathbf{p},\mathbf{q}), \tag{6.15a}$$

$$S^{\omega u}(\mathbf{p}|\mathbf{k}'|\mathbf{q}) + S^{\omega u}(\mathbf{p}|\mathbf{q}|\mathbf{k}') = S^{\omega u}(\mathbf{p}|\mathbf{k}',\mathbf{q}), \tag{6.15b}$$

$$S^{\omega u}(\mathbf{q}|\mathbf{k}'|\mathbf{p}) + S^{\omega u}(\mathbf{q}|\mathbf{p}|\mathbf{k}') = S^{\omega u}(\mathbf{q}|\mathbf{k}',\mathbf{p}). \tag{6.15c}$$

Note however that

$$S^{\omega u}(\mathbf{X}|\mathbf{Y}|\mathbf{Z}) + S^{\omega u}(\mathbf{Y}|\mathbf{X}|\mathbf{Z}) \neq 0 \tag{6.16}$$

due to the absence of back transfer $S^{u\omega}(\mathbf{Y}|\mathbf{X}|\mathbf{Z})$ from $\boldsymbol{\omega}(\mathbf{X})$ to $\mathbf{u}(\mathbf{Y})$.

We start with the assumptions (1–5) of Section 6.1.1 that yields $S^{\omega u}(\mathbf{k}'|\mathbf{p}|\mathbf{q})$ of the same form as Eq. (6.11). As described in Subsection 6.1.1, the first two terms are part of $S^{\omega\omega}(\mathbf{k}'|\mathbf{p}|\mathbf{q})$. Therefore,

$$\begin{aligned} S^{\omega u}(\mathbf{k}'|\mathbf{p}|\mathbf{q}) &= c_3 \Im \left[\{\mathbf{k}' \cdot \boldsymbol{\omega}(\mathbf{q})\}\{\boldsymbol{\omega}(\mathbf{p}) \cdot \mathbf{u}(\mathbf{k}')\}\right] \\ &+ c_4 \Im \left[\{\mathbf{k}' \cdot \boldsymbol{\omega}(\mathbf{q})\}\{\boldsymbol{\omega}(\mathbf{k}') \cdot \mathbf{u}(\mathbf{p})\}\right] \\ &+ c_5 \Im \left[\{\mathbf{k}' \cdot \boldsymbol{\omega}(\mathbf{p})\}\{\boldsymbol{\omega}(\mathbf{q}) \cdot \mathbf{u}(\mathbf{k}')\}\right] \\ &+ c_6 \Im \left[\{\mathbf{k}' \cdot \boldsymbol{\omega}(\mathbf{p})\}\{\boldsymbol{\omega}(\mathbf{k}') \cdot \mathbf{u}(\mathbf{q})\}\right]. \end{aligned} \tag{6.17a}$$

Since $\boldsymbol{\omega}(\mathbf{k}')$ is the receiver mode for the enstrophy (see Eq. (6.3b)), we can eliminate the first and third terms of Eq. (6.17a). Hence,

$$\begin{aligned} S^{\omega u}(\mathbf{k}'|\mathbf{p}|\mathbf{q}) &= c_4 \Im \left[\{\mathbf{k}' \cdot \boldsymbol{\omega}(\mathbf{q})\}\{\boldsymbol{\omega}(\mathbf{k}') \cdot \mathbf{u}(\mathbf{p})\}\right] \\ &+ c_6 \Im \left[\{\mathbf{k}' \cdot \boldsymbol{\omega}(\mathbf{p})\}\{\boldsymbol{\omega}(\mathbf{k}') \cdot \mathbf{u}(\mathbf{q})\}\right]. \end{aligned} \tag{6.18}$$

Now, note that in real space, the nonlinear term corresponding to $S^{\omega u}$ is $[(\boldsymbol{\omega} \cdot \nabla)\mathbf{u}] \cdot \boldsymbol{\omega}$, where $\mathbf{u}$ is the giver, the last $\boldsymbol{\omega}$ is the receiver, and the first $\boldsymbol{\omega}$ is the mediator. Hence, in Eq. (6.18), $\boldsymbol{\omega}(\mathbf{q})$ must be the mediator (see Fig. 6.2). Therefore, the first term represents the enstrophy transfer from $\mathbf{u}(\mathbf{p})$ to $\boldsymbol{\omega}(\mathbf{k}')$ with the mediation of $\boldsymbol{\omega}(\mathbf{q})$. Thus, $c_6 = 0$. In addition, Eqs. (6.3b, 6.15a) yields $c_4 = 1$. Hence,

$$S^{\omega u}(\mathbf{k}'|\mathbf{p}|\mathbf{q}) = \Im \left[\{\mathbf{k}' \cdot \boldsymbol{\omega}(\mathbf{q})\}\{\boldsymbol{\omega}(\mathbf{k}') \cdot \mathbf{u}(\mathbf{p})\}\right]. \tag{6.19}$$

We remark that for the wavenumber triad,

$$\sum_{\mathbf{k}} \sum_{\mathbf{p}} S^{\omega u}(\mathbf{k}|\mathbf{p}|\mathbf{q}) \neq 0 \tag{6.20}$$

due to vortex stretching. Also note that the mode-to-mode enstrophy transfer is zero in the absence of nonlinearity. A single mode with $\mathbf{u}(\mathbf{k}) \parallel \boldsymbol{\omega}(\mathbf{k})$ cannot stretch

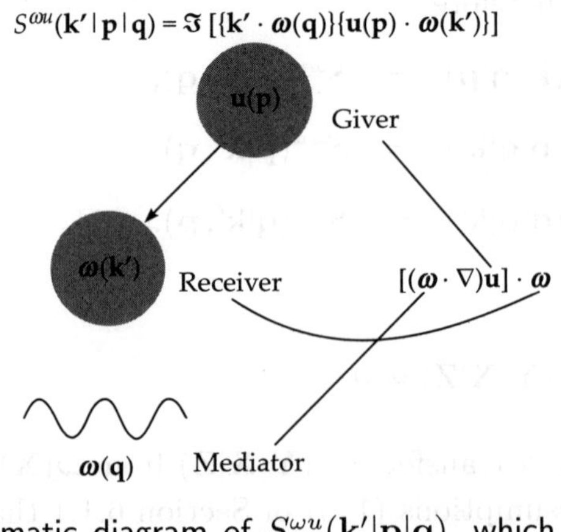

Figure 6.2 A schematic diagram of $S^{\omega u}(\mathbf{k}'|\mathbf{p}|\mathbf{q})$, which is the mode-to-mode enstrophy transfer from $\mathbf{u}(\mathbf{p})$ to $\boldsymbol{\omega}(\mathbf{k}')$ with the mediation of $\boldsymbol{\omega}(\mathbf{k}')$.

itself and increase the net enstrophy. In some sense, the other modes of the triad squeeze $\boldsymbol{\omega}(\mathbf{k})$ that leads to enstrophy enhancement. We will illustrate this aspect using several examples. Refer to Tennekes and Lumley (1972) for physical interpretation of vortex stretching in real space. It will be interesting to work out the connections between real and Fourier space interpretations.

Example 6.1: Consider the following 2D fluid flows with three components:

$$(a): \quad \mathbf{u} = 2(-\sin y - \sin(x+y)/\sqrt{2}, \sin x + \sin(x+y)/\sqrt{2}, \cos x + \cos y + \cos(x+y)).$$

$$(b): \quad \mathbf{u} = 2(-\sin y - \sin(x+y)/\sqrt{2}, \sin x + \sin(x+y)/\sqrt{2}, -\cos x - \cos y - \cos(x+y)).$$

$$(c): \quad \mathbf{u} = 2(-\sin y - \sin(x+y)/\sqrt{2}, \sin x + \sin(x+y)/\sqrt{2}, \cos x + \cos y - \cos(x+y)).$$

Compute the velocity and vorticity fields in Fourier space, as well as mode-to-mode kinetic energy and enstrophy transfers.

Solution: In Table 6.1, we list the active wavenumbers and their corresponding velocity and vorticity Fourier modes. The wavenumbers $\mathbf{k}', \mathbf{p}, \mathbf{q}$ form a triad and they satisfy a condition $\mathbf{k}' + \mathbf{p} + \mathbf{q} = \mathbf{0}$. Note that a complimentary triad $-\mathbf{k}'$, $-\mathbf{p}, -\mathbf{q}$ too exists in the system due to the reality condition.

Table 6.1 Example 6.1: Amplitudes of the velocity and vorticity fields of wavenumbers $\mathbf{k}'$, $\mathbf{p}$, $\mathbf{q}$.

Part	Mode	$\mathbf{k}$	$\mathbf{u}(\mathbf{k})$	$\boldsymbol{\omega}(\mathbf{k})$
(a)	$\mathbf{k}'$	$(-1,-1)$	$(-i/\sqrt{2}, i/\sqrt{2}, 1)$	$(-i, i, \sqrt{2})$
	$\mathbf{p}$	$(0,1)$	$(i,0,1)$	$(i,0,1)$
	$\mathbf{q}$	$(1,0)$	$(0,-i,1)$	$(0,-i,1)$
(b)	$\mathbf{k}'$	$(-1,-1)$	$(-i/\sqrt{2}, i/\sqrt{2}, -1)$	$(i, -i, \sqrt{2})$
	$\mathbf{p}$	$(0,1)$	$(i,0,-1)$	$(-i,0,1)$
	$\mathbf{q}$	$(1,0)$	$(0,-i,-1)$	$(0,i,1)$
(c)	$\mathbf{k}'$	$(-1,-1)$	$(-i/\sqrt{2}, i/\sqrt{2}, 1)$	$(-i, i, \sqrt{2})$
	$\mathbf{p}$	$(0,1)$	$(i,0,1)$	$(i,0,1)$
	$\mathbf{q}$	$(1,0)$	$(0,-i,-1)$	$(0,i,1)$

The mode-to-mode kinetic energy and enstrophy transfers are listed in Table 6.2. These transfers are same for cases (a) and (b). We list all the six enstrophy transfers, $S^{\omega u}(\mathbf{X}|\mathbf{Y}|\mathbf{Z})$, because

$$S^{\omega u}(\mathbf{X}|\mathbf{Y}|\mathbf{Z}) + S^{\omega u}(\mathbf{Y}|\mathbf{X}|\mathbf{Z}) \neq 0.$$

For all the cases,

$$\sum_{\mathbf{X}} \sum_{\mathbf{Y}} S^{\omega u}(\mathbf{X}|\mathbf{Y}|\mathbf{Z}) = 0.$$

Hence, there is no net gain of enstrophy in the aforementioned triadic interactions. It will be interesting to understand the vortex dynamics and its relation to the mode-to-mode enstrophy transfers computed here.

Table 6.2 Example 6.1: Transfers S^{uu}, $S^{\omega\omega}$, and $S^{\omega u} = (S^{\omega u}(\mathbf{X}|\mathbf{Y}|\mathbf{Z}), S^{\omega u}(\mathbf{X}|\mathbf{Z}|\mathbf{Y})$.

Part	Transfers	S^{uu}	$S^{\omega\omega}$	$S^{\omega u}$
(a)	$S(\mathbf{k}'\|\mathbf{p}\|\mathbf{q})$	$-1-1/\sqrt{2}$	$-1-\sqrt{2}$	$1+\sqrt{2}, -1-\sqrt{2}$
	$S(\mathbf{p}\|\mathbf{q}\|\mathbf{k}')$	$-1/\sqrt{2}$	$-1/\sqrt{2}$	$1, -1-1/\sqrt{2}$
	$S(\mathbf{q}\|\mathbf{k}'\|\mathbf{p})$	$-1-1/\sqrt{2}$	$-1-\sqrt{2}$	$1+1/\sqrt{2}, -1$
(b)	$S(\mathbf{k}'\|\mathbf{p}\|\mathbf{q})$	$-1-1/\sqrt{2}$	$-1-\sqrt{2}$	$1+\sqrt{2}, -1-\sqrt{2}$
	$S(\mathbf{p}\|\mathbf{q}\|\mathbf{k}')$	$-1/\sqrt{2}$	$-1/\sqrt{2}$	$1, -1-1/\sqrt{2}$
	$S(\mathbf{q}\|\mathbf{k}'\|\mathbf{p})$	$-1-1/\sqrt{2}$	$-1-\sqrt{2}$	$1+1/\sqrt{2}, -1$
(c)	$S(\mathbf{k}'\|\mathbf{p}\|\mathbf{q})$	$1-1/\sqrt{2}$	$1-\sqrt{2}$	$-1+\sqrt{2}, 1-\sqrt{2}$
	$S(\mathbf{p}\|\mathbf{q}\|\mathbf{k}')$	$-1/\sqrt{2}$	$-1/\sqrt{2}$	$-1, 1-1/\sqrt{2}$
	$S(\mathbf{q}\|\mathbf{k}'\|\mathbf{p})$	$1-1/\sqrt{2}$	$1-\sqrt{2}$	$-1+1/\sqrt{2}, 1$

In the next section, we will show how the enstrophy transfers in 2D have a relatively simpler form.

6.2 Mode-to-mode Enstrophy Transfers in 2D Hydrodynamics

As shown in Section 2.2, in 2D hydrodynamics with $\mathbf{u} = \hat{x}u_x + \hat{y}u_y$, the vorticity is along $\hat{z}$. As in Section 4.3, in Fourier space, we represent the 2D velocity field as

$$\mathbf{u}(\mathbf{p}) = u(\mathbf{p})(\hat{p} \times \hat{z}), \tag{6.21}$$

where $u(\mathbf{p})$ is the amplitude of the velocity Fourier mode. See Fig. 4.5 for an illustration. Therefore, the corresponding vorticity mode is

$$\boldsymbol{\omega}(\mathbf{p}) = i\mathbf{p} \times (\hat{p} \times \hat{z})u(\mathbf{p}) = -ipu(\mathbf{p})\hat{z}. \tag{6.22}$$

Hence, the vorticity field is along $\hat{z}$ with its amplitude being $\omega(\mathbf{p}) = -ipu(\mathbf{p})$.

Since the vorticity field and the velocity field are perpendicular to each other, using Eq. (6.19), we immediately deduce that

$$S^{\omega u}(\mathbf{k}'|\mathbf{p}|\mathbf{q}) = 0. \tag{6.23}$$

Physically, in 2D hydrodynamics, infinitely long vortex tubes cannot be stretched or compressed along $\hat{z}$. However, the enstrophy transfer $S^{\omega\omega}(\mathbf{k}|\mathbf{p}|\mathbf{q})$ is nonzero, and is given by

$$\begin{aligned} S^{\omega\omega}(\mathbf{k}'|\mathbf{p}|\mathbf{q}) &= -\Im\left[\{\mathbf{k}' \cdot \mathbf{u}(\mathbf{q})\}\{\omega(\mathbf{p})\omega(\mathbf{k}')\}\right] \\ &= -\Im\left[\{\mathbf{k}' \cdot (\hat{q} \times \hat{n})u(\mathbf{q})\}\{-ipu(\mathbf{p})\}\{-ik'u(\mathbf{k}')\}\right] \\ &= [k'^2 p(\hat{k} \times \hat{q}) \cdot \hat{z}]\Im\{u(\mathbf{q})u(\mathbf{p})u(\mathbf{k}')\} \\ &= k'^2 p \sin\beta \Im\{u(\mathbf{q})u(\mathbf{p})u(\mathbf{k}')\}. \end{aligned} \tag{6.24}$$

A comparison of Eq. (6.24) with Eq. (4.34) reveals that

$$\frac{S^{uu}(\mathbf{k}|\mathbf{p}|\mathbf{q})}{S^{\omega\omega}(\mathbf{k}|\mathbf{p}|\mathbf{q})} = \frac{\cos\gamma}{kp}, \tag{6.25}$$

where γ is the angle between $\mathbf{k}$ and $\mathbf{p}$. See Fig. 4.5 for an illustration of the aforementioned angles.

Example 6.2: Consider the two-dimensional fluid flow with two components:

$$\mathbf{u} \quad = \quad 2(-\sin y - \sin(x+y)/\sqrt{2}, \sin x + \sin(x+y)/\sqrt{2}).$$

Compute the velocity and vorticity fields in the Fourier space, as well as mode-to-mode kinetic energy and enstrophy transfers.

Solution: The flow field is formed by wavenumbers $\mathbf{k}' = (-1,-1)$, $\mathbf{p} = (0,1)$ and $\mathbf{q} = (1,0)$ and their negative counterparts $-\mathbf{k}', -\mathbf{p}, -\mathbf{q}$. Note that $(\mathbf{k}', \mathbf{p}, \mathbf{q})$ is an interacting triad. The amplitudes of the velocity and vorticity Fourier modes for these wavenumbers are listed in Table 6.3.

Table 6.3 Example 6.3: Mode amplitudes.

Mode	$\mathbf{k}$	$\mathbf{u}(\mathbf{k})$	$\boldsymbol{\omega}(\mathbf{k})$
$\mathbf{k}'$	$(-1,-1)$	$(-i/\sqrt{2}, i/\sqrt{2}, 0)$	$(0,0,\sqrt{2})$
$\mathbf{p}$	$(0,1)$	$(i,0,0)$	$(0,0,1)$
$\mathbf{q}$	$(1,0)$	$(0,-i,0)$	$(0,0,1)$

In Table 6.4, we list the mode-to-mode energy and enstrophy transfer rates. Since $\mathbf{u} \perp \boldsymbol{\omega}$ for the present flow, $S^{\omega u} = 0$.

Table 6.4 Example 6.3: Energy and enstrophy transfers among the modes.

Transfers	S^{uu}	$S^{\omega\omega}$	$S^{\omega u}$
$S(\mathbf{k}'\vert\mathbf{p}\vert\mathbf{q})$	$-1/\sqrt{2}$	$-\sqrt{2}$	0
$S(\mathbf{p}\vert\mathbf{q}\vert\mathbf{k}')$	0	$-1/\sqrt{2}$	0
$S(\mathbf{q}\vert\mathbf{k}'\vert\mathbf{p})$	$-1/\sqrt{2}$	$-\sqrt{2}$	0

A typical fluid flow involves a large number of interacting modes. In the following section, we describe interactions in such systems.

6.3 Enstrophy Transfers for Many Triads

In the presence of a large number of triads, the rate of change of modal enstrophy $E_\omega(\mathbf{k})$ is

$$\begin{aligned}\frac{d}{dt}E_\omega(\mathbf{k}) &= \sum_{\mathbf{p}} S^{\omega\omega}(\mathbf{k}|\mathbf{p}|\mathbf{q}) + \sum_{\mathbf{p}} S^{\omega u}(\mathbf{k}|\mathbf{p}|\mathbf{q}) + \mathcal{F}_\omega(\mathbf{k}) - 2\nu k^2 E_\omega(\mathbf{k}) \\ &= T_{\omega 1}(\mathbf{k}) + T_{\omega 2}(\mathbf{k}) + \mathcal{F}_\omega(\mathbf{k}) - D_\omega(\mathbf{k}),\end{aligned} \tag{6.26}$$

where $\mathbf{q} = \mathbf{k} - \mathbf{p}$, and $\mathcal{F}_\omega(\mathbf{k})$ is the enstrophy supply rate by the external force:

$$\mathcal{F}_\omega(\mathbf{k}) = \Re[i\mathbf{k} \times \mathbf{F}_u(\mathbf{k}) \cdot \boldsymbol{\omega}^*(\mathbf{k})]. \tag{6.27}$$

Thus, the modal enstrophy is affected by two nonlinear enstrophy transfer rates, the enstrophy transfer rate by the external force, and the enstrophy dissipation rate $D_\omega(\mathbf{k})$. When we sum over all $\mathbf{k}$'s, we obtain

$$\begin{aligned}\frac{d}{dt}E_\omega &= \sum_{\mathbf{k}}\sum_{\mathbf{p}} S^{\omega\omega}(\mathbf{k}|\mathbf{p}|\mathbf{q}) + \sum_{\mathbf{k}}\sum_{\mathbf{p}} S^{\omega u}(\mathbf{k}|\mathbf{p}|\mathbf{q}) \\ &\quad + \sum_{\mathbf{k}}[\mathcal{F}_\omega(\mathbf{k}) - D_\omega(\mathbf{k})]\end{aligned} \tag{6.28}$$

with

$$\sum_{\mathbf{k}}\sum_{\mathbf{p}} S^{\omega\omega}(\mathbf{k}|\mathbf{p}|\mathbf{q}) = 0. \tag{6.29}$$

Note however that

$$\sum_{\mathbf{k}}\sum_{\mathbf{p}} S^{\omega u}(\mathbf{k}|\mathbf{p}|\mathbf{q}) \neq 0, \tag{6.30}$$

which is due to the stretching of vorticity by the velocity field. Hence,

$$\frac{d}{dt}E_\omega = \sum_{\mathbf{k}}\sum_{\mathbf{p}} S^{\omega u}(\mathbf{k}|\mathbf{p}|\mathbf{q}) + \sum_{\mathbf{k}} \mathcal{F}_\omega(\mathbf{k}) - \sum_{\mathbf{k}} D_\omega(\mathbf{k}). \tag{6.31}$$

That is, the total enstrophy of the flow changes due to (a) the stretching of vortices by the velocity field, (b) external force, and (c) enstrophy dissipation.

Note that for any region A,

$$\sum_{\mathbf{k}\in A}\sum_{\mathbf{p}\in A} S^{\omega\omega}(\mathbf{k}|\mathbf{p}|\mathbf{q}) = 0. \tag{6.32}$$

However,

$$\sum_{\mathbf{k}\in A}\sum_{\mathbf{p}\in A} S^{\omega u}(\mathbf{k}|\mathbf{p}|\mathbf{q}) \neq 0 \tag{6.33}$$

due to vortex stretching.

In the next section, we will derive formulas for enstrophy fluxes and shell-to-shell enstrophy transfers.

6.4 Enstrophy Fluxes

The enstrophy flux for a wavenumber sphere has several parts. Here the superscript and subscript refer to the giver and receiver modes respectively. Moreover, the symbols $<$ and $>$ refer to the modes inside and outside the wavenumber sphere respectively.

1. $\Pi^{\omega<}_{\omega>}(k_0)$: Enstrophy transfers from all the vorticity modes inside the sphere of radius k_0 to all the vorticity modes outside the sphere with the velocity modes acting as mediator. This flux, which is via $S^{\omega\omega}(\mathbf{k}'|\mathbf{p}|\mathbf{q})$, is

$$\Pi^{\omega<}_{\omega>}(k_0) = \sum_{|\mathbf{p}|\leq k_0}\sum_{|\mathbf{k}'|>k_0} S^{\omega\omega}(\mathbf{k}'|\mathbf{p}|\mathbf{q}). \tag{6.34}$$

2. $\Pi^{u<}_{\omega>}(k_0)$: Enstrophy transfers from all the velocity modes inside the sphere to all the vorticity modes outside the sphere with the vorticity modes acting as mediator. This flux, which is via $S^{\omega u}(\mathbf{k}'|\mathbf{p}|\mathbf{q})$, is

$$\Pi^{u<}_{\omega>}(k_0) = \sum_{|\mathbf{p}|\leq k_0} \sum_{|\mathbf{k}'|>k_0} S^{\omega u}(\mathbf{k}|\mathbf{p}|\mathbf{q}). \tag{6.35}$$

3. $\Pi^{u<}_{\omega<}(k_0)$: For all the wavenumbers inside the sphere, enstrophy transfers from the velocity modes to the vorticity modes:

$$\Pi^{u<}_{\omega<}(k_0) = \sum_{|\mathbf{p}|\leq k_0} \sum_{|\mathbf{k}'|\leq k_0} S^{\omega u}(\mathbf{k}|\mathbf{p}|\mathbf{q}). \tag{6.36}$$

4. $\Pi^{u>}_{\omega>}(k_0)$: For all the wavenumbers outside the sphere, enstrophy transfers from the velocity modes to the vorticity modes:

$$\Pi^{u>}_{\omega>}(k_0) = \sum_{|\mathbf{p}|> k_0} \sum_{|\mathbf{k}'|> k_0} S^{\omega u}(\mathbf{k}|\mathbf{p}|\mathbf{q}). \tag{6.37}$$

5. $\Pi^{u>}_{\omega<}(k_0)$: Enstrophy transfers from all the velocity modes outside the sphere to all the vorticity modes inside the sphere:

$$\Pi^{u>}_{\omega<}(k_0) = \sum_{|\mathbf{p}|> k_0} \sum_{|\mathbf{k}'|< k_0} S^{\omega u}(\mathbf{k}|\mathbf{p}|\mathbf{q}). \tag{6.38}$$

In all the aforementioned cases, $\mathbf{q} = \mathbf{k} - \mathbf{p}$. These fluxes, except $\Pi^{u>}_{\omega<}(k_0)$, are illustrated in Fig. 6.3.

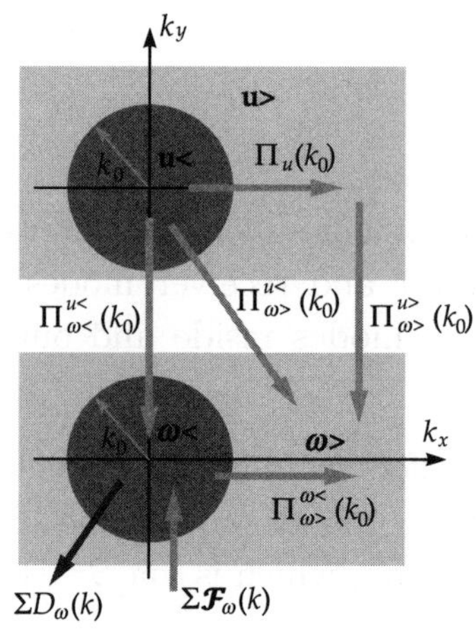

Figure 6.3 Various enstrophy fluxes in 3D hydrodynamics. See Section 6.4 for definitions.

When we sum Eq. (6.26) over all the modes outside the sphere of radius k_0, we obtain

$$\frac{d}{dt}\sum_{|\mathbf{k}|>k_0} E_\omega(\mathbf{k}) = \sum_{|\mathbf{k}|>k_0} [T_{\omega 1}(\mathbf{k}) + T_{\omega 2}(\mathbf{k})] - \sum_{|\mathbf{k}|>k_0} D_\omega(\mathbf{k}). \tag{6.39}$$

Since the enstrophy is conserved for interactions $S^{\omega\omega}$, following arguments similar to those in Section 4.4, we obtain

$$\begin{aligned}
\sum_{|\mathbf{k}|>k_0} T_{\omega 1}(\mathbf{k}) &= \sum_{|\mathbf{k}'|>k_0} \sum_{\mathbf{p}} S^{\omega\omega}(\mathbf{k}'|\mathbf{p}|\mathbf{q}) \\
&= \sum_{|\mathbf{k}'|>k_0} \sum_{|\mathbf{p}|<k_0} S^{\omega\omega}(\mathbf{k}'|\mathbf{p}|\mathbf{q}) + \sum_{|\mathbf{k}|>k_0} \sum_{|\mathbf{p}|>k_0} S^{\omega\omega}(\mathbf{k}'|\mathbf{p}|\mathbf{q}). \\
&= \Pi^{\omega<}_{\omega>}(k_0).
\end{aligned} \tag{6.40}$$

Note that the second term in the second line vanishes due to conservation laws given by Eq. (6.14). Similarly,

$$\begin{aligned}
\sum_{|\mathbf{k}|>k_0} T_{\omega 2}(\mathbf{k}) &= \sum_{|\mathbf{k}'|>k_0} \sum_{\mathbf{p}} S^{\omega u}(\mathbf{k}'|\mathbf{p}|\mathbf{q}) \\
&= \sum_{|\mathbf{k}'|>k_0} \sum_{|\mathbf{p}|<k_0} S^{\omega u}(\mathbf{k}'|\mathbf{p}|\mathbf{q}) + \sum_{|\mathbf{k}'|>k_0} \sum_{|\mathbf{p}|>k_0} S^{\omega u}(\mathbf{k}'|\mathbf{p}|\mathbf{q}) \\
&= \Pi^{u<}_{\omega>}(k_0) + \Pi^{u>}_{\omega>}(k_0).
\end{aligned} \tag{6.41}$$

Therefore, the net enstrophy flux from the modes inside the sphere to the modes outside the sphere is

$$\Pi_\omega(k_0) = \Pi^{\omega<}_{\omega>}(k_0) + \Pi^{u<}_{\omega>}(k_0) + \Pi^{u>}_{\omega>}(k_0). \tag{6.42}$$

Also note that $\Pi^{u>}_{\omega>}(k_0) \neq 0$ due to Eq. (6.20). In addition, we have

$$\Pi^{\omega<}_{\omega>}(k_0 = \infty) = \sum_{\mathbf{k}} \sum_{\mathbf{p}} S^{\omega\omega}(\mathbf{k}|\mathbf{p}|\mathbf{q}) = 0, \tag{6.43}$$

but

$$\Pi^{u<}_{\omega<}(k_0 = \infty) = \sum_{\mathbf{k}} \sum_{\mathbf{p}} S^{\omega u}(\mathbf{k}|\mathbf{p}|\mathbf{q}) \neq 0. \tag{6.44}$$

Since $S^{\omega u} = 0$ in 2D hydrodynamics, only $\Pi^{\omega<}_{\omega>}(k_0)$ flux survives in 2D turbulence. We will discuss this case in the next chapter.

When we sum Eq. (6.26) over the modes outside the wavenumber spheres of radii k and $k + dk$, and take the difference (as in Section 4.4), we obtain

$$\frac{\partial}{\partial t} E_\omega(k,t) = -\frac{\partial}{\partial k}[\Pi^{\omega<}_{\omega>}(k,t) + \Pi^{u<}_{\omega>}(k,t) + \Pi^{u>}_{\omega>}(k,t)] - D_\omega(k,t). \quad (6.45)$$

This equation needs careful analysis since enstrophy is not conserved in 3D hydrodynamics. For example, it is possible that $\partial E_\omega(k,t)/\partial t \neq 0$ under a steady state due to non-conservation of enstrophy.

6.5 Shell-to-shell Enstrophy Transfer

We define the shell-to-shell enstrophy transfers from shell m to shell n as

1. $T^{\omega,m}_{\omega,n}$: Enstrophy transfer from all the vorticity modes of shell m to all the vorticity modes of shell n with the velocity modes acting as mediator:

$$T^{\omega,m}_{\omega,n} = \sum_{\mathbf{p}\in m}\sum_{\mathbf{k}'\in n} S^{\omega\omega}(\mathbf{k}'|\mathbf{p}|\mathbf{q}). \quad (6.46)$$

2. $T^{u,m}_{\omega,n}$: Enstrophy transfer from all the velocity modes of shell m to all the vorticity modes of shell n with the vorticity modes acting as mediator:

$$T^{u,m}_{\omega,n} = \sum_{\mathbf{p}\in m}\sum_{\mathbf{k}'\in n} S^{\omega u}(\mathbf{k}|\mathbf{p}|\mathbf{q}). \quad (6.47)$$

In both the cases, $\mathbf{q} = -\mathbf{k}' - \mathbf{p}$.

In 2D hydrodynamics, $S^{\omega u}(\mathbf{k}|\mathbf{p}|\mathbf{q}) = 0$; hence, 2D hydrodynamics has only $T^{\omega,m}_{\omega,n}$.

6.6 Numerical Results on Enstrophy Fluxes

In this section we briefly describe several numerical results on the enstrophy fluxes. Sadhukhan et al. (2018) performed numerical simulation of hydrodynamic turbulence on a 512^3 grid. The Reynolds number of the flow was approximately 5700. They computed the enstrophy fluxes $\Pi^{\omega<}_{\omega>}(k)$ and $\Pi^{u<}_{\omega>}(k)$ for various k's. These fluxes are plotted in Fig. 6.4. They observed that $\Pi^{u<}_{\omega>}(k) \gg \Pi^{\omega<}_{\omega>}(k)$ in the forcing and inertial range, and

$$\Pi^{\omega<}_{\omega>}(k) \sim k^2, \quad (6.48a)$$

$$\Pi^{u<}_{\omega>}(k) \sim k. \quad (6.48b)$$

We can explain the scaling of Eq. (6.48a) using dimensional analysis. In hydrodynamic turbulence, the nonlinear interactions are local. Hence,

$$\begin{aligned} \Pi^{\omega<}_{\omega>}(k) &= \sum_{|\mathbf{p}|\le k} \sum_{|\mathbf{k}''|>k} -\Im\left[\{\mathbf{k}''\cdot\mathbf{u}(\mathbf{q})\}\{\boldsymbol{\omega}(\mathbf{p})\cdot\boldsymbol{\omega}(\mathbf{k}'')\}\right] \\ &\sim k^2 p u_k u_p u_q \sim k^2 \epsilon_u. \end{aligned} \tag{6.49}$$

Here we have used $k \sim p \sim q$ and $u_k \sim u_p \sim u_q \sim \epsilon_u^{1/3} k^{-1/3}$.

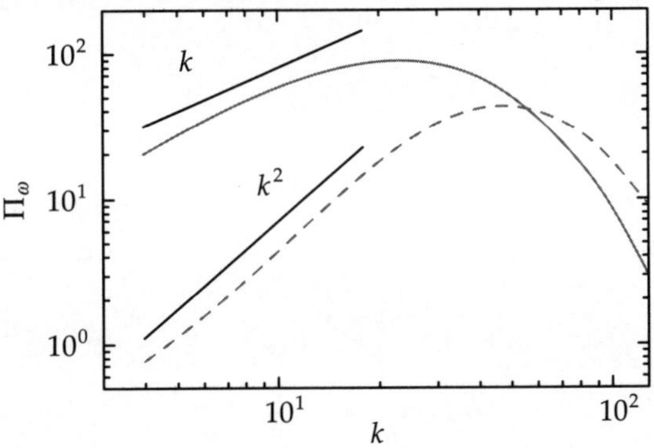

Figure 6.4 For hydrodynamic turbulence on a 512^3 grid, plots of enstrophy fluxes $\Pi^{u<}_{\omega>}$ (solid line) and $\Pi^{\omega<}_{\omega>}$ (dashed line). Note that $\Pi^{u<}_{\omega>} \gg \Pi^{\omega<}_{\omega>}$. Adopted from figures of Sadhukhan et al. (2018).

The scaling of $\Pi^{u<}_{\omega>}(k)$ is suppressed by a factor of k compared to $\Pi^{\omega<}_{\omega>}(k)$. The reason for this suppression is not clear at the moment. Note that the enstrophy spectrum $E_\omega(k)$ can be derived very easily from the energy spectrum $E(k)$:

$$E_\omega(k) = \frac{E_u(k)}{k^2} = K_{\mathrm{Ko}} \Pi_u^{2/3} k^{-11/3}. \tag{6.50}$$

With this, we end our discussion on the enstrophy transfers in hydrodynamic turbulence.

Further Reading

The textbooks by Lesieur (2008) and Davidson (2004) cover basic features of enstrophy. Mode-to-mode enstrophy transfer and enstrophy flux are described here for the first time. Refer to Sadhukhan et al. (2018) for further details on numerical computation of enstrophy fluxes.

Exercises

1. Consider the flows of Exercises 4.1 and 4.2. Compute the enstrophy transfers among the Fourier modes. Also, compute the enstrophy fluxes for them.
2. Consider the flow field of Example 3.6. Compute the enstrophy transfers among the modes.

Chapter 7

Two-dimensional Turbulence

On many occasions, flows become quasi-two-dimensional or two-dimensional (2D). For example, strong rotation suppresses the velocity component along the direction of rotation. Strong external magnetic field in magnetohydrodynamics exhibits similar behaviour. Thermal convection in viscous fluids too exhibits quasi-two-dimensional flow patterns. A cyclone is a practical example of a 2D flow. Due to these reasons, understanding 2D hydrodynamic turbulence becomes quite critical.

Properties of 2D hydrodynamic turbulence are very different from those of its 3D counterpart. In this chapter we describe characteristics of 2D hydrodynamic turbulence.

7.1 Conservation Laws; Energy and Enstrophy Transfers in 2D Hydrodynamics

In Chapter 2 we showed that in inviscid 2D hydrodynamics (with $\nu = 0$), the total kinetic energy (KE), $\int d\mathbf{r}u^2/2$, and the total enstrophy, $\int d\mathbf{r}\omega^2/2$, are conserved. These conservation laws play an important role in 2D hydrodynamics. Note that in 3D hydrodynamics, the total KE, $\int d\mathbf{r}u^2/2$, and the total kinetic helicity, $\int d\mathbf{r}(\mathbf{u}\cdot\boldsymbol{\omega})$, are conserved.

As argued in Sections 4.3 and 6.2, the mode-to-mode KE and enstrophy transfers for 2D hydrodynamics are

$$S^{uu}(\mathbf{k}'|\mathbf{p}|\mathbf{q}) \quad = \quad -\Im\left[\{\mathbf{k}'\cdot\mathbf{u}(\mathbf{q})\}\{\mathbf{u}(\mathbf{p})\cdot\mathbf{u}(\mathbf{k}')\}\right], \tag{7.1}$$

$$S^{\omega\omega}(\mathbf{k}'|\mathbf{p}|\mathbf{q}) = -\Im\left[\{\mathbf{k}'\cdot\mathbf{u}(\mathbf{q})\}\{\omega(\mathbf{p})\omega(\mathbf{k}')\}\right]. \tag{7.2}$$

Note that for 2D flows,

$$S^{\omega u}(\mathbf{k}'|\mathbf{p}|\mathbf{q}) = 0 \tag{7.3}$$

because $\mathbf{u}$ and $\boldsymbol{\omega}$ are perpendicular to each other. Also, the KE and enstrophy fluxes are computed using Eqs. (4.39) and (6.34) respectively. For brevity, we write

$$\Pi_\omega(k_0) = \Pi^{\omega<}_{\omega>}(k_0). \tag{7.4}$$

Following the arguments of Sections 4.4 and 6.4, we find that

$$\frac{\partial}{\partial t}E_u(k,t) = -\frac{\partial}{\partial k}\Pi_u(k,t) + \mathcal{F}_u(k,t) - D_u(k,t), \tag{7.5a}$$

$$\frac{\partial}{\partial t}E_\omega(k,t) = -\frac{\partial}{\partial k}\Pi_\omega(k,t) + \mathcal{F}_\omega(k,t) - D_\omega(k,t), \tag{7.5b}$$

where $E_u(k), E_\omega(k)$ are respectively the one-dimensional KE and enstrophy shell spectra; $D_u(k), D_\omega(k)$ are respectively the KE and enstrophy dissipation rates; and $\mathcal{F}_u(k), \mathcal{F}_\omega(k)$ are respectively the KE and enstrophy supply rates by the external force. These quantities have been defined in Chapters 3, 4, and 6. Under a steady state,

$$\frac{d}{dk}\Pi_u(k) = \mathcal{F}_u(k) - D_u(k), \tag{7.6a}$$

$$\frac{d}{dk}\Pi_\omega(k) = \mathcal{F}_\omega(k) - D_\omega(k). \tag{7.6b}$$

As remarked in Chapter 4, turbulent fluxes are not directly related to conservation laws; rather, they follow from energetics and nonlinear interactions. Yet, conservations laws provide useful clues towards the nature of fluxes. For 2D hydrodynamic turbulence, Fjørtoft (1953) showed that simultaneous conservation of KE and enstrophy leads to inverse cascade of kinetic energy and forward cascade of enstrophy. Nazarenko (2011) arrived at similar results based on centroids of KE and enstrophy. Using direct interaction approximation, Kraichnan (1967) deduced similar patterns for the KE and enstrophy fluxes. In Chapter 10, we will compute the KE transfers for 2D hydrodynamic turbulence using field-theoretic treatment and show similar results (also see Verma et al. (2005)).

Based on the directions of KE and enstrophy transfers, Kraichnan (1967) constructed a phenomenological theory of 2D hydrodynamic turbulence, which is described as follows.

7.2 Kraichnan's Theory for 2D Hydrodynamic Turbulence

Unlike 3D hydrodynamics where external force is employed at large scales, for 2D turbulence, Kraichnan (1967) considered forcing to be employed at intermediate scales, i.e., at $k_f \gg 1/L$. Using field-theoretic calculations, Kraichnan (1967) showed that for $k < k_f$, a constant energy flux $\Pi_u(k)$ cascades to small wavenumbers. For wavenumber range $k < k_f$, a dimensional analysis similar to that of Chapter 5 yields the following results:

$$E_u(k) = K_{2D}\Pi_u^{2/3}k^{-5/3}, \tag{7.7a}$$

$$\Pi_u = \text{const} < 0, \tag{7.7b}$$

where K_{2D} is a constant whose numerical value is approximately 5.5–7.0.

For $k > k_f$, Kraichnan (1967) argued that a constant enstrophy flux Π_ω flows from intermediate wavenumbers to larger wavenumbers. Since

$$[\Pi_\omega] = [\omega^2]/[T] = [T]^{-3}, \tag{7.8}$$

dimensional analysis yields

$$E_u(k) = K'_{2D}\Pi_\omega^{2/3}k^{-3}, \tag{7.9a}$$

$$\Pi_\omega = \text{const} > 0, \tag{7.9b}$$

where K'_{2D} is another constant whose numerical value is 1.3–1.7.

Note that 2D hydrodynamic turbulence exhibits inverse cascade of KE in contrast to forward cascade of KE in 3D hydrodynamic turbulence. These transfers lead to a growth of $E_u(k)$ at small wavenumbers or large length scales. This growth may saturate due to friction at large scales, for example, Eckmann damping. The growth of large-scale structures in 2D hydrodynamics, as in cyclones, is connected to the aforementioned inverse cascade of kinetic energy.

7.3 Subtleties in Energy and Enstrophy Fluxes

Under a steady state, in the inertial–dissipation range where the external force is absent, Eqs. (7.6) yield

$$\frac{d}{dk}\Pi_u(k) = -2\nu k^2 E_u(k), \tag{7.10a}$$

$$\frac{d}{dk}\Pi_\omega(k) = -2\nu k^2 E_\omega(k). \tag{7.10b}$$

Therefore, using $E_\omega(k) = k^2 E_u(k)$, we obtain

$$\frac{d}{dk}\Pi_\omega(k) = k^2 \frac{d}{dk}\Pi_u(k) \tag{7.11}$$

Hence, the relationship between $\Pi_u(k)$ and $\Pi_\omega(k)$ is somewhat tricky. We derive these fluxes in the $k < k_f$ and $k > k_f$ regimes using Eqs. (7.10).

In Section 5.5.1, we derived Pao's formula for $E_u(k)$ and $\Pi_u(k)$ in the inertial–dissipative range of 3D hydrodynamics as

$$\Pi_u(k) = \epsilon_u \exp\left(-\frac{3}{2}K_{\mathrm{Ko}}(k/k_d)^{4/3}\right), \tag{7.12a}$$

$$E_u(k) = K_{\mathrm{Ko}}\epsilon_u^{2/3}k^{-5/3}\exp\left(-\frac{3}{2}K_{\mathrm{Ko}}(k/k_d)^{4/3}\right). \tag{7.12b}$$

The above formulas are expected to work in the $k < k_f$ regime of steady 2D hydrodynamics turbulence with the following modifications:

$$\Pi_u(k) = -\epsilon_u \exp\left(\frac{3}{2}K_{2D}(k/k_d)^{4/3}\right), \tag{7.13a}$$

$$E_u(k) = K_{2D}\epsilon_u^{2/3}k^{-5/3}\exp\left(\frac{3}{2}K_{2D}(k/k_d)^{4/3}\right), \tag{7.13b}$$

where $k_d = (\epsilon_u/\nu^3)^{1/4}$ is Kolmogorov's wavenumber. Note the change in the sign of $\Pi_u(k)$ and in the argument of the exponential function. Consequently, $|\Pi_u(k)|$ is expected to increase with k, though marginally.

When we substitute $E_u(k)$ of Eq. (7.13b) in Eq. (7.10b), we obtain

$$\frac{d}{dk}\Pi_\omega(k) = -2\nu k^4 E_u(k). \tag{7.14}$$

Hence,

$$\begin{aligned}\Pi_\omega(k) &= -2\nu K_{2D}\epsilon_u^{2/3}\int^k k'^{7/3}\exp\left(\frac{3}{2}K_{2D}(k'/k_d)^{4/3}\right)dk' \\ &= -2\nu K_{2D}\epsilon_u^{2/3}k_d^{10/3}\int^x dx' x'^{7/3}\exp\left(\frac{3}{2}K_{2D}x'^{4/3}\right),\end{aligned} \tag{7.15}$$

where $x = k/k_d$. This integral needs to be computed numerically.

For the $k > k_f$ regime, in the inertial–dissipative range, the energy spectrum and flux of Eq. (7.9) need to be modified as follows. Since there are two unknowns and a single equation, following Pao (1965), we assume that $E_\omega(k)/\Pi_\omega(k)$ is independent of ν, and it depends only on the enstrophy dissipation rate, ϵ_ω, and k. Under this ansatz, $E_\omega(k)$ and $\Pi_\omega(k)$ that satisfy Eq. (7.10b) are

$$\Pi_\omega(k) = \epsilon_\omega \exp\left(-K'_{2D}(k/k_{d2D})^2\right), \tag{7.16a}$$
$$E_\omega(k) = K'_{2D}\epsilon_\omega^{2/3}k^{-1}\exp\left(-K'_{2D}(k/k_{d2D})^2\right), \tag{7.16b}$$
$$E_u(k) = K'_{2D}\epsilon_\omega^{2/3}k^{-3}\exp\left(-K'_{2D}(k/k_{d2D})^2\right), \tag{7.16c}$$

where

$$k_{d2D} = \frac{\epsilon_\omega^{1/6}}{\sqrt{\nu}} \tag{7.17}$$

is the enstrophy dissipation wavenumber.

When we substitute Eq. (7.16b) in Eq. (7.10a), we obtain

$$\begin{aligned}\Pi_u(k) &= -2\nu\epsilon_\omega^{2/3}\int^k \frac{1}{k'}\exp\left(-K'_{2D}(k'/k_{d2D})^2\right)dk' \\ &= -\frac{\epsilon_\omega}{k_{d2D}^2}\int^k \frac{1}{k'}\exp\left(-K'_{2D}(k'/k_{d2D})^2\right)dk' \\ &= -\frac{\epsilon_\omega}{k_{d2D}^2}\mathrm{Ei}(-K'_{2D}(k/k_{d2D})^2),\end{aligned} \tag{7.18}$$

where Ei is the exponential integral. Asymptotically, $-\mathrm{Ei}(-x) \sim \exp(-x)/x$. Hence,

$$\Pi_u(k) \approx \frac{\epsilon_\omega}{k^2}\exp\left(-K'_{2D}(k/k_{d2D})^2\right). \tag{7.19}$$

In the next section we attempt to verify the aforementioned spectra and fluxes using numerical simulations.

7.4 Verification of 2D Hydrodynamic Turbulence Models Using Numerical Simulations

In this section we report some of the numerical results related to $E_u(k)$, $\Pi_u(k)$, and $\Pi_\omega(k)$ of 2D hydrodynamic turbulence.

Gupta et al. (2018) performed numerical simulations of force 2D hydrodynamic turbulence on a 2048^2 grid. They employed forcing at wavenumber band (50,51). The viscosity was set at 3×10^{-3} that led to the Reynolds number to be approximately 1.2×10^4. Gupta et al. (2018) computed the spectra and fluxes of kinetic energy and enstrophy, as well as shell-to-shell energy transfers. These results are described below.

In Fig. 7.1 we plot the numerical computed KE spectrum using solid lines, and the predictions from Eqs. (7.13b, 7.16c) as dashed lines. We observe that these equations describe the numerical data quite well. Note that $E_u(k) \sim k^{-5/3}$ for $k < k_f$, and $E_u(k) \sim k^{-3}\exp(-(k/k_{d2D})^2)$ for $k > k_f$. We remark that Sharma et al. (2018b) observed $E_u(k) \sim k^{-3}\exp(-(k/k_{d2D})^2)$ in their simulation of rotating turbulence that yields a quasi-two-dimensional flow.

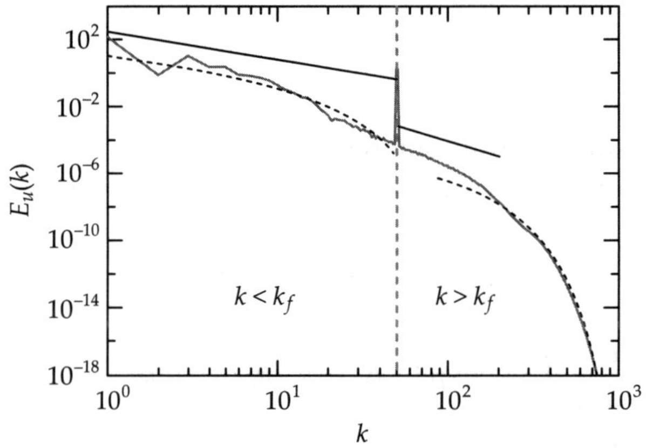

Figure 7.1 KE spectrum for 2D hydrodynamic turbulence with forcing at $k = (50, 51)$. The predictions of Eqs. (7.13b, 7.16c), shown as dashed lines, match with the numerical data quite well. The exponential fits are better than $k^{-5/3}$ and k^{-3} lines (solid) shown in the figure. Adopted from a figure of Gupta et al. (2018).

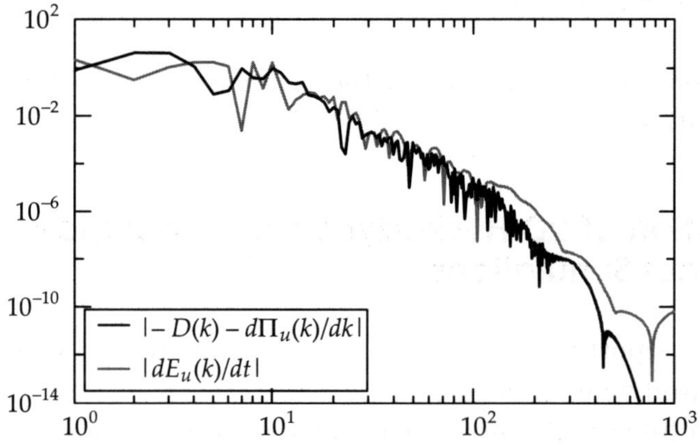

Figure 7.2 For numerical simulation of Gupta et al. (2018), comparison of $|\partial E(k,t)/\partial t|$ with $| - d\Pi/dk - D_u(k)|$ of the energy equation [Eq. (7.5a)]. Significant measures of $|\partial E(k,t)/\partial t|$ indicates the unsteady nature of the flow. Adapted from a figure of Gupta et al. (2018).

Two-dimensional turbulence involves inverse cascade of kinetic energy and continual build up of large-scale structures. Such dynamic processes are expected to make the flow unsteady. To verify whether 2D turbulence is steady or not, Gupta et al. (2018) computed each term of Eq. (7.5a) and compared them. In Fig. 7.2 we exhibit $|\partial E(k,t)/\partial t|$ and $|-d\Pi/dk - D_u(k)|$. Though the left-hand and right-hand sides of Eq. (7.5a) match each other, noticeably, $|\partial E(k,t)/\partial t|$ is significant, thus indicating an unsteady nature of the flow. This feature leads to strong fluctuations in $\Pi_u(k)$ and $\Pi_\omega(k)$, and invalidates the formulas derived in Section 7.3. However, as we show in the following, the time-averaged fluxes match with the formulas of the previous section reasonably well.

In Fig. 7.3 we illustrate the numerically computed KE and enstrophy fluxes along with the predictions from Eqs. (7.13a, 7.15, 7.16a, 7.19). Figures 7.3(a)

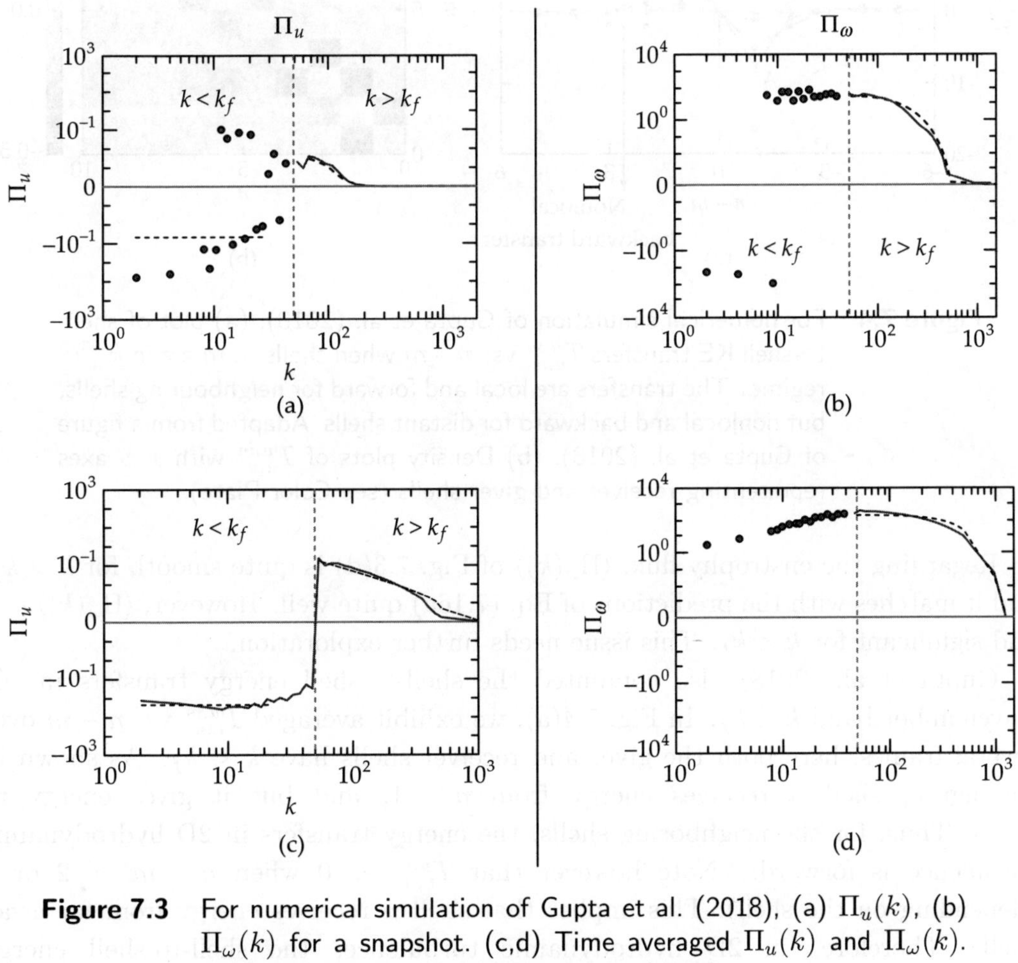

Figure 7.3 For numerical simulation of Gupta et al. (2018), (a) $\Pi_u(k)$, (b) $\Pi_\omega(k)$ for a snapshot. (c,d) Time averaged $\Pi_u(k)$ and $\Pi_\omega(k)$. Adapted from figures of Gupta et al. (2018).

and (b) exhibit the fluxes for a single snapshot. Clearly, both the fluxes show strong fluctuations, specially for $k < k_f$, due to the unsteady nature of the flow. Note that the fluctuations for every snapshot are different.

In Fig. 7.3(c,d) we exhibit the time-averaged $\Pi_u(k)$ and $\Pi_\omega(k)$. $\langle \Pi_u(k) \rangle$ of Fig. 7.3(c) is quite smooth. For $k < k_f$, $\langle \Pi_u(k) \rangle \approx$ const, which is consistent with Kraichnan's predictions, but the fluctuations are still strong. However, for $k > k_f$, $\langle \Pi_u(k) \rangle$ matches with the predictions of Eq. (7.19) quite well.

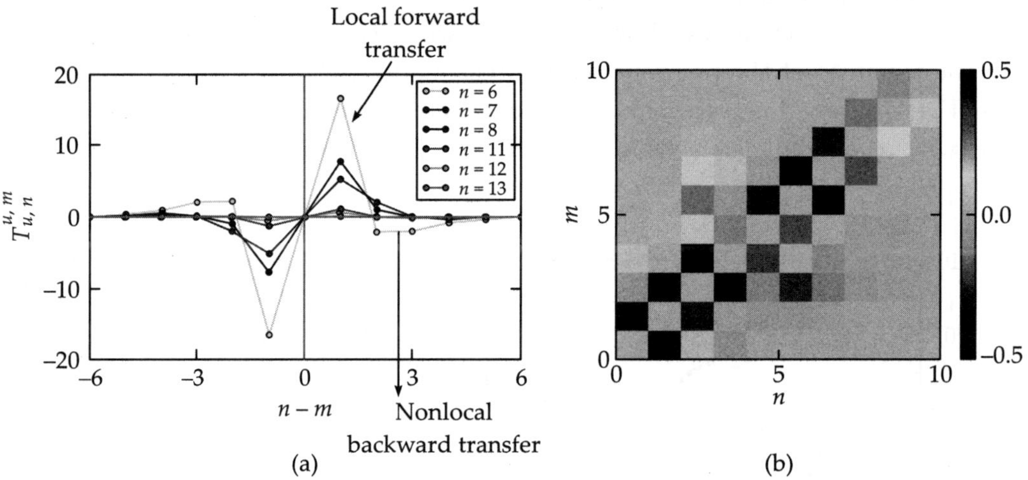

Figure 7.4 For numerical simulation of Gupta et al. (2018), (a) plot of shell-to-shell KE transfers $T^{u,m}_{u,n}$ vs. $n-m$ when shells n, m are in $k^{-5/3}$ regime. The transfers are local and forward for neighbouring shells, but nonlocal and backward for distant shells. Adapted from a figure of Gupta et al. (2018). (b) Density plots of $T^{u,m}_{u,n}$ with x, y axes representing receiver and giver shells (see Color Plate).

Regarding the enstrophy flux, $\langle \Pi_\omega(k) \rangle$ of Fig. 7.3(d) is quite smooth for $k > k_f$, and it matches with the predictions of Eq. (7.16a) quite well. However, $\langle \Pi_\omega(k) \rangle > 0$ and significant for $k < k_f$. This issue needs further exploration.

Gupta et al. (2018) also computed the shell-to-shell energy transfers in the wavenumber band $k < k_f$. In Fig. 7.4(a), we exhibit averaged $T^{u,m}_{u,n}$ vs. $n-m$ over several frames; here both the giver and receiver shells have $k < k_f$. As shown in the figure, shell n receives energy from $n-1$, and but it gives energy to $n+1$. Thus, for the neighboring shells, the energy transfers in 2D hydrodynamic turbulence is forward. Note however that $T^{u,m}_{u,n} < 0$ when $n-m > 2$ or 3 (depending on the shell). This implies that shell n receives energy from far away shells. Therefore, in 2D hydrodynamic turbulence, the shell-to-shell energy transfers to the neighboring shells are forward, but the energy transfers are backward for the distant shells.

In Fig. 7.4(b) we exhibit the density plot of $T_{u,n}^{u,m}$ with x, y axes representing the receiver (n) and giver (m) shells respectively. These plots show significant fluctuations, yet $n = m + 1$ diagonal is primarily positive. Hence, we conclude forward energy transfer to the nearest neighbor. The diagonals $n = m + 3$ and $n = m+4$ are negative that indicates nonlocal backward transfer. These observations are consistent with the description of the previous paragraphs.

We remark that for $k > k_f$, we do not observe k^{-3} scaling for $E_u(k)$, which is due to the dissipative effects in this wavenumber regime. We need higher resolution simulations with forcing at lower wavenumbers to observe k^{-3} scaling and a constant Π_ω in the $k > k_f$ regime. We refer the reader to Boffetta and Ecke (2012) for further details.

Verma et al. (2005) observed similar features in their field-theoretic computations. They argued that the backward KE transfers from the distant shells yield negative KE flux in the $k^{-5/3}$ regime. We will revisit these arguments in Chapter 10.

With this, we close our discussion on 2D hydrodynamic turbulence.

Further Reading

The model of Section 7.2 was proposed by Kraichnan (1967). For a detailed discussion on 2D hydrodynamic turbulence, we refer the reader to Lesieur (2008), Boffetta and Ecke (2012), and Tabeling (2002). Verma et al. (2005) discussed shell-to-shell energy transfers in 2D turbulence. The simulation results presented in this chapter have been taken from Gupta et al. (2018).

Exercises

1. Read the arguments of Fjørtoft (1953) and Nazarenko (2011) that determine the tendencies of energy and enstrophy transfers.
2. What kind of energy spectrum is expected in a decaying 2D hydrodynamic turbulence with large Re?

Chapter 8

Helical Turbulence

Kinetic helicity is an important quantity of hydrodynamic flows. In this chapter we will derive formulas for mode-to-mode kinetic helicity transfers and kinetic helicity flux. Later in the chapter we will describe the numerical results on kinetic helicity spectrum and flux.

8.1 Mode-to-mode Kinetic Helicity Transfers in Hydrodynamics

To derive an equation for mode-to-mode kinetic helicity transfers, we follow a similar procedure as that adopted for the derivation of mode-to-mode kinetic energy transfers (see Section 4.1). We start with Eq. (3.47), which is

$$\frac{d}{dt}H_K(\mathbf{k}) = \sum_{\mathbf{p}} \Re[\mathbf{u}(\mathbf{q}) \cdot \{\boldsymbol{\omega}(\mathbf{p}) \times \boldsymbol{\omega}^*(\mathbf{k})\}] + \mathcal{F}_{H_K}(\mathbf{k}) - \nu k^2 H_K(\mathbf{k}), \tag{8.1}$$

where $\mathbf{q} = \mathbf{k} - \mathbf{p}$, and $\mathcal{F}_{H_K}(\mathbf{k})$ is the kinetic helicity supply rate by the external force. We focus on a pair of triads, $(\mathbf{k}', \mathbf{p}, \mathbf{q})$ and $(-\mathbf{k}', -\mathbf{p}, -\mathbf{q})$, where $\mathbf{k}' = -\mathbf{k}$. Note that the modal kinetic helicity is defined as (see Section 3.3)

$$H_K(\mathbf{k}) = \frac{1}{2}\Re[\mathbf{u}(\mathbf{k}) \cdot \boldsymbol{\omega}^*(\mathbf{k})]. \tag{8.2}$$

Following Eq. (8.1), for a triad, the evolution equation for $H_K(\mathbf{k})$ with $\nu = 0$ and $\mathcal{F}_{H_K} = 0$ is

$$\begin{aligned}\frac{d}{dt}H_K(\mathbf{k}') &= S^{H_K}(\mathbf{k}'|\mathbf{p},\mathbf{q}) \\ &= \Re[\mathbf{u}(\mathbf{q})\cdot\{\boldsymbol{\omega}(\mathbf{p})\times\boldsymbol{\omega}(\mathbf{k}')\}] + \Re[\mathbf{u}(\mathbf{p})\cdot\{\boldsymbol{\omega}(\mathbf{q})\times\boldsymbol{\omega}(\mathbf{k}')\}]. \end{aligned} \tag{8.3}$$

Similar equations can be derived for $H_K(\mathbf{p})$ and $H_K(\mathbf{q})$.

Using the identity $\mathbf{A}\times\mathbf{B} = -\mathbf{B}\times\mathbf{A}$, we obtain

$$S^{H_K}(\mathbf{k}'|\mathbf{p},\mathbf{q}) + S^{H_K}(\mathbf{p}|\mathbf{k}',\mathbf{q}) + S^{H_K}(\mathbf{q}|\mathbf{k}',\mathbf{p}) = 0. \tag{8.4}$$

Hence,

$$H_K(\mathbf{k}') + H_K(\mathbf{p}) + H_K(\mathbf{q}) = \text{const.} \tag{8.5}$$

Or, the total kinetic helicity is conserved in a triadic interaction. This is the statement of the *detailed conservation of kinetic helicity* in a triad.

Next, we derive a formula for $S^{H_K}(\mathbf{k}'|\mathbf{p}|\mathbf{q})$ that represents the mode-to-mode kinetic helicity transfer from wavenumber $\mathbf{p}$ to wavenumber $\mathbf{k}'$ with wavenumber $\mathbf{q}$ acing as a mediator. The derivation described in the following is very similar to that for kinetic energy (see Section 4.1).

The mode-to-mode kinetic helicity transfer must satisfy the following properties. For convenience, we denote the wavenumbers as $\mathbf{X},\mathbf{Y},\mathbf{Z}$ with $\mathbf{X}+\mathbf{Y}+\mathbf{Z}=0$.

1. By definition, the sum of $S^{H_K}(\mathbf{X}|\mathbf{Y}|\mathbf{Z})$ and $S^{H_K}(\mathbf{X}|\mathbf{Z}|\mathbf{Y})$ is the combined kinetic helicity transfer $S^{H_K}(\mathbf{X}|\mathbf{Y},\mathbf{Z})$ defined in Eq. (8.3). Therefore,

$$S^{H_K}(\mathbf{k}'|\mathbf{p}|\mathbf{q}) + S^{H_K}(\mathbf{k}'|\mathbf{q}|\mathbf{p}) = S^{H_K}(\mathbf{k}'|\mathbf{p},\mathbf{q}), \tag{8.6a}$$

$$S^{H_K}(\mathbf{p}|\mathbf{k}'|\mathbf{q}) + S^{H_K}(\mathbf{p}|\mathbf{q}|\mathbf{k}') = S^{H_K}(\mathbf{p}|\mathbf{k}',\mathbf{q}), \tag{8.6b}$$

$$S^{H_K}(\mathbf{q}|\mathbf{k}'|\mathbf{p}) + S^{H_K}(\mathbf{q}|\mathbf{p}|\mathbf{k}') = S^{H_K}(\mathbf{q}|\mathbf{k}',\mathbf{p}). \tag{8.6c}$$

2. The kinetic helicity transfer from wavenumber $\mathbf{X}$ to wavenumber $\mathbf{Y}$ is equal and opposite to that from $\mathbf{Y}$ to $\mathbf{X}$. Therefore,

$$S^{H_K}(\mathbf{k}'|\mathbf{p}|\mathbf{q}) + S^{H_K}(\mathbf{p}|\mathbf{k}'|\mathbf{q}) = 0, \tag{8.7a}$$

$$S^{H_K}(\mathbf{k}'|\mathbf{q}|\mathbf{p}) + S^{H_K}(\mathbf{q}|\mathbf{k}'|\mathbf{p}) = 0, \tag{8.7b}$$

$$S^{H_K}(\mathbf{p}|\mathbf{q}|\mathbf{k}') + S^{H_K}(\mathbf{q}|\mathbf{p}|\mathbf{k}') = 0. \tag{8.7c}$$

Equations (8.6, 8.7) do not yield a unique solution to six unknowns $S^{H_K}(\mathbf{X}|\mathbf{Y}|\mathbf{Z})$ because the determinant of the solution matrix is zero. Hence, we need more constraints.

We employ tensor analysis to derive the formula for $S^{H_K}(\mathbf{X}|\mathbf{Y}|\mathbf{Z})$. Using the structure of Eq. (8.3), we infer that $S^{H_K}(\mathbf{k}'|\mathbf{p}|\mathbf{q})$ is a function of $\mathbf{u}(\mathbf{k}')$, $\mathbf{u}(\mathbf{p})$, $\mathbf{u}(\mathbf{q})$, $\boldsymbol{\omega}(\mathbf{k}')$, $\boldsymbol{\omega}(\mathbf{p})$, and $\boldsymbol{\omega}(\mathbf{q})$ with the following additional conditions:

1. $S^{H_K}(\mathbf{k}'|\mathbf{p}|\mathbf{q})$ is real.
2. $S^{H_K}(\mathbf{k}'|\mathbf{p}|\mathbf{q})$ has two $\boldsymbol{\omega}$'s and one $\mathbf{u}$, with each of them function of a distinct wavenumber among $\mathbf{k}', \mathbf{p}, \mathbf{q}$.
3. $S^{H_K}(\mathbf{k}'|\mathbf{p}|\mathbf{q})$ must have $\boldsymbol{\omega}(\mathbf{k}')$ in the expression.
4. $S^{H_K}(\mathbf{k}'|\mathbf{p}|\mathbf{q}) = S^{H_K}(-\mathbf{k}'|-\mathbf{p}|-\mathbf{q})$.

Using these conditions and the structure of Eq. (8.3), we find that

$$S^{H_K}(\mathbf{k}'|\mathbf{p}|\mathbf{q}) = c_1\Re[\mathbf{u}(\mathbf{q})\cdot\{\boldsymbol{\omega}(\mathbf{p})\times\boldsymbol{\omega}(\mathbf{k}')\}] + c_2\Re[\mathbf{u}(\mathbf{p})\cdot\{\boldsymbol{\omega}(\mathbf{q})\times\boldsymbol{\omega}(\mathbf{k}')\}]. \quad (8.8)$$

The imaginary part is zero due to condition (4). Now we employ Eq. (8.7a) that yields $c_2 = 0$. Therefore,

$$S^{H_K}(\mathbf{k}'|\mathbf{p}|\mathbf{q}) = c_1\Re[\mathbf{u}(\mathbf{q})\cdot\{\boldsymbol{\omega}(\mathbf{p})\times\boldsymbol{\omega}(\mathbf{k}')\}]. \quad (8.9)$$

Application of Eqs. (8.3, 8.6a) yields $c_1 = 1$. Therefore,

$$S^{H_K}(\mathbf{k}'|\mathbf{p}|\mathbf{q}) = \Re[\mathbf{u}(\mathbf{q})\cdot\{\boldsymbol{\omega}(\mathbf{p})\times\boldsymbol{\omega}(\mathbf{k}')\}] \quad (8.10)$$

is the mode-to-mode kinetic helicity transfer from wavenumber $\mathbf{p}$ to wavenumber $\mathbf{k}'$ with wavenumber $\mathbf{q}$ acting as a mediator.

The function $S^{H_K}(\mathbf{k}'|\mathbf{p}|\mathbf{q})$ satisfies many important relations. One among them is that for any wavenumber region A (which could be a single triad),

$$\sum_{\mathbf{k}'\in A}\sum_{\mathbf{p}\in A} S^{H_K}(\mathbf{k}'|\mathbf{p}|\mathbf{q}) = 0. \quad (8.11)$$

In the presence of a large number of triads, the rate of change of modal kinetic helicity $H_K(\mathbf{k})$ is

$$\begin{aligned}\frac{d}{dt}H_K(\mathbf{k}) &= \sum_{\mathbf{p}} S^{H_K}(\mathbf{k}|\mathbf{p}|\mathbf{q}) + \mathcal{F}_{H_K}(\mathbf{k}) - 2\nu k^2 H_K(\mathbf{k}), \\ &= T_{H_K}(\mathbf{k}) + \mathcal{F}_{H_K}(\mathbf{k}) - 2\nu k^2 H_K(\mathbf{k}), \end{aligned} \quad (8.12)$$

where $\mathbf{q} = \mathbf{k} - \mathbf{p}$.

In the next section, we define flux and shell-to-shell transfers of kinetic helicity for hydrodynamic turbulence.

8.2 Flux and Shell-to-shell Transfers of Kinetic Helicity

Using the formula of the mode-to-mode kinetic helicity transfer we derive expressions for the flux and shell-to-shell transfers of kinetic helicity. The derivation is similar to that for the kinetic energy flux and shell-to-shell transfers.

The kinetic helicity flux $\Pi_{H_K}(k_0)$ is defined as the rate of kinetic helicity transfer from all the modes inside the sphere of radius k_0 to all the modes outside the sphere. Using the mode-to-mode kinetic helicity transfer we deduce that

$$\Pi_{H_K}(k_0) = \sum_{|\mathbf{k}'|>k_0} \sum_{|\mathbf{p}|>k_0} S^{H_K}(\mathbf{k}'|\mathbf{p}|\mathbf{q}). \tag{8.13}$$

Similarly, we can derive the shell-to-shell transfer of kinetic helicity from shell m to shell n as

$$T^{H_K,m}_{H_K,n} = \sum_{\mathbf{k}'\in n} \sum_{\mathbf{p}\in m} S^{H_K}(\mathbf{k}'|\mathbf{p}|\mathbf{q}). \tag{8.14}$$

When we sum Eq. (8.12) over the modes in a sphere of radius k_0, we obtain

$$\begin{aligned}\frac{d}{dt}\sum_{|\mathbf{k}|\le k_0} H_K(\mathbf{k}) &= \sum_{|\mathbf{k}|\le k_0} T_{H_K}(\mathbf{k}) + \sum_{|\mathbf{k}|\le k_0} \mathcal{F}_{H_K}(\mathbf{k}) - \sum_{|\mathbf{k}|\le k_0} 2\nu k^2 H_K(\mathbf{k}),\\ &= -\Pi_{H_K}(k_0) + \sum_{|\mathbf{k}|\le k_0} \mathcal{F}_{H_K}(\mathbf{k}) - \sum_{|\mathbf{k}|\le k_0} 2\nu k^2 H_K(\mathbf{k}).\end{aligned} \tag{8.15}$$

In this derivation, the conservation of kinetic helicity plays an important role. Following the similar lines of arguments as in Section 4.4, we observe that 1D kinetic helicity spectrum follows the following evolution equation:

$$\frac{\partial}{\partial t} H_K(k,t) = -\frac{\partial}{\partial k}\Pi_{H_K}(k,t) + \mathcal{F}_{H_K}(\mathbf{k}) - D_{H_K}(k,t), \tag{8.16}$$

where $D_{H_K}(k)$ is the kinetic helicity dissipation rate in the shell. See Fig. 8.1 for an illustration. Under a steady state, $\partial H_K(k,t)/\partial t = 0$, and in the absence of an external force, we obtain

$$\frac{\partial}{\partial k}\Pi_{H_K}(k) = -D_{H_K}(k). \tag{8.17}$$

In the next section, we will discuss the spectrum of kinetic helicity in a fully-developed isotropic and homogeneous turbulence.

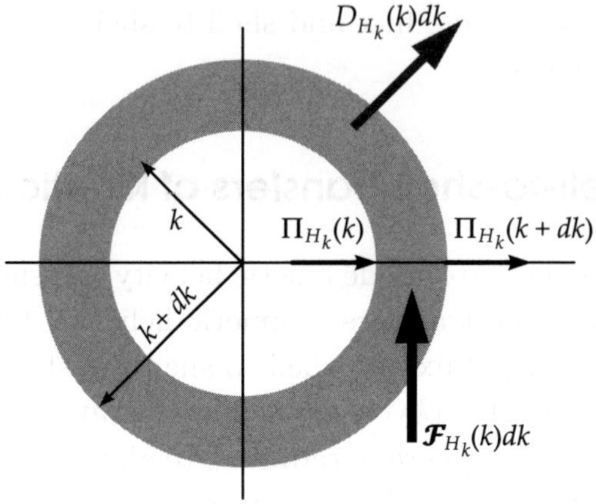

Figure 8.1 The rate of change of kinetic helicity in a shell is given by kinetic helicity flux difference $\Pi_{H_K}(k+dk) - \Pi_{H_K}(k)$, the kinetic helicity supply rate by the external force $\mathcal{F}_{H_K}(k)dk$, and the kinetic helicity dissipation rate $D_{H_K}(k)dk$.

8.3 Phenomenology of Helical Turbulence

In Chapter 5 we derived that fully-developed isotropic and homogeneous turbulence exhibits Kolmogorov spectrum ($E_u(k) \sim k^{-5/3}$). This derivation was silent on the kinetic helicity. A leading question is whether the kinetic energy spectrum depends on the kinetic helicity.

Zhou (1993) showed that the kinetic helicity does not alter the renormalized viscosity (see Appendix C). Hence we can argue that the kinetic energy spectrum and flux remain unaffected by the kinetic helicity. That is, Eqs. (5.6, 5.11) are applicable to a hydrodynamic turbulence irrespective of the amount of kinetic helicity.

After these arguments, we derive the kinetic helicity flux and spectrum in isotropic and homogeneous turbulence. Under a steady state, in the inertial range where $D_{H_K}(k) \to 0$, using Eq. (8.17) we deduce that

$$\Pi_{H_K} = \text{const.} \tag{8.18}$$

Now we can derive an expression for $H_K(k)$ using dimensional analysis and the following inputs:

1. $H_K(k)$ depends only on Π_u, Π_{H_K}, and k.

2. $[\Pi_u] = [L^2/T^3]$; $[\Pi_{H_K}] = [L/T^3]$; $[H_K(k)] = [L^2/T^2]$.

Using these inputs, based on dimensional analysis we postulate that

$$H_K(k) = (\Pi_{H_K})^\alpha (\Pi_u)^\beta k^\gamma. \tag{8.19}$$

Note that we have only two fundamental dimensions, $[L]$ and $[T]$, but three unknowns—α, β, γ. To reduce one of the parameters, we exploit the fact that both $H_K(k)$ and Π_{H_K} are proportional to $H_K = \mathbf{u} \cdot \boldsymbol{\omega}$, hence $H_K(k) \propto \Pi_{H_K}$.[1] Therefore, $\alpha = 1$.

Now matching the dimensions of [L] and [T] yields

$$\beta = -1/3; \quad \gamma = -5/3. \tag{8.20}$$

Hence,

$$H_K(k) = K_H \Pi_{H_K} (\Pi_u)^{-1/3} k^{-5/3}, \tag{8.21}$$

where K_H is a nondimensional constant. Using field-theoretic computation, Avinash et al. (2006) estimated K_H to be 2.47. Following similar steps as in the derivation of Eq. (5.8), we show that Π_{H_K} equals the dissipation rate of kinetic helicity, which is

$$\epsilon_{H_K} = \int_0^\infty 2\nu k^2 H_K(k) dk. \tag{8.22}$$

The formula of Eq. (8.21) is extended to the dissipative range using the same procedure as discussed in Section 5.5.1. The kinetic helicity spectrum and flux thus obtained are as follows:

$$\Pi_{H_K}(k) = \epsilon_{H_K} \exp\left(-\frac{3}{2} K_H (k/k_d)^{4/3}\right), \tag{8.23a}$$

$$H_K(k) = K_H \epsilon_{H_K} \epsilon_u^{-1/3} k^{-5/3} \exp\left(-\frac{3}{2} K_H (k/k_d)^{4/3}\right), \tag{8.23b}$$

where

$$k_d = \left(\frac{\epsilon_u}{\nu^3}\right)^{1/4} \tag{8.24}$$

is Kolmogorov's wavenumber. We will verify these formulas using numerical simulations.

A word of caution however is in order. Some of the recents works (Kessar et al., 2015; Sahoo and Biferale, 2018) show that hydrodynamic turbulence with near

[1]Using field-theoretic arguments, Avinash et al. (2006) showed that $\Pi_{H_K} \propto H_K(k)$.

maximum helicity have properties different from those described here. These topics however are beyond the scope of this book.

In the next section, we describe numerical results related to helical hydrodynamics.

8.4 Numerical Verification of Kinetic Helicity Spectrum and Flux

Sadhukhan et al. (2018) performed hydrodynamic turbulence simulation for the nonhelical case ($H_K = 0$) on a 1024^3 grid. They employed 0.1 nondimensional unit of kinetic energy injection rate at large length scales, that is $\epsilon_u = 0.1$. As expected, in the steady state, they observed $k^{-5/3}$ kinetic energy spectrum.

Sadhukhan et al. (2018) also performed a forced helical run with $\epsilon_u = 0.1$ and the kinetic helicity injection rate $\epsilon_h = 0.2$ (the forcing employed at large scales). The run was continued till the flow reached a steady state. In the following discussion, we compare the results of helical turbulence with those of nonhelical ones. The Reynolds number of both the flows are approximately 5700.

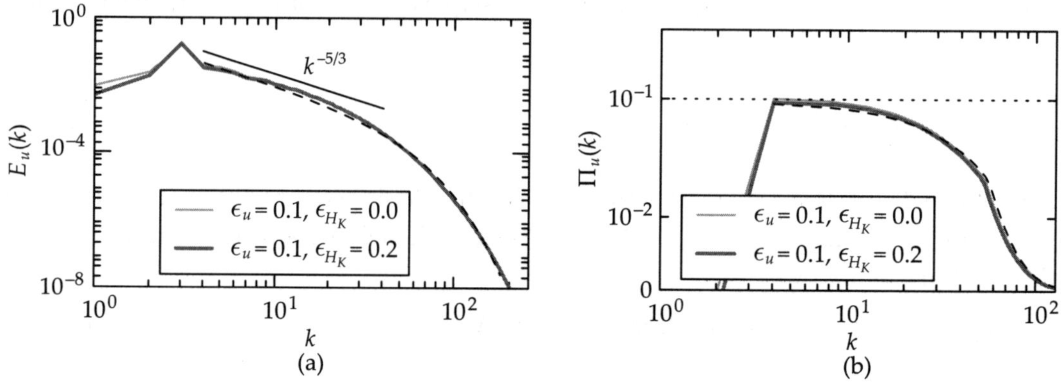

Figure 8.2 For nonhelical ($\epsilon_u = 0.1$, thin line) and helical ($\epsilon_u = 0.1$, $\epsilon_{H_k} = 0.2$, thick line) hydrodynamics: (a) $E_u(k)$ and (b) $\Pi_u(k)$. The plots show that kinetic helicity does not affect the kinetic energy spectrum and flux significantly. The predictions by Pao's formula (shown as dashed lines) fits with the numerical data quite well. Adopted from figures of Sadhukhan et al. (2018).

Using the steady state data, Sadhukhan et al. (2018) computed the spectra and fluxes of kinetic energy and kinetic helicity. In Fig. 8.2, we exhibit the kinetic energy spectra and fluxes for both helical and nonhelical runs. We observe that the kinetic energy spectra and fluxes for both helical and nonhelical runs are approximately equal, and they are described quite well by Pao's formula (Eqs. (5.34a, 5.34b)). In the figure, the analytical formula are shown by dashed lines. Note that $E_u(k) \sim k^{-5/3}$ and $\Pi_u(k) \approx$ const. in the inertial range, as expected. There is a minor difference at small wavenumbers, which is primarily due to the forcing. Thus, $E_u(k)$ and $\Pi_u(k)$ remain primarily unaffected by helicity (Lesieur, 2008; Avinash et al., 2006).

Sadhukhan et al. (2018) computed the kinetic helicity spectrum $H_K(k)$ as well as the kinetic helicity flux $\Pi_{H_K}(k)$ for the helical run; these quantities are plotted in Fig. 8.3. In the figure, we also plot the predictions of Eqs. (8.23). The predictions by the analytical formula match quite well with the numerical results, thus validating the phenomenology discussed in Section 8.3. Note that in the inertial range, $H_K(k) \sim k^{-5/3}$ and $\Pi_{H_K}(k) \sim$ const., consistent with the turbulence phenomenology described earlier.

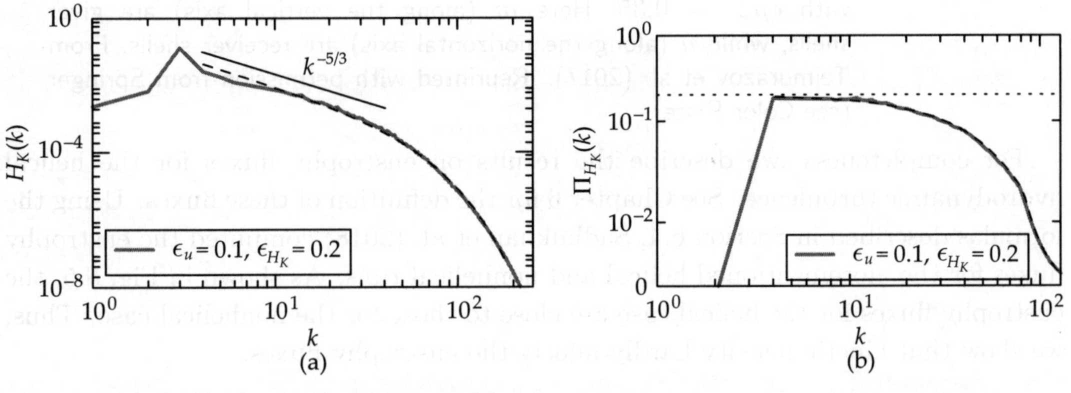

Figure 8.3 For helical hydrodynamics with $\epsilon_u = 0.1$, $\epsilon_{H_K} = 0.2$, (a) $H_K(k)$ and (b) $\Pi_{H_K}(k)$. The numerical results, shown as solid lines, match quite well with the predictions by formulas (dashed lines) of Eqs. (8.23). In the inertial range, $H_K(k) \sim k^{-5/3}$ and $\Pi_{H_K}(k) =$ const. Adopted from figures of Sadhukhan et al. (2018).

Teimurazov et al. (2017) computed the shell-to-shell kinetic energy and kinetic helicity transfers for helical and nonhelical hydrodynamics. We exhibit these quantities in Fig. 8.4. Here the giver shells m are along the vertical axis, while the receiver shells n are along the horizontal axis. The figures show that all the transfers are local and forward. These observations are consistent with the aforementioned result that hydrodynamic turbulence is not affected significantly by kinetic helicity. Note however that for helical turbulence, both kinetic energy and kinetic helicity exhibit slight nonlocal shell-to-shell transfers from the forcing band to distant shells. This issue needs to be explored further.

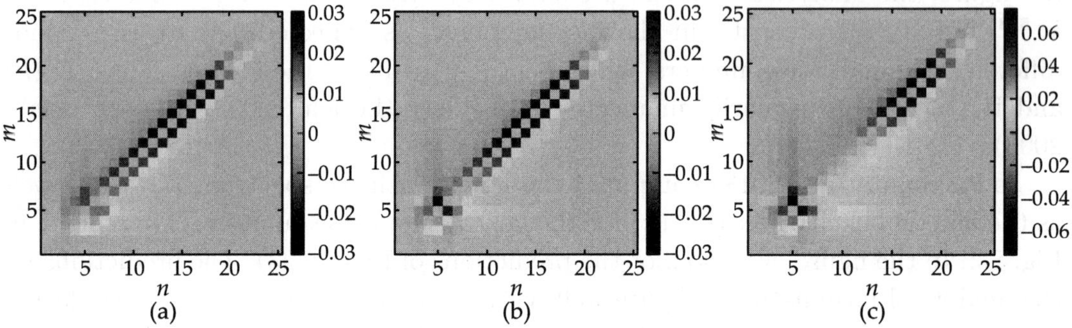

Figure 8.4 The shell-to-shell energy transfer for kinetic energy $T_{u,n}^{u,m}$ for (a) nonhelical turbulence, and (b) helical turbulence with $\epsilon_{H_K} = 0.35$. (c) Kinetic helicity shell-to-shell transfer $T_{H_K,n}^{H_K,m}$ for helical turbulence with $\epsilon_{H_K} = 0.35$. Here m (along the vertical axis) are giver shells, while n (along the horizontal axis) are receiver shells. From Teimurazov et al. (2017). Reprinted with permission from Springer (see Color Plate).

For completeness, we describe the results on enstrophy fluxes for the helical hydrodynamic turbulence. See Chapter 6 for the definition of these fluxes. Using the formulas described in Section 6.4, Sadhukhan et al. (2018) computed the enstrophy fluxes for the aforementioned helical and nonhelical runs. As shown in Fig. 8.5, the enstrophy fluxes for the helical case are close to those for the nonhelical case. Thus, we show that kinetic helicity hardly affects the enstrophy fluxes.

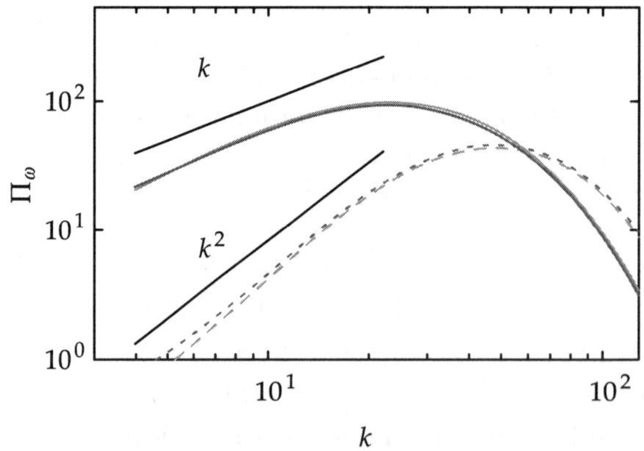

Figure 8.5 For nonhelical ($\epsilon_u = 0.1$, dark curves) and helical ($\epsilon_u = 0.1$, $\epsilon_{H_K} = 0.2$, light curves) hydrodynamics, plots of enstrophy fluxes $\Pi_{\omega>}^{u<}$ (solid line) and $\Pi_{\omega>}^{\omega<}$ (dashed line). Note that $\Pi_{\omega>}^{u<} \gg \Pi_{\omega>}^{\omega<}$. The fluxes for helical and nonhelical cases almost overlap on each other. Adopted from figures of Sadhukhan et al. (2018).

With this, we close our discussion on phenomenology of helical turbulence. We will revisit kinetic helicity in subsequent chapters.

Further Reading

Lesieur (2008) describes the scaling of kinetic helicity spectrum and flux of helical turbulence under eddy-damped quasi-normal Markovian (EDQNM) approximation. The mode-to-mode kinetic helicity transfer was discussed earlier by Avinash et al. (2006), but its proof is given here for the first time. Avinash et al. (2006) also discussed the field-theoretic aspects of helical turbulence. The numerical results described here are from Teimurazov et al. (2017) and Sadhukhan et al. (2018).

In recent times, Biferale et al. (2013) and coworkers, Stepanov et al. (2015), and Kessar et al. (2015) have studied the energy transfers and fluxes for strongly helical turbulence. These works are summarized in a recent review by Alexakis and Biferale (2018).

Chapter 9

Craya–Herring and Helical Basis

In this chapter we describe Craya–Herring and helical basis that simplify the energy transfer computations. In addition, it is easier to construct low-dimensional models in these basis. We start with the description of Craya–Herring basis.

9.1 Craya–Herring Basis for Hydrodynamics

Craya (1958) and Herring (1974) constructed the following basis vectors to represent the velocity Fourier modes:

$$\hat{e}_3(\mathbf{k}) = \hat{k}, \tag{9.1a}$$

$$\hat{e}_1(\mathbf{k}) = \frac{\hat{k} \times \hat{n}}{|\hat{k} \times \hat{n}|}, \tag{9.1b}$$

$$\hat{e}_2(\mathbf{k}) = \hat{e}_3(\mathbf{k}) \times \hat{e}_1(\mathbf{k}), \tag{9.1c}$$

where $\hat{k}$ is the unit vector along the wavenumber $\mathbf{k}$, and $\hat{n}$ is along any direction (also see Lesieur (2008); Sagaut and Cambon (2008)). For an illustration, see Fig. 9.1 in which $\hat{n}$ is along $\hat{z}$. It is customary to choose $\hat{n}$ along the anisotropy direction, for example, along the acceleration due to gravity, or rotation axis, or the mean magnetic field.

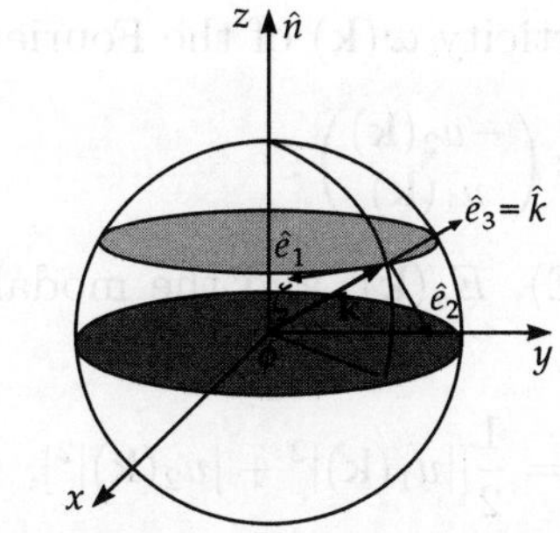

Figure 9.1 Illustration of Craya–Herring basis vectors. The unit vector $\hat{e}_3$ is along $\mathbf{k}$, while $\mathbf{u}(\mathbf{k})$ lies in the plane formed by $\hat{e}_1$ and $\hat{e}_2$.

For an incompressible flow, $\mathbf{k} \cdot \mathbf{u}(\mathbf{k}) = 0$, hence $u_3(\mathbf{k}) = 0$. Therefore, in Craya–Herring basis,

$$\mathbf{u}(\mathbf{k}) \quad = \quad u_1(\mathbf{k})\hat{e}_1 + u_2(\mathbf{k})\hat{e}_2. \tag{9.2}$$

Note that $u_{1,2}(\mathbf{k})$ are in general complex numbers.[1] We can compute the corresponding Cartesian components as

$$\begin{aligned} \mathbf{u}(\mathbf{k}) = \begin{pmatrix} u_x \\ u_y \\ u_z \end{pmatrix} &= \begin{pmatrix} u_1(\mathbf{k}) \sin\phi + u_2(\mathbf{k}) \cos\zeta \cos\phi \\ -u_1(\mathbf{k}) \cos\phi + u_2(\mathbf{k}) \cos\zeta \sin\phi \\ -u_2(\mathbf{k}) \sin\zeta \end{pmatrix} \\ &= \begin{pmatrix} u_1(\mathbf{k}) \frac{k_y}{k_\perp} + u_2(\mathbf{k}) \frac{k_z k_x}{k k_\perp} \\ -u_1(\mathbf{k}) \frac{k_x}{k_\perp} + u_2(\mathbf{k}) \frac{k_z k_y}{k k_\perp} \\ -u_2(\mathbf{k}) \frac{k_\perp}{k} \end{pmatrix}, \end{aligned} \tag{9.3}$$

where ζ, ϕ are respectively the polar and azimuthal angles of the wave vector $\mathbf{k}$, as shown in Fig. 9.2.

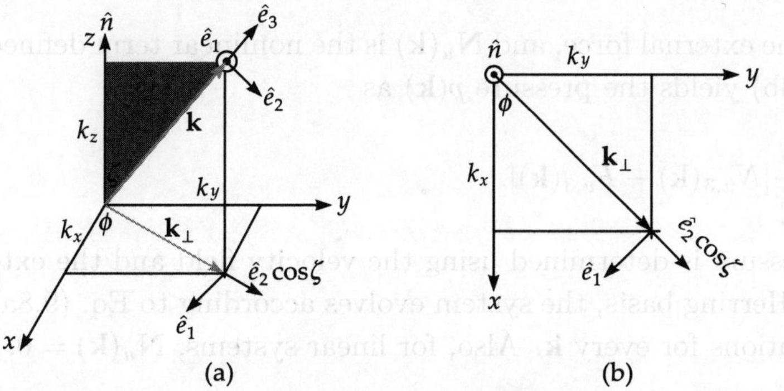

Figure 9.2 Craya–Herring basis vectors and their relationships with the Cartesian basis vectors: (a) in 3D, (b) in k_x-k_y plane.

[1] $u_{1,2}(\mathbf{k})$ are not the x and y components of $\mathbf{u}(\mathbf{k})$.

In Craya–Herring basis, vorticity $\boldsymbol{\omega}(\mathbf{k})$ of the Fourier mode is

$$\boldsymbol{\omega}(\mathbf{k}) = i\mathbf{k} \times \mathbf{u}(\mathbf{k}) = ik \begin{pmatrix} -u_2(\mathbf{k}) \\ u_1(\mathbf{k}) \end{pmatrix}. \tag{9.4}$$

The modal kinetic energy (KE), $E_u(\mathbf{k})$, and the modal kinetic helicity, $H_K(\mathbf{k})$, of the velocity mode are

$$E_u(\mathbf{k}) = \frac{1}{2}\mathbf{u}^*(\mathbf{k})\mathbf{u}(\mathbf{k}) = \frac{1}{2}[|u_1(\mathbf{k})|^2 + |u_2(\mathbf{k})|^2], \tag{9.5a}$$

$$H_K(\mathbf{k}) = \frac{1}{2}\Re[\mathbf{u}^*(\mathbf{k}) \cdot \boldsymbol{\omega}(\mathbf{k})] = k\Im[u_1^*(\mathbf{k})u_2(\mathbf{k})]. \tag{9.5b}$$

Therefore, for nonzero helicity, (a) both u_1 and u_2 must be nonzero, and (b) u_1 and u_2 cannot be pure real or pure imaginary simultaneously. Also, if

$$u_1(\mathbf{k}) = |u_1|\exp(i\phi_1); \quad u_2(\mathbf{k}) = |u_2|\exp(i\phi_2), \tag{9.6}$$

then

$$H_K(\mathbf{k}) = k\Im[u_1^*(\mathbf{k})u_2(\mathbf{k})] = k|u_1||u_2|\sin(\phi_2 - \phi_1). \tag{9.7}$$

Hence, the mode $\mathbf{u}(\mathbf{k})$ is maximally helical with $H_K(\mathbf{k}) = \pm k|u_1|u_2|$ when $\phi_2 - \phi_1 = \pm\pi/2$.

Now let us write down the momentum equation in the Craya–Herring basis. Since $u_3(\mathbf{k}) = 0$, Eq. (3.17a) transforms to

$$\frac{d}{dt}u_\alpha(\mathbf{k}) = -N_{u,\alpha}(\mathbf{k}) + F_{u,\alpha}(\mathbf{k}) - \nu k^2 u_\alpha(\mathbf{k}) \quad \text{for } \alpha = 1, 2, \tag{9.8a}$$

$$\frac{d}{dt}u_3(\mathbf{k}) = 0 = -N_{u,3}(\mathbf{k}) - ikp(\mathbf{k}) + F_{u,3}(\mathbf{k}), \tag{9.8b}$$

where $\mathbf{F}_u$ is the external force, and $\mathbf{N}_u(\mathbf{k})$ is the nonlinear term defined in Eq. (3.16). Equation (9.8b) yields the pressure $p(\mathbf{k})$ as

$$p(\mathbf{k}) = \frac{i}{k}[N_{u,3}(\mathbf{k}) - F_{u,3}(\mathbf{k})]. \tag{9.9}$$

Thus, the pressure is determined using the velocity field and the external force.

In Craya–Herring basis, the system evolves according to Eq. (9.8a) that contains a pair of equations for every $\mathbf{k}$. Also, for linear systems, $\mathbf{N}_u(\mathbf{k}) = 0$, and hence,

$$\frac{d}{dt}u_\alpha(\mathbf{k}) = F_{u,\alpha}(\mathbf{k}) - \nu k^2 u_\alpha(\mathbf{k}) \quad \text{for } \alpha = 1, 2, \tag{9.10a}$$

$$p(\mathbf{k}) = -\frac{i}{k}F_{u,3}(\mathbf{k}). \tag{9.10b}$$

Note that under *parity transformation* $\mathbf{k} \to -\mathbf{k}$,

$$\hat{e}_1(-\mathbf{k}) = -\hat{e}_1(\mathbf{k}), \tag{9.11a}$$
$$\hat{e}_2(-\mathbf{k}) = \hat{e}_2(\mathbf{k}), \tag{9.11b}$$
$$\hat{e}_3(-\mathbf{k}) = -\hat{e}_3(\mathbf{k}). \tag{9.11c}$$

See Fig. 9.3 for an illustration. Therefore, the reality condition $\mathbf{u}(-\mathbf{k}) = \mathbf{u}^*(\mathbf{k})$ yields the following relations:

$$u_1(-\mathbf{k}) = -u_1^*(\mathbf{k}) \tag{9.12a}$$
$$u_2(-\mathbf{k}) = u_2^*(\mathbf{k}). \tag{9.12b}$$

Under parity, the modal energy and modal kinetic helicity remain unchanged, that is,

$$E_u(-\mathbf{k}) = E_u(\mathbf{k}), \tag{9.13}$$
$$\begin{aligned} H_K(-\mathbf{k}) &= k\Im\langle u_1^*(-\mathbf{k})u_2(-\mathbf{k})\rangle = k\Im\langle -u_1(\mathbf{k})u_2^*(\mathbf{k})\rangle \\ &= k\Im\langle u_1^*(\mathbf{k})u_2(\mathbf{k})\rangle = H_K(\mathbf{k}). \end{aligned} \tag{9.14}$$

This is expected because E_u and H_K are real.

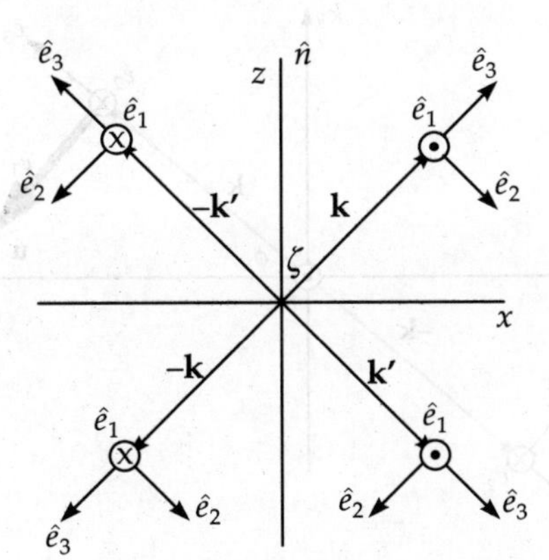

Figure 9.3 Transformations of $\hat{e}_1$ and $\hat{e}_2$ under parity: $\mathbf{k} \to -\mathbf{k}$.

The aforementioned representation of Craya–Herring basis is used for 3D systems. The basis functions become somewhat simpler for 2D flows, as shown in the following.

A 2D flow is a function only of coordinates x and y in real space, or of k_x and k_y in spectral space. Thus,

$$\mathbf{k} = k_x\hat{x} + k_y\hat{y}. \tag{9.15}$$

The unit vector $\hat{n}$ is chosen along $\hat{z}$. Hence, the basis vectors $\hat{e}_1$, $\hat{e}_2$, and $\hat{e}_3$ appear as shown in Fig. 9.4. A two-dimensional incompressible velocity field can be expressed as

$$\mathbf{u}(\mathbf{k}) = u_1 \hat{e}_1. \tag{9.16}$$

The Cartesian components of the velocity field are

$$\mathbf{u}(\mathbf{k}) = \begin{pmatrix} u_x \\ u_y \end{pmatrix} = \begin{pmatrix} u_1 \sin\phi \\ -u_1 \cos\phi \end{pmatrix} = \begin{pmatrix} (k_y/k)u_1 \\ -(k_x/k)u_1 \end{pmatrix}. \tag{9.17}$$

Note that the kinetic helicity is zero for a 2D flow since the vorticity $\boldsymbol{\omega}(\mathbf{k})$ is perpendicular to $\mathbf{u}(\mathbf{k})$.

A general 2D incompressible velocity field is

$$\mathbf{u}(\mathbf{k}) = u_1 \hat{e}_1 + u_2 \hat{e}_2 \tag{9.18}$$

with $u_2 \hat{e}_2$ along $-\hat{n}$. Such two-dimensional three-component (2D3C) fields are often encountered in magnetohydrodynamics with a strong external magnetic field.

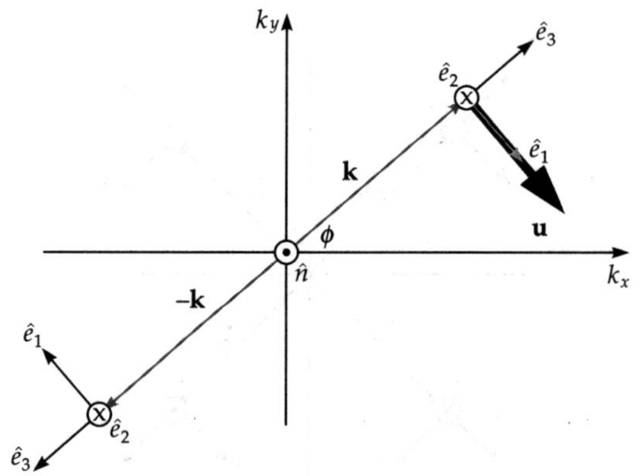

Figure 9.4 Illustration of Craya–Herring basis vectors and $\mathbf{u}(\mathbf{k})$ for a 2D flow.

As argued earlier, under parity transformation, $\mathbf{k} \to -\mathbf{k}$,

$$\hat{e}_1(-\mathbf{k}) = -\hat{e}_1(\mathbf{k}), \tag{9.19a}$$

$$u_1(-\mathbf{k}) = -u_1^*(\mathbf{k}). \tag{9.19b}$$

Therefore, if $u_1(\mathbf{k}) = A + B/i$ with A, B as real, then

$$u_1(-\mathbf{k}) = -A + B/i. \tag{9.20}$$

9.2 Equations of Motion in Craya–Herring Basis

As described in Chapter 4, the basic unit of nonlinear interactions in hydrodynamics is a wavenumber triad $(\mathbf{k}', \mathbf{p}, \mathbf{q})$ with $\mathbf{k}' + \mathbf{p} + \mathbf{q} = 0$. The derivation of equations of motion for the Fourier modes gets simplified considerably in the Craya–Herring basis.

Consider a plane formed by the wavenumber triad $(\mathbf{k}', \mathbf{p}, \mathbf{q})$, as shown in Fig. 9.5. Here the wavenumbers traverse in a clockwise direction. We choose

$$\hat{n} = \frac{\mathbf{q} \times \mathbf{p}}{|\mathbf{q} \times \mathbf{p}|}, \tag{9.21}$$

which is perpendicular to the aforementioned plane. Now the three Craya–Herring basis vectors are constructed using Eqs. (9.1). As a result, $\mathbf{k}', \mathbf{p}, \mathbf{q}$, $\hat{e}_1(\mathbf{k}')$, $\hat{e}_1(\mathbf{p})$, $\hat{e}_1(\mathbf{q})$ lie in the plane of the triad, $\hat{e}_2$'s are along $-\hat{n}$, and the vectors $\hat{e}_3$ are along the wavenumbers. Note that $\hat{e}_1$'s are to the right of the respective wavenumbers.

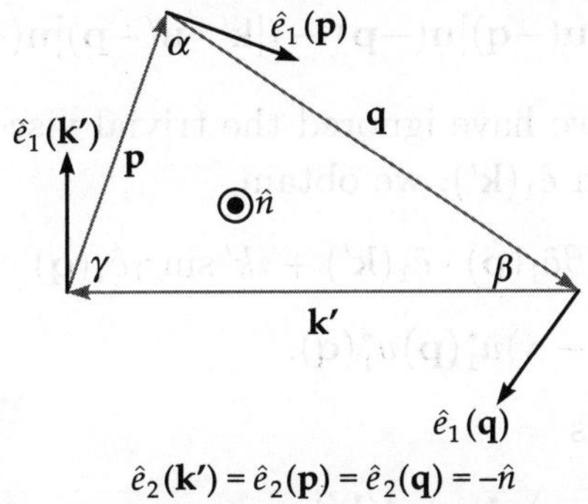

Figure 9.5 Craya–Herring basis vectors when the wavenumbers traverse in the clockwise direction. Here, $\hat{n} = \mathbf{q} \times \mathbf{p}/|\mathbf{q} \times \mathbf{p}|$.

In the following derivation, we will need quantities like $\hat{e}_i(\mathbf{k}') \cdot \hat{e}_j(\mathbf{p})$, where i, j take values 1,2,3. For illustration, we compute two of them:

$$\begin{aligned} \hat{e}_1(\mathbf{k}') \cdot \hat{e}_1(\mathbf{p}) &= (\hat{k}' \times \hat{n}) \cdot (\hat{p} \times \hat{n}) = \{(\hat{k}' \times \hat{n}) \times \hat{p}\} \cdot \hat{n} = \hat{k}' \cdot \hat{p} = -\cos\gamma, & (9.22a) \\ \hat{e}_3(\mathbf{k}') \cdot \hat{e}_1(\mathbf{p}) &= \hat{k}' \cdot (\hat{p} \times \hat{n}) = (\hat{k}' \times \hat{p}) \cdot \hat{n} = -\sin\gamma. & (9.22b) \end{aligned}$$

We list these products in Table 9.1. In addition,

$$\begin{aligned} \hat{e}_2(\mathbf{X}) \cdot \hat{e}_1(\mathbf{Y}) &= 0 & (9.23a) \\ \hat{e}_2(\mathbf{X}) \cdot \hat{e}_2(\mathbf{Y}) &= 1 & (9.23b) \\ \hat{e}_2(\mathbf{X}) \cdot \hat{e}_3(\mathbf{Y}) &= 0, & (9.23c) \end{aligned}$$

where $\mathbf{X}, \mathbf{Y}$ take values $\mathbf{k}', \mathbf{p}, \mathbf{q}$.

Table 9.1 Product rules for the Craya–Herring basis vectors when the wavenumbers traverse in the clockwise direction (See Fig. 9.5).

·	$\hat{e}_1(\mathbf{k}')$	$\hat{e}_1(\mathbf{p})$	$\hat{e}_1(\mathbf{q})$
$\hat{e}_3(\mathbf{k}')$	0	$-\sin\gamma$	$\sin\beta$
$\hat{e}_3(\mathbf{p})$	$\sin\gamma$	0	$-\sin\alpha$
$\hat{e}_3(\mathbf{q})$	$-\sin\beta$	$\sin\alpha$	0
$\hat{e}_1(\mathbf{k}')$	1	$-\cos\gamma$	$-\cos\beta$
$\hat{e}_1(\mathbf{p})$	$-\cos\gamma$	1	$-\cos\alpha$
$\hat{e}_1(\mathbf{q})$	$-\cos\beta$	$-\cos\alpha$	1

Craya–Herring basis provide a convenient framework to derive equations of motion for the Fourier modes. A major advantage of this scheme is that the pressure term gets eliminated from the equations automatically (see Eq. (9.8a)). In the following, we derive an equation for $\mathbf{u}(\mathbf{k}')$:

$$\frac{d}{dt}\mathbf{u}(\mathbf{k}') = -i[\mathbf{k}'\cdot\mathbf{u}(-\mathbf{q})]\mathbf{u}(-\mathbf{p}) - i[\mathbf{k}'\cdot\mathbf{u}(-\mathbf{p})]\mathbf{u}(-\mathbf{q}) - i\mathbf{k}p(\mathbf{k}) \tag{9.24}$$

since $\mathbf{k}' = -\mathbf{p}-\mathbf{q}$. Here we have ignored the trivial viscous term. By taking a dot product of Eq. (9.24) with $\hat{e}_1(\mathbf{k}')$, we obtain

$$\begin{aligned}\dot{u}_1(\mathbf{k}') &= [-ik'\sin\beta\hat{e}_1(\mathbf{p})\cdot\hat{e}_1(\mathbf{k}') + ik'\sin\gamma\hat{e}_1(\mathbf{q})\cdot\hat{e}_1(\mathbf{k}')]u_1^*(\mathbf{p})u_1^*(\mathbf{q}) \\ &= ik'\sin(\beta-\gamma)u_1^*(\mathbf{p})u_1^*(\mathbf{q}). \end{aligned} \tag{9.25}$$

Similar computation yields

$$\dot{u}_1(\mathbf{p}) = ip\sin(\gamma-\alpha)u_1^*(\mathbf{q})u_1^*(\mathbf{k}'), \tag{9.26a}$$

$$\dot{u}_1(\mathbf{q}) = iq\sin(\alpha-\beta)u_1^*(\mathbf{p})u_1^*(\mathbf{k}'), \tag{9.26b}$$

and

$$\dot{u}_2(\mathbf{k}') = ik'\{\sin\gamma u_1^*(\mathbf{p})u_2^*(\mathbf{q}) - \sin\beta u_1^*(\mathbf{q})u_2^*(\mathbf{p})\}, \tag{9.27a}$$

$$\dot{u}_2(\mathbf{p}) = ip\{\sin\alpha u_1^*(\mathbf{q})u_2^*(\mathbf{k}') - \sin\gamma u_1^*(\mathbf{k}')u_2^*(\mathbf{q})\}, \tag{9.27b}$$

$$\dot{u}_2(\mathbf{q}) = iq\{\sin\beta u_1^*(\mathbf{k}')u_2^*(\mathbf{p}) - \sin\alpha u_1^*(\mathbf{p})u_2^*(\mathbf{k}')\}. \tag{9.27c}$$

These formulas are applicable when the wavenumbers $\mathbf{k}', \mathbf{p}, \mathbf{q}$ traverse in the clockwise direction (see Fig. 9.5). However, when the wavenumbers traverse in the counterclockwise direction, as shown in Fig. 9.6, we choose

$$\hat{n} = \frac{\mathbf{p}\times\mathbf{q}}{|\mathbf{p}\times\mathbf{q}|} \tag{9.28}$$

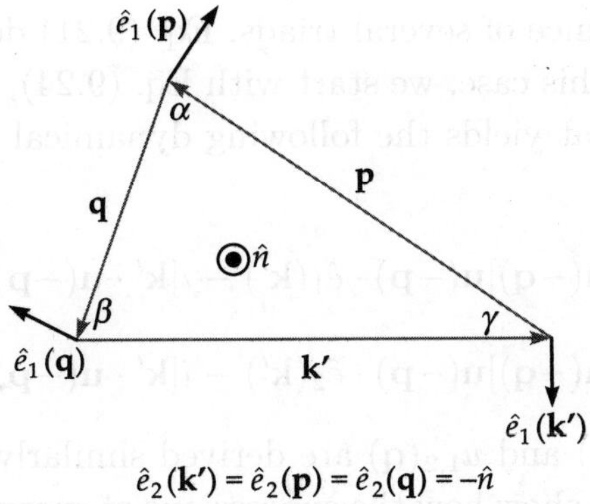

Figure 9.6 Craya–Herring basis vectors when the wavenumbers traverse in the counterclockwise direction. Here, $\hat{n} = \mathbf{p} \times \mathbf{q}/|\mathbf{p} \times \mathbf{q}|$.

so that $\hat{e}_1$s still point toward the right of the respective wavenumber. The basis vectors are shown in Fig. 9.6.

Under this scheme, $\hat{e}_1(\mathbf{X}) \cdot \hat{e}_1(\mathbf{Y})$ are the same as before, but $\hat{e}_3(\mathbf{X}) \cdot \hat{e}_1(\mathbf{Y})$ have opposite signs. For example,

$$\hat{e}_3(\mathbf{k}') \cdot \hat{e}_1(\mathbf{p}) \quad = \quad \hat{k'} \cdot (\hat{p} \times \hat{n}) = (\hat{k'} \times \hat{p}) \cdot \hat{n} = \sin\gamma. \tag{9.29}$$

See Table 9.2 for the multiplication table when the wavenumbers traverse in counterclockwise direction, as shown in Fig. 9.6.

Table 9.2 Product rules for the Craya–Herring basis vectors when the wavenumbers traverse in the counterclockwise direction (See Fig. 9.6).

$\cdot$	$\hat{e}_1(\mathbf{k}')$	$\hat{e}_1(\mathbf{p})$	$\hat{e}_1(\mathbf{q})$
$\hat{e}_3(\mathbf{k}')$	0	$\sin\gamma$	$-\sin\beta$
$\hat{e}_3(\mathbf{p})$	$-\sin\gamma$	0	$\sin\alpha$
$\hat{e}_3(\mathbf{q})$	$\sin\beta$	$-\sin\alpha$	0
$\hat{e}_1(\mathbf{k}')$	1	$-\cos\gamma$	$-\cos\beta$
$\hat{e}_1(\mathbf{p})$	$-\cos\gamma$	1	$\cos\alpha$
$\hat{e}_1(\mathbf{q})$	$-\cos\beta$	$-\cos\alpha$	1

Due to the change in sign of $\hat{e}_3(\mathbf{X})\cdot\hat{e}_1(\mathbf{Y})$, Eqs. (9.25, 9.26, 9.27) get an additional negative sign, for example,

$$ik'\sin(\beta - \gamma) \to -ik'\sin(\beta - \gamma); \quad ik'\sin\gamma \to -ik'\sin\gamma. \tag{9.30}$$

The above forms of equations are very useful. However, sometimes it is also useful to construct equations of motion using a specific $\hat{n}$, not that of Eq. (9.21) or Eq. (9.28).

For example, in the presence of several triads, Eq. (9.21) does not provide a unique $\hat{n}$ (see Example 9.5). In this case, we start with Eq. (9.24), and take its dot product with $\hat{e}_1(\mathbf{k}')$ and $\hat{e}_2(\mathbf{k}')$ that yields the following dynamical equations for $u_1(\mathbf{k}')$ and $u_2(\mathbf{k}')$:

$$\frac{d}{dt}u_1(\mathbf{k}') = -i[\mathbf{k}'\cdot\mathbf{u}(-\mathbf{q})]\mathbf{u}(-\mathbf{p})\cdot\hat{e}_1(\mathbf{k}') - i[\mathbf{k}'\cdot\mathbf{u}(-\mathbf{p})]\mathbf{u}(-\mathbf{q})\cdot\hat{e}_1(\mathbf{k}'), \tag{9.31a}$$

$$\frac{d}{dt}u_2(\mathbf{k}') = -i[\mathbf{k}'\cdot\mathbf{u}(-\mathbf{q})]\mathbf{u}(-\mathbf{p})\cdot\hat{e}_2(\mathbf{k}') - i[\mathbf{k}'\cdot\mathbf{u}(-\mathbf{p})]\mathbf{u}(-\mathbf{q})\cdot\hat{e}_2(\mathbf{k}'). \tag{9.31b}$$

The equations for $u_{1,2}(\mathbf{p})$ and $u_{1,2}(\mathbf{q})$ are derived similarly.

In the next section, we show how the expressions of energy transfers get simplified in Craya–Herring basis.

9.3 Energy Transfer Functions in Craya–Herring Basis

The expression of the mode-to-mode energy transfer $S^{uu}(\mathbf{k}'|\mathbf{p}|\mathbf{q})$ in a triad takes a simpler form in Craya–Herring basis, and it helps in many calculations.

We consider a triad $(\mathbf{k}', \mathbf{p}, \mathbf{q})$ with a condition that $\mathbf{k}' + \mathbf{p} + \mathbf{q} = 0$. To derive the mode-to-mode transfers, it is convenient to work with the triangle of the previous section. For a triad in which the wavenumbers traverse in a clockwise direction, the formula for the mode-to-mode transfer is as follows:

$$\begin{aligned}
S^{uu}(\mathbf{k}'|\mathbf{p}|\mathbf{q}) &= -\Im\left[\{\mathbf{k}'\cdot\mathbf{u}(\mathbf{q})\}\{\mathbf{u}(\mathbf{p})\cdot\mathbf{u}(\mathbf{k}')\}\right] \\
&= -\Im\left[\{k'\hat{e}_3(\mathbf{k}')\cdot\hat{e}_1(\mathbf{q})u_1(\mathbf{q})\}\{\hat{e}_1(\mathbf{p})\cdot\hat{e}_1(\mathbf{k}')u_1(\mathbf{p})u_1(\mathbf{k}')\right. \\
&\qquad\qquad \left.+\,\hat{e}_2(\mathbf{p})\cdot\hat{e}_2(\mathbf{k}')u_2(\mathbf{p})u_2(\mathbf{k}')\}\right] \\
&= k'\sin\beta\cos\gamma\,\Im\{u_1(\mathbf{q})u_1(\mathbf{p})u_1(\mathbf{k}')\} - k'\sin\beta\,\Im\{u_1(\mathbf{q})u_2(\mathbf{p})u_2(\mathbf{k}')\} \\
&= S^{u_1u_1}(\mathbf{k}'|\mathbf{p}|\mathbf{q}) + S^{u_2u_2}(\mathbf{k}'|\mathbf{p}|\mathbf{q}),
\end{aligned} \tag{9.32}$$

where

$$S^{u_1u_1}(\mathbf{k}'|\mathbf{p}|\mathbf{q}) = k'\sin\beta\cos\gamma\,\Im\{u_1(\mathbf{q})u_1(\mathbf{p})u_1(\mathbf{k}')\}, \tag{9.33a}$$

$$S^{u_2u_2}(\mathbf{k}'|\mathbf{p}|\mathbf{q}) = -k'\sin\beta\,\Im\{u_1(\mathbf{q})u_2(\mathbf{p})u_2(\mathbf{k}')\}. \tag{9.33b}$$

The sign would be reversed if the wavenumbers traverse in the counterclockwise direction.

In Eq. (9.32), the first term $S^{u_1u_1}(\mathbf{k}'|\mathbf{p}|\mathbf{q})$ represents the mode-to-mode kinetic energy transfer from $u_1(\mathbf{p})$ to $u_1(\mathbf{k}')$ with $\mathbf{u}(\mathbf{q})$ acting as a mediator, while the second term $S^{u_2u_2}(\mathbf{k}'|\mathbf{p}|\mathbf{q})$ represents the corresponding transfer from $u_2(\mathbf{p})$ to $u_2(\mathbf{k}')$ with $\mathbf{u}(\mathbf{q})$ acting as a mediator. Note that the full velocity mode $\mathbf{u}(\mathbf{q})$, not the components, is the mediator for both these transfers. Also, there is no energy exchange between u_1 and u_2 modes.

Following the same steps as in Section 4.1, we can derive the following identities for the wavenumbers $\mathbf{X}, \mathbf{Y}, \mathbf{Z}$ of the triad:

$$S^{u_1u_1}(\mathbf{X}|\mathbf{Y}|\mathbf{Z}) = -S^{u_1u_1}(\mathbf{Y}|\mathbf{X}|\mathbf{Z}), \tag{9.34a}$$

$$S^{u_2u_2}(\mathbf{X}|\mathbf{Y}|\mathbf{Z}) = -S^{u_2u_2}(\mathbf{Y}|\mathbf{X}|\mathbf{Z}). \tag{9.34b}$$

Hence,

$$\sum_{\mathbf{X},\mathbf{Y}} S^{u_1u_1}(\mathbf{X}|\mathbf{Y}|\mathbf{Z}) = 0, \tag{9.35a}$$

$$\sum_{\mathbf{X},\mathbf{Y}} S^{u_2u_2}(\mathbf{X}|\mathbf{Y}|\mathbf{Z}) = 0. \tag{9.35b}$$

These identities, which are based on energetics considerations, yields the following detailed energy conservation for each component of the velocity field in a triad:

$$\frac{d}{dt}\sum_{\mathbf{X}} \frac{1}{2}|\mathbf{u}_1(\mathbf{X})|^2 = 0; \quad \frac{d}{dt}\sum_{\mathbf{X}} \frac{1}{2}|\mathbf{u}_2(\mathbf{X})|^2 = 0. \tag{9.36}$$

Following the similar procedure as earlier, we derive

$$S^{u_1u_1}(\mathbf{p}|\mathbf{q}|\mathbf{k}') = p\sin\gamma\cos\alpha\Im\{u_1(\mathbf{q})u_1(\mathbf{p})u_1(\mathbf{k}')\}, \tag{9.37a}$$

$$S^{u_2u_2}(\mathbf{p}|\mathbf{q}|\mathbf{k}') = -p\sin\gamma\Im\{u_1(\mathbf{k}')u_2(\mathbf{p})u_2(\mathbf{q})\}, \tag{9.37b}$$

$$S^{u_1u_1}(\mathbf{q}|\mathbf{k}'|\mathbf{p}) = q\sin\alpha\cos\beta\Im\{u_1(\mathbf{q})u_1(\mathbf{p})u_1(\mathbf{k}')\}, \tag{9.37c}$$

$$S^{u_2u_2}(\mathbf{q}|\mathbf{k}'|\mathbf{p}) = -q\sin\alpha\Im\{u_1(\mathbf{p})u_2(\mathbf{q})u_2(\mathbf{k}')\}. \tag{9.37d}$$

A recipe to construct the above formulas is as follows: $S^{u_1u_1}$ have $\Im\{u_1(\mathbf{q}) u_1(\mathbf{p})u_1(\mathbf{k}')\}$ as a common factor. The remaining parts of the factor are the wavenumber magnitude and sin / cos of the two angles at the two sides of the receiver wavenumber. The angle at the head of the receiver wavenumber arrow is the argument of the cos function, while that at the tail of the wavenumber is the argument of the sin function.

The energy transfer formula $S^{u_2u_2}$ has a factor which is an imaginary part of the product of the u_1 component of the mediator mode with u_2 components of the receiver and giver modes. The other factors is a product of the magnitude of the receiver wavenumber with sin function of the angle at the tail of the receiver wavenumber.

Following similar arguments as earlier, we derive the combined energy transfer to the mode $\mathbf{u}(\mathbf{k}')$ from the modes $\mathbf{u}(\mathbf{p})$ and $\mathbf{u}(\mathbf{q})$ as

$$\begin{aligned} S^{u_1u_1}(\mathbf{k}'|\mathbf{p},\mathbf{q}) &= S^{u_1u_1}(\mathbf{k}'|\mathbf{p}|\mathbf{q}) + S^{u_1u_1}(\mathbf{k}'|\mathbf{q}|\mathbf{p}) \\ &= k'\sin(\gamma-\beta)\Im\{u_1(\mathbf{q})u_1(\mathbf{p})u_1(\mathbf{k}')\} \end{aligned} \tag{9.38}$$

and

$$\begin{aligned} S^{u_2u_2}(\mathbf{k}'|\mathbf{p},\mathbf{q}) &= S^{u_2u_2}(\mathbf{k}'|\mathbf{p}|\mathbf{q}) + S^{u_2u_2}(\mathbf{k}'|\mathbf{q}|\mathbf{p}) \\ &= +k'\sin\gamma\Im\{u_1(\mathbf{q})u_2(\mathbf{p})u_2(\mathbf{k}')\} - k'\sin\beta\Im\{u_1(\mathbf{p})u_2(\mathbf{q})u_2(\mathbf{k}')\} \end{aligned} \tag{9.39}$$

In 2D flows, $u_2 = 0$, hence the terms involving u_2 vanish automatically. Therefore,

$$S^{uu}(\mathbf{k}'|\mathbf{p}|\mathbf{q}) = S^{u_1u_1}(\mathbf{k}'|\mathbf{p}|\mathbf{q}) = k'\sin\beta\cos\gamma\Im\{u_1(\mathbf{q})u_1(\mathbf{p})u_1(\mathbf{k}')\}, \tag{9.40a}$$

$$S^{uu}(\mathbf{k}'|\mathbf{p},\mathbf{q}) = S^{u_1u_1}(\mathbf{k}'|\mathbf{p},\mathbf{q}) = k'\sin(\gamma-\beta)\Im\{u_1(\mathbf{q})u_1(\mathbf{p})u_1(\mathbf{k}')\}. \tag{9.40b}$$

For such flows, in a triad with $p = q$, $\beta = \gamma$, hence $S^{uu}(\mathbf{k}'|\mathbf{p},\mathbf{q}) = 0$. Thus, the aforementioned formulas provide intuitive understanding of the energy transfers.

Following similar procedure as before, we can derive expressions for the mode-to-mode enstrophy transfers. In 3D, the enstrophy transfer from $\omega(\mathbf{p})$ to $\omega(\mathbf{k}')$ with $\mathbf{u}(\mathbf{q})$ acting as a mediator is

$$\begin{aligned} S^{\omega\omega}(\mathbf{k}'|\mathbf{p}|\mathbf{q}) &= -\Im\left[\{\mathbf{k}'\cdot\mathbf{u}(\mathbf{q})\}\{\boldsymbol{\omega}(\mathbf{p})\cdot\boldsymbol{\omega}(\mathbf{k}')\}\right] \\ &= k'^2 p\sin\beta\Im\{u_1(\mathbf{q})u_1(\mathbf{p})u_1(\mathbf{k}')\} \\ &\quad -k'^2 p\sin\beta\cos\gamma\Im\{u_1(\mathbf{q})u_2(\mathbf{p})u_2(\mathbf{k}')\} \\ &= S^{\omega_1\omega_1}(\mathbf{k}'|\mathbf{p}|\mathbf{q}) + S^{\omega_2\omega_2}(\mathbf{k}'|\mathbf{p}|\mathbf{q}), \end{aligned} \tag{9.41}$$

where the first term $S^{\omega_1\omega_1}(\mathbf{k}'|\mathbf{p}|\mathbf{q})$ has contribution only from u_1 modes, while the second term $S^{\omega_2\omega_2}(\mathbf{k}'|\mathbf{p}|\mathbf{q})$ transfers via $u_2(\mathbf{p})$ to $u_2(\mathbf{k}')$. In 2D, $u_2 = 0$. Therefore,

$$S^{\omega\omega}(\mathbf{k}'|\mathbf{p}|\mathbf{q}) = S^{\omega_1\omega_1}(\mathbf{k}'|\mathbf{p}|\mathbf{q}) = k'^2 p\sin\gamma\Im\{u_1(\mathbf{q})u_1(\mathbf{p})u_1(\mathbf{k}')\}. \tag{9.42}$$

Hence,

$$\frac{S^{uu}(\mathbf{k}'|\mathbf{p}|\mathbf{q})}{S^{\omega\omega}(\mathbf{k}'|\mathbf{p}|\mathbf{q})} = \frac{\cos\gamma}{kp}. \tag{9.43}$$

The formula for the enstrophy transfer from $\mathbf{u}(\mathbf{p})$ to $\omega(\mathbf{k}')$ with $\omega(\mathbf{q})$ acting as a mediator is

$$\begin{aligned} S^{\omega u}(\mathbf{k}'|\mathbf{p}|\mathbf{q}) &= \Im\left[\{\mathbf{k}'\cdot\boldsymbol{\omega}(\mathbf{q})\}\{\mathbf{u}(\mathbf{p})\cdot\boldsymbol{\omega}(\mathbf{k}')\}\right] = k'^2 q\sin\beta\cos\gamma\Im\{u_2(\mathbf{q})u_1(\mathbf{p})u_2(\mathbf{k}')\} \\ &\quad +k'^2 q\sin\beta\Im\{u_2(\mathbf{q})u_2(\mathbf{p})u_1(\mathbf{k}')\}. \end{aligned} \tag{9.44}$$

In 2D, $u_2 = 0$. Therefore,

$$S^{\omega u}(\mathbf{k}'|\mathbf{p}|\mathbf{q}) = 0, \tag{9.45}$$

Similarly, using Eq. (8.10), we rederive the formula for the mode-to-mode kinetic helicity transfer as

$$\begin{aligned} S^{H_K}(\mathbf{k}'|\mathbf{p}|\mathbf{q}) &= \Re\left[\mathbf{u}(\mathbf{q}) \cdot \boldsymbol{\omega}(\mathbf{p}) \times \boldsymbol{\omega}(\mathbf{k}')\}\right] \\ &= -k'p[\sin\alpha\Re\{u_1(\mathbf{q})u_2(\mathbf{p})u_1(\mathbf{k}')\} + \sin\beta\Re\{u_1(\mathbf{q})u_1(\mathbf{p})u_2(\mathbf{k}')\}] \\ &\quad +k'p\sin\gamma\Re\{u_2(\mathbf{q})u_2(\mathbf{p})u_2(\mathbf{k}')\}. \end{aligned} \tag{9.46}$$

Note that the transfers $S^{\omega u}$ and S^{H_K} vanish when $u_2 = 0$.

These derivations demonstrate the usefulness of Craya–Herring basis toward computation of various transfers and derivation of truncated nonlinear equations. In the next section we will use these formulas to derive energy fluxes in Craya–Herring basis.

9.4 Fluxes in Craya–Herring Basis

A general flow involves many triads. We can generalize the energy transfers discussed in the previous section to many triads. Since energy is an additive quantity, we can compute the energy transferred to $u_1(\mathbf{k})$ from $u_1(\mathbf{p})$'s with $\mathbf{p} \in B$, where B is a Fourier space region, as

$$\sum_{\mathbf{p}\in B} S^{u_1 u_1}(\mathbf{k}'|\mathbf{p}|\mathbf{q}) = \sum_{\mathbf{p}\in B} k' \sin\beta\cos\gamma\Im\{u_1(\mathbf{q})u_1(\mathbf{p})u_1(\mathbf{k}')\}. \tag{9.47}$$

Similarly, we can compute the energy transferred to $u_2(\mathbf{k})$ from a set of $u_2(\mathbf{p})$ modes. An important thing to keep in mind is that the components u_1's and u_2's would need to be computed for each triad individually. This is because the unit vectors $\hat{e}_i$'s change for each triad.

Using this recipe, we can compute the energy fluxes $\Pi_{u_1}(k_0)$ and $\Pi_{u_2}(k_0)$ for u_1 and u_2 fields separately. They are as follows:

$$\Pi_{u_1}(k_0) = \sum_{|\mathbf{p}|\le k_0} \sum_{|\mathbf{k}|>k_0} S^{u_1 u_1}(\mathbf{k}|\mathbf{p}|\mathbf{q}), \tag{9.48a}$$

$$\Pi_{u_2}(k_0) = \sum_{|\mathbf{p}|\le k_0} \sum_{|\mathbf{k}|>k_0} S^{u_2 u_2}(\mathbf{k}|\mathbf{p}|\mathbf{q}). \tag{9.48b}$$

Similarly, the shell-to-shell energy transfers for u_1 and u_2 fields from shell m to shell n can be defined as

$$T^{u_1,m}_{u_1,n} = \sum_{\mathbf{p}\in m}\sum_{\mathbf{k}\in n} S^{u_1u_1}(\mathbf{k}|\mathbf{p}|\mathbf{q}), \tag{9.49a}$$

$$T^{u_2,m}_{u_2,n} = \sum_{\mathbf{p}\in m}\sum_{\mathbf{k}\in n} S^{u_2u_2}(\mathbf{k}|\mathbf{p}|\mathbf{q}). \tag{9.49b}$$

Similar formulas can be written down for the enstrophy transfers.

Example 9.1: Consider the 2D fluid flows with three components:

$$\begin{aligned}
(a): \quad \mathbf{u} &= 2(-\sin y - \sin(x+y)/\sqrt{2}, \sin x + \sin(x+y)/\sqrt{2}, \\
&\quad \cos x + \cos y + \cos(x+y)). \\
(b): \quad \mathbf{u} &= 2(-\sin y - \sin(x+y)/\sqrt{2}, \sin x + \sin(x+y)/\sqrt{2}, \\
&\quad -\cos x - \cos y - \cos(x+y)). \\
(c): \quad \mathbf{u} &= 2(-\sin y - \sin(x+y)/\sqrt{2}, \sin x + \sin(x+y)/\sqrt{2}, \\
&\quad \cos x + \cos y - \cos(x+y)).
\end{aligned}$$

Compute the Craya–Herring basis vectors and the components along the basis vectors. Also compute the mode-to-mode kinetic energy, enstrophy, and kinetic helicity transfers.

Solution: We compute the Craya–Herring basis vectors using Eqs. (9.1). Here we choose $\hat{n} = \hat{z}$. The associated components are computed using

$$u_1(\mathbf{k}') = \mathbf{u}(\mathbf{k}') \cdot \hat{e}_1(\mathbf{k}'); \qquad u_2(\mathbf{k}') = \mathbf{u}(\mathbf{k}') \cdot \hat{e}_2(\mathbf{k}').$$

For the flow fields of (a,b,c), the Craya–Herring basis vectors and the components along these vectors are listed in Table 9.3. Note that $\hat{e}_2(\mathbf{k}) = -\hat{z}$ for all the modes.

The table also lists the modal KE and modal kinetic helicity, which are

$$\begin{aligned}
E_u(\mathbf{k}) &= \frac{1}{2}(|u_1(\mathbf{k}|^2 + |u_2(\mathbf{k}|^2), \\
H_K(\mathbf{k}) &= k\Im[u_1^*(\mathbf{k})u_2(\mathbf{k})].
\end{aligned}$$

A careful observations of the Fourier modes indicate the following two interacting triads for all the three cases:

$$\begin{aligned}
(-1,-1) &= -(0,1) \bigoplus -(1,0), \\
(1,1) &= -(0,-1) \bigoplus -(-1,0).
\end{aligned}$$

Table 9.3 Example 9.1: The Craya–Herring basis vectors, the velocity components, modal KE and kinetic helicity for the velocity fields (a,b,c).

Part	Mode	$\hat{e}_1(\mathbf{a})$	$\hat{e}_2(\mathbf{a})$	$(u_1(\mathbf{a}), u_2(\mathbf{a}))$	$E_u(\mathbf{a})$	$H_K(\mathbf{a})$
(a)	(1,1)	$(\hat{x}-\hat{y})/\sqrt{2}$	$-\hat{z}$	$(i,-1)$	1	1
	$(-1,-1)$	$(-\hat{x}+\hat{y})/\sqrt{2}$	$-\hat{z}$	$(i,-1)$	1	1
	(0,1)	$\hat{x}$	$-\hat{z}$	$(i,-1)$	1	1
	$(0,-1)$	$-\hat{x}$	$-\hat{z}$	$(i,-1)$	1	1
	(1,0)	$-\hat{y}$	$-\hat{z}$	$(i,-1)$	1	1
	$(-1,0)$	$\hat{y}$	$-\hat{z}$	$(i,-1)$	1	1
(b)	(1,1)	$(\hat{x}-\hat{y})/\sqrt{2}$	$-\hat{z}$	$(i,1)$	1	-1
	$(-1,-1)$	$(-\hat{x}+\hat{y})/\sqrt{2}$	$-\hat{z}$	$(i,1)$	1	-1
	(0,1)	$\hat{x}$	$-\hat{z}$	$(i,1)$	1	-1
	$(0,-1)$	$-\hat{x}$	$-\hat{z}$	$(i,1)$	1	-1
	(1,0)	$-\hat{y}$	$-\hat{z}$	$(i,1)$	1	-1
	$(-1,0)$	$\hat{y}$	$-\hat{z}$	$(i,1)$	1	-1
(c)	(1,1)	$(\hat{x}-\hat{y})/\sqrt{2}$	$-\hat{z}$	$(i,1)$	1	-1
	$(-1,-1)$	$(-\hat{x}+\hat{y})/\sqrt{2}$	$-\hat{z}$	$(i,1)$	1	-1
	(0,1)	$\hat{x}$	$-\hat{z}$	$(i,-1)$	1	1
	$(0,-1)$	$-\hat{x}$	$-\hat{z}$	$(i,-1)$	1	1
	(1,0)	$-\hat{y}$	$-\hat{z}$	$(i,-1)$	1	1
	$(-1,0)$	$\hat{y}$	$-\hat{z}$	$(i,-1)$	1	1

Table 9.4 Example 9.1: The mode-to-mode transfers of kinetic helicity transfer in a triad with wavenumbers $\mathbf{k}' = (-1,-1)$, $\mathbf{q} = (0,1)$ and $\mathbf{p} = (1,0)$.

Part	Transfers	S^{H_K}
(a)	$S(\mathbf{k}'\|\mathbf{p}\|\mathbf{q})$	$-2-\sqrt{2}$
	$S(\mathbf{p}\|\mathbf{q}\|\mathbf{k}')$	$-1-\sqrt{2}$
	$S(\mathbf{q}\|\mathbf{k}'\|\mathbf{p})$	$-2-\sqrt{2}$
(b)	$S(\mathbf{k}'\|\mathbf{p}\|\mathbf{q})$	$2+\sqrt{2}$
	$S(\mathbf{p}\|\mathbf{q}\|\mathbf{k}')$	$1+\sqrt{2}$
	$S(\mathbf{q}\|\mathbf{k}'\|\mathbf{p})$	$2+\sqrt{2}$
(c)	$S(\mathbf{k}'\|\mathbf{p}\|\mathbf{q})$	$2-\sqrt{2}$
	$S(\mathbf{p}\|\mathbf{q}\|\mathbf{k}')$	$1-\sqrt{2}$
	$S(\mathbf{q}\|\mathbf{k}'\|\mathbf{p})$	$2-\sqrt{2}$

We assume $\mathbf{k}' = (-1,-1)$, $\mathbf{q} = (0,1)$, and $\mathbf{p} = (1,0)$, and compute the mode-to-mode transfers of KE, enstrophy, and kinetic helicity using the formulas described in this section. The KE and enstrophy transfers computed using the Craya–Herring basis are exactly the same as those computed in Example 6.1. The only extra computation is for the kinetic helicity, which is listed in Table 9.4.

Example 9.2: Consider the following flow field:

$$\mathbf{u} = \hat{x}2B\cos y + \hat{y}2C\cos x + (\hat{x} - \hat{y})2A\sin(x+y).$$

Compute the components of the velocity field in Craya–Herring basis. Also derive equations of motion for A, B, and C using Craya–Herring decomposition.

Solution: The velocity field is composed with wavenumbers $(1,0)$, $(0,1)$, $(1,1)$, $(-1,0)$, $(0,-1)$, and $(-1,-1)$. We choose $\hat{n} = \hat{z}$. Using the entries of Table 3.4 and $u_1(-\mathbf{k}) = -u_1^*(\mathbf{k})$, $\hat{e}_1$, we deduce the u_1 component of these modes as $-C$, B, $A\sqrt{2}/i$, C, $-B$, $-A\sqrt{2}/i$ respectively. We consider a wavenumber triad $\{(1,0),(0,1),(-1,-1)\}$ as shown in Fig. 9.7(a). The equations of motion for the Fourier modes of the triad are

$$\begin{aligned}
\dot{u}_1(-1,-1) &= i\sqrt{2}\sin(45-45)u_1^*(1,0)u_1^*(0,1) = 0, \\
\dot{u}_1(0,1) &= i\sin(45-90)u_1^*(-1,-1)u_1^*(1,0), \\
\dot{u}_1(1,0) &= i\sin(90-45)u_1^*(-1,-1)u_1^*(0,1).
\end{aligned}$$

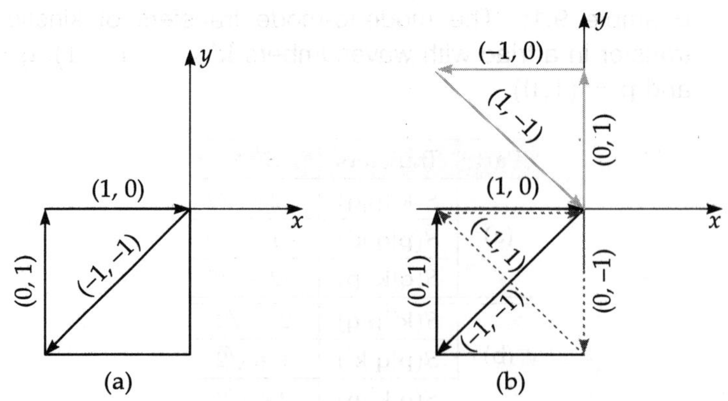

Figure 9.7 (a) Example 9.2: The interacting wavenumber triads of the flow. (b) Example 9.4: The three interacting wavenumber triads are in black, gray, and dotted lines.

Substitution of the amplitudes of the above modes in the aforementioned equations yields

$$\begin{aligned}\dot{A} &= 0,\\ \dot{B} &= -AC,\\ \dot{C} &= AB,\end{aligned}$$

which are the same as those derived in Example 3.4. Note however that the present derivation is much simpler than that of Example 3.4.

Example 9.3: How does the result in Example 9.2 change if the velocity field under consideration is

$$\mathbf{u} = \hat{x}2B\cos y + \hat{y}2C\cos x + (\hat{x} - \hat{y})2A\cos(x + y)?$$

Solution: We follow the same procedure as in Example 9.2. Note however that the Carya-Herring amplitudes of the wavenumbers $(1, 1)$ and $(-1, -1)$ are $A\sqrt{2}$ and $-A\sqrt{2}$ respectively. When we substitute them in the equations of motion, we obtain the following equations:

$$\begin{aligned}\dot{A} &= 0,\\ \dot{B} &= -iAC,\\ \dot{C} &= iAB.\end{aligned}$$

Since A, B, C are real, we obtain $\dot{A} = \dot{B} = \dot{C} = 0$, which are the same as those of Example 3.5.

Example 9.4: Consider the flow field

$$\mathbf{u} = \hat{x}2B\cos y + \hat{y}2C\cos x + 4A(\hat{x}\sin x\cos y - \hat{y}\cos x\sin y).$$

Perform the same analysis as in Example 9.2.

Solution: This flow field contains Fourier modes with wavenumbers $(0, \pm1)$, $(\pm1, 0)$, and $(\pm1, \pm1)$. When $\hat{n} = \hat{z}$, the amplitudes of the u_1 component of the four modes are

$$\begin{aligned}u_1(1,0) &= -C,\\ u_1(0,1) &= B,\\ u_1(1,1) &= \sqrt{2}A/i\\ u_1(1,-1) &= -\sqrt{2}A/i.\end{aligned}$$

The remaining modes are obtained using the relation $u_1(-\mathbf{k}) = -u_1^*(\mathbf{k})$. Note that $u_2(\mathbf{k}) = 0$.

The wavenumbers and the three interacting triads are shown in Fig. 9.7(b) using which we deduce that

$$
\begin{aligned}
(-1,-1) &= -(1,0) \bigoplus -(0,1), \\
(1,-1) &= -(-1,0) \bigoplus -(0,1), \\
(0,1) &= [-(1,0) \bigoplus -(-1,-1)] + [-(-1,0) \bigoplus -(1,-1)], \\
(1,0) &= [-(0,1) \bigoplus -(-1,-1)] + [-(0,-1) \bigoplus -(-1,1)],
\end{aligned}
$$

where $\bigoplus$ represents the nonlinear interaction. It can be shown that the modes $u(1,0)$, $u(0,1)$, $u(1,1)$, $u(1,-1)$ are generated by the aforementioned nonlinear interactions.[2] Therefore, following Eqs. (4.37, 9.26), we deduce that

$$
\begin{aligned}
\dot{u}_1(-1,-1) &= i\sqrt{2}\sin(45-45)u_1^*(1,0)u_1^*(0,1) = 0, \\
\dot{u}_1(1,-1) &= i\sqrt{2}\sin(45-45)u_1^*(-1,0)u_1^*(0,1) = 0, \\
\dot{u}_1(0,1) &= i\sin(45-90)u_1^*(-1,-1)u_1^*(1,0) - i\sin(45-90)u_1^*(1,-1)u_1^*(-1,0), \\
\dot{u}_1(1,0) &= i\sin(90-45)u_1^*(-1,-1)u_1^*(0,1) + i\sin(45-90)u_1^*(-1,1)u_1^*(0,-1).
\end{aligned}
$$

The negative sign in the second term of $\dot{u}_1(0,1)$ is due to the counterclockwise sense of traversal of the wavenumbers in the triad $\{(0,1),(-1,0),(1,-1)\}$.

From these equations, we immediately deduce that

$$
\begin{aligned}
u_1(1,1) &= \text{const} = \sqrt{2}A/i, \\
u_1(1,-1) &= \text{const} = -\sqrt{2}A/i.
\end{aligned}
$$

Hence,

$$\dot{A} = 0.$$

[2] If $k_x, k_y = \pm 1$, then the nonlinear interactions occur only through the following triads:

$$
\begin{aligned}
(k_x,k_y) &= (k_x,0) \bigoplus (0,k_y), \\
(k_x,0) &= [(k_x,k_y) \bigoplus (0,-k_y)] + [(k_x,-k_y) \bigoplus (0,k_y)], \\
(0,k_y) &= [(k_x,k_y) \bigoplus (-k_x,0)] + [(-k_x,k_y) \bigoplus (k_x,0)].
\end{aligned}
$$

Substitution of the amplitudes of the Fourier modes in the other two equations yields

$$\begin{aligned}\dot{B} &= 0,\\ \dot{C} &= 0,\end{aligned}$$

which implies that B and C are constants. Hence, in the absence of viscosity, the flow is time independent. When we contrast the present example with Example 9.2, we find that inclusion of an additional triad has led to cancelation of the nonlinear interactions. Hence, having more Fourier modes does not necessarily imply stronger nonlinear interactions. Such kind of interactions may be present in the square patterns of thermal convection.

We compute the energy transfers to each of the Fourier modes using the formulas for the mode-to-mode energy transfers, and observe them to be zero for all the modes. We leave this computation as an exercise. These energy transfers are consistent with the constancy of $A, B,$ and C.

Example 9.5: Consider the field configuration of Example 3.6. Using $\hat{n} = \hat{z}$, compute the Craya–Herring basis vectors and the components along the basis vectors. Derive the nonlinear terms and equations of motion.

Solution: We choose $\hat{n} = \hat{z}$ and compute the Craya–Herring basis vectors using the formulas of Eqs. (9.1). The associated components are computed using

$$u_1(\mathbf{k}') = \mathbf{u}(\mathbf{k}') \cdot \hat{e}_1(\mathbf{k}'); \quad u_2(\mathbf{k}') = \mathbf{u}(\mathbf{k}') \cdot \hat{e}_2(\mathbf{k}').$$

These quantities are listed in Table 9.5. The righthand side of Eqs. (9.31a, 9.31b) are the components $-N_{u,1}, -N_{u,2}$ of the nonlinear term. The computed values are also listed in the table.

As illustrated in Example 3.6, the nonlinear interactions have the following triads:

$$\begin{aligned}(-1,-1,-2) &= -(1,0,1) \bigoplus -(0,1,1),\\ (1,0,1) &= [-(0,1,1) \bigoplus -(-1,-1,-2)] + [-(0,-1,1) \bigoplus -(-1,1,-2)],\\ (0,1,1) &= [-(1,0,1) \bigoplus -(-1,-1,-2)] + [-(-1,0,1) \bigoplus -(1,-1,-2)],\end{aligned}$$

where $\bigoplus$ represents the nonlinear interaction. From the entries of the nonlinear term of the table, we deduce that

$$\begin{aligned}\dot{u}_1(-1,-1,-2) &= 0,\\ \dot{u}_2(-1,-1,-2) &= i\frac{4}{\sqrt{3}}BC,\end{aligned}$$

$$\begin{aligned}
\dot{u}_1(1,0,1) &= (i-i) = 0, \\
\dot{u}_2(1,0,1) &= -i2\sqrt{2}AB, \\
\dot{u}_1(0,1,1) &= (i-i) = 0, \\
\dot{u}_2(0,1,1) &= -i2\sqrt{2}AC.
\end{aligned}$$

Table 9.5 Example 9.5: The Craya–Herring basis vectors $\hat{e}_{1,2}$, the components of the velocity (u_1, u_2), and negative of the nonlinear term $(N_{u,1}, N_{u,2})$. The three sets in the table are for three triads.

Mode	e_1	e_2	(u_1, u_2)	$-(N_{u,1}, N_{u,2})$
$(-1,-1,-2)$	$(-1/\sqrt{2}, 1/\sqrt{2}, 0)$	$(1/\sqrt{3}, 1/\sqrt{3}, -1/\sqrt{3})$	$(0, -iA\sqrt{3})$	$(0, i(4/\sqrt{3})BC)$
(1,0,1)	$(0,-1,0)$	$(1/\sqrt{2}, 0, -1/\sqrt{2})$	$(0, -iC\sqrt{2})$	$(i, -i\sqrt{2}AB)$
(0,1,1)	$(1,0,0)$	$(0, 1/\sqrt{2}, -1/\sqrt{2})$	$(0, -iB\sqrt{2})$	$(-i, -i\sqrt{2}AC)$
$(-1,1,-2)$	$(1/\sqrt{2}, 1/\sqrt{2}, 0)$	$(1/\sqrt{3}, -1/\sqrt{3}, -1/\sqrt{3})$	$(0, -iA\sqrt{3})$	$(0, i(4/\sqrt{3})BC)$
(1,0,1)	$(0,-1,0)$	$(1/\sqrt{2}, 0, -1/\sqrt{2})$	$(0, -iC\sqrt{2})$	$(-i, -i\sqrt{2}AB)$
$(0,-1,1)$	$(-1,0,0)$	$(0, -1/\sqrt{2}, -1/\sqrt{2})$	$(0, -iB\sqrt{2})$	$(i, -i\sqrt{2}AC)$
$(1,-1,-2)$	$(1/\sqrt{2}, -1/\sqrt{2}, 0)$	$(-1/\sqrt{3}, 1/\sqrt{3}, -1/\sqrt{3})$	$(0, -iA\sqrt{3})$	$(0, i(4/\sqrt{3})BC)$
$(-1,0,1)$	$(0,1,0)$	$(-1/\sqrt{2}, 0, -1/\sqrt{2})$	$(0, -iC\sqrt{2})$	$(-i, -i\sqrt{2}AB)$
(0,1,1)	$(1,0,0)$	$(0, 1/\sqrt{2}, -1/\sqrt{2})$	$(0, -iB\sqrt{2})$	$(i, -i\sqrt{2}AC)$

Substitution of $u_2(-1,-1,-2) = -iA\sqrt{3}$, $u_2(1,0,1) = -iC\sqrt{2}$, and $u_2(0,1,1) = -iB\sqrt{2}$ in these equations yields

$$\begin{aligned}
\dot{A} &= -\frac{4}{3}BC, \\
\dot{B} &= 2AC, \\
\dot{C} &= 2AB,
\end{aligned}$$

which are identical to those derived in Example 3.6. The computations in Craya–Herring basis function are relatively more convenient because the pressure gets eliminated automatically in this basis.

Example 9.6: Consider the following flow field:

$$\mathbf{u} = 2C\sin(x+z)(\hat{x}-\hat{z}) + 2B\sin(y+z)(\hat{y}-\hat{z}) + 2A\sin(x+y+2z)(-\hat{x}-\hat{y}+\hat{z}).$$

By considering $\hat{n}$ perpendicular to the plane formed by the interacting wavenumbers, construct the Craya–Herring basis vectors and the components along the basis vectors. Derive the nonlinear terms and equations of motion. Compute the mode-to-mode kinetic energy, enstrophy, and kinetic helicity transfers.

Solution: The flow field is composed of Fourier modes with wavenumbers $\mathbf{k}' = (-1,-1,-2)$, $\mathbf{p} = (1,0,1)$, $\mathbf{q} = (0,1,1)$, and its negative counterparts. Note that these wavenumbers form a triad ($\mathbf{k}' + \mathbf{p} + \mathbf{q} = 0$). The wavenumbers $(1,0,1)$, $(0,1,1)$, and $(-1,-1,-2)$ form a plane whose normal is along

$$\hat{n} = \frac{\mathbf{q} \times \mathbf{p}}{|\mathbf{q} \times \mathbf{p}|} = (1/\sqrt{3}, 1/\sqrt{3}, -1/\sqrt{3}).$$

Now we construct the Craya–Herring basis vectors for a wavenumber $\mathbf{p}$ using

$$\hat{e}_3(\mathbf{p}) = \hat{p}; \quad \hat{e}_1(\mathbf{p}) = \frac{\hat{p} \times \hat{n}}{|\hat{p} \times \hat{n}|}, \quad \hat{e}_2(\mathbf{p}) = \hat{e}_3(\mathbf{p}) \times \hat{e}_1(\mathbf{p}).$$

The Craya–Herring basis vectors for other wavenumbers are computed similarly. Note that all the wavenumbers have

$$\hat{e}_2 = -\hat{n} = (1/\sqrt{3}, 1/\sqrt{3}, -1/\sqrt{3}).$$

We also compute the components along the basis vectors using

$$u_1(\mathbf{p}) = \mathbf{u}(\mathbf{p}) \cdot e_1(\mathbf{p}); \quad u_2(\mathbf{p}) = \mathbf{u}(\mathbf{p}) \cdot e_2(\mathbf{p}).$$

All these quantities are listed in Table 9.6. Note that the kinetic helicity of each of the modes is zero.

Table 9.6 Example 9.6: The Craya–Herring basis vectors $\hat{e}_{1,2}$, the components of the velocity field (u_1, u_2), and negative of the nonlinear term $(N_{u,1}, N_{u,2})$.

Mode	$\hat{e}_1$	(u_1, u_2)	$-(N_{u,1}, N_{u,2})$
$(-1,-1,-2)$	$(-1/\sqrt{2}, 1/\sqrt{2}, 0)$	$(0, -iA\sqrt{3})$	$(0, i(4/\sqrt{3})BC)$
(0,1,1)	$(\sqrt{6}/3, -\sqrt{6}/6, \sqrt{6}/6)$	$i(\sqrt{6}/3, -2/\sqrt{3})B$	$(0, -i\sqrt{3}AC)$
(1,0,1)	$(\sqrt{6}/6, -\sqrt{6}/3, -\sqrt{6}/6)$	$i(-\sqrt{6}/3, -2/\sqrt{3})C$	$(0, -i\sqrt{3}AB)$

We compute the nonlinear terms for each of the wavenumbers using Eqs. (9.25, 9.26, 9.27); these terms are also listed in the table. Combining these results, we obtain

$$\frac{d}{dt}u_1(-1,-1,-2) = \frac{d}{dt}u_1(0,1,1) = \frac{d}{dt}u_1(1,0,1) = 0.$$

Hence,

$$u_1(-1,-1,-2) = 0; \quad u_1(0,1,1) = i\frac{\sqrt{6}}{3}; \quad u_1(1,0,1) = i\frac{\sqrt{6}}{3}.$$

Using the results of the u_2 components, we obtain

$$\begin{aligned}\dot{A} &= -\frac{4}{3}BC,\\ \dot{B} &= \frac{3}{2}AC,\\ \dot{C} &= \frac{3}{2}AB.\end{aligned}$$

It is important to note that

$$\sum_{\mathbf{k}} \frac{1}{2}|\mathbf{u}_2(\mathbf{k})|^2 = \frac{1}{2}\left(3A^2 + \frac{4}{3}B^2 + \frac{4}{3}C^2\right),$$

and

$$\sum_{\mathbf{k}} \frac{1}{2}|\mathbf{u}_1(\mathbf{k})|^2 = \frac{1}{3}\left(B^2 + C^2\right)$$

are separately conserved. These conservation laws are described in Eq. (9.36).

Table 9.7 Example 9.6: The mode-to-mode transfers of KE, enstrophy, and H_K in a triad $\mathbf{k}' = (-1,-1,-2)$, $\mathbf{p} = (0,1,1)$, and $\mathbf{q} = (1,0,1)$. Note that $S^{\omega u} = (S^{\omega u}(\mathbf{X}|\mathbf{Y}|\mathbf{Z}), S^{\omega u}(\mathbf{X}|\mathbf{Z}|\mathbf{Y}))$.

Transfers/ABC	S^{uu}	$S^{\omega\omega}$	$S^{\omega u}$	S^{H_K}
$S(\mathbf{k}'|\mathbf{p}|\mathbf{q})$	-2	-6	$(-6,-6)$	0
$S(\mathbf{p}|\mathbf{q}|\mathbf{k}')$	0	0	$(-6,4)$	0
$S(\mathbf{q}|\mathbf{k}'|\mathbf{p})$	2	6	$(4,-6)$	0

We compute the mode-to-mode transfers among the Fourier modes. These transfers are listed in Table 9.7. It is interesting to note that S^{H_K} is zero for all the transfers. Also,

$$\sum_{\mathbf{a}}\sum_{\mathbf{b}} S^{\omega u}(\mathbf{a}|\mathbf{b}|\mathbf{c}) = -16ABC.$$

Thus, there is a net enstrophy transfer due to vortex stretching.

Another set of basis vectors, called helical basis vectors, are useful for representing helical fields. We describe them in the next section.

9.5 Helical Decomposition

The helical basis vectors are expressed using Craya–Herring basis vectors as (Waleffe, 1992; Sagaut and Cambon, 2008):

$$\hat{e}_{s_k}(\mathbf{k}) = \frac{1}{\sqrt{2}}[\hat{e}_2(\mathbf{k}) - is_k\hat{e}_1(\mathbf{k})], \tag{9.50}$$

where s_k takes values 1 and -1. Note that the vectors $\hat{e}_\pm(\mathbf{k})$ are complex unlike Craya–Herring basis vectors that are real, and

$$\hat{e}^*_\pm(\mathbf{k}) = \hat{e}_\mp(\mathbf{k}). \tag{9.51}$$

It is important to note that our normalization differs from that of Waleffe (1992).[3] Our basis vectors are normalized so as to yield a unit magnitude. The vectors $\hat{e}_\pm(\mathbf{k})$ have several interesting and queer properties: (Waleffe, 1992; Sagaut and Cambon, 2008):

$$\hat{e}_{s_k}(\mathbf{k}) \cdot \hat{e}_{s_k}(\mathbf{k}) = 0, \tag{9.52a}$$

$$\hat{e}_{s_k}(\mathbf{k}) \cdot \hat{e}^*_{s_k}(\mathbf{k}) = 1, \tag{9.52b}$$

$$\hat{e}_+(\mathbf{k}) \cdot \hat{e}_-(\mathbf{k}) = \hat{e}_-(\mathbf{k}) \cdot \hat{e}_+(\mathbf{k}) = 1, \tag{9.52c}$$

$$i\mathbf{k} \times \hat{e}_{s_k}(\mathbf{k}) = ik\hat{e}_3(\mathbf{k}) \times \frac{1}{\sqrt{2}}[\hat{e}_2(\mathbf{k}) - is_k\hat{e}_1(\mathbf{k})] = s_k k\hat{e}_{s_k}(\mathbf{k}). \tag{9.52d}$$

Under parity transformation, $\mathbf{k} \to -\mathbf{k}$. Using $\hat{e}_2(-\mathbf{k}) = \hat{e}_2(\mathbf{k})$ and $\hat{e}_1(-\mathbf{k}) = -\hat{e}_1(\mathbf{k})$, we derive that

$$\begin{aligned} \hat{e}_{s_k}(-\mathbf{k}) &= \frac{1}{\sqrt{2}}[\hat{e}_2(-\mathbf{k}) - is_k\hat{e}_1(-\mathbf{k})] \\ &= \frac{1}{\sqrt{2}}[\hat{e}_2(\mathbf{k}) + is_k\hat{e}_1(\mathbf{k})] \\ &= \hat{e}^*_{s_k}(\mathbf{k}). \end{aligned} \tag{9.53}$$

In helical basis, the velocity field is represented as

$$\mathbf{u}(\mathbf{k}) = u_+(\mathbf{k})\hat{e}_+(\mathbf{k}) + u_-(\mathbf{k})\hat{e}_-(\mathbf{k}). \tag{9.54}$$

[3] Waleffe's notation: $\hat{e}_{s_k}(\mathbf{k}) = \hat{e}_2(\mathbf{k}) - is_k\hat{e}_1(\mathbf{k})$, hence $\hat{e}_{s_k}(\mathbf{k}) \cdot \hat{e}^*_{s_k}(\mathbf{k}) = 2$.

A comparison of this equation with $\mathbf{u}(\mathbf{k}) = u_1(\mathbf{k})\hat{e}_1(\mathbf{k}) + u_2(\mathbf{k})\hat{e}_2(\mathbf{k})$ yields the following relationships between $u_\pm(\mathbf{k})$ and $u_{1,2}(\mathbf{k})$:

$$u_{s_k}(\mathbf{k}) = \frac{1}{\sqrt{2}}[u_2(\mathbf{k}) + is_k u_1(\mathbf{k})], \tag{9.55a}$$

$$u_1(\mathbf{k}) = -\frac{i}{\sqrt{2}}[u_+(\mathbf{k}) - u_-(\mathbf{k})], \tag{9.55b}$$

$$u_2(\mathbf{k}) = \frac{1}{\sqrt{2}}[u_+(\mathbf{k}) + u_-(\mathbf{k})]. \tag{9.55c}$$

The vorticity field is conveniently represented in helical basis as

$$\boldsymbol{\omega}(\mathbf{k}) = i\mathbf{k} \times \mathbf{u}(\mathbf{k}) = k[u_+(\mathbf{k})\hat{e}_+(\mathbf{k}) - u_-(\mathbf{k})\hat{e}_-(\mathbf{k})]. \tag{9.56}$$

Note that $\boldsymbol{\omega}(\mathbf{k})$ lies in the plane of $\hat{e}_1(\mathbf{k})$ and $\hat{e}_2(\mathbf{k})$, and it has no component along $\hat{e}_3(\mathbf{k})$. The modal kinetic energy and kinetic helicity are

$$E_u(\mathbf{k}) = \frac{1}{2}\mathbf{u}^*(\mathbf{k}) \cdot \mathbf{u}(\mathbf{k}) = \frac{1}{2}[|u_+(\mathbf{k})|^2 + |u_-(\mathbf{k})|^2], \tag{9.57a}$$

$$H_K(\mathbf{k}) = \frac{1}{2}\Re[\mathbf{u}^*(\mathbf{k}) \cdot \boldsymbol{\omega}(\mathbf{k})] = \frac{1}{2}k[|u_+(\mathbf{k})|^2 - |u_-(\mathbf{k})|^2]. \tag{9.57b}$$

Thus, $u_\pm(\mathbf{k})$ are the maximal helical mode with $H_K(\mathbf{k})/(kE(\mathbf{k})) = \pm 1$ respectively.

The helical modes have interesting properties in real space, which will be discussed in the next sections.

9.6 Helical Modes

In this section we describe pure and mixed helical modes.

9.6.1 The helical mode $\mathbf{u}_+$

In real space, the helical mode $\mathbf{u}_+$ appears as

$$\begin{aligned} \mathbf{u}_+(\mathbf{r}) = \begin{pmatrix} u_1(\mathbf{r}) \\ u_2(\mathbf{r}) \end{pmatrix} &= \Re[\hat{e}_+(\mathbf{k})u_+(\mathbf{k}) \exp i(\mathbf{k}\cdot\mathbf{r})] \\ &= \frac{1}{\sqrt{2}}|u_+(\mathbf{k})|\Re[(\hat{e}_2(\mathbf{k}) - i\hat{e}_1(\mathbf{k})) \exp i(kz' + \phi_{k+})] \\ &= \frac{1}{\sqrt{2}}|u_+(\mathbf{k})| \begin{pmatrix} \sin(kz' + \phi_{k+}) \\ \cos(kz' + \phi_{k+}) \end{pmatrix}, \end{aligned} \tag{9.58}$$

where $\Re[.]$ stands for the real part of the argument, $u_+(\mathbf{k}) = |u_+(\mathbf{k})| \exp(i\phi_{k+})$ with ϕ_{k+} as the phase of the mode, and z' is the coordinate along $\hat{e}_3(\mathbf{k})$. See Fig. 9.8 for an illustration of $\mathbf{u}_+(\mathbf{r})$. When we hold the z' axis in the *left hand*, the tip of the vector $\mathbf{u}_+$ turns along the fingers. Therefore, $\mathbf{u}_+(\mathbf{r})$ is said to be *left-handed.*

Pure $\mathbf{u}_+(\mathbf{k})$ mode in the Craya–Herring basis is

$$\frac{1}{\sqrt{2}}\begin{pmatrix}-i\\1\end{pmatrix} = \frac{1}{\sqrt{2}}\begin{pmatrix}\exp(-i\pi/2)\\1\end{pmatrix}. \tag{9.59}$$

Therefore, phase of u_2 is ahead of u_1 by $\pi/2$. This is also reflected in the real space description because in Eq. (9.58), $\cos(kz' + \phi_{k+}) = \sin(kz' + \phi_{k+} + \pi/2)$. Also,

$$|u_1(\mathbf{r})|^2 + |u_2(\mathbf{r})|^2 = \frac{1}{2}|u_+(\mathbf{k})|^2 = \text{const.} \tag{9.60}$$

Therefore, the mode $\mathbf{u}_+$ is said to be *circularly polarized.*

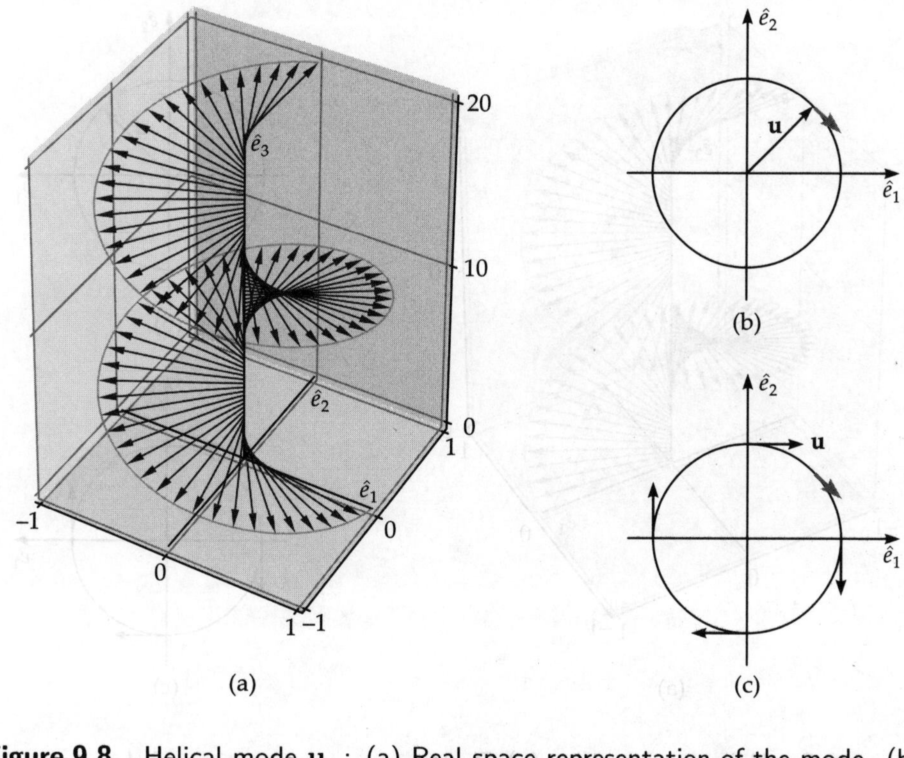

Figure 9.8 Helical mode $\mathbf{u}_+$: (a) Real space representation of the mode. (b,c) The tip of $\mathbf{u}_+(\mathbf{r})$ vector rotates in a clockwise direction as we traverse along z'.

9.6.2 The helical mode $\mathbf{u}_-$

In real space, the helical mode $\mathbf{u}_-$ appears as

$$\begin{aligned}
\mathbf{u}_-(\mathbf{r}) = \begin{pmatrix} u_1(\mathbf{r}) \\ u_2(\mathbf{r}) \end{pmatrix} &= \Re[\hat{e}_-(\mathbf{k}) u_-(\mathbf{k}) \exp i(\mathbf{k} \cdot \mathbf{r})] \\
&= \frac{1}{\sqrt{2}} |u_-(\mathbf{k})| \Re[(\hat{e}_2(\mathbf{k}) + i\hat{e}_1(\mathbf{k})) \exp i(kz' + \phi_{k-})] \\
&= \frac{1}{\sqrt{2}} |u_-(\mathbf{k})| \begin{pmatrix} -\sin(kz' + \phi_{k-}) \\ \cos(kz' + \phi_{k-}) \end{pmatrix}, \qquad (9.61)
\end{aligned}$$

where $u_-(\mathbf{k}) = |u_-(\mathbf{k})| \exp(i\phi_{k-})$. We illustrate $\mathbf{u}_-(\mathbf{r})$ in Fig. 9.9. When we hold the z' axis in the *right hand*, the tip of $\mathbf{u}_-(\mathbf{r})$ turns along the fingers. Therefore, $\mathbf{u}_-(\mathbf{r})$ is said to be *right-handed.* From the definition of the pure $\mathbf{u}_-$ mode, we deduce that the phase of u_1 is ahead of u_2 by $\pi/2$. In addition, $|u_1(\mathbf{r})|^2 + |u_2(\mathbf{r})|^2 = |u_-(\mathbf{k})|^2/2$, hence we conclude that $\mathbf{u}_-$ is *circularly polarized.*

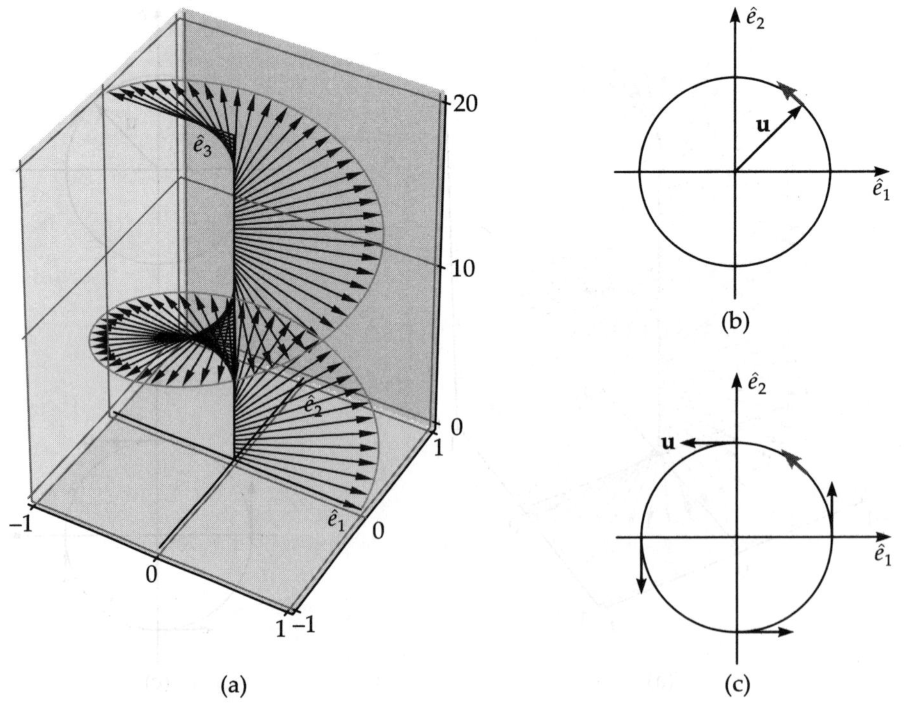

Figure 9.9 Helical mode $\mathbf{u}_-$: (a) Real space representation of the mode. (b, c) The tip of $\mathbf{u}_-(\mathbf{r})$ rotates in a counterclockwise direction as we traverse along z'.

9.6.3 Mixture of $\mathbf{u}_+$ and $\mathbf{u}_-$

A combination of $\mathbf{u}_+$ and $\mathbf{u}_-$ modes yields

$$\begin{aligned}\mathbf{u}(\mathbf{r}) &= \Re[\{\hat{e}_+(\mathbf{k})u_+(\mathbf{k}) + \hat{e}_-(\mathbf{k})u_-(\mathbf{k})\}\exp i(\mathbf{k}\cdot\mathbf{r})] \\ &= \Re[\{u_1(\mathbf{k})\hat{e}_1(\mathbf{k}) + u_2(\mathbf{k})\hat{e}_2(\mathbf{k})\}\exp i(kz')] \\ &= \Re[\{|u_1(\mathbf{k})|\exp(i\phi_1)\hat{e}_1 + |u_2(\mathbf{k})|\exp(i\phi_2)\hat{e}_2\}\exp i(kz')] \\ &= \begin{pmatrix} |u_1(\mathbf{k})|\cos(kz'+\phi_1) \\ |u_2(\mathbf{k})|\cos(kz'+\phi_2)\end{pmatrix}. \end{aligned} \tag{9.62}$$

Hence,

$$u_1(\mathbf{r}) = |u_1(\mathbf{k})|\cos(kz'+\phi_1) = |u_1(\mathbf{k})|\cos\Phi, \tag{9.63a}$$

$$u_2(\mathbf{r}) = |u_2(\mathbf{k})|\cos(kz'+\phi_2) = |u_2(\mathbf{k})|\cos(\Phi+\Delta\Phi), \tag{9.63b}$$

where $\Phi = kz'+\phi_1$ and $\Delta\Phi = \phi_2-\phi_1$. Elimination of Φ from these equations yields

$$\frac{u_1^2(\mathbf{r})}{|u_1(\mathbf{k})|^2} + \frac{u_2^2(\mathbf{r})}{|u_2(\mathbf{k})|^2} - \frac{2u_1(\mathbf{r})u_2(\mathbf{r})}{|u_1(\mathbf{k})||u_2(\mathbf{k})|}\cos(\phi_2-\phi_1) = \sin^2(\phi_2-\phi_1), \tag{9.64}$$

which is an equation of an ellipse as shown in Fig. 9.10(a). This is the reason why $\mathbf{u}(\mathbf{r})$ of Eq. (9.62) is said to be elliptic polarized.

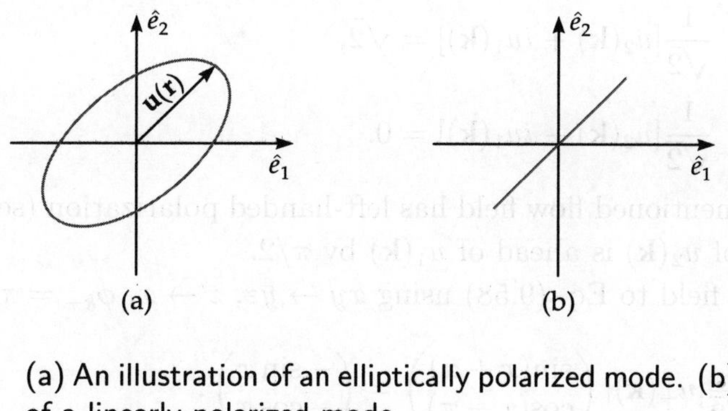

Figure 9.10 (a) An illustration of an elliptically polarized mode. (b) An illustration of a linearly polarized mode.

When $\phi_1 = \phi_2$, Eq. (9.64) yields

$$u_2(\mathbf{r}) = \frac{|u_2(\mathbf{k})|}{|u_1(\mathbf{k})|}u_1(\mathbf{r}) \tag{9.65}$$

indicating that $u_2(\mathbf{r})$ is proportional to $u_1(\mathbf{r})$; this configuration corresponds to a plane polarised wave (see Fig. 9.10(b)). When $\phi_2 - \phi_1 = \pm\pi/2$, they correspond to left-handed $\mathbf{u}_+$ and right-handed $\mathbf{u}_-$ fields respectively. For these cases,

$$u_1^2(\mathbf{r}) + u_2^2(\mathbf{r}) = |u_1(\mathbf{k})|^2 + |u_2(\mathbf{k})|^2 = \text{const.} \tag{9.66}$$

Hence, they are circularly polarized modes. The aforementioned linear and circular polarization are special cases of elliptic polarization.

Example 9.7: Consider the flow field

$$\mathbf{u} = -\hat{y}2\sin x - \hat{z}2\cos x.$$

Analyze the flow field in the light of helical basis functions.

Solution: The flow field consists of $\mathbf{u}(\mathbf{k})$ and $\mathbf{u}(-\mathbf{k})$ with $\mathbf{k} = (1,0,0)$. Hence, $\hat{e}_3(\mathbf{k}) = \hat{x}$. We choose $\hat{n} = \hat{z}$. Using Eqs. (9.1), we obtain the other two Craya–Herring basis vectors as

$$\hat{e}_1(\mathbf{k}) = -\hat{y}; \quad \hat{e}_2(\mathbf{k}) = -\hat{z}.$$

Hence, $u_1(\mathbf{k}) = 1/i$ and $u_2(\mathbf{k}) = 1$. Using the conversion formulas described in Section 9.1, we deduce the velocity Fourier modes and its components in the helical basis as

$$\begin{aligned}
\mathbf{u}(\mathbf{k}) &= \frac{1}{i}\hat{e}_1(\mathbf{k}) + \hat{e}_2(\mathbf{k}), \\
u_+(\mathbf{k}) &= \frac{1}{\sqrt{2}}[u_2(\mathbf{k}) + iu_1(\mathbf{k})] = \sqrt{2}, \\
u_-(\mathbf{k}) &= \frac{1}{\sqrt{2}}[u_2(\mathbf{k}) - iu_1(\mathbf{k})] = 0.
\end{aligned}$$

Thus, the aforementioned flow field has left-handed polarization (see Fig. 9.8). Note that the phase of $u_2(\mathbf{k})$ is ahead of $u_1(\mathbf{k})$ by $\pi/2$.

We map this field to Eq. (9.58) using $xy \to yz$, $z \to x$, $\phi_{k+} = \pi$ that yields

$$\mathbf{u}_+(\mathbf{r}) = \frac{1}{\sqrt{2}}|u_+(\mathbf{k})| \begin{pmatrix} \sin(x+\pi) \\ \cos(x+\pi) \end{pmatrix} = \begin{pmatrix} -\sin x \\ -\cos x \end{pmatrix}.$$

The modal energy and kinetic helicity are

$$\begin{aligned}
E_u(\mathbf{k}) &= \frac{1}{2}|\mathbf{u}(\mathbf{k})|^2 = 1, \\
H_K(\mathbf{k}) &= k\Im(u_1^* u_2) = k = 1.
\end{aligned}$$

Since the flow contains two wavenumbers—$(1,0,0)$ and $(-1,0,0)$, the total energy $E_u = 2$, and the total kinetic helicity $H_K = 2$. The flow field has maximal helicity.

Example 9.8: Analyze the helicity of the following flow:

$$\mathbf{u} = \hat{y}2\sin x - \hat{z}2\cos x.$$

Solution: Following similar steps as the earlier example, we can show that

$$\mathbf{u}(\mathbf{k}) = -\frac{1}{i}\hat{e}_1(\mathbf{k}) + \hat{e}_2(\mathbf{k}).$$

Hence, $u_1(\mathbf{k}) = i$, $u_2(\mathbf{k}) = 1$, $u_+(\mathbf{k}) = 0$, and $u_-(\mathbf{k}) = \sqrt{2}$. The map of the aforementioned field to Eq. (9.61) is as follows: $xy \to yz$, $z \to x$, $\phi_{k-} = \pi$, and

$$\mathbf{u}_-(\mathbf{r}) = \frac{1}{\sqrt{2}}|u_-(\mathbf{k})| \begin{pmatrix} -\sin(x+\pi) \\ \cos(x+\pi) \end{pmatrix} = \begin{pmatrix} \sin x \\ -\cos x \end{pmatrix}.$$

For this case, the modal energy and kinetic helicity of the mode (1,0,0) are 1 and −1 respectively. The flow field is right-handed and it has maximal helicity with negative sign.

In the next section we will derive equations of motion for $\mathbf{u}(\mathbf{k})$ in the helical basis.

9.7 Equations of Motion in Helical Basis

The equation of motion of hydrodynamics can also be derived in helical basis.[4] Following the procedure of Section 9.2 and Waleffe (1992), we consider the plane formed by the wavenumbers of the triad $(\mathbf{k}', \mathbf{p}, \mathbf{q})$, and take

$$\hat{n} = \frac{\mathbf{q} \times \mathbf{p}}{|\mathbf{q} \times \mathbf{p}|}. \tag{9.67}$$

Consequently, for all the wavenumbers, $\hat{e}_1$ of Craya–Herring basis lie in the plane, and $\hat{e}_2 = -\hat{n}$. Note that we consider clockwise orientation for the wavenumbers of the triad, and that $\hat{e}_\pm$ do not lie in the plane of the triad. See Fig. 9.5 for an illustration.

To derive the equation of motion for $\mathbf{u}(\mathbf{k})$, we start with Navier–Stokes equation:

$$\frac{d}{dt}\mathbf{u}(\mathbf{k}') = -i\mathbf{k}'p(\mathbf{k}') + \mathbf{u}(-\mathbf{q}) \times \boldsymbol{\omega}(-\mathbf{p}) + \mathbf{u}(-\mathbf{p}) \times \boldsymbol{\omega}(-\mathbf{q}). \tag{9.68}$$

We expand the velocity and vorticity fields in helical basis, and take dot product of Eq. (9.68) with $\hat{e}^*_{s_{k'}}(\mathbf{k}')$. As a result, we obtain

[4] This analysis was first done by Waleffe (1992)

$$
\begin{aligned}
\dot{u}_{s_k}(\mathbf{k}') &= [\mathbf{u}^*(\mathbf{q}) \times \boldsymbol{\omega}^*(\mathbf{p}) + \mathbf{u}^*(\mathbf{p}) \times \boldsymbol{\omega}^*(\mathbf{q})] \cdot \hat{e}^*_{s_{k'}}(\mathbf{k}') \\
&= \sum_{s_p, s_q} [p s_p u^*_{s_q}(\mathbf{q}) u^*_{s_p}(\mathbf{p}) \{\hat{e}^*_{s_q}(\mathbf{q}) \times \hat{e}^*_{s_p}(\mathbf{p}) \cdot \hat{e}^*_{s_{k'}}(\mathbf{k}')\}] \\
&\quad + \sum_{s_p, s_q} [q s_q u^*_{s_q}(\mathbf{q}) u^*_{s_p}(\mathbf{p}) \{\hat{e}^*_{s_p}(\mathbf{p}) \times \hat{e}^*_{s_q}(\mathbf{q}) \cdot \hat{e}^*_{s_{k'}}(\mathbf{k}')\}] \\
&= \sum_{s_p, s_q} (p s_p - q s_q) u^*_{s_q}(\mathbf{q}) u^*_{s_p}(\mathbf{p}) \{\hat{e}^*_{s_q}(\mathbf{q}) \times \hat{e}^*_{s_p}(\mathbf{p}) \cdot \hat{e}^*_{s_{k'}}(\mathbf{k}')\},
\end{aligned}
\tag{9.69}
$$

or

$$
\dot{u}_{s_k}(\mathbf{k}') = \sum_{s_p, s_q} g(p s_p - q s_q) u^*_{s_q}(\mathbf{q}) u^*_{s_p}(\mathbf{p}), \tag{9.70}
$$

where

$$
\begin{aligned}
g &= \hat{e}^*_{s_q}(\mathbf{q}) \times \hat{e}^*_{s_p}(\mathbf{p}) \cdot \hat{e}^*_{s_{k'}}(\mathbf{k}') \\
&= \frac{1}{2\sqrt{2}} s_{k'} s_p s_q \{s_{k'} \sin\alpha + s_p \sin\beta + s_q \sin\gamma\}.
\end{aligned}
\tag{9.71}
$$

Here, α, β, γ are the angles of the triangle formed by the wavenumbers $\mathbf{k}', \mathbf{p}, \mathbf{q}$. See Fig. 9.5 for an illustration.

If a triad $(\mathbf{k}', \mathbf{p}, \mathbf{q})$ consists only of three modes—$(u_{s_{k'}}, u_{s_p}, u_{s_q})$, then the equations of motion for these modes would be

$$
\dot{u}_{s_{k'}}(\mathbf{k}') = g(p s_p - q s_q) u^*_{s_p}(\mathbf{p}) u^*_{s_q}(\mathbf{q}), \tag{9.72a}
$$

$$
\dot{u}_{s_p}(\mathbf{p}) = g(q s_q - k' s_{k'}) u^*_{s_q}(\mathbf{q}) u^*_{s_{k'}}(\mathbf{k}'), \tag{9.72b}
$$

$$
\dot{u}_{s_q}(\mathbf{q}) = g(k' s_{k'} - p s_p) u^*_{s_{k'}}(\mathbf{k}') u^*_{s_p}(\mathbf{p}). \tag{9.72c}
$$

It is easy to verify that

$$
\frac{1}{2}[|u_{s_{k'}}(\mathbf{k}')|^2 + |u_{s_p}(\mathbf{p})|^2 + |u_{s_q}(\mathbf{q})|^2] = \text{const.} \tag{9.73}
$$

that corresponds to the *detailed conservation of kinetic energy* for a helical triad. We could also arrive at the same conclusion from the energy transfer formulas.

The aforementioned set of equations can be mapped to Euler's equation for rigid body rotation (Landau and Lifshitz, 1976; Verma, 2016). From the above we can make several simple deductions. If $p s_p = q s_q$, then $u_{s_{k'}}(\mathbf{k}') = $ const., and the other two modes oscillate around zero such that $|u_{s_p}(\mathbf{p})|^2 + |u_{s_q}(\mathbf{q})|^2 = $ const. The dynamics is similar to that of a symmetric top (Landau and Lifshitz, 1976; Verma, 2016).

Let us consider stability of the solution—$(u_{s_{k'}}, u_{s_p}, u_{s_q}) = (A, 0, 0)$. When we take another time derivative of Eq. (9.72b), we obtain

$$\ddot{u}_{s_p}(\mathbf{p}) = (qs_q - k's_{k'})(k's_{k'} - ps_p)|gA|^2 u_{s_p}(\mathbf{p}). \tag{9.74}$$

Hence, $u_{s_p}(\mathbf{p})$ will grow when

$$(qs_q - k's_{k'})(k's_{k'} - ps_p) > 0. \tag{9.75}$$

This is the condition for instability, which is achieved when $qs_q < k's_{k'} < ps_p$ or $qs_q > k's_{k'} > ps_p$. In other words, the system becomes unstable when the value of $k's_{k'}$ lies between ps_p and qs_q. The instability condition is akin to the instability condition of an asymmetric top (Landau and Lifshitz, 1976; Verma, 2016). Note however that the linear growth is observed only for a short time until $u_{s_p}(\mathbf{p})$ and $u_{s_q}(\mathbf{q})$ are small. When these modes become comparable to A, we need to consider the full set of nonlinear equations (Eqs. (9.72)).

In the next section we will derive formulas for the mode-to-mode transfers in helical basis.

9.8 Mode-to-mode Transfer Functions in Helical Basis

To derive a formula for the mode-to-mode KE transfer in helical basis, we write the giver mode $\mathbf{u}(\mathbf{p})$ and the receiver mode $\mathbf{u}(\mathbf{k'})$ as

$$\mathbf{u}(\mathbf{p}) = \sum_{s_p} u_{s_p}(\mathbf{p})\hat{e}_{s_p}(\mathbf{p}), \tag{9.76a}$$

$$\mathbf{u}(\mathbf{k'}) = \sum_{s_{k'}} u_{s_{k'}}(\mathbf{k'})\hat{e}_{s_{k'}}(\mathbf{k'}). \tag{9.76b}$$

Therefore, the mode-to-mode kinetic energy transfer is

$$S^{uu}(\mathbf{k'}|\mathbf{p}|\mathbf{q}) = \sum_{s_p, s_{k'}} -\Im\left[\{\mathbf{k'}\cdot\mathbf{u}(\mathbf{q})\} u_{s_p}(\mathbf{p}) u_{s_{k'}}(\mathbf{k'})\{\hat{e}_{s_p}(\mathbf{p})\cdot\hat{e}_{s_{k'}}(\mathbf{k'})\}\right]. \tag{9.77}$$

Note that

$$\begin{aligned}\hat{e}_{s_p}(\mathbf{p})\cdot\hat{e}_{s_{k'}}(\mathbf{k'}) &= \frac{1}{2}[\hat{e}_2(\mathbf{p}) - is_p\hat{e}_1(\mathbf{p})]\cdot[\hat{e}_2(\mathbf{k'}) - is_{k'}\hat{e}_1(\mathbf{k'})] \\ &= \frac{1}{2}(1 + s_p s_{k'}\cos\gamma).\end{aligned} \tag{9.78}$$

Therefore,

$$S^{uu}(\mathbf{k'}|\mathbf{p}|\mathbf{q}) = \sum_{s_p, s_{k'}} S^{uu}_{s_{k'}s_p}(\mathbf{k'}|\mathbf{p}|\mathbf{q}), \tag{9.79}$$

where $S^{uu}_{s_{k'}s_p}(\mathbf{k}'|\mathbf{p}|\mathbf{q})$ is the KE transfer from mode $u_{s_p}(\mathbf{p})$ to $u_{s_{k'}}(\mathbf{k}')$:

$$S^{uu}_{s_{k'}s_p}(\mathbf{k}'|\mathbf{p}|\mathbf{q}) = -\frac{k'}{2}\sin\beta(1+s_p s_{k'}\cos\gamma)\Im\{u_1(\mathbf{q})u_{s_p}(\mathbf{p})u_{s_{k'}}(\mathbf{k}')\}. \tag{9.80}$$

We use Eq. (9.55b) to expand $u_1(\mathbf{q})$ in terms of $u_\pm$ that yields

$$\Im\{u_1(\mathbf{q})u_{s_p}(\mathbf{p})u_{s_{k'}}(\mathbf{k}')\} = -\frac{1}{\sqrt{2}}\Re[u_+(\mathbf{q})u_{s_p}(\mathbf{p})u_{s_{k'}}(\mathbf{k}')] + \frac{1}{\sqrt{2}}\Re[u_-(\mathbf{q})u_{s_p}(\mathbf{p})u_{s_{k'}}(\mathbf{k}')]. \tag{9.81}$$

The other energy transfers in the triad are

$$S^{uu}_{s_p s_q}(\mathbf{p}|\mathbf{q}|\mathbf{k}') = -\frac{p}{2}\sin\gamma(1+s_p s_q\cos\alpha)\Im\{u_1(\mathbf{k}')u_{s_q}(\mathbf{q})u_{s_p}(\mathbf{p})\}, \tag{9.82a}$$

$$S^{uu}_{s_{k'}s_p}(\mathbf{q}|\mathbf{k}'|\mathbf{p}) = -\frac{q}{2}\sin\alpha(1+s_q s_{k'}\cos\beta)\Im\{u_1(\mathbf{p})u_{s_q}(\mathbf{q})u_{s_{k'}}(\mathbf{k}')\}. \tag{9.82b}$$

In helical basis, the mode-to-mode kinetic helicity transfer via $u_{s_p}(\mathbf{p})$ and $u_{s_{k'}}(\mathbf{k}')$ channel is derived as follows. Using

$$\boldsymbol{\omega}_{s_p}(\mathbf{p}) = i\mathbf{p}\times[u_{s_p}(\mathbf{p})\hat{e}_{s_p}(\mathbf{p})] = p s_p u_{s_p}\hat{e}_{s_p}(\mathbf{p}), \tag{9.83}$$

we obtain

$$\begin{aligned} S^{H_K}(\mathbf{k}'|\mathbf{p}|\mathbf{q}) &= \Re[\mathbf{u}(\mathbf{q})\cdot\boldsymbol{\omega}(\mathbf{p})\times\boldsymbol{\omega}(\mathbf{k}')\}] \\ &= \sum_{s_p,s_{k'}} s_{k'}s_p p k'\Re[\{\mathbf{u}(\mathbf{q})\cdot\hat{e}_{s_p}(\mathbf{p})\times\hat{e}_{s_p}(\mathbf{k}')\}u_{s_p}(\mathbf{p})u_{s_{k'}}(\mathbf{k}')] \\ &= \sum_{s_p,s_{k'}} S^{H_K}_{s_{k'}s_p}(\mathbf{k}'|\mathbf{p},\mathbf{q}). \end{aligned} \tag{9.84}$$

Using

$$\hat{e}_1(\mathbf{q})\cdot\hat{e}_{s_p}(\mathbf{p})\times\hat{e}_{s_{k'}}(\mathbf{k}') = \frac{i}{2}[s_{k'}\sin\beta + s_p\sin\alpha], \tag{9.85a}$$

$$\hat{e}_2(\mathbf{q})\cdot\hat{e}_{s_p}(\mathbf{p})\times\hat{e}_{s_{k'}}(\mathbf{k}') = \frac{1}{2}s_{k'}s_p\sin\gamma, \tag{9.85b}$$

we derive that

$$\begin{aligned} S^{H_K}_{s_{k'}s_p}(\mathbf{k}'|\mathbf{p},\mathbf{q}) &= -\frac{1}{2}pk'[s_{k'}\sin\beta + s_p\sin\alpha]\Im\{u_1(\mathbf{q})u_{s_p}(\mathbf{p})u_{s_{k'}}(\mathbf{k}')\} \\ &\quad +\frac{1}{2}pk'\sin\gamma\Re\{u_2(\mathbf{q})u_{s_p}(\mathbf{p})u_{s_{k'}}(\mathbf{k}')\}. \end{aligned} \tag{9.86}$$

Following similar arguments as earlier, we derive the following enstrophy transfer formulas.

$$\begin{aligned} S^{\omega\omega}_{s_{k'}s_p}(\mathbf{k'}|\mathbf{p}|\mathbf{q}) &= -\Im\left[\{\mathbf{k'}\cdot\mathbf{u}(\mathbf{q})\}\{\boldsymbol{\omega}_{s_p}(\mathbf{p})\cdot\boldsymbol{\omega}_{s_{k'}}(\mathbf{k'})\}\right] \\ &= -\frac{1}{2}pk'^2\sin\beta[\cos\gamma + s_p s_{k'}]\Im\{u_1(\mathbf{q})u_{s_p}(\mathbf{p})u_{s_{k'}}(\mathbf{k'})\}. \end{aligned} \quad (9.87)$$

For

$$S^{\omega u}_{s_{k'}s_p}(\mathbf{k'}|\mathbf{p}|\mathbf{q}) = \Im\left[\{\mathbf{k'}\cdot\boldsymbol{\omega}(\mathbf{q})\}\{\boldsymbol{\omega}_{s_{k'}}(\mathbf{k'})\cdot\mathbf{u}_{s_p}(\mathbf{p})\}\right], \quad (9.88)$$

we employ

$$\begin{aligned} \mathbf{k'}\cdot\boldsymbol{\omega}(\mathbf{q}) &= i\mathbf{k'}\cdot(\mathbf{q}\times\mathbf{u}(\mathbf{q})) \\ &= i(\mathbf{k'}\times\mathbf{q})\cdot\mathbf{u}(\mathbf{q}) = -ik'qu_2(\mathbf{q})\sin\beta, \end{aligned} \quad (9.89)$$

and

$$\begin{aligned} \boldsymbol{\omega}_{s_{k'}}(\mathbf{k'})\cdot\mathbf{u}_{s_p}(\mathbf{p}) &= s_{k'}k'\hat{e}_{s_{k'}}(\mathbf{k'})\cdot\hat{e}_{s_p}(\mathbf{p})u_{s_{k'}}(\mathbf{k'})u_{s_p}(\mathbf{p}) \\ &= k'\frac{1}{2}(s_{k'} + s_p\cos\gamma)u_{s_{k'}}(\mathbf{k'})u_{s_p}(\mathbf{p}). \end{aligned} \quad (9.90)$$

Therefore,

$$S^{\omega u}_{s_{k'}s_p}(\mathbf{k'}|\mathbf{p}|\mathbf{q}) = -\frac{1}{2}qk'^2\sin\beta(s_{k'} + s_p\cos\gamma)\Re\{u_2(\mathbf{q})u_{s_{k'}}(\mathbf{k'})u_{s_p}(\mathbf{p}\}. \quad (9.91)$$

In the next section we will employ the aforementioned mode-to-mode transfer functions to compute fluxes in helical basis.

9.9 Fluxes and Shell-to-shell Energy Transfers in Helical Basis

Following the same lines of arguments as in Section 9.4, we define energy and other fluxes in helical basis. Using Eq. (9.80), the kinetic energy flux from helical modes u_{s_g} to u_{s_r}, where s_g and s_r are the signs of giver and receiver modes respectively, can be written as

$$\Pi^{u_{s_g}<}_{u_{s_r}>}(k_0) = \sum_{|\mathbf{p}|\le k_0}\sum_{|\mathbf{k}|>k_0} S^{uu}_{s_g s_r}(\mathbf{k'}|\mathbf{p}|\mathbf{q}). \quad (9.92)$$

Using this set of formulas we can compute the kinetic energy fluxes from u_+ modes to u_+ modes (Π^{u+}_{u+}), u_+ modes to u_- modes (Π^{u+}_{u-}), u_- modes to u_+ modes (Π^{u-}_{u+}), and u_- modes to u_- modes (Π^{u-}_{u-}).

Similarly, the shell-to-shell energy transfers from u_{s_g} modes of shell m to u_{s_r} modes of shell n can be defined as

$$T_{u_{s_r},n}^{u_{s_g},m} = \sum_{\mathbf{p}\in m}\sum_{\mathbf{k}\in n} S_{s_r s_g}^{uu}(\mathbf{k}'|\mathbf{p}|\mathbf{q}). \tag{9.93}$$

Similar formulas can be written down for the enstrophy and helicity transfers. For example, the kinetic helicity flux along the + to + channel is

$$\Pi_{H_K+}^{H_K+}(k_0) = \sum_{|\mathbf{p}|\le k_0}\sum_{|\mathbf{k}|>k_0} S_{++}^{H_K}(\mathbf{k}'|\mathbf{p}|\mathbf{q}). \tag{9.94}$$

The above discussion shows that helical basis provides interesting and useful formulas, especially for helical flows.

Example 9.9: Consider the flow fields of Example 9.1. Compute the helical basis vectors and the components along the basis vectors of these fields.

Solution: Refer to the solution of Example 9.1. Using $\hat{e}_{1.2}$ and $u_{1,2}$, we compute

$$\begin{aligned}
\hat{e}_{\pm}(\mathbf{k}) &= \frac{1}{\sqrt{2}}[\hat{e}_2(\mathbf{k}) \mp i\hat{e}_1(\mathbf{k})], \\
\boldsymbol{\omega}(\mathbf{k}) &= i\mathbf{k}\times\mathbf{u}(\mathbf{k}) = k[u_+(\mathbf{k})\hat{e}_+(\mathbf{k}) - u_-(\mathbf{k})\hat{e}_-(\mathbf{k})], \\
E_u(\mathbf{k}) &= \frac{1}{2}\mathbf{u}^*(\mathbf{k})\cdot\mathbf{u}(\mathbf{k}) = \frac{1}{2}[|u_+(\mathbf{k})|^2 + |u_-(\mathbf{k})|^2], \\
H_K(\mathbf{k}) &= \frac{1}{2}\Re[\mathbf{u}^*(\mathbf{k})\cdot\boldsymbol{\omega}(\mathbf{k})] = \frac{1}{2}k[|u_+(\mathbf{k})|^2 - |u_-(\mathbf{k})|^2].
\end{aligned}$$

The resulting values are listed in Table 9.8.

Table 9.8 Example 9.9: The helical basis vectors $\hat{e}_{\pm}$ and their associated components, modal kinetic energy and kinetic helicity. Note that $\hat{e}_+(\mathbf{k}') = (-\hat{z} - i(\hat{x} - \hat{y})/\sqrt{2})/\sqrt{2}$, $\hat{e}_-(\mathbf{k}') = (-\hat{z} + i(\hat{x} - \hat{y})/\sqrt{2})/\sqrt{2}$.

Part	Mode	$\hat{e}_+(\mathbf{k})$	$\hat{e}_-(\mathbf{k})$	$(u_+(\mathbf{k}), u_-(\mathbf{k}))$	$E_u(\mathbf{k})$	$H_K(\mathbf{k})$
(a)	(1,1)	$\hat{e}_+^*(\mathbf{k}')$	$\hat{e}_-^*(\mathbf{k}')$	$(-\sqrt{2}, 0)$	1	1
	$(-1,-1)$	$\hat{e}_+(\mathbf{k}')$	$\hat{e}_-(\mathbf{k}')$	$(-\sqrt{2}, 0)$	1	1
	(0,1)	$(-\hat{z} - i\hat{x})/\sqrt{2}$	$(-\hat{z} + i\hat{y})/\sqrt{2}$	$(-\sqrt{2}, 0)$	1	1
	$(0,-1)$	$(-\hat{z} + i\hat{x})/\sqrt{2}$	$(-\hat{z} - i\hat{y})/\sqrt{2}$	$(-\sqrt{2}, 0)$	1	1
	(1,0)	$(-\hat{z} + i\hat{y})/\sqrt{2}$	$(-\hat{z} - i\hat{y})/\sqrt{2}$	$(-\sqrt{2}, 0)$	1	1
	$(-1,0)$	$(-\hat{z} - i\hat{y})/\sqrt{2}$	$(-\hat{z} + i\hat{y})/\sqrt{2}$	$(-\sqrt{2}, 0)$	1	1

Table 9.8 Contd...

	(1,1)	$\hat{e}^*_+(\mathbf{k}')$	$\hat{e}^*_-(\mathbf{k}')$	$(0, \sqrt{2})$	1	−1
	$(-1,-1)$	$\hat{e}_+(\mathbf{k}')$	$\hat{e}_-(\mathbf{k}')$	$(0, \sqrt{2})$	1	−1
(b)	(0,1)	$(-\hat{z} - i\hat{x})/\sqrt{2}$	$(-\hat{z} + i\hat{y})/\sqrt{2}$	$(0, \sqrt{2})$	1	−1
	$(0,-1)$	$(-\hat{z} + i\hat{x})/\sqrt{2}$	$(-\hat{z} - i\hat{y})/\sqrt{2}$	$(0, \sqrt{2})$	1	−1
	(1,0)	$(-\hat{z} + i\hat{y})/\sqrt{2}$	$(-\hat{z} - i\hat{y})/\sqrt{2}$	$(0, \sqrt{2})$	1	−1
	$(-1,0)$	$(-\hat{z} - i\hat{y})/\sqrt{2}$	$(-\hat{z} + i\hat{y})/\sqrt{2}$	$(0, \sqrt{2})$	1	−1
	(1,1)	$\hat{e}^*_+(\mathbf{k}')$	$\hat{e}^*_-(\mathbf{k}')$	$(0, \sqrt{2})$	1	−1
	$(-1,-1)$	$\hat{e}_+(\mathbf{k}')$	$\hat{e}_-(\mathbf{k}')$	$(0, \sqrt{2})$	1	−1
(c)	(0,1)	$(-\hat{z} - i\hat{x})/\sqrt{2}$	$(-\hat{z} + i\hat{y})/\sqrt{2}$	$(-\sqrt{2}, 0)$	1	1
	$(0,-1)$	$(-\hat{z} + i\hat{x})/\sqrt{2}$	$(-\hat{z} - i\hat{y})/\sqrt{2}$	$(-\sqrt{2}, 0)$	1	1
	(1,0)	$(-\hat{z} + i\hat{y})/\sqrt{2}$	$(-\hat{z} - i\hat{y})/\sqrt{2}$	$(-\sqrt{2}, 0)$	1	1
	$(-1,0)$	$(-\hat{z} - i\hat{y})/\sqrt{2}$	$(-\hat{z} + i\hat{y})/\sqrt{2}$	$(-\sqrt{2}, 0)$	1	1

Example 9.10: Consider an arbitrary flow field constructed with wavenumbers $\mathbf{k}' = (-1,-1,0)$, $\mathbf{q} = (0,1,0)$, $\mathbf{p} = (1,0,0)$, and their negative counterparts. Derive the equations for the modes $\mathbf{u}_\pm(\mathbf{k}')$, $\mathbf{u}_\pm(\mathbf{p})$, $\mathbf{u}_\pm(\mathbf{q})$. Also compute the energy transfers among the modes.

Solution: The interacting wavenumbers form a right-angle triangle. The associated magnitudes and the angles are (see Fig. 9.5(a))

$$k' = \sqrt{2}; \quad p = q = 1;$$

$$\alpha = \pi/2; \quad \beta = \gamma = \pi/4.$$

Following Eqs. (9.72), we derive the equations of motion for the helical modes as

$$\begin{aligned}
\dot{u}_{s_{k'}}(\mathbf{k}') &= g(s_p - s_q) u^*_{s_p}(\mathbf{p}) u^*_{s_q}(\mathbf{q}), \\
\dot{u}_{s_p}(\mathbf{p}) &= g(s_q - \sqrt{2} s_{k'}) u^*_{s_q}(\mathbf{q}) u^*_{s_{k'}}(\mathbf{k}'), \\
\dot{u}_{s_q}(\mathbf{q}) &= g(\sqrt{2} s_{k'} - s_p) u^*_{s_{k'}}(\mathbf{k}') u^*_{s_p}(\mathbf{p}),
\end{aligned}$$

where

$$\begin{aligned}
g &= \frac{1}{2\sqrt{2}} s_{k'} s_p s_q \{ s_{k'} \sin\alpha + s_p \sin\beta + s_q \sin\gamma \} \\
&= \frac{1}{2\sqrt{2}} s_{k'} s_p s_q \{ s_{k'} + \frac{1}{\sqrt{2}} (s_p + s_q) \}.
\end{aligned}$$

If $s_p = s_q$, then

$$\dot{u}_{s_{k'}}(\mathbf{k}') = 0 \quad \Longrightarrow u_{s_{k'}}(\mathbf{k}') = C,$$

and

$$\begin{aligned} \dot{u}_{s_p}(\mathbf{p}) &= -aC^* u^*_{s_q}(\mathbf{q}), \\ \dot{u}_{s_q}(\mathbf{q}) &= aC^* u^*_{s_p}(\mathbf{p}), \end{aligned}$$

where $a = g(\sqrt{2}s_{k'} - s_p)$. By taking another derivative of these equations, we obtain

$$\ddot{u}_{s_p} = -|aC|^2 u_{s_p}; \quad \ddot{u}_{s_q} = -|aC|^2 u_{s_q}.$$

Hence, both these modes oscillate with a frequency of $\omega = |aC|$.

The energy transfers among the modes can be computed following the procedure of the previous example.

With this, we close our discussion on the Craya–Herring and helical basis. We will revisit them in the next chapter itself when we compute the energy transfers using field theory. Later, we will employ these basis functions to thermal convection, magnetohydrodynamic (MHD) turbulence, and dynamo.

Further Reading

For introduction to Craya–Herring and helical basis, refer to Lesieur (2008), Sagaut and Cambon (2008), and Waleffe (1992). Waleffe (1992) discusses the energy transfers in the helical basis. The mode-to-mode transfers in Craya–Herring and helical basis have been discussed in this chapter for the first time. Recently, Biferale et al. (2013) and coworkers studied energy transfers and fluxes in truncated Navier–Stokes equations with u^+ modes only.

Exercises

1. In Craya–Herring basis, the velocity field $\mathbf{u}(\mathbf{k}_1) = \hat{e}_2(\mathbf{k}_1)+i\hat{e}_1(\mathbf{k}_1)$ and $\mathbf{u}(\mathbf{k}_2) = \hat{e}_2(\mathbf{k}_2)+2\hat{e}_1(\mathbf{k}_2)$, where $\mathbf{k}_1 = (1,0,1)$ and $\mathbf{k}_2 = (1,1,1)$. The reality condition on the velocity field demands that $\mathbf{u}(-\mathbf{k}_1) = \mathbf{u}^*(\mathbf{k}_1)$ and $\mathbf{u}(-\mathbf{k}_2) = \mathbf{u}^*(\mathbf{k}_2)$. Write down the velocity field in Cartesian coordinates, in both Fourier space and real space.
2. Generalize Exercise 4.3 to make the velocity field quasi-2D. Write down the equations of motion for the components of Craya–Herring basis.

3. Consider a 2D flow field with Fourier modes with wavenumbers $(0,1)$, (p_x, p_y), $(-p_x, -1-p_y)$, and their complex conjugates. The kinematic viscosity of the fluid is ν, and the flow field is driven by an external force $\mathbf{F} = \hat{x}\cos y$. Write down the equation of motion for the Fourier modes.
4. Compute the total energy transferred to each of the Fourier modes of Example 9.4. Show that these transfers are zeros.
5. For the flow field of Example 9.5, compute the energy and enstrophy transfers among the modes in Craya–Herring and helical basis.
6. Consider the flow fields of Exercise 4.1. Compute the components of the velocity field in Craya–Herring basis. Derive equations of motion for A, B, and C.
7. Repeat Exercise 9.6 for Exercise 4.2. Derive equations of motion for A, B, C, and D.

Chapter 10

Field-theoretic Treatment of Energy Transfers

In earlier chapters we derived analytic formulas for kinetic energy transfers and flux (see Chapter 4). These energy transfers have universal properties during a steady state of hydrodynamic turbulence. We can compute these quantities using experimental and/or numerical data of steady state turbulence. In addition, there is an approximate yet powerful formalism based on field theory that provides estimates for these transfers; this is the topic of the present chapter.

In some field-theoretic treatments (McComb, 1990; Leslie, 1973; Verma, 2004), energy transfers are computed perturbatively to first order. The resulting expressions involve Green's function and the correlation function. Kolmogorov's spectrum $E_u(k) \sim k^{-5/3}$ is substituted for the correlation function, and Green's function is modeled similarly. By construction, these formulas are applicable to the inertial range of homogeneous and isotropic hydrodynamic turbulence. We will detail these computations in this chapter.

First, we describe the correlation functions of homogeneous and isotropic turbulence.

10.1 Correlation Functions in Homogeneous and Isotropic Turbulence

In homogenous and isotropic turbulence, the velocity–velocity correlation is of the following form:

$$\langle u_i(\mathbf{k})u_j(\mathbf{k}')\rangle = \left[P_{ij}(\mathbf{k})C_u(k) - i\epsilon_{ijl}k_l\frac{H_K(k)}{k^2}\right]\delta(\mathbf{k}+\mathbf{k}'), \tag{10.1}$$

where

$$P_{ij}(\mathbf{k}) = \delta_{ij} - \frac{k_ik_j}{k^2} \tag{10.2}$$

is the projection operator, and $C_u(k)$ and $H_K(k)$ are real functions of k. Note that the correlation $\langle u_i(\mathbf{k})u_j(\mathbf{k}')\rangle$ vanishes unless $\mathbf{k}' = -\mathbf{k}$.

In most field-theoretic treatment, the components are taken to be along Cartesian axis—x, y, z. In this chapter, however, we will consider the components in Craya-Herring basis (see Figs. 9.1 and 9.2). Hence, in 3D, for a wavenumber $\mathbf{k}$,

$$k_1 = k_2 = 0; \quad k_3 = k; \quad u_3(\mathbf{k}) = 0. \tag{10.3}$$

From Eq. (10.1), we deduce that in 3D,

$$\langle |u_1(\mathbf{k})|^2\rangle = \langle |u_2(\mathbf{k})|^2\rangle = C_u(k). \tag{10.4}$$

Note that the modal kinetic energy (KE) is defined as

$$C_u(\mathbf{k}) = \frac{1}{2}(|u_1(\mathbf{k})|^2 + |u_2(\mathbf{k})|^2). \tag{10.5}$$

Therefore,

$$\langle C_u(\mathbf{k})\rangle = \frac{1}{2}\langle |u_1(\mathbf{k})|^2 + |u_2(\mathbf{k})|^2\rangle = C_u(k). \tag{10.6}$$

Thus, statistical average of modal KE equals $C_u(k)$, which is a statement of isotropic turbulence.

Similarly, using Eq. (10.1) we deduce that

$$\langle u_1(\mathbf{k})u_2(-\mathbf{k})\rangle = \langle u_1(\mathbf{k})u_2^*(\mathbf{k})\rangle = -i\epsilon_{123}\frac{k}{k^2}H_K(k) = -\frac{i}{k}H_K(k). \tag{10.7}$$

Therefore,

$$H_K(k) = ik\langle u_1(\mathbf{k})u_2^*(\mathbf{k})\rangle. \tag{10.8}$$

Since $H_K(k)$ is purely real, $\langle u_1(\mathbf{k})u_2^*(\mathbf{k})\rangle$ must be pure imaginary. Hence,

$$H_K(k) = k\Im\langle u_1^*(\mathbf{k})u_2(\mathbf{k})\rangle. \tag{10.9}$$

Using Eq. (9.5b) we deduce that the righthand side of Eq. (10.9) is the statistical average of the modal kinetic helicity $H_K(\mathbf{k})$. Hence,

$$\langle H_K(\mathbf{k}) \rangle = H_K(k). \tag{10.10}$$

This is also a statement of statistical isotropy. Note that at a particular instance, for a given $\mathbf{k}$, $u_1^*(\mathbf{k})u_2(\mathbf{k})$ may have a real part, but it will vanish on statistical averaging.

In 3D, one-dimensional energy spectrum $E_u(k)$ is related to $C_u(\mathbf{k})$ in the following manner:

$$\begin{aligned} E_u &= \int E_u(k)dk = \frac{1}{2}\langle u^2 \rangle = \frac{1}{2}[\langle u_1^2 \rangle + \langle u_2^2 \rangle] \\ &= \int C_u(\mathbf{k})d\mathbf{k} = \int 4\pi k^2 C_u(\mathbf{k})dk. \end{aligned} \tag{10.11}$$

Hence,

$$E_u(k) = 4\pi k^2 C_u(\mathbf{k}). \tag{10.12}$$

For a 2D flow in the xy coordinate system, in Craya–Herring basis,

$$k_1 = 0; \quad k_2 = k; \quad k_3 = 0; \quad u_2(\mathbf{k}) = 0, \tag{10.13}$$

and

$$C_u(\mathbf{k}) = \langle |u_1(\mathbf{k})|^2 \rangle = C_u(k). \tag{10.14}$$

Here,

$$E_u = \int E_u(k)dk \;=\; \frac{1}{2}\langle u_1^2 \rangle = \frac{1}{2}\int C_u(\mathbf{k})d\mathbf{k} = \int \pi k C_u(\mathbf{k})dk. \tag{10.15}$$

Hence,

$$E_u(k) = \pi k C_u(k). \tag{10.16}$$

Note that there is no kinetic helicity associated with 2D flows.

Now we are ready to compute expressions for the mode-to-mode energy transfers in homogeneous and isotropic turbulent flows.

10.2 Field-theoretic Treatment of Mode-to-mode Kinetic Energy Transfers and Flux

In this section we derive an expression for the mode-to-mode kinetic energy transfers in Craya-Herring basis. We remark that earlier field theoretic derivations are based on Cartesian coordinates. We will show below that Craya-Herring basis makes the derivations simpler and more intuitive.

We start with Eq. (9.32). The mode-to-mode kinetic energy transfer from $\mathbf{u}(\mathbf{p})$ to $\mathbf{u}(\mathbf{k}')$ with $\mathbf{u}(\mathbf{q})$ acting as a mediator is

$$S^{uu}(\mathbf{k}'|\mathbf{p}|\mathbf{q}) = S^{u_1u_1}(\mathbf{k}'|\mathbf{p}|\mathbf{q}) + S^{u_2u_2}(\mathbf{k}'|\mathbf{p}|\mathbf{q}), \tag{10.17}$$

where

$$\begin{aligned} S^{u_1u_1}(\mathbf{k}'|\mathbf{p}|\mathbf{q}) &= k'\sin\beta\cos\gamma\,\Im\{u_1(\mathbf{q})u_1(\mathbf{p})u_1(\mathbf{k}')\}, &(10.18a)\\ S^{u_2u_2}(\mathbf{k}'|\mathbf{p}|\mathbf{q}) &= -k'\sin\beta\,\Im\{u_1(\mathbf{q})u_2(\mathbf{p})u_2(\mathbf{k}')\}. &(10.18b)\end{aligned}$$

Here, β, γ are two angles of the triangle formed by the wavenumbers $\mathbf{k}', \mathbf{p}, \mathbf{q}$ (see Fig. 9.5). The quantity $S^{uu}(\mathbf{k}'|\mathbf{p}|\mathbf{q})$ depends on the amplitudes of the Fourier modes. However, a statistical average prescribed by field theory provides very useful results. We estimate average values of two triple correlation of Eq. (10.17) to first order in perturbation expansion. Here, u_1, u_2 are assumed to be quasi-Gaussian with $\langle XYZ\rangle = 0$ to zeroth order, but nonzero to first order. The computation is sketched in the following discussion.

10.2.1 Computation of $\Im\langle u_1(\mathbf{q},t)u_1(\mathbf{p},t)u_1(\mathbf{k}',t)\rangle$

Under Gaussian approximation, $\Im\langle u_1(\mathbf{q},t)u_1(\mathbf{p},t)u_1(\mathbf{k}',t)\rangle$ vanishes to zeroth order. Hence, we go to the next order. Using Eq. (9.25), we write $u_1(\mathbf{k}',t)$ in terms of Green's function as

$$u_1(\mathbf{k}',t) = ik'\int_{-\infty}^{t} dt' G(\mathbf{k}',t-t')\sin(\beta-\gamma)u_1^*(\mathbf{p},t')u_1^*(\mathbf{q},t'), \tag{10.19}$$

substitution of which in the aforementioned triple correlation yields

$$\begin{aligned}\Im\langle u_1(\mathbf{q},t)u_1(\mathbf{p},t)u_1(\mathbf{k}',t)\rangle &= k'\int_{-\infty}^{t} dt' G(\mathbf{k}',t-t')\sin(\beta-\gamma)\\ &\quad\times\langle u_1^*(\mathbf{p},t')u_1(\mathbf{p},t)\rangle\langle u_1^*(\mathbf{q},t')u_1(\mathbf{q},t)\rangle. \end{aligned}\tag{10.20}$$

Note that $\langle u_1^*(\mathbf{p},t')u_1(\mathbf{q},t)\rangle = 0$ when $\mathbf{p} \neq \mathbf{q}$ (see Eq. (10.1)). We assume that the average quantities are homogeneous in time; hence,

$$\langle u_1^*(\mathbf{p},t')u_1(\mathbf{p},t)\rangle = C(\mathbf{p},t-t'), \tag{10.21a}$$

$$\langle u_1^*(\mathbf{q},t')u_1(\mathbf{q},t)\rangle = C(\mathbf{q},t-t'). \tag{10.21b}$$

To simplify the above expression, we assume that Green's function and the correlation function relax with a time scale of $1/(\nu(k)k^2)$, where $\nu(k)$ is the renormalized viscosity at wavenumber k, that is,

$$G(\mathbf{k}',t-t') = \exp[-\nu(k')k^2(t-t')], \tag{10.22a}$$

$$C(\mathbf{k}',t-t') = C(\mathbf{k}')\exp[-\nu(k)k^2(t-t')]. \tag{10.22b}$$

Under the aforementioned assumptions, we obtain

$$\Im\langle u_1(\mathbf{q},t)u_1(\mathbf{p},t)u_1(\mathbf{k}',t)\rangle = \frac{1}{\text{denr}}k'\sin(\beta-\gamma)C(\mathbf{p})C(\mathbf{q}), \tag{10.23}$$

where

$$\text{denr} = \nu(k)k^2 + \nu(p)p^2 + \nu(q)q^2. \tag{10.24}$$

But this is not all. We need to also expand $u_1(\mathbf{q},t)$ and $u_1(\mathbf{p},t)$ of the correlation function $\Im\langle u_1(\mathbf{q},t)u_1(\mathbf{p},t)u_1(\mathbf{k}',t)\rangle$. Following a similar procedure as earlier, we obtain

$$\begin{aligned}\Im\langle u_1(\mathbf{q},t)u_1(\mathbf{p},t)u_1(\mathbf{k}',t)\rangle = &\frac{1}{\text{denr}}[k'\sin(\beta-\gamma)C(\mathbf{p})C(\mathbf{q})\\ &+p\sin(\gamma-\alpha)C(\mathbf{k}')C(\mathbf{q})\\ &+q\sin(\alpha-\beta)C(\mathbf{k}')C(\mathbf{p})],\end{aligned} \tag{10.25}$$

where denr is given in Eq. (10.24). We compute $S^{u_1u_1}(\mathbf{k}'|\mathbf{p}|\mathbf{q})$ using the correlation function described earlier.

10.2.2 Computation of $\Im\langle u_1(\mathbf{q},t)u_2(\mathbf{p},t)u_2(\mathbf{k}',t)\rangle$

Using Eq. (9.27a), we write $u_2(\mathbf{k}',t)$ in terms of Green's function as

$$u_2(\mathbf{k}',t) = ik'\int_{-\infty}^{t} dt' G(\mathbf{k}',t-t')[\sin\gamma u_1^*(\mathbf{p},t')u_2^*(\mathbf{q},t') - \sin\beta u_1^*(\mathbf{q},t')u_2^*(\mathbf{p},t')], \tag{10.26}$$

substitution of which in the triple correlation and following similar steps as in the previous subsection yields

$$\begin{aligned}\Im\langle u_1(\mathbf{q},t)u_2(\mathbf{p},t)u_2(\mathbf{k}',t)\rangle &= \frac{1}{\mathrm{denr}}[k'\sin\beta C(\mathbf{q})(C(\mathbf{k}')-C(\mathbf{p})) \\ &\quad +k'\sin\gamma\frac{H_K(\mathbf{p})}{p}\frac{H_K(\mathbf{q})}{q} - p\sin\gamma\frac{H_K(\mathbf{k}')}{k'}\frac{H_K(\mathbf{q})}{q} \\ &\quad +q\sin(\alpha-\beta)\frac{H_K(\mathbf{k}')}{k'}\frac{H_K(\mathbf{p})}{p}], \end{aligned} \quad (10.27)$$

where denr is given in Eq. (10.24). We now compute $S^{u_2u_2}(\mathbf{k}'|\mathbf{p}|\mathbf{q})$ using the correlation function described earlier. A sum of $S^{u_1u_1}(\mathbf{k}'|\mathbf{p}|\mathbf{q})$ and $S^{u_2u_2}(\mathbf{k}'|\mathbf{p}|\mathbf{q})$ yields $S^{uu}(\mathbf{k}'|\mathbf{p}|\mathbf{q})$. For subsequent computation, we assume that the flow is nonhelical, that is, $H_K = 0$.

To understand the properties of $\langle S^{uu}(\mathbf{k}'|\mathbf{p}|\mathbf{q})\rangle$ in the inertial range, we assume that k', p, q lie in the inertial range, and that

$$C_u(\mathbf{k}) = \frac{E_u(k)}{4\pi k^2} = \frac{K_{\mathrm{Ko}}}{4\pi}\Pi_u^{2/3}k^{-11/3}. \quad (10.28)$$

In Fig. 10.1 we present the density plot of $\langle S^{uu}(\mathbf{k}'|\mathbf{p}|\mathbf{q})\rangle$ in logarithmic scale. Here the x and y axes are p/k and q/k respectively. We observe that in most regions,

$$\langle S^{uu}(\mathbf{k}'|\mathbf{p}|\mathbf{q})\rangle > 0 \quad \text{for } p < k;$$
$$\langle S^{uu}(\mathbf{k}'|\mathbf{p}|\mathbf{q})\rangle < 0 \quad \text{for } p > k.$$

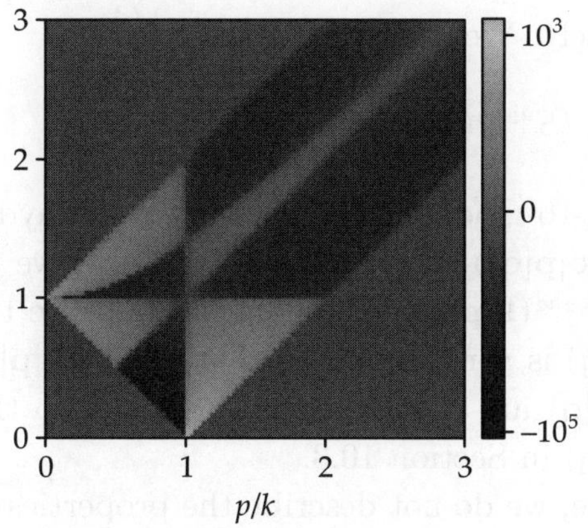

Figure 10.1 For 3D hydrodynamic turbulence, density plot of mode-to-mode KE transfer $\langle S^{uu}(\mathbf{k}'|\mathbf{p}|\mathbf{q})\rangle$ computed using field theory. Figure indicates significant KE transfer from small p ($p \ll k$) to k.

Hence, wavenumbers p with $p < k$ transfer KE to k, while wavenumbers p with $p > k$ receive KE from k. This indicates that the mode-to-mode KE transfer in 3D hydrodynamic turbulence is forward. In addition, as shown in Fig. 10.1, $\langle S^{uu}(\mathbf{k}'|\mathbf{p}|\mathbf{q})\rangle$ is strong when $p/k \ll 1$, hence the mode-to-mode KE transfer is predominantly nonlocal (see Fig. 10.2(a)). Also, in an equilateral triangle for which $\alpha = \beta = \gamma = \pi/3$, there is no KE exchange among the modes of the triad. See Fig. 10.2(b) for an illustration. Domaradzki and Rogallo (1990) and Verma et al. (2005) showed that 3D hydrodynamic turbulence exhibits nonlocal mode-to-mode energy transfers but local shell-to-shell transfers. Refer to the original papers for further details.

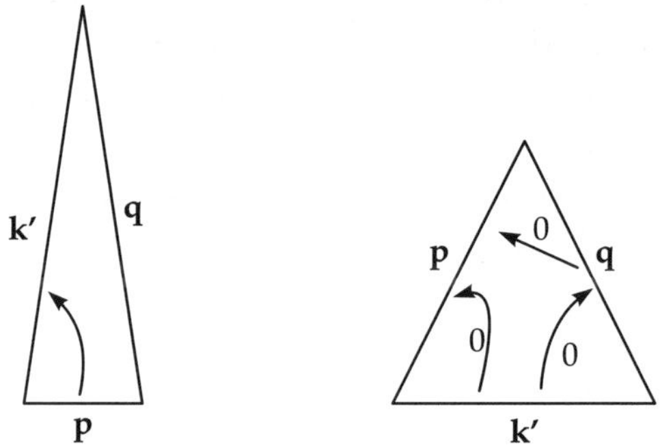

Figure 10.2 For 3D hydrodynamic turbulence, (a) $\langle S^{uu}(\mathbf{k}'|\mathbf{p}|\mathbf{q})\rangle$ is dominant when $p \ll k$, and (b) $\langle S^{uu}(\mathbf{k}'|\mathbf{p}|\mathbf{q})\rangle = 0$ when $k = p = q$.

To diagnose the energy transfers further, we compute $\langle S^{u_1u_1}(\mathbf{k}|\mathbf{p}|\mathbf{q})\rangle$ and $\langle S^{u_2u_2}(\mathbf{k}|\mathbf{p}|\mathbf{q})\rangle$ separately. We observe that

$$\langle S^{u_1u_1}(\mathbf{k}|\mathbf{p}|\mathbf{q})\rangle \ll \langle S^{u_2u_2}(\mathbf{k}|\mathbf{p}|\mathbf{q})\rangle. \tag{10.29}$$

Hence, the total mode-to-mode energy transfer in 3D hydrodynamic turbulence is dominated by $\langle S^{u_2u_2}(\mathbf{k}'|\mathbf{p}|\mathbf{q})\rangle$. In Fig. 10.3(a) and (b) we present the density plots of $S^{u_1u_1}(\mathbf{k}|\mathbf{p}|\mathbf{q})$ and $S^{u_2u_2}(\mathbf{k}|\mathbf{p}|\mathbf{q})$. These figures indicate that the nature of energy transfer of $S^{u_2u_2}(\mathbf{k}|\mathbf{p}|\mathbf{q})$ is very similar to that of $S^{uu}(\mathbf{k}|\mathbf{p}|\mathbf{q})$. However, the energy transfers of $S^{u_1u_1}(\mathbf{k}|\mathbf{p}|\mathbf{q})$ are approximately opposite to that of $S^{u_2u_2}(\mathbf{k}|\mathbf{p}|\mathbf{q})$. We will revisit $S^{u_1u_1}(\mathbf{k}|\mathbf{p}|\mathbf{q})$ in Section 10.3.

Due to lack of space, we do not describe the properties of $\langle S^{uu}(\mathbf{k}'|\mathbf{p}|\mathbf{q})\rangle$ in more detail. The aforementioned results however show that important deductions can be made using field-theoretic computations.

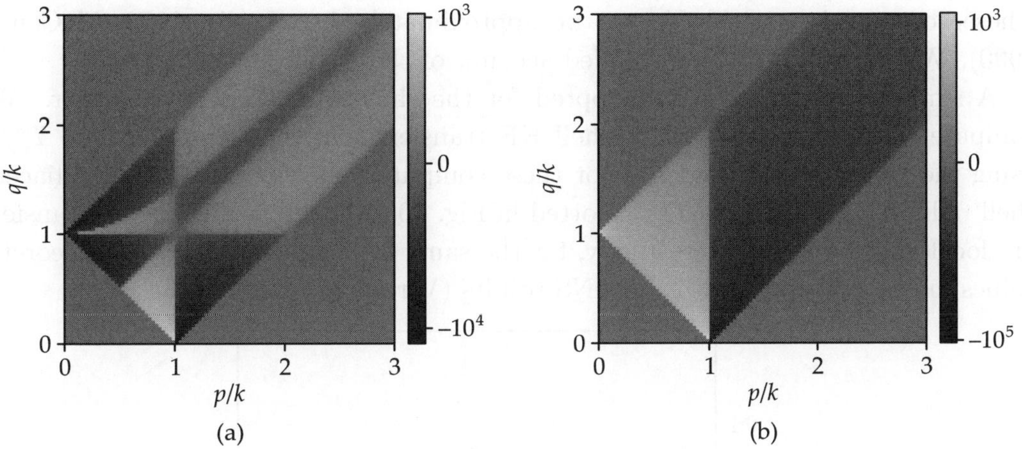

Figure 10.3 For 3D hydrodynamic turbulence, density plots of mode-to-mode KE transfers: (a) $\langle S^{u_1u_1}(\mathbf{k}'|\mathbf{p}|\mathbf{q})\rangle$ and (b) $\langle S^{u_2u_2}(\mathbf{k}'|\mathbf{p}|\mathbf{q})\rangle$. See Eqs. (10.25, 10.27).

10.2.3 Computation of kinetic energy flux and shell-to-shell kinetic energy transfer

Using $\langle S^{uu}(\mathbf{k}'|\mathbf{p}|\mathbf{q})\rangle$ computed in the previous subsection, we can compute average kinetic energy flux, $\Pi_u(k_0)$, as

$$\langle \Pi_u(k_0)\rangle = \sum_{|\mathbf{p}|\le k_0} \sum_{|\mathbf{k}|>k_0} \langle S^{uu}(\mathbf{k}|\mathbf{p}|\mathbf{q})\rangle. \tag{10.30}$$

For the computation of this integral (or sum), refer to Leslie (1973) and McComb (1990). They obtain an expression of the following form:

$$\langle \Pi_u(k_0)\rangle = \Pi_u \frac{K_{\mathrm{Ko}}^{-3/2}}{\nu_*} I, \tag{10.31}$$

where I is the value of the integral, a nondimensional number, and K_{Ko} is Kolmogorov's constant. The constant ν_* is computed analytically using renormalization group approach (see Appendix C; McComb (1990, 2014); McComb and Shanmugasundaram (1983, 1984); Zhou et al. (1988); Zhou (2010); Verma (2004)). Also, $\langle \Pi_u(k_0)\rangle = \Pi_u$, the kinetic energy flux. These relations help us deduce K_{Ko} as

$$K_{\mathrm{Ko}} = \left(\frac{\nu_*}{I}\right)^{2/3}. \tag{10.32}$$

These computations yield K_{Ko} to be approximately 1.6 (Leslie, 1973; McComb, 1990). We will not provide a detailed account of these computations here.

A similar approach can be adopted for the shell-to-shell energy transfer. We compute the normalized shell-to-shell KE transfer from shell m to shell n, $T^{u,m}_{u,n}$, using the values of K_{Ko} and ν_*. For these computations, the ratio of the adjacent shell radii, $k_{n+1}/k_n = 2^{1/4}$. $T^{u,m}_{u,n}$ plotted in Fig. 10.4 show that the energy transfers are local and forward. Interestingly, for the same k_{n+1}/k_n ratio, the field-theoretic values match quite well with the DNS results (Verma et al., 2005).

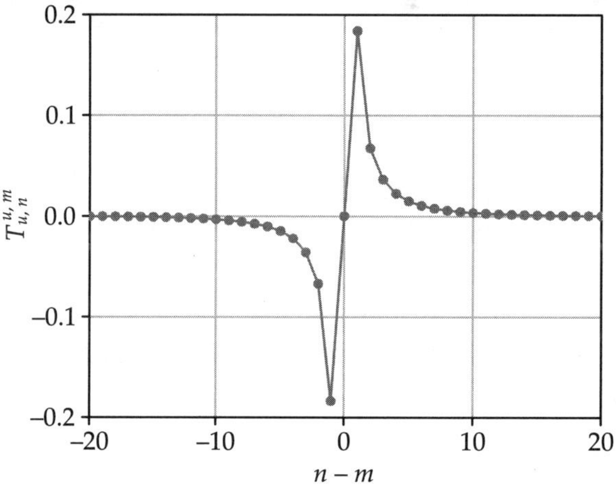

Figure 10.4 For 3D hydrodynamic turbulence, shell-to-shell KE transfer $T^{u,m}_{urn}$ from shell m to shell n. The plot indicates local and forward energy transfer.

10.2.4 Energy transfers for absolute equilibrium turbulence or Euler turbulence

In Sections 4.5 and 5.4 we briefly discussed the energy transfers for equilibrium or Euler turbulence. For such flows, the modal energies for all the Fourier modes are equal, that is, $C_u(\mathbf{k}) = C$, where C is a constant. When we substitute the above form of modal energy in Eq. (10.25), we obtain

$$\begin{aligned} \Im\langle u_1(\mathbf{q},t)u_1(\mathbf{p},t)u_1(\mathbf{k}',t)\rangle &= \frac{C^2}{\mathrm{denr}}[k'\sin(\beta-\gamma)+p\sin(\gamma-\alpha)+q\sin(\alpha-\beta) \\ &= 0 \end{aligned} \tag{10.33}$$

because

$$\frac{\sin\alpha}{k} = \frac{\sin\beta}{p} = \frac{\sin\gamma}{q}. \tag{10.34}$$

In addition, Eq. (10.27) with $H_K = 0$ yields

$$\Im\langle u_1(\mathbf{q},t)u_2(\mathbf{p},t)u_2(\mathbf{k}',t)\rangle = \frac{1}{\mathrm{denr}}[k'\sin\beta C(\mathbf{q})(C(\mathbf{k}') - C(\mathbf{p}))] = 0.$$

Therefore, for an equilibrium configuration, $S^{uu}(\mathbf{k}'|\mathbf{p}|\mathbf{q}) = 0$, or, there is no energy exchange from one mode to another. Therefore, $\Pi_u(k) = 0$. This is what is expected for an equilibrium system, and it is captured quite nicely by field-theoretic calculations.

In the next section, we will briefly describe the energy and enstrophy transfers in 2D hydrodynamic turbulence.

10.3 Energy and Enstrophy Transfers in 2D Hydrodynamic Turbulence

In 2D hydrodynamics, $u_2 = 0$, hence,

$$\langle S^{uu}(\mathbf{k}|\mathbf{p}|\mathbf{q})\rangle = \langle S^{u_1u_1}(\mathbf{k}|\mathbf{p}|\mathbf{q})\rangle. \tag{10.35}$$

In a 2D turbulence forced at $k = k_f$, the kinetic energy exhibits inverse cascade in the wavenumber regime $k < k_f$ where $E_u(k) \sim k^{-5/3}$. In the wavenumber regime $k > k_f$, the enstrophy cascades is in the forward direction and $E_u(k) \sim k^{-3}$. See Chapter 7 for details. In this section, we will discuss the properties of $\langle S^{uu}(\mathbf{k}|\mathbf{p}|\mathbf{q})\rangle$ in the $k < k_f$ regime where kinetic energy cascades from small scales to large scales.

We compute $\langle S^{uu}(\mathbf{k}|\mathbf{p}|\mathbf{q})\rangle$ in 2D hydrodynamic turbulence using Eq. (10.25). In this expression, we substitute Kolmogorov's spectrum for the kinetic energy. In Fig. 10.5, we exhibit the density plot of $\langle S^{uu}(\mathbf{k}|\mathbf{p}|\mathbf{q})\rangle$ as a function of p/k and q/k. As shown in the figure, $\langle S^{uu}(\mathbf{k}|\mathbf{p}|\mathbf{q})\rangle$ is strongly negative when $p/k \to 0$. Hence, small p modes receive energy from intermediate scale wavenumbers k ($k \gg p$) via nonlocal transfer (see the left oval of Fig. 10.5). These energy transfers are responsible for the inverse energy cascade.

Now we focus on the bottom oval of Fig. 10.5. In this regime, $S^{uu}(\mathbf{k}|\mathbf{p}|\mathbf{q}) > 0$ when $p < k$, and vice versa when $p > k$. It implies forward energy transfers for neighboring wavenumber modes. Thus, in 2D hydrodynamic turbulence, we have local forward energy transfers and nonlocal inverse energy transfers.

We also perform energy flux and shell-to-shell energy transfer computations for 2D turbulence in the $k^{-5/3}$ regime. We compute Kolmogorov's constant for 2D hydrodynamic turbulence following the same scheme as described in the previous section. For $\nu_* = -0.6$, we find that $K_{\mathrm{Ko}} \approx 6.3$ (Verma, 2004). We plot the shell-to-shell energy transfers in Fig. 10.6 that shows several interesting features. The shell-to-shell energy transfers from shell m to $m+1, m+2$ are forward and

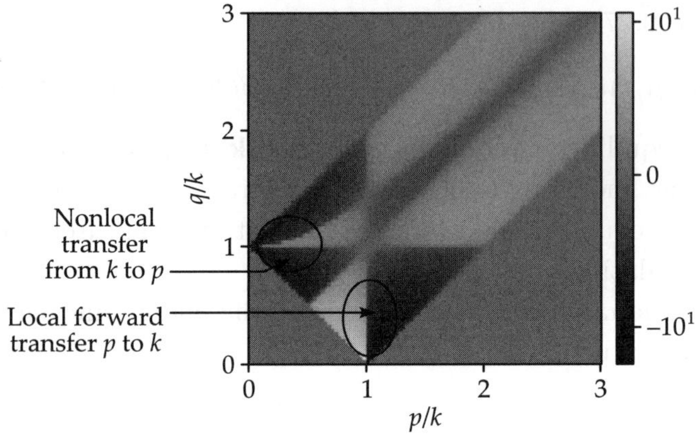

Figure 10.5 For 2D hydrodynamic turbulence, density plot of mode-to-mode KE transfer $\langle S^{uu}(\mathbf{k}'|\mathbf{p}|\mathbf{q})\rangle$ computed using field theory. The figure indicates that small p modes receive KE from modes with $k \gg p$ via nonlocal transfer; these transfers are responsible for the inverse cascade. There are forward and local transfers for modes with $p/k \approx 1$.

significant. But those from $m + i$ with $i > 2$ are negative (inverse), but small (see Fig. 10.6). These negative transfers are responsible for the inverse energy cascade in the $k^{-5/3}$ regime, and they are consistent with the mode-to-mode energy transfers described in the previous paragraph. Similar results were observed in direct numerical simulations of 2D hydrodynamic turbulence (see Chapter 7 and Gupta et al. (2018)).

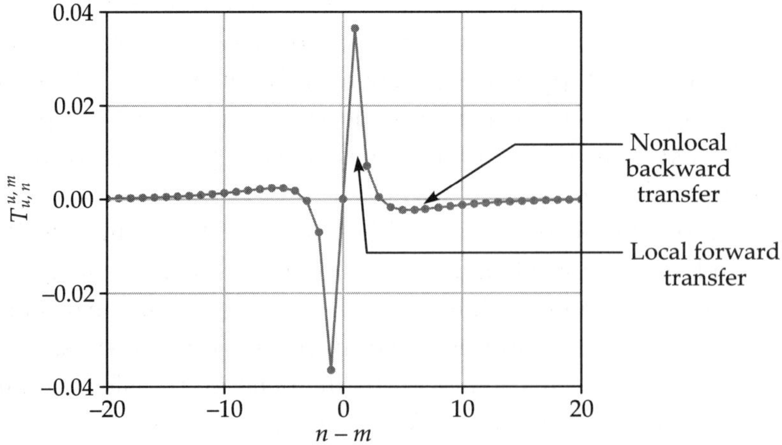

Figure 10.6 For 2D hydrodynamic turbulence, plot of shell-to-shell KE transfer $T^{u,m}_{u,n}$ from shell m to shell n vs. $n - m$. The plot indicates local forward KE transfers (from m to $m + 1, m + 2$), and nonlocal backward energy transfers.

Moreover, the energy flux is a cumulative sum of many shell-to-shell energy transfers. Consider a wavenumber sphere that contains all the shells up to shell m. If we assume that energy transfers are scale invariant, that is, $T^{u,m}_{u,n} = T^{u,m+i}_{u,n+i}$, then the energy flux crossing the aforementioned sphere is

$$\Pi_u(k) = \sum_{n=m+1}^{\infty} (n-m) T^{u,m}_{u,n}. \tag{10.36}$$

Verma et al. (2005) computed $T^{u,m}_{u,n}$ in the $k^{-5/3}$ regime of 2D hydrodynamic turbulence, and showed that

$$\sum_{n=m+1}^{\infty} (n-m) \frac{T^{u,m}_{u,n}}{\Pi} = -1, \tag{10.37}$$

which is consistent with the inverse cascade of kinetic energy in this regime. The nonlocal and backward shell-to-shell transfers contribute to the negative sum (see Fig. 10.6).

It is straightforward to compute the mode-to-mode enstrophy transfer in 2D. Using Eq. (6.25), we deduce that

$$S^{\omega\omega}(\mathbf{k}|\mathbf{p}|\mathbf{q}) = \frac{kp}{\cos\gamma} S^{uu}(\mathbf{k}|\mathbf{p}|\mathbf{q}). \tag{10.38}$$

In Fig. 10.7, we exhibit the density plot of $S^{\omega\omega}(\mathbf{k}|\mathbf{p}|\mathbf{q})$ in the $k < k_f$ regime. The enstrophy transfer comes with both positive and negative signs in approximately the same measure, and they tend to cancel each other. The strong fluctuations

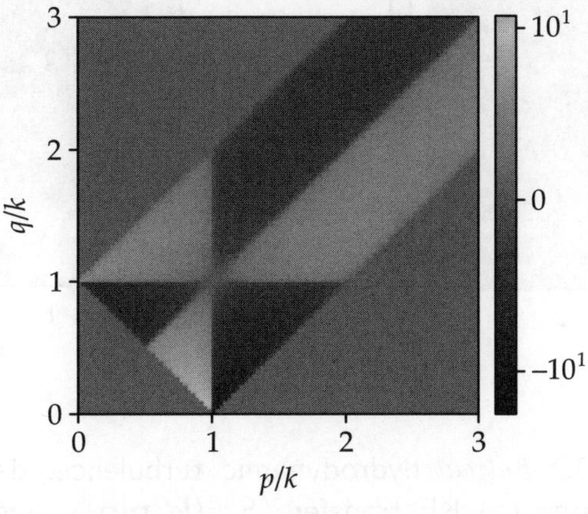

Figure 10.7 For 2D hydrodynamic turbulence, density plot of mode-to-mode enstrophy transfer $\langle S^{\omega\omega}(\mathbf{k}'|\mathbf{p}|\mathbf{q})\rangle$ computed using field theory.

in numerically-computed enstrophy flux, which was reported in Section 7.4, may be due to the aforementioned reasons. Also refer to Boffetta and Ecke (2012) for further details on Π_ω. The forward enstrophy cascade regime of 2D turbulence is interesting as well. However, we do not perform field-theoretic computation for this regime here.

In the next section we describe the mode-to-mode kinetic energy and kinetic helicity transfers in helical hydrodynamic turbulence.

10.4 Kinetic Energy and Helicity Transfers in Helical Turbulence

First, we estimate the effects of kinetic helicity on the kinetic energy flux. Using Eqs. (10.25, 10.27), we deduce that the kinetic helicity can influence $\langle S^{uu}(\mathbf{k}'|\mathbf{p},\mathbf{q})\rangle$ and kinetic energy flux only through $\langle S^{u_2u_2}(\mathbf{k}'|\mathbf{p},\mathbf{q})\rangle$. Following the discussion of helical turbulence of Chapter 8, in the expressions of Eq. (10.27), we substitute

$$\frac{H_K(k)}{k} \sim k^{-8/3}; \quad \frac{H_K(k)}{kE_u(k)} = r_K, \tag{10.39}$$

where r_K is a constant. For the computation of $\langle S^{uu}(\mathbf{k}'|\mathbf{p},\mathbf{q})\rangle$ of helical turbulence, we take $r_K = 0.3$. The computed $\langle S^{uu}(\mathbf{k}'|\mathbf{p},\mathbf{q})\rangle$'s are plotted in Fig. 10.8(a). When we compare this transfer with its nonhelical counterpart shown in Fig. 10.1, we observe that the plots are roughly similar, except that the negative regions are

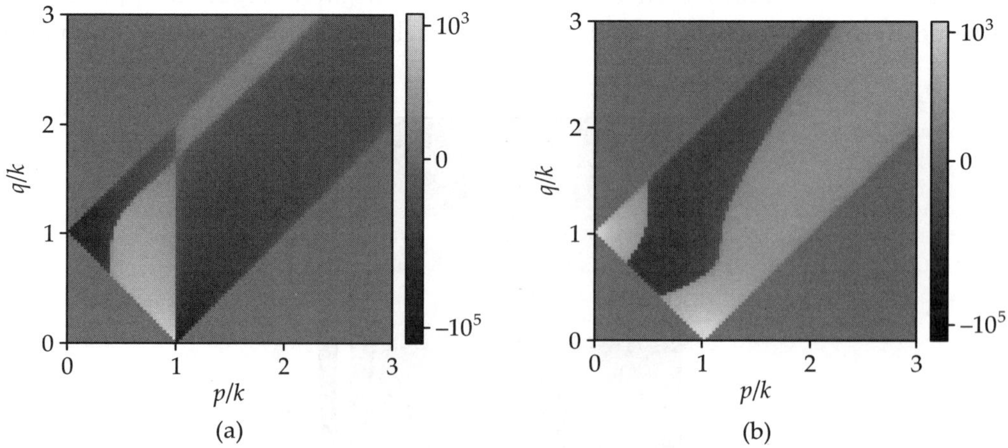

Figure 10.8 For 3D *helical* hydrodynamic turbulence, density plot of mode-to-mode (a) KE transfer $\langle S^{uu}(\mathbf{k}'|\mathbf{p}|\mathbf{q})\rangle$, and (b) kinetic helicity transfer $\langle S^{H_K}(\mathbf{k}'|\mathbf{p}|\mathbf{q})\rangle$. For these computations we choose $H_K(k)/kE_u(k) = 0.3$.

more prominent in the helical turbulence. These results are consistent with the fact that kinetic helicity does not affect the kinetic energy flux significantly (see Chapter 8).

After the discussion on energy flux, we derive an expression for the mode-to-mode kinetic helicity transfer using field-theoretic arguments. We start with Eq. (9.46):

$$\begin{aligned} S^{H_K}(\mathbf{k}'|\mathbf{p}|\mathbf{q}) &= -\Im\left[\mathbf{u}(\mathbf{q}) \cdot \boldsymbol{\omega}(\mathbf{p}) \times \boldsymbol{\omega}(\mathbf{k}')\}\right] \\ &= -k'p\sin\alpha\Re\{u_1(\mathbf{q})u_2(\mathbf{p})u_1(\mathbf{k}')\} - k'p\sin\beta\Re\{u_1(\mathbf{q})u_1(\mathbf{p})u_2(\mathbf{k}')\} \\ &\quad +k'p\sin\sin\gamma\Re\{u_2(\mathbf{q})u_2(\mathbf{p})u_2(\mathbf{k}')\}. \end{aligned} \tag{10.40}$$

To zeroth order, $S^{H_K}(\mathbf{k}'|\mathbf{p},\mathbf{q}) = 0$, hence we go to the next order. For the same, we expand one of the arguments using Green's function. Following similar steps as in Section 10.2, we obtain the following formula for the mode-to-mode kinetic helicity transfer:

$$\begin{aligned} \langle S^{H_K}(\mathbf{k}'|\mathbf{p}|\mathbf{q})\rangle &= -kp\sin\alpha[k\sin(\gamma-\beta)\frac{H_K(p)}{p}C(q) - q\sin(\beta-\alpha)\frac{H_K(p)}{p}C(k) \\ &\quad +\sin\beta\{H_K(k)C(q) - H_K(q)C(k)\}] \\ &\quad -kp\sin\beta[p\sin(\gamma-\alpha)\frac{H_K(k)}{k}C(q) + q\sin(\alpha-\beta)\frac{H_K(k)}{k}C(p) \\ &\quad +\sin\alpha\{H_K(p)C(q) - H_K(q)C(p)\}] \\ &\quad +kp\sin\gamma[-k\sin\gamma\frac{H_K(p)}{p}C(q) \\ &\quad +p\sin\gamma\frac{H_K(k)}{k}C(q) - q\sin\beta\frac{H_K(k)}{k}C(p)]. \end{aligned} \tag{10.41}$$

We plot $\langle S^{H_K}(\mathbf{k}'|\mathbf{p}|\mathbf{q})\rangle$ computed using $H_K(k)$ prescribed in Eq. (10.39). The resulting kinetic helicity transfer is plotted in Fig. 10.8(b). The figure shows a mixture of positive and negative transfers.

Field-theoretic treatment of turbulence is an extensive topic. Here we only sketch some of the main results of the field. A novelty of the treatment presented here is an introduction of mode-to-mode energy transfers, and an application of Craya-Herring basis for these computations. These computations provide valuable insights to the physics of turbulence.

Further Reading

There is an extensive literature on the field-theoretic treatment of turbulence. Here we list only a small fraction of them. McComb (1990, 2014); Leslie (1973);

Adzhemyan et al. (1999) are textbooks on this subject. For research papers, refer to Kraichnan (1959); Yakhot and Orszag (1986); McComb et al. (1992); Zhou (2010); Verma (2004); Verma et al. (2005), and references therein.

Exercises

1. Derive the transfer functions of Eqs. (10.25, 10.27, 10.41) yourself.
2. Derive $S^{\omega\omega}(\mathbf{k}'|\mathbf{p}|\mathbf{q})$ and $S^{\omega u}(\mathbf{k}'|\mathbf{p}|\mathbf{q})$ for 3D hydrodynamic turbulence using field-theoretic arguments.
3. List the major assumptions of the field-theoretic treatment described in this chapter. Are they likely to breakdown?

Chapter 11

Energy Transfers in Anisotropic Flows

Many natural and engineering flows become anisotropic on introduction of a strong external force and/or walls. Consequently, the structures in such flows exhibit anisotropy. A terrestrial hurricane which is created due to the strong rotation and gravity of the Earth is one such example.

In the earlier chapters we discussed the shell spectrum, energy flux, and shell-to-shell energy transfers that are averaged quantities over the polar angle. Hence, these measures do not capture angular anisotropy. For the description of scale-by-scale anisotropy, we divide the Fourier space along the polar angle, and construct measures that capture the anisotropic energy distribution and energy transfers.

11.1 Ring Spectrum for Spherical Rings

First, we discuss the energy contents of an anisotropic flow in various regions of Fourier space. We assume the asymmetry direction to be along $\hat{z}$. Hence, the flow is symmetric azimuthally (along ϕ), but asymmetric along the polar angle ζ.[1] Hence, the energy contents in Fourier space depend on both k and ζ. Therefore, we divide a wavenumber shell into rings as shown in Fig. 11.1. A ring is

[1]In this book, we denote the polar angle by ζ. The usual symbol θ is used to denote the passive scalar and temperature fluctuations.

characterized by two indices—the shell index k and the sector index β (Teaca et al., 2009; Reddy and Verma, 2014; Verma, 2017).

Note that the anisotropy direction is along $\zeta = 0$. For convenience, we call the rings containing $\zeta = 0$ and $\zeta = \pi/2$ as the *polar* and the *equatorial* rings respectively. The energy spectrum of a ring, called the *ring spectrum*, is defined as

$$E_u(k, \beta) = \frac{1}{\mathrm{Norm}(k, \beta)} \sum_{\substack{k-1<k'\leq k; \\ \angle \mathbf{k}' \in (\zeta_{\beta-1}, \zeta_\beta]}} \frac{1}{2}|\mathbf{u}(\mathbf{k}')|^2, \tag{11.1}$$

where $\angle \mathbf{k}'$ is the angle between $\mathbf{k}'$ and the unit vector $\hat{z}$ (anisotropy direction). The sector β contains the modes between the polar angles $\zeta_{\beta-1}$ to ζ_β, as shown in Fig. 11.1(b), with a caveat. The band $(\zeta_{\beta-1}, \zeta_\beta]$ means open interval from the left and closed interval from the right, that is, modes with $\zeta = \zeta_\beta$ are included, but the modes with $\zeta = \zeta_{\beta-1}$ are not.

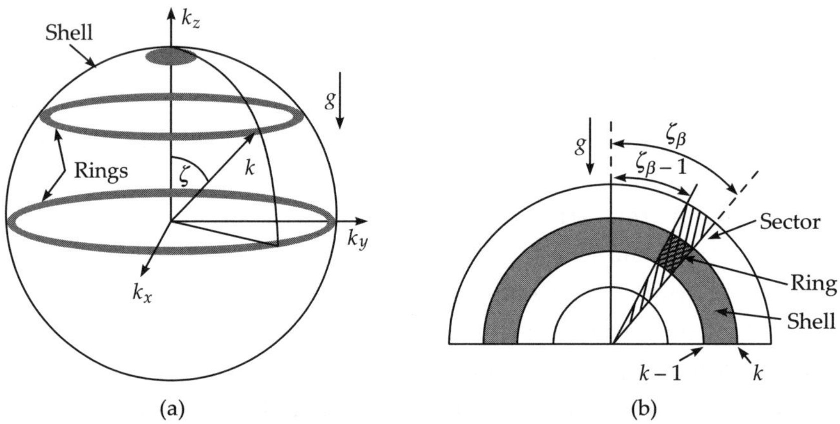

Figure 11.1 A schematic diagram exhibiting spherical rings. (b) A vertical cross section of a wavenumber sphere exhibiting shells, sectors, and rings. From Nath et al. (2016). Reprinted with permission from APS.

For a uniform $\Delta\zeta = \zeta_\beta - \zeta_{\beta-1}$, the sectors near the equator contain more modes than those near the poles. In addition, the number of Fourier modes in a ring of radius k is proportional to $k^2 \sin\zeta$, whose integral for the ring (k, β) yields the following approximate number of modes in the ring:

$$\text{No of modes} = 2\pi k^2 |\cos(\zeta_{\beta-1}) - \cos(\zeta_\beta)|. \tag{11.2}$$

Therefore, we employ one of the two normalization in Eq. (11.1):

$$\mathrm{Norm}_1(k, \beta) \quad = \quad 2\pi k^2 |\cos(\zeta_{\beta-1}) - \cos(\zeta_\beta)|. \tag{11.3a}$$

$$\mathrm{Norm}_2(k,\beta) = |\cos(\zeta_{\beta-1}) - \cos(\zeta_\beta)|. \tag{11.3b}$$

Under the choice of $\mathrm{Norm}_1(k,\beta)$, $E_u(k,\beta)$ is an average energy content of a Fourier mode in a ring. On the other hand, the second normalization only compensates for the polar angle dependence; under this definition, $E_u(k,\beta)$ has similar k dependence as $E_u(k)$. Note that for both the definitions,

$$E(k) = \sum_\beta \mathrm{Norm}(k,\beta) E_u(k,\beta). \tag{11.4}$$

We also define the ring spectra for the perpendicular and parallel components of the velocity as

$$E_{u,\perp}(k,\beta) = \frac{1}{\mathrm{Norm}(k,\beta)} \sum_{\substack{k-1<k'\le k;\\ \angle(\mathbf{k}')\in(\zeta_{\beta-1},\zeta_\beta]}} \frac{1}{2}|\mathbf{u}_\perp(\mathbf{k}')|^2, \tag{11.5a}$$

$$E_{u,\parallel}(k,\beta) = \frac{1}{\mathrm{Norm}(k,\beta)} \sum_{\substack{k-1<k'\le k;\\ \angle(\mathbf{k}')\in(\zeta_{\beta-1},\zeta_\beta]}} \frac{1}{2}|u_\parallel(\mathbf{k}')|^2, \tag{11.5b}$$

where

$$u_\parallel(\mathbf{k}') = \mathbf{u}(\mathbf{k}')\cdot\hat{z}, \tag{11.6a}$$

$$\mathbf{u}_\perp(\mathbf{k}') = \mathbf{u}(\mathbf{k}') - \hat{z}u_\parallel(\mathbf{k}'). \tag{11.6b}$$

Note that the total energy

$$E_u(k,\beta) = E_{u,\perp}(k,\beta) + E_{u,\parallel}(k,\beta). \tag{11.7}$$

We can define energy contents of a sector β as

$$E_u(\beta) = \sum_k E_u(k,\beta), \tag{11.8a}$$

$$E_{u,\perp,\parallel}(\beta) = \sum_k E_{u,\perp,\parallel}(k,\beta). \tag{11.8b}$$

In the next section we extend the above definition of the ring spectrum to cylindrical rings.

11.2 Ring Spectrum for Cylindrical Rings

Cylindrical rings, illustrated in Fig. 11.2, is another type of ring decomposition. We divide $k_\perp$ into shells $(k_\perp - 1, k_\perp)$, and $k_\parallel$ into vertical segments. A cylindrical ring is specified by a shell index $k_\perp$ and a height index i, where the ith segment contains $k_\parallel$ in the range $(H_{i-1} : H_i]$ (open interval in the left and closed interval in the right).

The energy spectrum of this cylindrical ring is defined as

$$E_u(k_\perp, i) = \frac{1}{\mathrm{Norm}(k_\perp, i)} \sum_{\substack{k-1<k'_\perp \le k \\ H_{i-1}<k'_\parallel \le H_i}} \frac{1}{2}|\mathbf{u}(\mathbf{k}')|^2. \tag{11.9}$$

Since the number of modes in a cylindrical ring is approximately $2\pi k_\perp(H_i - H_{i-1})$, we could employ two types of normalization:

$$\mathrm{Norm}_1(k_\perp, i) = 2\pi k_\perp(H_i - H_{i-1}), \tag{11.10a}$$

$$\mathrm{Norm}_2(k_\perp, i) = (H_i - H_{i-1}). \tag{11.10b}$$

With the first normalization, $E_u(k_\perp, i)$ yields an average energy content of a Fourier mode in a ring. However, the second normalization does not compensate for $k_\perp$ factor. We can define $E_{u,\perp}(k, i)$ and $E_{u,\parallel}(k, i)$ in a similar manner.

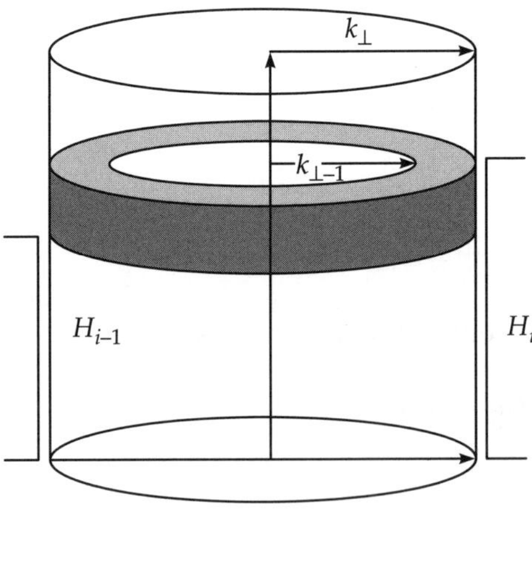

Figure 11.2 A schematic diagram of a cylindrical ring. The inner and outer radii of the ring are $k_\perp - 1$ and $k_\perp$ respectively, while its vertical extent is from H_{i-1} to H_i. From Verma (2018). Reprinted with permission from World Sci.

In an anisotropic flow,

$$\langle|u_x(\mathbf{k})|^2\rangle = \langle|u_y(\mathbf{k})|^2\rangle \neq \langle|u_z(\mathbf{k})|^2\rangle. \tag{11.11}$$

Therefore, as discussed in Chapter 3, the anisotropic parameter

$$A(k) = \frac{E_{u,\perp}(k)}{2E_{u,\parallel}(k)}. \tag{11.12}$$

is another measure of anisotropy. See Section 3.2 for details.

In the next section we quantify the anisotropic energy transfers using ring-to-ring transfers.

11.3 Ring-to-ring Energy Transfers

In Chapter 4, we introduced the energy flux that captures the energy transfers from the modes inside the sphere to the modes outside the sphere. Similarly, the shell-to-shell energy transfers tell us about the energy transfers from a shell to another shell. Unfortunately these diagnostics are averaged over the polar angle; hence, they do not describe the anisotropic energy transfers. In this section we introduce ring-to-ring energy transfers that capture the anisotropic effects.

As described in Section 11.1, we divide the Fourier space into a set of rings. The energy transfers among these rings are defined as follows. The modes in a ring (m, α), where m and α represent the shell and sector indices of the ring, interacts with all other rings. The ring-to-ring kinetic energy transfer from ring (m, α) to ring (n, β) is

$$T^{(u,m,\alpha)}_{(u,n,\beta)} = \sum_{\mathbf{k}\in(n,\beta)} \sum_{\mathbf{p}\in(m,\alpha)} S^{uu}(\mathbf{k}|\mathbf{p}|\mathbf{q}), \tag{11.13}$$

where $\mathbf{q} = \mathbf{k} - \mathbf{p}$, and $S^{uu}(\mathbf{k}|\mathbf{p}|\mathbf{q})$ is the mode-to-mode energy transfer defined in Section 4.1. $T^{(u,m,\alpha)}_{(u,n,\beta)}$ is computed using the procedure described in Section 4.11. Note that numerical computation of the ring-to-ring energy transfers for all the rings is very time-consuming since it involves a large number of rings.

For cylindrical rings, the energy transfer from ring (m, i_1) to ring (n, i_2) is given by

$$T^{(u,m,i_1)}_{(u,n,i_2)} = \sum_{\mathbf{k}\in(n,i_2)} \sum_{\mathbf{p}\in(m,i_1)} S^{uu}(\mathbf{k}|\mathbf{p}|\mathbf{q}), \tag{11.14}$$

where $\mathbf{q} = \mathbf{k} - \mathbf{p}$, $S^{uu}(\mathbf{k}|\mathbf{p}|\mathbf{q})$ is the mode-to-mode energy transfer; m, n are shell indices; and i_1, i_2 are height indices.

In anisotropic flows, the velocity components parallel and perpendicular to the anisotropy direction have different energies. Therefore, there is a preferential energy transfer either from the equatorial region to the polar region or vice versa. In quasi-static magnetohydrodynamics (QS MHD) with large interaction parameters, $u_\perp \gg u_\parallel$, and hence the kinetic energy tends to flow from the equator to the polar region (Reddy et al., 2014; Verma, 2017). A similar trend has been observed in magnetohydrodynamic (MHD) turbulence (Teaca et al., 2009; Sundar et al., 2017) and in rotating turbulence. However, in turbulent thermal convection, the anisotropic energy transfer is in the reverse direction, that is, from the polar region to the equator region (Nath et al., 2016) because $u_\parallel > u_\perp$ in turbulent thermal convection. We will revisit these systems in later parts of this book.

A corollory of the aforementioned properties: In isotropic turbulence, there is no energy transfer among the spherical rings of the same radius k due to the rotation symmetry. That is, for all α, β,

$$T^{(u,n,\alpha)}_{(u,n,\beta)} = 0 \tag{11.15}$$

on an average. However, in an anisotropic flow, these transfers will be nonzero. In strong MHD and QS MHD turbulence, $E_u(m,\beta) > E_u(m,\alpha)$ for $\beta > \alpha$. Hence, we expect that

$$T^{(u,n,\alpha)}_{(u,n,\beta)} < 0 \tag{11.16}$$

for such systems. However, the inequality is reversed in turbulent thermal convection because the rings closer to the poles have more energy than those away from it.

In the following section, we will describe how energy is exchanged between $u_\parallel$ and $u_\perp$ in anisotropic turbulence.

11.4 Anisotropic Energy Fluxes, and $u_\parallel \leftrightarrow \mathbf{u}_\perp$ Energy Exchange

In an anisotropic turbulence, $E_\perp \neq 2E_\parallel$; hence, we expect energy transfers from $E_\perp$ to $E_\parallel$, or vice versa. In this section we compute these transfers and show them to be facilitated by the pressure.

In Section 4.1, we showed that pressure does not participate in the mode-to-mode energy in an incompressible flow. However, the pressure redistributes the energy between $E_\parallel(\mathbf{k}')$ and $E_\perp(\mathbf{k}')$ in an anisotropic turbulence. We derive these results in the following.

Using Eq. (3.17) we obtain the following equations for $\mathbf{u}_\perp(\mathbf{k})$ and $u_\parallel(\mathbf{k})$:

$$\frac{d}{dt}u_\parallel(\mathbf{k}) = i\sum_{\mathbf{q}}[\mathbf{k}\cdot\mathbf{u}(\mathbf{q})]u_\parallel(\mathbf{p}) - ik_\parallel p(\mathbf{k}) + F_u(\mathbf{k}) - \nu k^2 u_\parallel(\mathbf{k}), \tag{11.17a}$$

$$\frac{d}{dt}\mathbf{u}_\perp(\mathbf{k}) = i\sum_{\mathbf{q}}[\mathbf{k}\cdot\mathbf{u}(\mathbf{q})]\mathbf{u}_\perp(\mathbf{p}) - i\mathbf{k}_\perp p(\mathbf{k}) - \nu k^2\mathbf{u}_\perp(\mathbf{k}'), \tag{11.17b}$$

where $\mathbf{q} = \mathbf{k} - \mathbf{p}$, p is the pressure, and $\mathbf{F}_u$ is along the anisotropy direction. Note that the external force is assumed to act along $u_\parallel$. As in Section 3.2, we define modal kinetic energy for $u_\parallel(\mathbf{k})$ and $u_\perp(\mathbf{k})$, which are

$$\frac{\partial E_\parallel(\mathbf{k}')}{\partial t} = \sum_{\mathbf{p}} S_\parallel(\mathbf{k}'|\mathbf{p}|\mathbf{q}) + \mathcal{P}_\parallel(\mathbf{k}') - 2\nu k'^2 E_\parallel(\mathbf{k}') + \Re(u_\parallel^*(\mathbf{k}')F_u(\mathbf{k}')), \tag{11.18a}$$

$$\frac{\partial E_\perp(\mathbf{k}')}{\partial t} = \sum_{\mathbf{p}} S_\perp(\mathbf{k}'|\mathbf{p}|\mathbf{q}) + \mathcal{P}_\perp(\mathbf{k}') - 2\nu k'^2 E_\perp(\mathbf{k}'), \tag{11.18b}$$

where $\mathbf{k}' + \mathbf{p} + \mathbf{q} = 0$, $\mathbf{k}' = -\mathbf{k}$, and

$$S_\perp(\mathbf{k}'|\mathbf{p}|\mathbf{q}) = -\Im\{[\mathbf{k}'\cdot\mathbf{u}(\mathbf{q})]\,[\mathbf{u}_\perp(\mathbf{k}')\cdot\mathbf{u}_\perp(\mathbf{p})]\}, \tag{11.19a}$$

$$S_\parallel(\mathbf{k}'|\mathbf{p}|\mathbf{q}) = -\Im\{[\mathbf{k}'\cdot\mathbf{u}(\mathbf{q})]\,[u_\parallel(\mathbf{k}')u_\parallel(\mathbf{p})]\}, \tag{11.19b}$$

$$\mathcal{P}_\perp(\mathbf{k}') = -\Im\{[\mathbf{k}'_\perp\cdot\mathbf{u}_\perp(\mathbf{k}')]\,p(\mathbf{k}')\}, \tag{11.19c}$$

$$\mathcal{P}_\parallel(\mathbf{k}') = -\Im\left\{\left[k'_\parallel u_\parallel(\mathbf{k})\right]p(\mathbf{k}')\right\}. \tag{11.19d}$$

Note that $S_\perp(\mathbf{k}'|\mathbf{p}|\mathbf{q})$ and similar formulas for other modes satisfy Eqs. (4.10, 4.11) as long as the giver and receiver modes are $\mathbf{u}_\perp$, while the mediator mode is full $\mathbf{u}$. Hence, following the same arguments as in Section 4.1, we can argue that $S_\perp(\mathbf{k}'|\mathbf{p}|\mathbf{q})$ represents the energy transfer from $\mathbf{u}_\perp(\mathbf{p})$ to $\mathbf{u}_\perp(\mathbf{k}')$ with $\mathbf{u}(\mathbf{q})$ acting as a mediator. Following similar arguments we can show that $S_\parallel(\mathbf{k}'|\mathbf{p}|\mathbf{q})$ represents the energy transfer from $u_\parallel(\mathbf{p})$ to $u_\parallel(\mathbf{k}')$ with $\mathbf{u}(\mathbf{q})$ acting as a mediator.

Following the arguments of Section 4.1, it is easy to show that

$$\sum_{\mathbf{k}'}\sum_{\mathbf{p}} S_\perp(\mathbf{k}'|\mathbf{p}|\mathbf{q}) = 0; \quad \sum_{\mathbf{k}'}\sum_{\mathbf{p}} S_\parallel(\mathbf{k}'|\mathbf{p}|\mathbf{q}) = 0 \tag{11.20}$$

for the modes in a triad $(\mathbf{k}', \mathbf{p}, \mathbf{q})$, as well as for the whole system. In addition, using $\mathbf{k}'\cdot\mathbf{u}(\mathbf{k}') = 0$, we can show that

$$\mathcal{P}_\perp(\mathbf{k}') + \mathcal{P}_\parallel(\mathbf{k}') = 0. \tag{11.21}$$

Now let us assume that $\nu = 0$ and $\mathbf{F}_u = 0$. For this case, we sum the terms of Eqs. (11.18) over all $\mathbf{k}'$ of the triad that yields

$$\frac{d}{dt}\sum_{\mathbf{k}''} E_{u,\perp}(\mathbf{k}'') = \sum_{\mathbf{k}''}\mathcal{P}_\perp(\mathbf{k}''), \tag{11.22a}$$

$$\frac{d}{dt}\sum_{\mathbf{k}''} E_{u,\|}(\mathbf{k}'') = \sum_{\mathbf{k}''} \mathcal{P}_{\|}(\mathbf{k}''). \tag{11.22b}$$

Therefore, using Eq. (11.21), we can show that for a triad

$$\sum_{\mathbf{k}''} [E_{u,\perp}(\mathbf{k}'') + E_{u,\|}(\mathbf{k}'')] = \text{constant}, \tag{11.23}$$

and that

1. $\mathcal{P}_{\perp}(\mathbf{k}'')$ is the energy transfer from $u_{\|}(\mathbf{k}'')$ to $\mathbf{u}_{\perp}(\mathbf{k}'')$.
2. $\mathcal{P}_{\|}(\mathbf{k}'')$ is the energy transfer from $\mathbf{u}_{\perp}(\mathbf{k}'')$ to $u_{\|}(\mathbf{k}'')$.

Thus, Eq. (11.18b) reveals that $E_{u,\perp}(\mathbf{k})$ receives energy from $E_{u,\|}(\mathbf{k})$ with a rate of $\mathcal{P}_{\perp}(\mathbf{k})$. The energy gained by $\mathbf{u}_{\perp}(\mathbf{k})$ via pressure equals the energy lost by $u_{\|}(\mathbf{k})$. Thus, pressure plays an important role in the energy exchange between $u_{\|}$ and $\mathbf{u}_{\perp}$. It is important to note that there is no direct energy transfer between $\mathbf{u}_{\perp}(\mathbf{k})$ and $u_{\|}(\mathbf{k})$. It follows from the fact that the nonlinear transfer does not have a term of the type $[\mathbf{u}_{\perp}^*(\mathbf{k}) \cdot \mathbf{u}_{\|}(\mathbf{p})]$ (which would anyway vanish since they are perpendicular to each other)

Following the arguments similar to those in Section 4.4, we define the energy flux for $\mathbf{u}_{\perp}$ using $S_{\perp}(\mathbf{k}'|\mathbf{p}|\mathbf{q})$. The energy flux $\Pi_{u,\perp}(k_0)$ is defined as the transfer of $u_{\perp}^2/2$ from all the modes residing inside the wavenumber sphere of radius k_0 to the modes outside the sphere, and is given by

$$\Pi_{u,\perp}(k_0) = \sum_{|\mathbf{k}|>k_0} \sum_{|\mathbf{p}|\le k_0} S_{\perp}(\mathbf{k}'|\mathbf{p}|\mathbf{q}). \tag{11.24}$$

Similarly, for $u_{\|}$, the energy flux $\Pi_{u,\|}(k_0)$ is defined as

$$\Pi_{u,\|}(k_0) = \sum_{|\mathbf{k}|>k_0} \sum_{|\mathbf{p}|\le k_0} S_{\|}(\mathbf{k}'|\mathbf{p}|\mathbf{q}). \tag{11.25}$$

The total kinetic energy flux Π_u is a sum of $\Pi_{u,\perp}$ and $\Pi_{u,\|}$:

$$\Pi_u(k_0) = \Pi_{u,\perp}(k_0) + \Pi_{u,\|}(k_0). \tag{11.26}$$

We can make the following interesting corollary. Using Eqs. (11.20), we derive the temporal evolution of $E_{u,\perp}$ as

$$\frac{d}{dt}E_{u,\perp} = \sum_{\mathbf{k}'} \mathcal{P}_{\perp}(\mathbf{k}') - \sum_{\mathbf{k}'} 2\nu k'^2 E_{u,\perp}(\mathbf{k}). \tag{11.27}$$

Thus, the viscous term dissipates $E_{u,\perp}$. Hence, for a steady state, $E_{u,\perp}$ needs an energy source, which is provided by $\mathcal{P}_\perp$; $E_{u,\perp}$ would vanish without $\mathcal{P}_\perp$. The term $\mathcal{P}_\perp$ is the energy supply to $\mathbf{u}_\perp$ by $u_\parallel$. This mechanism is present in turbulent thermal convection. Here, buoyancy feeds $E_\parallel$, which in turn feeds $E_\perp$.

In the next chapter, we will describe the energy transfers in real space.

Further Reading

For a discussion on the ring spectrum and ring-to-ring energy transfers in MHD turbulence, refer to Teaca et al. (2009); Sundar et al. (2017). These ideas have been extended to liquid metals by Reddy and Verma (2014); Verma (2017); these works also discuss energy transfers among $u_\perp$ and $u_\parallel$ modes via pressure. Nath et al. (2016) discuss the anisotropy in turbulent thermal convection.

Chapter 12

Turbulence Properties in Real Space and K41 Theory

In Chapter 5 we described spectral energy flux and spectrum of three-dimensional hydrodynamic turbulence. We showed that in the inertial range:

$$\Pi_u(k) = \epsilon_u; \quad E_u(k) = K_{\mathrm{Ko}} \epsilon_u^{2/3} k^{-5/3}, \tag{12.1}$$

where ϵ_u is the energy dissipation rate, and K_{Ko} is Kolmogorov's constant.

In 1941, Kolmogorov (1941c) and Kolmogorov (1941a) derived the laws of turbulence in real space. Using similarity hypothesis, Kolmogorov (1941c) showed that for a homogeneous, isotropic, incompressible, and steady hydrodynamic turbulence, the second order correlation varies as

$$\langle \mathbf{u}(\mathbf{r}+\mathbf{l}) \cdot \mathbf{u}(\mathbf{r}) \rangle = u_{\mathrm{rms}}^2 - c\epsilon_u^{2/3} l^{2/3}, \tag{12.2}$$

where c is a constant. Throughout this chapter, $\langle \cdot \rangle$ represents ensemble averaging. The Fourier transform of this correlation yields Kolmogorov's $k^{-5/3}$ energy spectrum. In a companion paper, Kolmogorov (1941a) showed that in the limit of infinite Reynolds number, the third order structure function

$$\left\langle (\triangle \mathbf{u})_{\|}^3 \right\rangle = -\frac{4}{5} \epsilon_u l, \tag{12.3}$$

where $(\triangle u)_{\|}$ is the component of $\mathbf{u}(\mathbf{x}+\mathbf{l}) - \mathbf{u}(\mathbf{x})$ along $\mathbf{l}$. This is an exact relation.

In this chapter we present Kolmogorov's original results. In the later part of the chapter, we will describe the properties of higher order structure functions.

12.1 Second Order Correlation Functions

We consider two real space points at $\mathbf{r}$ and $\mathbf{r}+\mathbf{l}$ where the velocity fields are $\mathbf{u}(\mathbf{r})$ and $\mathbf{u}(\mathbf{r}+\mathbf{l})$ respectively (see Fig. 12.1). For brevity, we employ symbols $\mathbf{r}'$, u_i, u_i', ∂_i, ∂_i' to denote $\mathbf{r}+\mathbf{l}$, $u_i(\mathbf{r})$, $u_i(\mathbf{r}+\mathbf{l})$, $\partial/\partial x_i$, $\partial/\partial x_i'$ respectively.

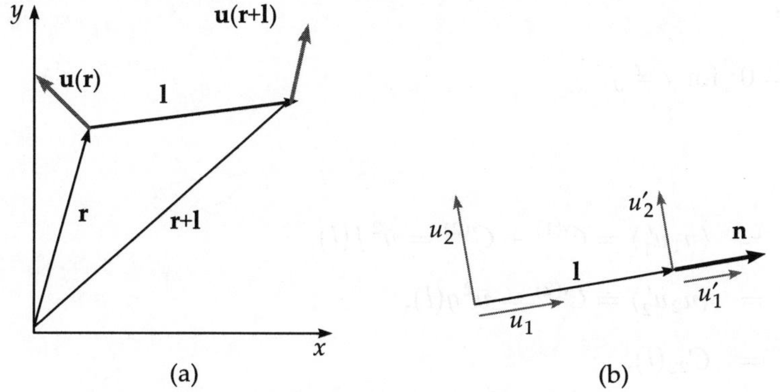

Figure 12.1 (a) We consider two points located at $\mathbf{r}$ and $\mathbf{r}+\mathbf{l}$. The velocity fields at the two points are $\mathbf{u}(\mathbf{r})$ and $\mathbf{u}(\mathbf{x}+\mathbf{r})$ respectively. (b) The components of velocity fields along $\mathbf{l}$ are u_1 and u_1' respectively. The perpendicular components are u_2, u_3, u_2' and u_3' respectively. $\mathbf{n}=\mathbf{l}/l$ is the unit vector along $\mathbf{l}$.

Kolmogorov assumed the velocity field to be random. In addition, the flow is assumed to be statistically homogeneous, isotropic, and steady. The homogeneity property implies that the correlation functions are invariant under space translation, and they depend only on the separation between the points. For example, for the two points of Fig. 12.1,

$$\langle u_i(\mathbf{r},t)u_j(\mathbf{r}+\mathbf{l},t)\rangle = C_{ij}(\mathbf{l},t). \tag{12.4}$$

For a steady flow, the correlation function is independent of time (t). For isotropic flows, the correlation functions are invariant under rotation of the reference points.

Using the homogeneity property we deduce that

$$\partial_{l_j}\langle f(\mathbf{r})g(\mathbf{r}+\mathbf{l})\rangle = \partial_{l_j}\langle f(\mathbf{r}-\mathbf{l})g(\mathbf{r})\rangle = -\partial_{(\mathbf{r}-\mathbf{l})_j}\langle f(\mathbf{r}-\mathbf{l})g(\mathbf{r})\rangle. \tag{12.5}$$

Therefore,

$$\partial'_j \langle f g' \rangle = -\partial_j \langle f g' \rangle. \tag{12.6}$$

Using the isotropy of the flow, we deduce that the second order correlation function is (Batchelor, 1953)

$$C_{ij}(\mathbf{l}) = \langle u_i u'_j \rangle = \langle u_i(\mathbf{r}) u_j(\mathbf{r}+\mathbf{l}) \rangle = C^{(1)}(l) n_i n_j + C^{(2)}(l) \delta_{ij}, \tag{12.7}$$

where $\mathbf{n} = \mathbf{l}/l$ is the unit vector along $\mathbf{l}$. For the choice of coordinate system of Fig. 12.1(b),

$$C_{ij}(\mathbf{l}) = 0 \ \text{ for } i \neq j, \tag{12.8}$$

and

$$C_{11}(l) = \langle u_1 u'_1 \rangle = C^{(1)} + C^{(2)} = \bar{u}^2 f(l), \tag{12.9a}$$

$$C_{22}(l) = \langle u_2 u'_2 \rangle = C^{(2)} = \bar{u}^2 g(l), \tag{12.9b}$$

$$C_{33}(l) = C_{22}(l). \tag{12.9c}$$

Here,

$$C_{11}(0) = C_{22}(0) = C_{33}(0) = \bar{u}^2 = \frac{u^2}{3} = \frac{2}{3} u_{\mathrm{rms}}^2, \tag{12.10}$$

and hence $f(0) = g(0) = 1$. In this chapter we assume that the flow respects mirror symmetry. Hence, in Eq. (12.7) we have ignored the mirror asymmetric term $C^{(3)} \epsilon_{ijm}(l) n_m$.

For the following discussion, we need the following identities:

$$\partial_j n_j = \partial_j (l_j/l) = \frac{1}{l} \partial_j l_j + l_j \partial_j (1/l) = \frac{2}{l}, \tag{12.11a}$$

$$\partial_j n_i = \partial_j (l_i/l) = \frac{1}{l} \delta_{ij} - \frac{1}{l^3} l_i l_j = \frac{1}{l} P_{ij}(\mathbf{n}), \tag{12.11b}$$

$$\partial_j (n_i n_j) = n_i \partial_j n_j + n_j \partial_j n_i = \frac{2}{l} n_i, \tag{12.11c}$$

where

$$P_{ij}(\mathbf{n}) = \delta_{ij} - n_i n_m \tag{12.12}$$

is the projection tensor. The incompressibility condition, $\partial_j u_j = 0$, implies that

$$\partial_i C_{ij}(l) = \partial_j C_{ij}(l) = 0. \tag{12.13}$$

Application of this relation to Eq. (12.7) yields

$$\frac{2}{l}C^{(1)}(l) + \frac{d}{dl}(C^{(1)}(l) + C^{(2)}(l)) = 0. \tag{12.14}$$

Equation (12.14) yields the following relation for $f(l)$ and $g(l)$:

$$g = f + \frac{l}{2}\frac{df}{dl}. \tag{12.15}$$

Cauchy–Schwarz inequality leads to

$$\langle u_1 u_1' \rangle \le |u_1|^2; \quad \langle u_2 u_2' \rangle \le |u_2|^2. \tag{12.16}$$

Hence, near $l = 0$, $f(l)$ and $g(l)$ must be convex functions with maxima at $l = 0$. Therefore, for small l,

$$f(l) \;=\; 1 + \frac{1}{2}f''(0)l^2 + .. \approx 1 - \frac{1}{2\lambda^2}l^2, \tag{12.17a}$$

$$g(l) \;=\; f + \frac{l}{2}f' \approx 1 - \frac{1}{\lambda^2}l^2, \tag{12.17b}$$

where

$$\lambda = \frac{1}{\sqrt{-f''(0)}} \tag{12.18}$$

is Taylor's microscale. We combine all the aforementioned properties to derive the following relation:

$$\begin{aligned} C_{ii}(l) &\approx \bar{u}^2[f(l) + 2g(l)] \\ &\approx \bar{u}^2\left[3 - \frac{5}{2\lambda^2}l^2\right]. \end{aligned} \tag{12.19}$$

Note that this correlation is valid for the dissipation range because l is assumed to be small.

The dissipation tensor is

$$\begin{aligned} \langle \omega_i \omega_i' \rangle &= \epsilon_{ijm}\epsilon_{i\alpha\beta}(\partial_j u_m)(\partial_\alpha' u_\beta') \\ &= (\delta_{j\alpha}\delta_{m\beta} - \delta_{j\beta}\delta_{m\alpha})(\partial_j u_m)(\partial_\alpha' u_\beta') \\ &= \partial_j \partial_j' \langle u_m u_m' \rangle - \partial_m' \partial_j \langle u_m u_j' \rangle \end{aligned}$$

$$
\begin{aligned}
&= -\partial'_j \partial'_j \langle u_m u'_m \rangle + \cancel{\partial'_m \partial'_j \langle u_m u'_j \rangle} \\
&= -\nabla^2 C_{ii}(l).
\end{aligned} \tag{12.20}
$$

Due to isotropy of the flow, it is convenient to use spherical coordinates. Therefore,

$$
\langle \omega_i \omega'_i \rangle = -\nabla^2 C_{ii}(l) = -\frac{1}{l^2}\frac{d}{dl} l^2 \frac{d}{dl} C_{ii}(l). \tag{12.21}
$$

We compute the dissipation rate by substituting $C_{ii}(l)$ from Eq. (12.19) in Eq. (12.21):

$$
\epsilon_u = \nu \langle \omega_i \omega_i \rangle = 15 \frac{\nu \bar{u}^2}{\lambda^2}. \tag{12.22}
$$

We employ the identity:

$$
\begin{aligned}
\frac{\partial^2}{\partial l_x^2} \langle u_1 u'_1 \rangle &= \langle u_1 \frac{\partial^2}{\partial l_x^2} u'_1 \rangle = \langle u_1 \frac{\partial}{\partial l_x} u'_1 |_{\text{surface}} \rangle - \langle (\partial u_1 / \partial x_1)^2 \rangle \\
&= -\langle (\partial u_1 / \partial x_1)^2 \rangle
\end{aligned} \tag{12.23}
$$

because the surface term is zero due to periodic or vanishing boundary condition. Therefore,

$$
\langle (\partial u_1 / \partial x_1)^2 \rangle = \bar{u}^2 / \lambda^2. \tag{12.24}
$$

The Reynolds number based on Taylor's microscale is defined as

$$
\mathrm{Re}_\lambda = \frac{\bar{u}\lambda}{\nu}. \tag{12.25}
$$

Equations (12.22, 12.10) and the definition $\mathrm{Re} = u_{\mathrm{rms}} L / \nu$ yield

$$
\lambda = \bar{u} \sqrt{\frac{15\nu}{\epsilon_u}} = u_{\mathrm{rms}} \sqrt{2/3} \sqrt{\frac{15 \nu L}{u_{\mathrm{rms}}^3}} = L \sqrt{\frac{10}{\mathrm{Re}}}. \tag{12.26}
$$

Therefore,

$$
\mathrm{Re}_\lambda = \frac{\bar{u}\lambda}{\nu} = \sqrt{\frac{20\mathrm{Re}}{3}}. \tag{12.27}
$$

In the next section, we derive the third order correlation and structure functions.

12.2 Third Order Correlation and Structure Functions

We define the third order correlation function as

$$C_{ij,m}(\mathbf{l}) = \langle u_i u_j u'_m \rangle. \tag{12.28}$$

For isotropic flows, we express $C_{ij,m}(\mathbf{l})$ using the isotropic tensor:

$$C_{ij,m}(\mathbf{l}) = A(l)\delta_{ij}n_m + D(l)(\delta_{im}n_j + \delta_{jm}n_i) + F(l)n_i n_j n_m. \tag{12.29}$$

The incompressibility condition yields

$$\partial_m C_{ij,m}(\mathbf{l}) = 0, \tag{12.30}$$

application of which to Eq. (12.29) yields

$$lA' + 2(A + D) = 0, \tag{12.31a}$$

$$(2D' + F')l + 2F - 2D = 0. \tag{12.31b}$$

In the above expressions, $'$ is shorthand for the derivative with respect to l. Using the aforementioned equations, we derive

$$[l^2(3A + 2D + F)]' = 0. \tag{12.32}$$

Hence,

$$3A + 2D + F = \frac{\text{const.}}{l^2}. \tag{12.33}$$

The correlation is finite for $l = 0$, which is possible only if the constant in Eq. (12.33) vanishes. Therefore,

$$3A + 2D + F = 0. \tag{12.34}$$

Equations (12.31, 12.34) yield

$$D = -(A + lA'/2), \tag{12.35a}$$

$$F = lA' - A. \tag{12.35b}$$

Hence, the final form for $C_{ij,m}$ tensor is

$$C_{ij,m}(\mathbf{l}) = A(l)\delta_{ij}n_m - (A + lA'/2)(\delta_{im}n_j + \delta_{jm}n_i) + (lA' - A)n_i n_j n_m. \tag{12.36}$$

The third order structure function is defined as

$$Q_{ijm}(\mathbf{l}) = \langle (u'_i - u_i)(u'_j - u_j)(u'_m - u'_m) \rangle. \tag{12.37}$$

When $i = j$,

$$\begin{aligned} Q_{iij}(\mathbf{l}) &= \langle (u'_i - u_i)(u'_i - u_i)(u'_j - u_j) \rangle \\ &= -\langle u'_i u'_i u_j \rangle + \langle u_i u_i u'_j \rangle - 2\langle u_i u'_i u'_j \rangle + 2\langle u_i u'_i u_j \rangle. \end{aligned} \tag{12.38}$$

In this equation, we set

$$\langle u'_i u'_i u'_j \rangle - \langle u_i u_i u_j \rangle = 0 \tag{12.39}$$

due to the homogeneity of the flow.

By taking a divergence of Eq. (12.38) with relative to $\mathbf{l}$ and using the property that $\partial_j u_j = 0$, we obtain

$$\partial_{l_j} Q_{iij}(\mathbf{l}) = \nabla_l \cdot \langle (u' - u)^2 (\mathbf{u}' - \mathbf{u}) \rangle = -2\partial'_j \langle u_i u'_i u'_j \rangle + 2\partial'_j \langle u_i u'_i u_j \rangle. \tag{12.40}$$

We express the third order structure function of Eq. (12.37) in terms of the aforementioned correlation function as

$$Q_{ijm} = 2(C_{ij,m} + C_{jm,i} + C_{im,j}). \tag{12.41}$$

Using the formula of Eq. (12.36), we deduce that

$$Q_{ijm}(l) = -2(lA' + A)(\delta_{ij} n_m + \delta_{im} n_j + \delta_{jm} n_i) + 6(lA' - A) n_i n_j n_m. \tag{12.42}$$

Hence,

$$S_3(l) = Q_{111}(l) = \langle [(\Delta u)_{\parallel}(l)]^3 \rangle = -12A \tag{12.43}$$

and

$$Q_{iim}(l) = [-4lA' - 16A] n_m, \tag{12.44}$$

or

$$\langle (u' - u)^2 (\mathbf{u}' - \mathbf{u}) \rangle = [-4lA' - 16A] \frac{\mathbf{l}}{l}. \tag{12.45}$$

Therefore, Eq. (12.40) can be rewritten as

$$
\begin{aligned}
\partial_{l_j} Q_{iij}(\mathbf{l}) &= \nabla_l \cdot \langle (u'-u)^2(\mathbf{u}'-\mathbf{u})\rangle = -4\frac{1}{l^2}\frac{d}{dl}l^2(lA'+4A) \\
&= -4\frac{1}{l^2}\frac{d}{dl}\left[\frac{1}{l}\frac{d}{dl}(l^4 A)\right] = \frac{1}{3}\frac{1}{l^2}\frac{d}{dl}\left[\frac{1}{l}\frac{d}{dl}(l^4 S_3(l))\right]. \qquad (12.46)
\end{aligned}
$$

In the above derivation, we employ the divergence operator in spherical coordinates system.

In the next section we derive Kolmogorov's theory of turbulence for third order structure function.

12.3 Kolmogorov's Theory of Turbulence: Four-fifth Law

Kolmogorov (1941c) and Kolmogorov (1941a) derived the four-fifth law of turbulence by relating the constant energy flux to the third order structure function. Also refer to Landau and Lifshitz (1987), Frisch (1995), and Brachet (2000) for the derivation.

The details of the derivation are as follows. First, we derive a dynamical equation for the second order correlation function:

$$
\begin{aligned}
\frac{\partial}{\partial t}\frac{1}{2}\langle u_i u_i'\rangle &= \frac{1}{2}\langle u_i'\frac{\partial}{\partial t}u_i\rangle + \frac{1}{2}\langle u_i\frac{\partial}{\partial t}u_i'\rangle \\
&= \frac{1}{2}\left[-\partial_j\langle u_i'(u_j u_i)\rangle - \partial_j'\langle u_i(u_j' u_i')\rangle - \cancel{\partial_i\langle p u_i'\rangle} - \cancel{\partial_i'\langle p' u_i\rangle}\right. \\
&\quad \left. +\langle u_i' F_i\rangle + \langle u_i F_i'\rangle + \nu\langle u_i'\nabla^2 u_i\rangle + \nu\langle u_i \nabla'^2 u_i'\rangle\right] \\
&= \frac{1}{2}\left[\partial_j'\langle u_i'(u_j u_i)\rangle - \partial_j'\langle u_i(u_j' u_i')\rangle + 2\langle u_i' F_i\rangle + 2\nu\nabla'^2\langle u_i u_i'\rangle\right] \\
&= \frac{1}{4}\nabla_l\cdot\langle (u'-u)^2(\mathbf{u}'-\mathbf{u})\rangle + \langle F_i u_i'\rangle + \nu\nabla'^2\langle u_i u_i'\rangle \\
&= T_u(\mathbf{l}) + \mathcal{F}_u(\mathbf{l}) - D_u(\mathbf{l}), \qquad (12.47)
\end{aligned}
$$

where $T_u(\mathbf{l})$ corresponds to the spectral energy transfer term $T_u(\mathbf{k})$, $\mathcal{F}_u(\mathbf{l})$ is the correlation corresponding to the energy feed by the external force, and $D_u(\mathbf{l})$ corresponds to the dissipation rate. In the above expression, $\langle p u_i'\rangle = \langle p' u_i\rangle = 0$ due to isotropy. In addition, in the second last step, we have used Eq. (12.40). Since the flow is isotropic, for any function f,

$$
f(\mathbf{l}) = f(l). \qquad (12.48)
$$

Using the aforementioned property, we deduce that

$$
\frac{\partial}{\partial t}C(l) = T_u(l) + \mathcal{F}_u(l) - D_u(l), \qquad (12.49)
$$

where

$$C(l) = \frac{1}{2}C_{ii}(l) = \frac{1}{2}\langle u_i u_i' \rangle. \tag{12.50}$$

To derive Kolmogorov's four-fifth law, we make the following assumptions:

1. The flow is steady, hence $\partial\langle u_i u_i' \rangle/\partial t = 0$.
2. $\nu \to 0$, hence the dissipation wavenumber is at infinity.
3. We assume that the flow is forced at large scales; hence, $\mathcal{F}_u(l)$ is approximately same at all scales. In spectral language, we assume that the forcing is employed at a narrow band of small wavenumbers with an injection rate of ϵ_u. For illustration, when $\mathcal{F}_u(\mathbf{k}_0) = \mathcal{F}_u(-\mathbf{k}_0) = \epsilon_u/2$ with $k_0 \to 0$,

$$\begin{aligned} \mathcal{F}_u(\mathbf{l}) &= \langle \mathbf{F}_u(\mathbf{r}) \cdot \mathbf{u}(\mathbf{r}+\mathbf{l}) \rangle = \sum_{\mathbf{k}} \mathcal{F}_u(\mathbf{k}) \exp(i\mathbf{k}\cdot\mathbf{l}) \\ &= \epsilon_u \cos(\mathbf{k}_0 \cdot \mathbf{l}) \approx \epsilon_u. \end{aligned} \tag{12.51}$$

4. We take the following limit: $l \ll L$ or $l \to 0$. In the inertial range, we ignore the dissipation term $\nu\nabla'^2\langle u_i u_i' \rangle$, and hence

$$\mathcal{F}_u(l) \approx \epsilon_u \approx -T_u(l). \tag{12.52}$$

The above assumptions and an application of Eq. (12.46) yield

$$\epsilon_u = -T_u(l) = -\frac{1}{4}\nabla_l \cdot \langle (u'-u)^2(\mathbf{u}'-\mathbf{u}) \rangle = -\frac{1}{12}\frac{1}{l^2}\frac{d}{dl}\left[\frac{1}{l}\frac{d}{dl}(l^4 S_3(l))\right]. \tag{12.53}$$

The first integration of Eq. (12.53) yields

$$\frac{1}{l}\frac{d}{dl}(l^4 S_3(l)) = -4\epsilon_u l^3. \tag{12.54}$$

Multiplication of both sides of the equation by l and a subsequent integration yields[1]

$$S_3(l) = -\frac{4}{5}\epsilon_u l. \tag{12.55}$$

This is the four-fifth law of Kolmogorov.

[1]The constants of integration vanish due to the condition that $S_3(l)$ is finite when $l \to 0$.

In the above equation we can include the viscous effects in the following manner. In Eq. (12.47), the dissipation term is

$$\begin{aligned} D_u(l) &= -\nu\nabla'^2\langle u_i u_i'\rangle = -\nu\bar{u}^2\nabla^2(f+2g) \\ &= -\nu\bar{u}^2\frac{d}{l^2}\frac{d}{dl}\left[l^2\frac{d}{dl}(f+2g)\right] \\ &= -\frac{\nu\bar{u}^2}{l^2}\frac{d}{dl}\left[\frac{1}{l}\frac{d}{dl}(l^4 f')\right]. \end{aligned} \tag{12.56}$$

Under a steady state, an inclusion of $D_u(l)$ in Eq. (12.49) leads to

$$D_u - T_u = \epsilon_u, \tag{12.57}$$

or

$$-\frac{1}{l^2}\frac{d}{dl}\left[\frac{1}{l}\frac{d}{dl}\{l^4(\nu\bar{u}^2 f' + \frac{1}{12}S_3(l))\}\right] = \epsilon_u. \tag{12.58}$$

This is a variant of Kármán–Howarth equation. Following similar steps as that involved in the derivation of Kolmogorov's four-fifth law, we derive that

$$\nu\bar{u}^2 f' + \frac{1}{12}S_3(l) = -\frac{1}{15}\epsilon_u l. \tag{12.59}$$

Note that ϵ_u traverses to the inertial range and then to the dissipation range. In the inertial range where $\nu\bar{u}^2 f'$ is negligible, we obtain Kolmogorov's four-fifth law. However, the viscous term dominates in the dissipation range. Hence, using Eq. (12.22) and the boundary condition $f(0) = 1$, an integration of

$$\nu\bar{u}^2 f' = -\frac{1}{15}\epsilon_u l \tag{12.60}$$

yields

$$f(l) = 1 - \frac{1}{2}\frac{l^2}{\lambda^2}, \tag{12.61}$$

which is the same as Eq. (12.17a).

From Eq. (12.55), using dimensional analysis, we can extrapolate that in the inertial range

$$\langle|\mathbf{u}(\mathbf{r}+\mathbf{l}) - \mathbf{u}(\mathbf{r})|^2\rangle \sim \epsilon_u^{2/3} l^{2/3}. \tag{12.62}$$

Hence,

$$C(\mathbf{l}) = \frac{1}{2}\langle \mathbf{u}(\mathbf{r}+\mathbf{l}) \cdot \mathbf{u}(\mathbf{r})\rangle \sim u_{\mathrm{rms}}^2 (1 - \mathrm{const} \times \epsilon_u^{2/3} l^{2/3}). \tag{12.63}$$

Fourier transform of this correlation function yields $k^{-5/3}$ spectrum that was discussed in Chapter 5. In Fig. 12.2, we illustrate the velocity correlation function in the dissipative and inertial regimes. Note that the correlation vanishes at large l.

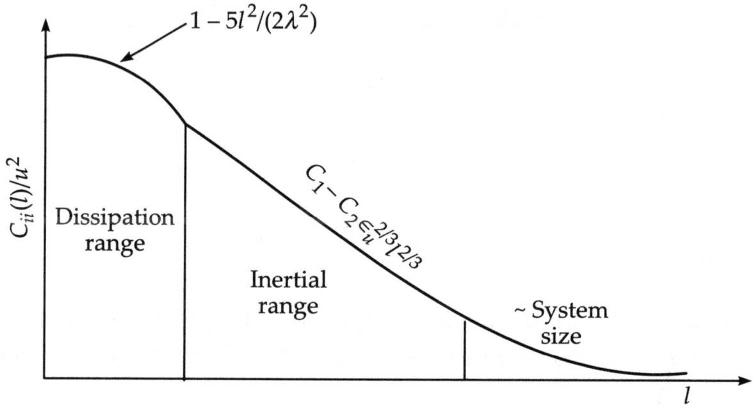

Figure 12.2 A schematic plot of normalized correlation function $C_{ii}(l)/\bar{u}^2$ vs. l.

12.4 Another Derivation of Four-fifth Law—Frisch (1995)

In the derivation of the previous section, it was assumed that $-T_u(l) \approx \epsilon_u$, which is not strictly valid. To place the derivation in a stronger footing, Frisch (1995) provided a variant of the aforementioned derivation of the four-fifth law in terms of energy flux. Also refer to Brachet (2000).

We start with the definition of the energy flux for a wavenumber sphere of radius K (see Section 4.4):

$$\Pi_u(K) = -\int_0^K T_u(k)dk = -\int_{|\mathbf{k}|<K} T_u(\mathbf{k})d\mathbf{k}. \tag{12.64}$$

Since $T_u(\mathbf{k})$ is the Fourier transform of $T_u(\mathbf{l})$ that appears in Eq. (12.47), we obtain

$$\begin{aligned} \Pi_u(K) &= -\int_{|\mathbf{k}|<K} d\mathbf{k} \int \frac{d\mathbf{l}}{(2\pi)^3} \exp(-i\mathbf{k}\cdot\mathbf{l}) T_u(l) \\ &= -\frac{1}{(2\pi)^3} \int d\mathbf{l} T_u(l) \int_{|\mathbf{k}|<K} d\mathbf{k} \exp(-i\mathbf{k}\cdot\mathbf{l}). \end{aligned} \tag{12.65}$$

For the $\mathbf{k}$ integral, we choose $\mathbf{l}$ along $\hat{z}$. We denote the angle between $\mathbf{l}$ and $\mathbf{k}$ by ζ. With this, the integral becomes

$$\begin{aligned}\Pi_u(K) &= -\frac{1}{2\pi^2}\int d\mathbf{l}T_u(l)\int k^2 dk d\cos\zeta \exp(-ikl\cos\zeta)\\ &= -\frac{1}{2\pi^2}\int d\mathbf{l}\frac{\sin Kl - Kl\cos Kl}{l^3}T_u(l)\\ &= -\frac{2}{\pi}\int_0^\infty dl\frac{\sin Kl}{l}(1+l\partial_l)T_u(l). \qquad (12.66)\end{aligned}$$

We make a change of variable $Kl = x$ and take the limit $K \to \infty$. In this limit, $\Pi_u(K)$ gets maximal contribution from $l \to 0$.

For simplification, we denote

$$-(1+l\partial_l)T_u(l) = G(l). \qquad (12.67)$$

With this, under the limit $K \to \infty$,

$$\Pi_u(K) = G(0)\frac{2}{\pi}\int_0^\infty \frac{\sin x}{x} = G(0). \qquad (12.68)$$

In the inertial range, $\Pi_u(K) = \epsilon_u$, and hence, in the limit of small l, we obtain

$$-(1+l\partial_l)T_u(l) = \epsilon_u. \qquad (12.69)$$

This equation has the following solution:

$$T_u(l) = -\epsilon_u + \frac{A}{l}, \qquad (12.70)$$

where A is a constant. Note however, $T_u(l)$ is finite when $l \to 0$, hence $A = 0$. Therefore,

$$T_u(l) = -\epsilon_u, \qquad (12.71)$$

which is same as Eq. (12.53) that yields the four-fifth law. Thus, we obtain the same relation when we start with the spectral definition of energy flux.

12.5 Comparison with Spectral Theory

When we compare Kolmogorov's arguments in real space with the energetics arguments of Fourier space (Chapter 4), we observe that both the theories assume that (a) $\nu \to 0$; (b) the flow is steady; (c) the external force field is applied at large scales. The two derivations (in real and Fourier space) are similar due to the energetics considerations. The energy supplied by the external force cascades from large scales to the dissipation range via the intermediate range of wavenumbers where the energy flux remains constant.

More crucially, the forms of $-T_u(l)$ and $\Pi_u(k)$ have close connections as illustrated in Fig. 12.3. The $\delta\mathbf{u}$'s of the product $(u' - u)^2$ are the giver and receiver fields, while the third $\delta\mathbf{u}$ of $-T_u(l)$ is the mediator field. In $\Pi_u(k)$, we sum over $\mathbf{p}$ and $\mathbf{k}$ modes because $\delta\mathbf{u}$ is a linear superposition of many Fourier modes. Also note that $K = 1/l$ is the radius of the wavenumber sphere for which energy flux is being computed. These arguments also indicate that $-T_u(l)$ is a sum of many $S^{uu}(\mathbf{k}|\mathbf{p}|\mathbf{q})$'s. Note that the giver and receiver fields cannot be contrasted in real space because $-T_u(l)$ is a sum of all the mode-to-mode energy transfers.

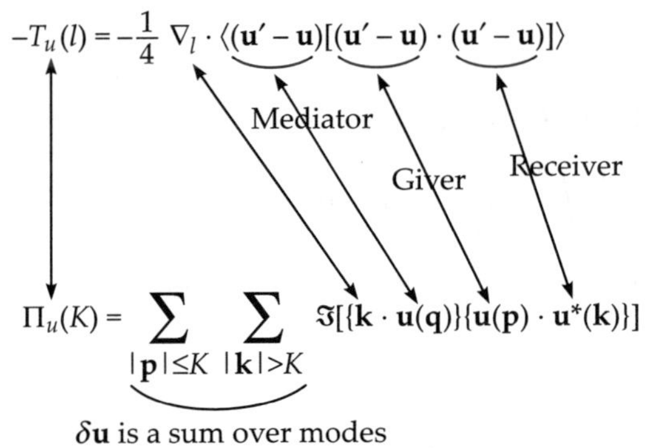

Figure 12.3 Connections between the formulas $-T_u(l)$ and the energy flux $\Pi_u(K)$. We identify the receiver, giver, and mediator fields in the figure.

Inspired by the above connections, we deduce the following relationships between the structure functions and the energy fluxes $\Pi_\perp$ and $\Pi_\parallel$ of Section 11.4. We consider an anisotropy direction, say $\hat{z}$, and split the velocity field into parallel and perpendicular components as:

$$\mathbf{u}(\mathbf{r}) = \mathbf{u}_\perp(\mathbf{r}) + \mathbf{u}_\parallel(\mathbf{r}). \tag{12.72}$$

Now, we split $T_u(\mathbf{l})$ into two parts:

$$T_{u,\perp}(\mathbf{l}) = \frac{1}{4}\nabla_l \cdot \langle(\mathbf{u}'_\perp - \mathbf{u}_\perp)^2(\mathbf{u}' - \mathbf{u})\rangle, \tag{12.73a}$$

$$T_{u,\parallel}(\mathbf{l}) = \frac{1}{4}\nabla_l \cdot \langle(\mathbf{u}'_\parallel - \mathbf{u}_\parallel)^2(\mathbf{u}' - \mathbf{u})\rangle. \tag{12.73b}$$

In $T_{u,\perp}(\mathbf{l})$, $(\mathbf{u}' - \mathbf{u})$ mediates energy transfers among $(\mathbf{u}'_\perp - \mathbf{u}_\perp)$ field components, while in $T_{u,\|}(\mathbf{l})$, the energy transfer is among the $(\mathbf{u}'_{\|} - \mathbf{u}_{\|})$ field components. Therefore, it is reasonable to expect that

$$\Pi_{\perp}(K) = -T_{u,\perp}(l), \tag{12.74a}$$
$$\Pi_{\|}(K) = -T_{u,\|}(l). \tag{12.74b}$$

The above formulas would be useful for quantifying anisotropy in turbulence with strong rotation, or magnetic field, or gravity. However, we need to verify the above conjecture.

Kolmogorov's theory assumes homogeneity and isotropy of the flow, but this assumption is not required for the statement $\Pi_u(K) = \text{const.}$ (see Section 4.5). This is because $\Pi_u(K)$ is a combined effect of the energy transfers among the modes inside and outside the sphere of radius K. Consequently, energy flux is meaningful for anisotropic flows as well. Hence, the energetics arguments of Fourier space, in some sense, are more general, and they could be employed to anisotropic turbulence, for example, to rotating turbulence, magnetohydrodynamic turbulence with external magnetic field, and buoyancy-driven turbulence. We will discuss these topics in later parts of the book.

In the next section, we describe the properties of higher order structure functions.

12.6 Higher Order Structure Functions of Hydrodynamic Turbulence

In literature, the structure function of order q is defined as

$$S_q(l) = \langle [\{\mathbf{u}(\mathbf{r}+\mathbf{l}) - \mathbf{u}(\mathbf{r})\} \cdot \mathbf{n}]^q \rangle, \tag{12.75}$$

where q is a real number, but is typically taken to be an integer. Customarily, $S_q(l)$ is modeled as the following:

$$S_q(l) \sim (\langle \epsilon_u \rangle l)^{\zeta_q}, \tag{12.76}$$

where ζ_q is the exponent of $S_q(l)$. In this section, we briefly describe several models for $S_q(l)$.

As described in the earlier sections, Kolmogorov derived an exact relation between $S_3(l)$ and the viscous dissipation rate. However, all attempts to derive exact relations for the structure functions of other orders have failed. There are several approximate theories, some of which are listed below.

A trivial generalization of Eq. (12.55) leads to

$$S_q(l) = \left\langle (\Delta u)_{\parallel}^q \right\rangle \sim (\langle \epsilon_u \rangle l)^{q/3}. \tag{12.77}$$

However, experiments and numerical simulations reveal that $\zeta_q \neq q/3$, and the correction, $\zeta_q - q/3$, is called the *intermittency correction* (Meneveau and Sreenivasan, 1987; She and Leveque, 1994; Sreenivasan, 1991; Lesieur, 2008). For example, numerical simulations and experiments show that $\zeta_2 \approx 0.71 \neq 2/3$ and $\zeta_{10} \approx 2.6 \neq 10/3$ (Meneveau and Sreenivasan, 1987; She and Leveque, 1994; Sreenivasan, 1991). Note that $\zeta_2 \approx 0.71$ yields $E(k) \sim k^{-1.71}$, thus the real spectral exponent is closer to -1.71, not $-5/3$.

These properties of ζ_q are quite interesting but complex, and they are beyond the scope of this book. Here we provide a brief description of several popular models of ζ_q.

1. The derivation of Eq. (12.77) is based on an assumption that in the inertial range, the energy flux $\Pi_u(k)$ is constant, that is, independent of k. This assumption however breaks down. The energy flux fluctuates around a mean value that makes $\zeta_q \neq q/3$.
2. *Kolmogorov's log-normal model*: Kolmogorov (1962) and Obukhov (1962) argued that ϵ_u fluctuates around its mean, and the probability distribution function of ϵ_u is log-normal. That is, the dispersion of the logarithm of ϵ_u follows:

$$\sigma^2_{\log \epsilon_u} = A + B \log L/l, \tag{12.78}$$

where A and B are constants. The log-normal model yields

$$\zeta_q = \frac{q}{3} - \frac{\mu}{18} q(q-1). \tag{12.79}$$

This model fits with numerical simulations and experiments quite well when $\mu = 0.2$.
3. *The β-model*: Frisch et al. (1978) and Novikov et al. (1997) proposed that the energy dissipation process is fractal-like. They argued that when we go from l_n to l_{n+1}, the dissipation rate D_n spilts into $2^{3-\delta}$ eddies of length l_{n+1}. Here $\beta = 2^{-\delta}$. In this model,

$$\zeta_q = \frac{q}{3} - \frac{\delta}{3}(q-3). \tag{12.80}$$

4. *The multifractal models*: Frisch and Parisi (1985) and Meneveau and Sreenivasan (1987) modeled the variations in the energy dissipation at

various scales using a multifractal model. Frisch and Parisi (1985) derived the generalized dimensions in their model. In Meneveau and Sreenivasan (1987)'s model, the viscous dissipation $D_u(l)$ at scale l is divided into two eddies of size $l/2$ as $pD_u(l)$ and $(1-p)D_u(l)$. It was derived that

$$\zeta_q = \left(\frac{q}{3} - 1\right) D_q + 1, \tag{12.81}$$

where the generalized dimension D_q is given by

$$D_q = \log_2 \left(p^q + (1-p^q)\right)^{1/(1-q)}. \tag{12.82}$$

Several experimental data fit well with the aforementioned model when $p = 0.7$.

5. *The log-Poisson model*: She and Leveque (1994) argued that the ratio $\epsilon_l^{(q)} = \langle \epsilon_l^{q+1} \rangle / \langle \epsilon_l^q \rangle$ forms a hierarchy with $\epsilon_l^\infty \sim l^{-2/3}$. Using various arguments, She and Leveque (1994) showed that

$$\zeta_q = \frac{q}{3} + 2\left[1 - \left(\frac{2}{3}\right)^{q/3}\right]. \tag{12.83}$$

We illustrate the model predictions in Fig. 12.4. It is observed that She and Leveque (1994)'s model provides best fit to experimental and numerical data. We reiterate that there is no analytical theory to derive ζ_q starting from Navier–Stokes equations.

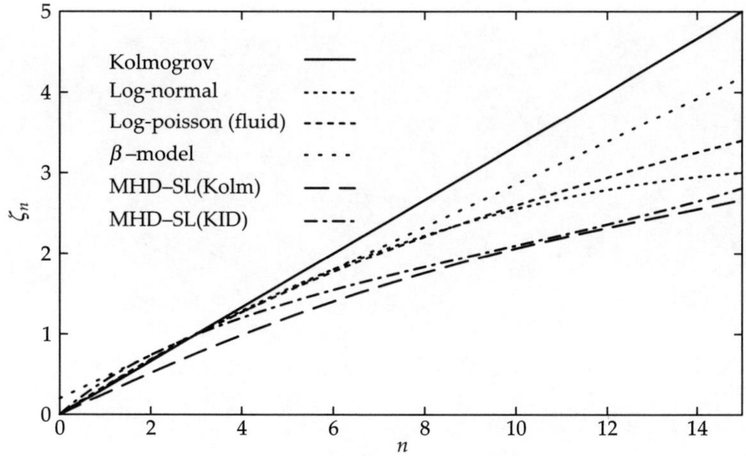

Figure 12.4 Plots of ζ_n vs. n for various intermittency models. The numerical simulation and experimental data appear to fit best with She and Leveque (1994)'s model. The functions for the MHD would be discussed in Chapter 22. From Verma (2004). Reprinted with permission from Elsevier.

With this, we end our discussion on turbulence properties in real space. In subsequent chapters of the book we will describe turbulence properties of scalars and vectors in a turbulent flow. They include passive scalar, temperature field in convection, fluid density in stably stratified flows, magnetohydrodynamics, rotating turbulence, etc.

Further Reading

Refer to the original papers by Kolmogorov (1941c) and Kolmogorov (1941a) for the derivation of the four-fifth law. Frisch (1995) and Brachet (2000) discuss these derivations in more detail. For a more detailed discussion on intermittency, refer to the textbook by Frisch (1995), as well as original papers cited in this chapter.

Part II

FLOWS WITH SCALARS

Chapter 13

Energy Transfers in Flows with Scalars

In the present and the next four chapters of the book we will discuss flows with scalars. Typical examples of scalar fields are density of dust particles, pollution, temperature, material density, etc. These scalars may or may not affect the velocity field. For example, massless or light dust particles do not backreact on the velocity field, while heavy or energy-releasing particles/fields often do. The temperature field also affects the flow via buoyancy. This property is used to classify a scalar field either as *active or passive*; scalars that back-react on the velocity field are called active, while those that do not are called passive.

In the present chapter we describe the governing equations and the energy transfers for a scalar flow. In Chapter 14, we will describe the spectral properties of a passive scalar, while in Chapters 15 and 16, we will describe buoyancy-driven flows where the fluid density and temperature act as active scalars. A more complex example called binary fluid mixture will be considered in Chapter 17.

In the next section we describe the governing equations of a flow containing a scalar.

13.1 Governing Equations

First, we describe the equations of a flow with a scalar. The evolution equation for the velocity field $\mathbf{u}$ and the scalar field θ are described respectively by the incompressible Navier–Stokes equations and an advection equation:

$$\frac{\partial \mathbf{u}}{\partial t} + (\mathbf{u} \cdot \nabla)\mathbf{u} = -\nabla(p/\rho) + \nu \nabla^2 \mathbf{u} + \mathbf{F}_u, \tag{13.1a}$$

$$\frac{\partial \theta}{\partial t} + (\mathbf{u} \cdot \nabla)\theta = \kappa \nabla^2 \theta + F_\theta, \tag{13.1b}$$

$$\nabla \cdot \mathbf{u} = 0, \tag{13.1c}$$

where p is the pressure field, ρ is the density which is assumed to be unity, ν is the kinematic viscosity of the fluid, κ is the diffusion coefficient of the passive scalar, and $\mathbf{F}_u, F_\theta$ are respectively the force fields for the velocity and scalar fields. The ratio of the viscosity and the diffusion coefficient is called the *Schmidt number* (Sc) in passive scalar literature, and the *Prandtl number* (Pr) for buoyancy-driven flows:

$$\mathrm{Sc} = \frac{\nu}{\kappa}; \quad \mathrm{Pr} = \frac{\nu}{\kappa}. \tag{13.2}$$

As we will show in subsequent chapters, this ratio plays an important role in scalar turbulence. For the scalar equation, the ratio of the nonlinear term and the diffusion term is the *Péclet number*:

$$\mathrm{Pe} = \frac{UL}{\kappa}, \tag{13.3}$$

where L is the system size, and U is the large-scale velocity of the flow.

A scalar is said to be an *active scalar* when it affects the force field $\mathbf{F}_u$, otherwise it is called a *passive scalar*. In addition, the force field of the scalar, F_θ, could either be a function of the velocity field or $F_\theta = 0$. In this book, we will cover four kinds of $\mathbf{F}_u$ and F_θ, which are related to important physical systems:

1. Passive scalar flow: $\mathbf{F}_u$ is independent of θ.
2. Stably stratified flow: $\mathbf{F}_u \propto -\rho$ and $F_\rho \propto u_z$; here, the fluid density ρ acts as an active scalar.
3. Thermal convection: $\mathbf{F}_u \propto \theta$ and $F_\theta \propto u_z$; here, the temperature θ acts as an active scalar.
4. Binary fluids: $\mathbf{F}_u \propto (\nabla\theta)(\nabla^2\theta)$ and $F_\theta = 0$, where θ is the difference in densities of the two fluid components. Here too θ is an active scalar.

In Eqs. (13.1), the equations for the velocity field are the Navier–Stokes equations discussed in Part I of the book. Hence, we need not repeat their properties. An immediate consequence of the aforementioned connection is that when $\mathbf{F}_u = 0$ and $\nu = 0$ (force-free and inviscid), the conservation laws for the velocity field remains the same as in hydrodynamics—in 3D hydrodynamics, the kinetic energy and kinetic helicity are conserved, but in 2D hydrodynamics, the kinetic energy and enstrophy are conserved. See Chapter 2 for details.

For the scalar field, we define the scalar energy density as

$$E_\theta(\mathbf{r}) = \frac{1}{2}\theta^2, \tag{13.4}$$

whose evolution equation is:

$$\frac{\partial}{\partial t}\frac{\theta^2}{2} + \nabla \cdot \left(\frac{1}{2}\theta^2 \mathbf{u}\right) = F_\theta \theta + \kappa \theta \nabla^2 \theta. \tag{13.5}$$

Clearly, in the absence of F_θ and κ, for periodic or vanishing boundary condition,

$$\int \frac{1}{2}\theta^2 d\mathbf{r} = \text{const.} \tag{13.6}$$

This is a statement of *conservation of scalar energy* in a nondiffusive scalar flow.

The equations for the velocity and scalar fields in Fourier space are as follows:

$$\frac{d}{dt}\mathbf{u}(\mathbf{k}) + \mathbf{N}_u(\mathbf{k}) = -i\mathbf{k}p(\mathbf{k}) + \mathbf{F}_u(\mathbf{k}) - \nu k^2 \mathbf{u}(\mathbf{k}), \tag{13.7a}$$

$$\frac{d}{dt}\theta(\mathbf{k}) + N_\theta(\mathbf{k}) = F_\theta(\mathbf{k}) - \kappa k^2 \theta(\mathbf{k}), \tag{13.7b}$$

$$\mathbf{k} \cdot \mathbf{u}(\mathbf{k}) = 0, \tag{13.7c}$$

where the nonlinear terms are

$$\mathbf{N}_u(\mathbf{k}) = i\sum_{\mathbf{p}} \{\mathbf{k} \cdot \mathbf{u}(\mathbf{q})\}\mathbf{u}(\mathbf{p}), \tag{13.8a}$$

$$N_\theta(\mathbf{k}) = i\sum_{\mathbf{p}} \{\mathbf{k} \cdot \mathbf{u}(\mathbf{q})\}\theta(\mathbf{p}), \tag{13.8b}$$

with $\mathbf{q} = \mathbf{k} - \mathbf{p}$. The equations for the kinetic energy and scalar energy are as follows:

$$\frac{d}{dt}E_u(\mathbf{k}) = \sum_{\mathbf{p}} \Im\left[\{\mathbf{k} \cdot \mathbf{u}(\mathbf{q})\}\{\mathbf{u}(\mathbf{p}) \cdot \mathbf{u}^*(\mathbf{k})\}\right] + \Re[\mathbf{F}_u(\mathbf{k}) \cdot \mathbf{u}^*(\mathbf{k})] - 2\nu k^2 E_u(\mathbf{k}), \tag{13.9a}$$

$$\frac{d}{dt}E_\theta(\mathbf{k}) = \sum_{\mathbf{p}} \Im\left[\{\mathbf{k} \cdot \mathbf{u}(\mathbf{q})\}\{\theta(\mathbf{p})\theta^*(\mathbf{k})\}\right] + \Re[F_\theta(\mathbf{k})\theta^*(\mathbf{k})] - 2\kappa k^2 E_\theta(\mathbf{k}). \tag{13.9b}$$

The first term in the RHS of Eq. (13.9b) represents the nonlinear transfer of scalar energy to $\theta(\mathbf{k})$ from all other modes in the system. The second and third terms of the RHS are respectively the scalar energy supply rate by the external source F_θ, and the decay rate of $E_\theta(\mathbf{k})$ by scalar diffusion.

In the following section we describe the mode-to-mode scalar energy transfers in a scalar flow.

13.2 Mode-to-mode Scalar Energy Transfers

In this section we derive an expression for the basic unit of scalar energy transfers. The derivation is similar to that in Section 4.1 where we derive formulas for the mode-to-mode kinetic energy transfers. As in Section 4.1, we focus on a triad $(\mathbf{k}', \mathbf{p}, \mathbf{q})$, and its negative counterpart, $(-\mathbf{k}', -\mathbf{p}, -\mathbf{q})$; the wavenumbers satisfy the condition $\mathbf{k}' + \mathbf{p} + \mathbf{q} = 0$. Note that $\mathbf{k}' = -\mathbf{k}$. Also, we set $F_\theta = 0$ and $\kappa = 0$. See Fig. 13.1 for a schematic diagram of the mode-to-mode energy transfers in a flow with a scalar.

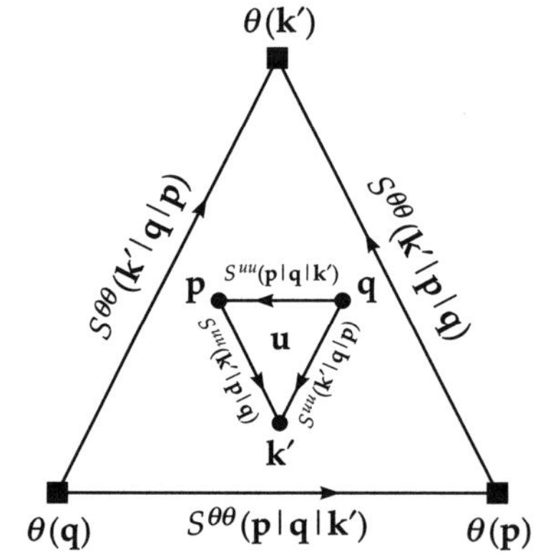

Figure 13.1 The circles represent the velocity modes, while the squares represent the scalar modes. The mode-to-mode kinetic energy transfer $S^{uu}(\mathbf{X}|\mathbf{Y}|\mathbf{Z})$ is from $\mathbf{u}(\mathbf{Y})$ to $\mathbf{u}(\mathbf{X})$ with $\mathbf{u}(\mathbf{Z})$ acting as a mediator. The mode-to-mode scalar energy transfer $S^{\theta\theta}(\mathbf{X}|\mathbf{Y}|\mathbf{Z})$ is from $\theta(\mathbf{Y})$ to $\theta(\mathbf{X})$ with $\mathbf{u}(\mathbf{Z})$ acting as a mediator.

We remark that the mode-to-mode kinetic energy transfer, whose origin is $[\mathbf{u} \cdot \nabla \mathbf{u}] \cdot \mathbf{u}$, is the same as that derived in Section 4.1. That is,

$$S^{uu}(\mathbf{k}'|\mathbf{p}|\mathbf{q}) = -\Im\left[\{\mathbf{k}' \cdot \mathbf{u}(\mathbf{q})\}\{\mathbf{u}(\mathbf{p}) \cdot \mathbf{u}(\mathbf{k}')\}\right]. \tag{13.10}$$

For the scalar field we start with Eq. (13.9b):

$$
\begin{aligned}
\frac{d}{dt}E_\theta(\mathbf{k}') &= S^{\theta\theta}(\mathbf{k}'|\mathbf{p},\mathbf{q}) \\
&= -\Im\left[\{\mathbf{k}'\cdot\mathbf{u}(\mathbf{q})\}\{\theta(\mathbf{p})\theta(\mathbf{k}')\}\right] - \Im\left[\{\mathbf{k}'\cdot\mathbf{u}(\mathbf{p})\}\{\theta(\mathbf{q})\theta(\mathbf{k}')\}\right]. \quad (13.11)
\end{aligned}
$$

The term $S^{\theta\theta}(\mathbf{k}'|\mathbf{p},\mathbf{q})$ is the *combined scalar energy transfer* to the mode $\theta(\mathbf{k}')$ from $\theta(\mathbf{p})$ and $\theta(\mathbf{q})$. It is easy to show that

$$
S^{\theta\theta}(\mathbf{k}'|\mathbf{p},\mathbf{q}) + S^{\theta\theta}(\mathbf{p}|\mathbf{q},\mathbf{k}') + S^{\theta\theta}(\mathbf{q}|\mathbf{k}',\mathbf{p}) = 0, \quad (13.12)
$$

which is a statement of the *detailed energy conservation via* $S^{\theta\theta}$ *channel.*

In this section, we derive a different quantity, $S^{\theta\theta}(\mathbf{k}'|\mathbf{p}|\mathbf{q})$, which is the mode-to-mode scalar energy transfer from $\theta(\mathbf{p})$ to $\theta(\mathbf{k}')$ with the mediation of $\mathbf{u}(\mathbf{q})$. For convenience, we denote the wavenumbers by variables $\mathbf{X},\mathbf{Y},\mathbf{Z}$ with a constraint that $\mathbf{X}+\mathbf{Y}+\mathbf{Z}=0$.

The functions $S^{\theta\theta}(\mathbf{X}|\mathbf{Y}|\mathbf{Z})$ satisfy the following relations:

1. By definition, the sum of $S^{\theta\theta}(\mathbf{X}|\mathbf{Y}|\mathbf{Z})$ and $S^{\theta\theta}(\mathbf{Y}|\mathbf{X}|\mathbf{Z})$ is the combined scalar energy transfer $S^{\theta\theta}(\mathbf{X}|\mathbf{Y},\mathbf{Z})$. Therefore,

$$
\begin{aligned}
S^{\theta\theta}(\mathbf{k}'|\mathbf{p}|\mathbf{q}) + S^{\theta\theta}(\mathbf{k}'|\mathbf{q}|\mathbf{p}) &= S^{\theta\theta}(\mathbf{k}'|\mathbf{p},\mathbf{q}), && (13.13a)\\
S^{\theta\theta}(\mathbf{p}|\mathbf{k}'|\mathbf{q}) + S^{\theta\theta}(\mathbf{p}|\mathbf{q}|\mathbf{k}') &= S^{\theta\theta}(\mathbf{p}|\mathbf{k}',\mathbf{q}), && (13.13b)\\
S^{\theta\theta}(\mathbf{q}|\mathbf{k}'|\mathbf{p}) + S^{\theta\theta}(\mathbf{q}|\mathbf{p}|\mathbf{k}') &= S^{\theta\theta}(\mathbf{q}|\mathbf{k}',\mathbf{p}). && (13.13c)
\end{aligned}
$$

2. The scalar energy transfer from $\theta(\mathbf{X})$ to $\theta(\mathbf{Y})$ is equal and opposite to the scalar energy transfer from $\theta(\mathbf{Y})$ to $\theta(\mathbf{X})$. Therefore,

$$
\begin{aligned}
S^{\theta\theta}(\mathbf{k}'|\mathbf{p}|\mathbf{q}) + S^{\theta\theta}(\mathbf{p}|\mathbf{k}'|\mathbf{q}) &= 0, && (13.14a)\\
S^{\theta\theta}(\mathbf{k}'|\mathbf{q}|\mathbf{p}) + S^{\theta\theta}(\mathbf{q}|\mathbf{k}'|\mathbf{p}) &= 0, && (13.14b)\\
S^{\theta\theta}(\mathbf{p}|\mathbf{q}|\mathbf{k}') + S^{\theta\theta}(\mathbf{q}|\mathbf{p}|\mathbf{k}') &= 0. && (13.14c)
\end{aligned}
$$

We find that

$$
S^{\theta\theta}(\mathbf{k}'|\mathbf{p}|\mathbf{q}) = -\Im\left[\{\mathbf{k}'\cdot\mathbf{u}(\mathbf{q})\}\{\theta(\mathbf{p})\theta(\mathbf{k}')\}\right] \quad (13.15)
$$

and similar formulas for other $S^{\theta\theta}$'s satisfy Eqs. (13.13a-13.14c). However, this set is not a unique solution because the determinant of the matrix formed by Eqs. (13.13, 13.14) is zero (see Section 4.1).

Following similar arguments as in Section 4.1, we provide the following physical and mathematical arguments to prove that Eq. (13.15) is indeed the recipe for the mode-to-mode scalar energy transfers in a triad.

13.2.1 A physical argument

In Eq. (13.1b), the nonlinear term $(\mathbf{u}\cdot\nabla)\theta$ represents an advection of the scalar field θ by the velocity field $\mathbf{u}$. Therefore, as argued in Section 4.1.1, $\mathbf{u}$ only mediates the scalar energy transfer. In the equation for the scalar energy, $[(\mathbf{u}\cdot\nabla)\theta]\theta$, the field $\mathbf{u}$ acts as a mediator for the scalar energy transfer between θ's that are present in the right of ∇ operator. In Fourier space, these fields correspond to the modes $\mathbf{u}(\mathbf{q})$, $\theta(\mathbf{p})$, and $\theta(\mathbf{k}')$ respectively; the scalar energy transfer occurs between mode $\theta(\mathbf{p})$ and mode $\theta(\mathbf{k}')$ with the mediation of $\mathbf{u}(\mathbf{q})$. See Fig. 13.2 for an illustration.

Hence, $S^{\theta\theta}$ of Eq. (13.15) is indeed the formula for the scalar energy transfer. Note that in Eq. (13.15), the giver and receiver scalar modes are multiplied together, while the mediator velocity mode and receiver wavenumber appear as a scalar product. Using this recipe we can write down the formulas for other mode-to-mode scalar energy transfers. Note that this formula is applicable to active as well as passive scalars because the nonlinear terms for them are the same.

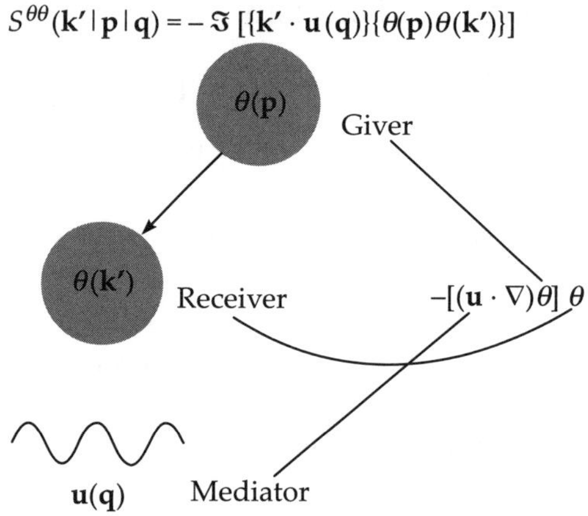

Figure 13.2 A schematic diagram exhibiting mode-to-mode scalar energy transfer from mode $\theta(p)$ to mode $\theta(k')$ with the mediation of mode $\mathbf{u}(\mathbf{q})$. For this energy transfer, the modes $\mathbf{u}(\mathbf{q})$ advects the scalar modes $\theta(p)$ and $\theta(k')$, who exchange energy among themselves.

13.2.2 A mathematical argument

We construct arguments similar to those in Section 4.1.2. Using the structure of Eq. (13.11) and tensor analysis, we demand that $S^{\theta\theta}(\mathbf{k}|\mathbf{p}|\mathbf{q})$ satisfies the following properties:

1. $S^{\theta\theta}(\mathbf{k}'|\mathbf{p}|\mathbf{q})$ is real.

2. $S^{\theta\theta}(\mathbf{k}'|\mathbf{p}|\mathbf{q})$ is a linear function of wave vector $\mathbf{k}'$ and $\theta(\mathbf{k}')$.

3. $S^{\theta\theta}(\mathbf{k}'|\mathbf{p}|\mathbf{q})$ is a linear function of one of the Fourier modes $\mathbf{u}(\mathbf{k}')$, $\mathbf{u}(\mathbf{p})$, and $\mathbf{u}(\mathbf{q})$. That is, the expressions of $S^{\theta\theta}(\mathbf{k}'|\mathbf{p}|\mathbf{q})$ includes one of the three velocity modes.

4. $S^{\theta\theta}(\mathbf{k}'|\mathbf{p}|\mathbf{q})$ is a linear function of two of the three scalar modes $\theta(\mathbf{k}')$, $\theta(\mathbf{p})$, and $\theta(\mathbf{q})$.

5. The arguments of the participating $\mathbf{u}$ and θ modes are distinct among wavenumbers $\mathbf{k}, \mathbf{p}, \mathbf{q}$.

6. Due to the equivalence of the triads $(\mathbf{k}, \mathbf{p}, \mathbf{q})$ and $(-\mathbf{k}, -\mathbf{p}, -\mathbf{q})$,

$$S^{\theta\theta}(-\mathbf{k}'|-\mathbf{p}|-\mathbf{q}) = S^{\theta\theta}(\mathbf{k}'|\mathbf{p}|\mathbf{q}). \tag{13.16}$$

Given these properties, the tensor analysis yields the following form for $S^{uu}(\mathbf{k}'|\mathbf{p}|\mathbf{q})$:

$$\begin{aligned} S^{\theta\theta}(\mathbf{k}'|\mathbf{p}|\mathbf{q}) &= c_1 \Im\left[\{\mathbf{k}' \cdot \mathbf{u}(\mathbf{q})\}\{\theta(\mathbf{p})\theta(\mathbf{k}')\}\right] \\ &+ c_2 \Im\left[\{\mathbf{k}' \cdot \mathbf{u}(\mathbf{p})\}\{\theta(\mathbf{q})\theta(\mathbf{k}')\}\right]. \end{aligned} \tag{13.17}$$

Here we have dropped the term involving $\mathbf{k}' \cdot \mathbf{u}(\mathbf{k}')$ because it is zero. Also note that $\Re[.]$ does not appear in Eq. (13.17) due to Eq. (13.16) (see Section 4.1). Now, an application of Eq. (13.14a) leads to $c_2 = 0$, while that of Eqs. (13.13a, 13.11) yields $c_1 = -1$. Thus, we deduce that

$$S^{\theta\theta}(\mathbf{k}|\mathbf{p}|\mathbf{q}) = -\Im\left[\{\mathbf{k}' \cdot \mathbf{u}(\mathbf{q})\}\{\theta(\mathbf{p})\theta(\mathbf{k}')\}\right] \tag{13.18}$$

is the formula for the scalar energy transfer from $\theta(\mathbf{p})$ to $\theta(\mathbf{k}')$ with the mediation of $\mathbf{u}(\mathbf{q})$.

Scalar flows may involve a large number of triads. Hence, a typical scalar mode $\theta(\mathbf{k})$ is affected by many modes belonging to various triads. Therefore, the evolution equation for the scalar energy spectrum $E_\theta(\mathbf{k})$ is

$$\begin{aligned} \frac{d}{dt}E_\theta(\mathbf{k}) &= \sum_{\mathbf{p}} S^{\theta\theta}(\mathbf{k}|\mathbf{p}|\mathbf{q}) + \mathcal{F}_\theta(\mathbf{k}) - 2\kappa k^2 E_\theta(\mathbf{k}), \\ &= T_\theta(\mathbf{k}) + \mathcal{F}_\theta(\mathbf{k}) - D_\theta(\mathbf{k}), \end{aligned} \tag{13.19}$$

where $\mathbf{q} = \mathbf{k} - \mathbf{p}$, and

$$\mathcal{F}_\theta(\mathbf{k}) = \Re[F_\theta(\mathbf{k})\theta^*(\mathbf{k})] \tag{13.20}$$

is the scalar energy supply rate by the external scalar force F_θ.

In the next section, we define scalar energy flux and shell-to-shell transfers for scalar turbulence.

13.3 Flux and Shell-to-shell Transfers for Scalar Turbulence

In scalar turbulence, the formulas for the kinetic energy flux and the shell-to-shell kinetic energy transfers are the same as those described in Chapter 4. In this section we derive the flux and shell-to-shell transfers for the scalar field.

We consider a wavenumber sphere of radius k_0. The scalar energy flux $\Pi_\theta(k_0)$ is the net scalar energy transferred from the modes inside the wavenumber sphere to the modes outside the sphere. Using the definition of the mode-to-mode scalar energy transfers, we define the scalar energy flux $\Pi_\theta(k_0)$ as

$$\Pi_\theta(k_0) = \sum_{|\mathbf{k}'|>k_0} \sum_{|\mathbf{p}|\le k_0} S^{\theta\theta}(\mathbf{k}'|\mathbf{p}|\mathbf{q}). \tag{13.21}$$

Similarly, we derive the shell-to-shell scalar energy transfer from shell m to shell n as

$$T^{\theta,m}_{\theta,n} = \sum_{\mathbf{k}'\in n} \sum_{\mathbf{p}\in m} S^{\theta\theta}(\mathbf{k}'|\mathbf{p}|\mathbf{q}). \tag{13.22}$$

We remark that the above formulas for the flux and the shell-to-shell energy transfers are the same for active and passive scalars. We can also extend these formulas to ring-to-ring scalar energy transfers.

Following arguments similar to those in Section 4.4, we derive the evolution equation for the scalar energy spectrum $E_\theta(k)$ as

$$\frac{\partial}{\partial t}E_\theta(k,t) = -\frac{\partial}{\partial k}\Pi_\theta(k,t) + \mathcal{F}_\theta(k,t) - D_\theta(k,t), \tag{13.23}$$

where $\mathcal{F}_\theta(k)$ is the scalar energy supply rate by F_θ, and $D_\theta(k)$ is the diffusion rate of the scalar energy in shell k, that is,

$$\mathcal{F}_\theta(k)dk = \sum_{k<k'\le k+dk} \Re[F_\theta(\mathbf{k}')\theta^*(\mathbf{k}')], \tag{13.24}$$

$$D_\theta(k)dk = \sum_{k<k'\le k+dk} 2\kappa k^2 E_\theta(\mathbf{k}'). \tag{13.25}$$

Under a steady state, $\partial E_\theta(k)/\partial t \approx 0$, we obtain

$$\frac{d}{dk}\Pi_\theta(k) = \mathcal{F}_\theta(k) - D_\theta(k). \tag{13.26}$$

In the next section, we will use this relation to describe variable scalar energy flux.

13.4 Variable Scalar Energy Flux

In Section 4.5 we described how the kinetic energy flux $\Pi_u(k)$ is affected by energy injection rate $\mathcal{F}_u(k)$. Equation (13.26) is similar to Eq. (4.49); hence, we expect similar behavior for the scalar energy flux $\Pi_\theta(k)$.

We focus on the inertial range of the scalar field; hence, we can assume that $D_\theta(k) \approx 0$. Therefore,

$$\frac{d}{dk}\Pi_\theta(k) = \mathcal{F}_{\theta,\text{in}}(k), \tag{13.27}$$

where $\mathcal{F}_{\theta,\text{in}}(k)$ is the scalar energy injection rate by F_θ in the inertial range. Using this equation, the behavior of the scalar energy flux can be classified into the following four categories. If $\mathcal{F}_{\theta,\text{in}}(k) = 0$, then $\Pi_\theta(k) = C$, where $C = 0$ or > 0; if $\mathcal{F}_{\theta,\text{in}}(k) < 0$, then $d\Pi_\theta(k)/dk < 0$; if $\mathcal{F}_{\theta,\text{in}}(k) > 0$, then $d\Pi_\theta(k)/dk > 0$. See Fig. 13.3 for an illustration. Following similar lines of arguments as in Section 4.5, we detail them as follows:

1. $\mathcal{F}_{\theta,\text{in}}(k) = 0$ and $\Pi_\theta(k) = 0$: This case, illustrated in Fig. 13.3(a), corresponds to an equilibrium configuration for which $\langle S^{\theta\theta}(\mathbf{k}'|\mathbf{p}|\mathbf{q})\rangle = 0$. That is, there is no net scalar energy transfer from any mode to any other mode, and the system respects detailed balance of energy exchange. According to the absolute equilibrium theory, such behavior is expected when $\kappa = 0$.

   In the absence of mode-to-mode energy transfers ($\langle S^{uu}(\mathbf{k}'|\mathbf{p}|\mathbf{q})\rangle = \langle S^{\theta\theta}(\mathbf{k}'|\mathbf{p}|\mathbf{q})\rangle = 0$), the Fourier modes $\mathbf{u}(\mathbf{k}')$, $\mathbf{u}(\mathbf{p})$, $\mathbf{u}(\mathbf{q})$, $\theta(\mathbf{k}')$, $\theta(\mathbf{p})$, and $\theta(\mathbf{q})$ are expected to have random and uncorrelated phases. For pure hydrodynamic turbulence, absolute equilibrium theory predicts that the modal kinetic energy of all the velocity modes are equal on an average. Hence,

   $$E_u(\mathbf{k}) \approx \text{const.} \implies E_u(k) \sim k^2. \tag{13.28}$$

   An extension of the aforementioned theory to scalar turbulence yields

   $$E_\theta(k) \sim k^2. \tag{13.29}$$

2. $\mathcal{F}_{\theta,\text{in}}(k) = 0$ and $\Pi_\theta(k) = C > 0$: As in Kolmogorov's picture of hydrodynamic turbulence, in scalar turbulence, the scalar energy would cascade from large scales (small k's) to small scales (large k's). See Fig. 13.3(b) for an illustration in which $\Pi_\theta(k) = C$ for $1/L < k < k_{\text{DI}}$, where L is the box size, and k_{DI} is the wavenumber from which diffusion becomes

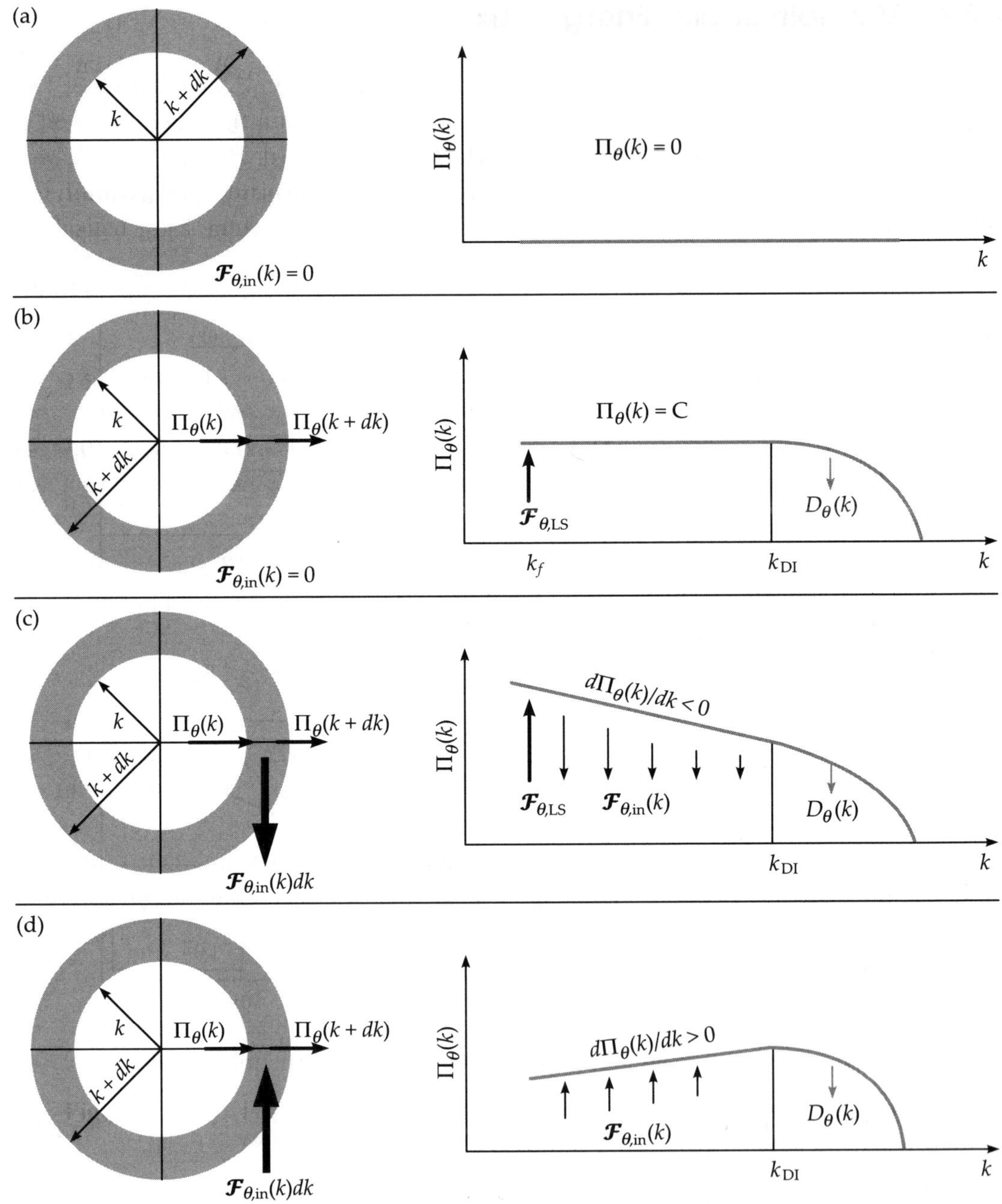

Figure 13.3 Schematic diagrams illustrating behavior of scalar energy flux $\Pi_\theta(k)$ for various types of scalar energy injection rates $\mathcal{F}_{\theta,\text{in}}(k)$: (a) $\mathcal{F}_{\theta,\text{in}}(k) = 0$ and $\Pi_\theta(k) = 0$; (b) $\mathcal{F}_{\theta,\text{in}}(k) = 0$ and $\Pi_\theta(k) = C > 0$; (c) $\mathcal{F}_{\theta,\text{in}}(k) < 0$ and $d\Pi_\theta(k)/dk < 0$; (d) $\mathcal{F}_{\theta,\text{in}}(k) > 0$ and $d\Pi_\theta(k)/dk > 0$.

significant. Here "DI" stands for diffusion. Here, κ is finite but small, and the nonlinear term, $(\mathbf{u} \cdot \nabla)\theta$, is much larger than the diffusion term, $\kappa\nabla^2\theta$.

For this case, which is detailed in Chapter 14 of this book, both $E_u(k)$ and $E_\theta(k)$ follow $k^{-5/3}$ spectrum.

3. $\mathcal{F}_{\theta,\text{in}}(k) < 0$ and $d\Pi_\theta(k)/dk < 0$: In this case, $\Pi_\theta(k)$ decreases with k as shown in Fig. 13.3(c). Stably stratified flows exhibit such behavior. The fluid density is to be treated as the scalar field for this case. See Chapter 15 for more details.

4. $\mathcal{F}_{\theta,\text{in}}(k) > 0$ and $d\Pi_\theta(k)/dk > 0$: In this case, $\Pi_\theta(k)$ increases with k as shown in Fig. 13.3(d). With temperature as the scalar field, the scalar energy flux of turbulent thermal convection shows such behavior.

The aforementioned discussion illustrates how variable energy flux captures various kinds of scalar turbulence in a single framework. We will revisit these arguments when we describe passive scalar turbulence and buoyancy-driven turbulence.

13.5 Scalar Field in Craya–Herring Basis

As described in Chapter 9, Craya–Herring basis provides a compact representation for hydrodynamic flows. In Section 9.2, we showed how to derive equations for $u_{1,2}$, the components of $\mathbf{u}$ in Craya–Herring basis.

In the following discussion, following Section 9.2, we focus on a single triad $(\mathbf{k}', \mathbf{p}, \mathbf{q})$ with a constraint that $\mathbf{k}' + \mathbf{p} + \mathbf{q} = 0$, and derive equations for the scalar modes corresponding to these wavenumbers. The equations of the velocity field are the same as those derived in Section 9.2.

For the scalar field, the equation for the Fourier mode $\theta(\mathbf{k}')$ is as follows:

$$\frac{d}{dt}\theta(\mathbf{k}') = -i[\mathbf{k}' \cdot \mathbf{u}(-\mathbf{q})]\theta(-\mathbf{p}) - i[\mathbf{k}' \cdot \mathbf{u}(-\mathbf{p})].\theta(-\mathbf{q}) - \kappa k^2\theta(\mathbf{k}') \tag{13.30}$$

with $\mathbf{k}' = -\mathbf{p} - \mathbf{q}$. The equations for $\theta(\mathbf{p})$ and $\theta(\mathbf{q})$ are written in a similar manner. Following the same steps as in Section 9.2, we simplify the aforementioned equations and those for the wavenumbers $\mathbf{p}$ and $\mathbf{q}$ as

$$\dot{\theta}(\mathbf{k}') = ik'\{\sin\gamma u_1^*(\mathbf{p})\theta^*(\mathbf{q}) - \sin\beta u_1^*(\mathbf{q})\theta^*(\mathbf{p})\} - \kappa k^2\theta(\mathbf{k}'), \tag{13.31a}$$

$$\dot{\theta}(\mathbf{p}) = ip\{\sin\alpha u_1^*(\mathbf{q})\theta^*(\mathbf{k}') - \sin\gamma u_1^*(\mathbf{k}')\theta^*(\mathbf{q})\} - \kappa p^2\theta(\mathbf{p}), \tag{13.31b}$$

$$\dot{\theta}(\mathbf{q}) = iq\{\sin\beta u_1^*(\mathbf{k}')\theta^*(\mathbf{p}) - \sin\alpha u_1^*(\mathbf{p})\theta^*(\mathbf{k}')\} - \kappa q^2\theta(\mathbf{q}). \tag{13.31c}$$

Note that the structure of the equations for the scalar field is same as that for u_2. Also note that the above equations conserve the total triadic scalar energy:

$$E_\theta = \frac{1}{2}\left(|\theta(\mathbf{k}')|^2 + |\theta(\mathbf{p})|^2 + |\theta(\mathbf{q})|^2\right). \tag{13.32}$$

The formulas for the mode-to-mode scalar energy transfers too get simplified in Craya–Herring basis. Following similar steps as in Section 9.3, we derive

$$S^{\theta\theta}(\mathbf{k}'|\mathbf{p}|\mathbf{q}) = -k' \sin\beta \Im\{u_1(\mathbf{q})\theta(\mathbf{p})\theta(\mathbf{k}')\}, \tag{13.33a}$$

$$S^{\theta\theta}(\mathbf{p}|\mathbf{q}|\mathbf{k}') = -p \sin\gamma \Im\{u_1(\mathbf{k}')\theta(\mathbf{p})\theta(\mathbf{q})\}, \tag{13.33b}$$

$$S^{\theta\theta}(\mathbf{q}|\mathbf{k}'|\mathbf{p}) = -q \sin\alpha \Im\{u_1(\mathbf{p})\theta(\mathbf{q})\theta(\mathbf{k}')\}. \tag{13.33c}$$

On some occasions (e.g. with many triads), it is practical to use the common direction as $\hat{n}$ for all the triads. For this case, the derivations of equations of motion require explicit computations of $[\mathbf{k}' \cdot \mathbf{u}(-\mathbf{q})]$ and $[\mathbf{k}' \cdot \mathbf{u}(-\mathbf{p})]$ in Eq. (13.30). Similar calculations are needed for the scalar energy transfers.

With this, we end our discussion on Craya–Herring basis for scalar flows.

Example 13.1: In a periodic box $[2\pi, 2\pi]$, consider the following force-free scalar flow with $\nu = \kappa = 0$:

$$\mathbf{u} = \hat{x}2B\cos y + \hat{y}2C\cos x + (\hat{x} - \hat{y})2A\sin(x + y),$$

$$\theta = 2B'\cos y + 2C'\cos x + 2A'\sin(x + y).$$

Derive equations for A', B', and C' as in Example 3.4.

Solution: The Fourier transforms of velocity and scalar fields yield Fourier modes with wavenumbers $(1,0)$, $(0,1)$, $(1,1)$, $(-1,0)$, $(0,-1)$, and $(-1,-1)$. We choose $\hat{n} = \hat{z}$. For the interacting triad $\mathbf{k}' = (-1,-1)$, $\mathbf{p} = (0,1)$, and $\mathbf{q} = (1,0)$, the amplitudes of the Fourier modes in the Craya–Herring basis are listed in Table 13.1. Note that $u_2 = 0$ since the field is two-dimensional.

Table 13.1 Example 13.1: The amplitudes of the velocity and scalar Fourier modes.

Mode	u_1	θ
$\mathbf{k}' = (-1,-1)$	$-\frac{A\sqrt{2}}{i}$	$-\frac{A'}{i}$
$\mathbf{p} = (0,1)$	B	B'
$\mathbf{q} = (1,0)$	$-C$	C'

The equations for the velocity field in a scalar flow are the Navier–Stokes equations. Hence, the equations of motion for A, B, and C are same as those in Example 3.4 or Example 9.2. The solution of these quantities are

$$\begin{aligned} A &= \text{constant}, \\ B &= c\cos(At), \\ C &= c\sin(At). \end{aligned}$$

For the scalar field, we consider the interacting triad shown in Fig. 9.7(a), whose internal angles are $\alpha = 90°$, $\beta = 45°$, and $\gamma = 45°$. Substitution of these in Eq. (13.31) yields

$$\begin{aligned} \dot{\theta}(-1,-1) &= -\frac{\dot{A}'}{i} = i\frac{\sqrt{2}}{\sqrt{2}}(BC' + CB'), \\ \dot{\theta}(0,1) &= \dot{B}' = i(-C(A'/i) - (A/i)C'), \\ \dot{\theta}(1,0) &= \dot{C}' = ((A/i)B' - B(A'/i)), \end{aligned}$$

or,

$$\begin{aligned} \dot{A}' &= BC' + CB', \\ \dot{B}' &= -CA' - AC', \\ \dot{C}' &= AB' - BA'. \end{aligned}$$

We can solve for A', B', and C' using the solution of A, B, and C. Also note that

$$A'^2 + B'^2 + C'^2 = \text{const.},$$

as expected from the conservation of scalar energy for the nondiffusive case.

Example 13.2: Compute the mode-to-mode scalar energy transfers for the flow field of Example 13.1.

Solution: We employ Eqs. (13.33) for these computations:

$$\begin{aligned} S^{\theta\theta}(\mathbf{k}'|\mathbf{p}|\mathbf{q}) &= -k'\sin\beta\Im\{u_1(\mathbf{q})\theta(\mathbf{p})\theta(\mathbf{k}')\} \\ &= -\frac{\sqrt{2}}{\sqrt{2}}\Im\{(-C)B'\left(-\frac{A'}{i}\right)\} = A'B'C. \\ S^{\theta\theta}(\mathbf{p}|\mathbf{q}|\mathbf{k}') &= -p\sin\gamma\Im\{u_1(\mathbf{k}')\theta(\mathbf{p})\theta(\mathbf{q})\} \\ &= -\frac{1}{\sqrt{2}}\Im\{\left(-\frac{A\sqrt{2}}{i}\right)B'C'\} = -AB'C'. \\ S^{\theta\theta}(\mathbf{q}|\mathbf{k}'|\mathbf{p}) &= -q\sin\alpha\Im\{u_1(\mathbf{p})\theta(\mathbf{q})\theta(\mathbf{k}')\} \\ &= -\Im\{BC'\left(-\frac{A'}{i}\right)\} = -A'BC'. \end{aligned}$$

We illustrate these scalar energy transfers in Fig. 13.4.

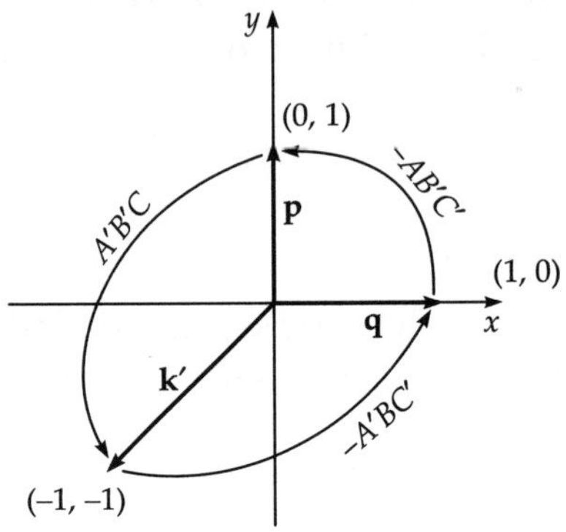

Figure 13.4 Scalar energy transfers among the modes of Example 13.2.

With this, we end our discussion on scalar flows. In the next four chapters, we will discuss various examples of passive and active scalars, in particular, passive scalar, stably stratified turbulence, turbulent thermal convection, and binary fluid mixture.

Exercises

1. In a closed box $[\pi, \pi, \pi]$, consider the following force-free scalar flow with $\nu = \kappa = 0$:

$$\begin{aligned} \mathbf{u} &= 4C(\hat{x}\sin x\cos z - \hat{z}\cos x\sin z) + 4B(\hat{y}\sin y\cos z - \hat{z}\cos y\sin z) \\ &\quad +8A(-\hat{x}\sin x\cos y\cos 2z - \hat{y}\cos x\sin y\cos 2z + \hat{z}\cos x\cos y\sin 2z). \\ \theta &= 4C'\cos x\sin z + 4B'\cos y\sin z + 8A'\hat{z}\cos x\cos y\sin 2z. \end{aligned}$$

   Derive equations for A, B, C, A', B', and C'.

2. In a periodic box $[\pi, \pi]$, consider the following force-free scalar flow with $\nu = \kappa = 0$:

$$\begin{aligned} \mathbf{u} &= 4C(\hat{x}\sin 3x\cos y - \hat{y}3\cos 3x\sin y) + 4B(\hat{y}\sin 2x\cos 2y - \hat{y}\cos 2x\sin 2y) \\ &\quad +4A(\hat{x}\sin x\cos y - \hat{y}\cos x\sin y). \\ \theta &= 4C'\cos 3x\sin y + 4B'\cos 2x\sin 2y + 4A'\cos x\sin y. \end{aligned}$$

   Derive equations for A, B, C, A', B', and C'.

Chapter 14

Flows with a Passive Scalar

In the previous chapter, we provided a framework for describing a fluid flow with a scalar, which could be either active or passive. In the present chapter, we will describe spectral properties of a passive scalar flow. We start with the governing equations of the flow.

14.1 Governing Equations

The equation of an incompressible passive scalar is a special case of Eqs. (13.1) in which $\mathbf{F}_u$ is independent of θ. Hence,

$$\frac{\partial \mathbf{u}}{\partial t} + (\mathbf{u} \cdot \nabla)\mathbf{u} = -\nabla(p/\rho) + \nu\nabla^2\mathbf{u} + \mathbf{F}_u, \tag{14.1a}$$

$$\frac{\partial \theta}{\partial t} + (\mathbf{u} \cdot \nabla)\theta = F_\theta + \kappa\nabla^2\theta, \tag{14.1b}$$

$$\nabla \cdot \mathbf{u} = 0, \tag{14.1c}$$

where $\mathbf{u}, \theta, p$ are the velocity, scalar, and pressure fields respectively; ρ is density which is assumed to be unity; $\mathbf{F}_u, F_\theta$ are the force fields for the velocity and scalar fields; ν is the kinematic viscosity for the flow; and κ is the diffusion coefficient for the scalar. The ratio of the viscosity and the diffusivity coefficient is the *Schmidt number*:

$$\mathrm{Sc} = \frac{\nu}{\kappa}. \tag{14.2}$$

The equation for the kinetic energy is the same as that for pure hydrodynamics; hence, it is not repeated here. As described in Chapter 13, the equation for the scalar energy density $\theta^2/2$ is

$$\frac{\partial}{\partial t}\frac{\theta^2}{2} + \nabla \cdot \left(\frac{1}{2}\theta^2 \mathbf{u}\right) = F_\theta \theta + \kappa \theta \nabla^2 \theta. \tag{14.3}$$

Clearly, for $\kappa = 0$ and $F_\theta = 0$, and for periodic or vanishing boundary condition,

$$\int \frac{1}{2}\theta^2 d\mathbf{r} = \text{const.} \tag{14.4}$$

This is a statement of *conservation of scalar energy* in a nondiffusive flow.

The equations for the velocity and scalar fields in Fourier space are the same as Eqs. (13.7). The equations for the mode-to-mode scalar energy transfers, scalar energy flux, and the shell-to-shell scalar energy transfers are the same as those described in Sections 13.2 and 13.3.

The evolution equation for the passive scalar energy spectrum $E_\theta(k)$ is

$$\frac{\partial}{\partial t} E_\theta(k,t) = -\frac{\partial}{\partial k}\Pi_\theta(k,t) + \mathcal{F}_\theta - D_\theta(k,t). \tag{14.5}$$

Under a steady state, $\partial E_\theta(k)/\partial t \approx 0$; hence, we obtain

$$\frac{d}{dk}\Pi_\theta(k) = \mathcal{F}_\theta - 2\kappa k^2 E_\theta(k). \tag{14.6}$$

In the next section, we will use this relation to derive the scalar energy spectrum and flux.

14.2 Phenomenology of Passive Scalar Turbulence

In passive scalar flows, the velocity field $\mathbf{u}$ is unaffected by the passive scalar; hence, for 3D and 2D systems, the kinetic energy spectra and fluxes would be same as those described in Chapters 5 and 7 respectively. In this section, we will derive the scalar energy spectrum and flux for 3D passive scalar turbulence. We assume that F_θ acts at large scales of the flow.

Under a steady state, in the inertial range where $\mathcal{F}_\theta = 0$ and $D_\theta(k) \to 0$, using Eq. (14.6), we deduce that

$$\frac{d}{dk}\Pi_\theta(k) = 0 \implies \Pi_\theta = \text{const.} \tag{14.7}$$

Based on scaling arguments, we expect that $E_\theta(k)$ depends on Π_u, Π_θ, and k, whose dimensions are respectively

$$[\Pi_u] = [L^2/T^3]; \quad [\Pi_\theta] = [\Theta^2/T]; \quad [E_\theta(k)] = [\Theta^2 L]. \tag{14.8}$$

Using dimensional analysis we postulate that

$$E_\theta(k) = (\Pi_\theta)^\alpha (\Pi_u)^\beta k^\gamma. \tag{14.9}$$

Now matching the dimensions of $[\Theta]$, $[L]$, and $[T]$ yields

$$\alpha = 1; \quad \beta = -1/3; \quad \gamma = -5/3. \tag{14.10}$$

Therefore,

$$E_\theta(k) = K_{\mathrm{OC}} \Pi_\theta (\Pi_u)^{-1/3} k^{-5/3}, \tag{14.11}$$

where K_{OC} is the *Obukhov–Corrsin constant.* Using field-theoretic computations, Yakhot and Orszag (1986) reported that $K_{\mathrm{OC}} = 1.16$, while Verma (2001b) observed that $K_{\mathrm{OC}} = 1.25$. Following the derivation of Eq. (5.8), using Eq. (14.6) we deduce that Π_θ equals the dissipation rate of scalar energy, that is,

$$\Pi_\theta \approx \epsilon_\theta = \int_0^\infty 2\kappa k^2 E_\theta(k) dk. \tag{14.12}$$

We can also derive Eq. (14.11) from Eq. (13.9b) by equating the nonlinear term with the scalar energy flux:

$$k\theta_k^2 u_k = \Pi_\theta. \tag{14.13}$$

Now, substitution of $u_k = \Pi_u^{1/3} k^{-1/3}$ immediately yields Eq. (14.11).

In the next section we will describe the spectral properties of scalar flows for various limiting cases of Reynolds and Péclet numbers.

14.3 Various Regimes of a Passive Scalar Flow

The spectral properties of a scalar flow depends on two parameters—Re and Pe. In the present section we will describe the properties of a passive scalar flow for various ranges of these parameters. In Fig. 14.1, we plot the kinetic and scalar energy spectra for various combinations of Re and Pe. For the plot, we take $U = 1$ and $\theta_{\mathrm{rms}} = 0.1$ (except in (c), where they are both 1).

In the following subsection, we explain these cases in detail.

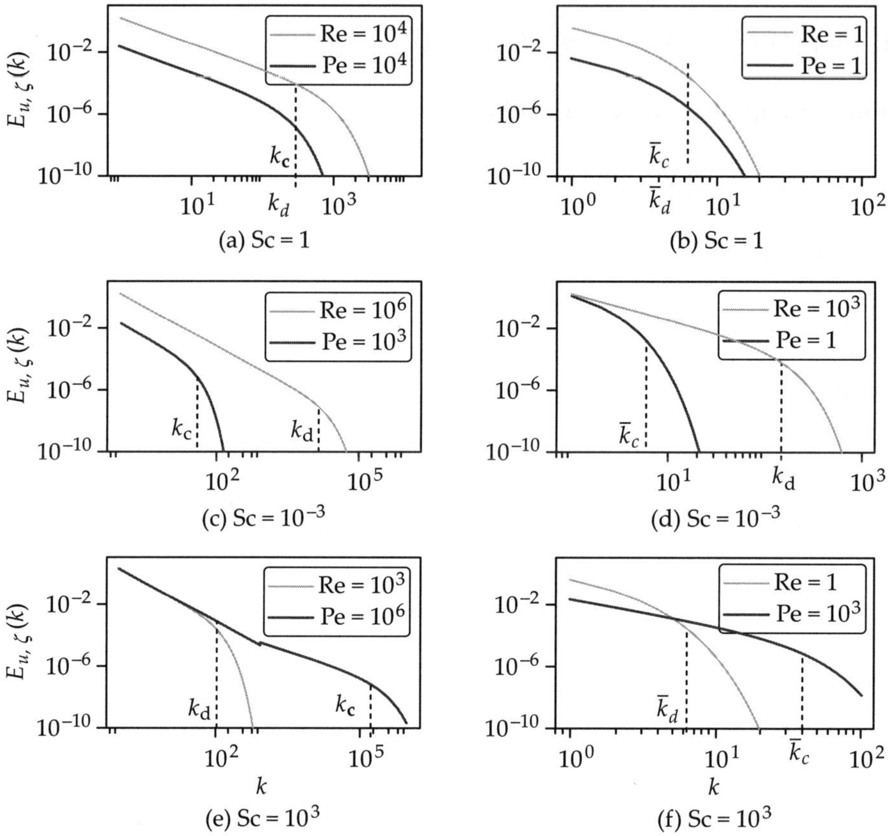

Figure 14.1 The kinetic energy spectrum $E_u(k)$ (thin curve) and scalar energy spectrum $E_\theta(k)$ (thick curve) for (a,b) $\mathrm{Sc} = 1$; (c,d) $\mathrm{Sc} = 10^{-3}$; (e,f) $\mathrm{Sc} = 10^3$. We take $U = 1$ and $\theta_{\mathrm{rms}} = 0.1$ (except in (c)). See Section 14.3 for details.

14.3.1 Turbulent regime I: $\mathrm{Re} \gg 1$; $\mathrm{Pe} \gg 1$; $\mathrm{Sc} \lesssim 1$

In the previous section we described the kinetic and scalar energy spectra and fluxes in the inertial range. Here we generalize the aforementioned results to inertial–dissipation range.

In Section 5.5.1 we showed how Pao (1965) derived the following kinetic energy spectrum and flux for the inertial–dissipation regime of hydrodynamic turbulence:

$$\Pi_u(k) = \epsilon_u \exp\left(-\frac{3}{2} K_{\mathrm{Ko}} (k/k_d)^{4/3}\right), \tag{14.14a}$$

$$E_u(k) = K_{\mathrm{Ko}} \epsilon_u^{2/3} k^{-5/3} \exp\left(-\frac{3}{2} K_{\mathrm{Ko}} (k/k_d)^{4/3}\right), \tag{14.14b}$$

where ϵ_u is the viscous dissipation rate, and

$$k_d = \left(\frac{\epsilon_u}{\nu^3}\right)^{1/4} \tag{14.15}$$

is Kolmogorov's wavenumber. For passive scalar turbulence, the kinetic energy spectrum and flux remain the same as that of Eq. (14.14) because the scalar field does not affect the velocity field.

Now we extend Pao (1965)'s arguments to passive scalar turbulence. We assume that in the inertial–dissipation range, $E_\theta(k)/\Pi_\theta(k)$ is independent of ν and κ, and depends on ϵ_u and k. Under this assumption, using Eq. (14.11), we obtain

$$\frac{E_\theta(k)}{\Pi_\theta(k)} = K_{\mathrm{OC}} \epsilon_u^{-1/3} k^{-5/3}, \tag{14.16}$$

substitution of which in Eq. (14.6) yields the following solution:

$$\Pi_\theta(k) = \epsilon_\theta \exp\left(-\frac{3}{2} K_{\mathrm{OC}} (k/k_c)^{4/3}\right), \tag{14.17a}$$

$$E_\theta(k) = K_{\mathrm{OC}} \epsilon_\theta \epsilon_u^{-1/3} k^{-5/3} \exp\left(-\frac{3}{2} K_{\mathrm{OC}} (k/k_c)^{4/3}\right), \tag{14.17b}$$

where ϵ_θ is the diffusion rate of the scalar, and

$$k_c = \left(\frac{\epsilon_u}{\kappa^3}\right)^{1/4} \tag{14.18}$$

is Kolmogorov's diffusion wavenumber. Therefore,

$$\frac{k_c}{k_d} = \left(\frac{\nu}{\kappa}\right)^{3/4} = \mathrm{Sc}^{3/4}. \tag{14.19}$$

Since $\mathrm{Sc} \leqq 1$, we deduce that $k_c \leqq k_d$. The other limit when $k_c \gg k_d$ will be discussed in Section 14.3.5. We illustrate these cases in Fig. 14.1(a, c); the former corresponds to $\mathrm{Sc} = 1$, and the latter to $\mathrm{Sc} \ll 1$.

In the next subsection, we describe the scaling for a passive scalar flow which is laminar.

14.3.2 Laminar regime: $\mathrm{Re} \lessapprox 1$; $\mathrm{Pe} \lessapprox 1$

For laminar flows, as derived in Section 5.6, the kinetic energy spectrum and flux are

$$E_u(k) = A\frac{1}{k}\exp(-k/\bar{k}_d), \tag{14.20a}$$

$$\Pi_u(k) = A2\nu\bar{k}_d^2(1 + (k/\bar{k}_d))\exp(-k/\bar{k}_d), \tag{14.20b}$$

where $\bar{k}_d$ is the dissipation wavenumber scale for laminar flows [see Eq. (5.41)].

When the nonlinear term of the scalar equation is smaller than the diffusion term, or when $\mathrm{Pe} < 1$, the scalar energy flux is nonzero but small. Note that the flux $\Pi_\theta(k) \neq 0$ as long as the nonlinear term $\mathbf{u} \cdot \nabla\theta$ is finite. When $\mathrm{Pe} \lessapprox 1$, there is no inertial range where $\Pi_\theta(k) \sim$ const. Rather, as shown in Fig. 14.1(b), $\Pi_\theta(k)$ and $E_\theta(k)$ decrease with k as $\exp(-k)$. For $\mathrm{Pe} \lessapprox 1$, the solution of Eq. (14.6) is

$$E_\theta(k) = B\frac{1}{k}\exp(-k/\bar{k}_c), \tag{14.21a}$$

$$\Pi_\theta(k) = 2B\kappa\bar{k}_c^2(1 + k/\bar{k}_c)\exp(-k/\bar{k}_c), \tag{14.21b}$$

where

$$\bar{k}_c = \frac{1}{L}\sqrt{\frac{UL}{\kappa}} = \frac{\sqrt{\mathrm{Pe}}}{L}. \tag{14.22}$$

Using Eqs. (5.41, 14.22) we deduce that for the laminar regime,

$$\frac{\bar{k}_c}{\bar{k}_d} \approx \sqrt{\frac{\mathrm{Pe}}{\mathrm{Re}}} = \sqrt{\mathrm{Sc}}. \tag{14.23}$$

Since $E_\theta \sim \theta^2$, an integration of Eq. (14.21a) yields

$$B \sim \theta_{\mathrm{rms}}^2. \tag{14.24}$$

This case is illustrated in Fig. 14.1(b).

14.3.3 Mixed regime I: $\mathrm{Re} \gg 1$; $\mathrm{Pe} \lessapprox 1$

Since $\mathrm{Re} \gg 1$, the kinetic energy spectrum and flux would be given by the formulas for the turbulent regime, which are Eqs. (14.14). For the passive scalar, since $\mathrm{Pe} \lessapprox 1$, the scalar energy spectrum and flux would be given by Eqs. (14.21). See Fig. 14.1(d) for an illustration. Note that $\mathrm{Sc} \ll 1$ for this case.

For this regime, Batchelor et al. (1959) proposed that

$$E_\theta(k) = \frac{1}{3}K_{\mathrm{Ko}}\epsilon_\theta\epsilon_u^{2/3}\kappa^{-3}k^{-17/3}. \tag{14.25}$$

A brief derivation of this scaling is as follows. Matching the diffusion term with the nonlinear term yields:

$$\kappa k^2 \theta_k \sim u_k k \theta_k \sim u_k (\nabla\theta)_k. \tag{14.26}$$

By squaring Eq. (14.26), and using

$$\kappa (\nabla\theta)^2 = \epsilon_\theta, \tag{14.27}$$

Batchelor et al. (1959) obtained

$$\kappa^3 k^4 \theta_k^2 \sim u_k^2 \epsilon_\theta \tag{14.28}$$

that leads to Eq. (14.25).

Though several authors, for example, Yeung and Sreenivasan (2013), have argued in favor of Batchelor et al. (1959)'s scaling, we believe that the exponential functions of Eqs. (14.21) would model the scalar energy spectrum and flux better.

14.3.4 Mixed regime II: $\mathrm{Re} \lessapprox 1$; $\mathrm{Pe} \gg 1$

There are situations when the velocity field is laminar, but the scalar field is turbulent, that is, $\mathrm{Re} \lessapprox 1$ and $\mathrm{Pe} \gg 1$. Since $\mathrm{Re} \lessapprox 1$, and the velocity field is unaffected by the scalar field, the formulas for the kinetic spectrum and flux would be same as Eqs. (14.20).

Since $\mathrm{Pe} \gg 1$, the scalar energy spectrum extends much beyond $\bar{k}_d$. For $k > \bar{k}_d$, since $u_k \to 0$, we expect that the effective $\mathrm{Pe} \approx u_{k_d} L/\kappa \ll 1$. Therefore, the scalar turbulence for $\bar{k}_d < k < k_c$ would be described by Eq. (14.21). In the wavenumber range $\bar{k}_d < k < k_c$, $\mathbf{u}(\bar{k}_d)$ would act as a mediator for nonlinear energy transfers of scalar energy. Hence, scalar energy transfers involve triads with $k, p \gg q \approx k_d$ that differs from typical local and forward transfers in 3D hydrodynamics that involves $k \approx p \approx q$. See Fig. 14.1(f) for an illustration.

The wavenumber band $k < \bar{k}_d$ is quite narrow. In this band, the kinetic energy flux decays exponentially. Hence, we expect $E_\theta(k)$ to have a laminar form even though $\mathrm{Pe} \gg 1$.

14.3.5 Turbulent regime II: $\mathrm{Re} \gg 1$; $\mathrm{Pe} \gg 1$; $\mathrm{Sc} \gg 1$

Since the velocity field is not affected by the passive scalar and $\mathrm{Re} \gg 1$, the kinetic energy spectrum and flux would be given by Eqs. (14.14). Note that $k_c \gg k_d$ because $\mathrm{Sc} \gg 1$. For $k < k_d$, the scalar energy spectrum would be given by Eq. (14.11).

However, the situation is somewhat tricky for $k_d < k < k_c$. In this wavenumber band, $\mathbf{u}(k_d)$ acts as a mediator for the nonlinear scalar energy transfers. As argued in Subsection 14.3.4, we expect that the scalar energy spectrum and flux would be described by Eq. (14.21). See Fig. 14.1(e) for an illustration.

Along with the above scaling arguments, we briefly describe two other popular models of passive scalar turbulence for $\mathrm{Sc} \gg 1$. Batchelor (1959) proposed that for $k_d < k < k_c$,

$$E_\theta(k) = K_{\mathrm{Ba}} \epsilon_\theta (\nu/\epsilon_u)^{1/2} k^{-1} \exp(-K_{\mathrm{Ba}} \kappa k^2 (\nu/\epsilon_u)^{1/2}), \tag{14.29}$$

but Kraichnan (1968) argued that

$$E_\theta(k) = K_{\mathrm{Ba}} \epsilon_\theta (\nu/\epsilon_u)^{1/2} (1 + \sqrt{6K_{\mathrm{Ba}}}) k^{-1} \exp(-\sqrt{6K_{\mathrm{Ba}}}(k/\bar{k}_c)). \tag{14.30}$$

The difference between Batchelor's and Kraichnan's models is in the form of the exponential function—Batchelor predicts $\exp(-k^2)$ form, while Kraichnan argues for $\exp(-k)$.

Note that Eq. (14.21a) derived using the flux equation has the same form as Eq. (14.30), proposed by Kraichnan. For $k_d < k < k_c$, the prefactors of all the three formulas are approximately the same:

$$\epsilon_\theta (\nu/\epsilon_u)^{1/2} = \frac{\theta_{\mathrm{rms}}^2 U}{L} \left[\frac{\nu L^2}{\nu U^2} \right]^{1/2} = \theta_{\mathrm{rms}}^2. \tag{14.31}$$

Our formula derived using variable energy flux is consistent with the $\exp(-k)$ form. Refer to Gotoh and Yeung (2013) for further details on this topic.

We summarize the aforementioned scaling in Table 14.1.

Table 14.1 Spectral properties for various regimes of scalar flow. Here Pao's spectra refer to Eqs. (14.14, 14.17), while laminar spectra refer to Eqs. (5.40, 14.21). Generalized spectrum is a combination of Eq. (14.11) and Eq. (14.21a), as described in Section 14.3.5.

Regime	Re	Pe	Sc	$E_u(k)$	$E_\theta(k)$	Fig. reference
Turb I	$\gg 1$	$\gg 1$	≈ 1	Pao	Pao	14.1(a)
Turb I	$\gg 1$	$\gg 1$	$\ll 1$	Pao	Pao	14.1(c)
Turb II	$\gg 1$	$\gg 1$	$\gg 1$	Pao	Generalized	14.1(e)
Laminar	$\lessapprox 1$	$\lessapprox 1$	Any	Laminar	Laminar	14.1(b)
Mix I	$\gg 1$	$\lessapprox 1$	$\ll 1$	Pao	Laminar	14.1(d)
Mix II	$\lessapprox 1$	$\gg 1$	$\gg 1$	Laminar	Laminar	14.1(f)

In the next section, we describe several numerical simulations that attempt to verify some of the aforementioned scaling.

14.4 Numerical Simulations of Passive Scalar Turbulence

Flows with passive scalar are simulated by solving Eqs. (14.1). There is a large body of work in this field, and it is impossible to detail them in a short chapter. In the following, we list main results for the three cases—$\mathrm{Sc} \approx 1$, $\mathrm{Sc} \ll 1$, $\mathrm{Sc} \gg 1$.

14.4.1 $\mathrm{Sc} \approx 1$

Yeung et al. (2005) simulated passive scalar turbulence for Taylor-scale Reynolds number, $\mathrm{Re}_\lambda \approx 700$, and for $\mathrm{Sc} = 1, 1/8$. These parameters corresponds to Turbulent regime I of Subsection 14.3.1 and Fig. 14.1(a). Yeung et al. (2005) demonstrated that in the inertial range, the scalar energy spectrum follows Eq. (14.11). See Fig. 14.2 for an illustration. Yeung et al. (2005) reported that $K_{\mathrm{OC}} = 0.67$.

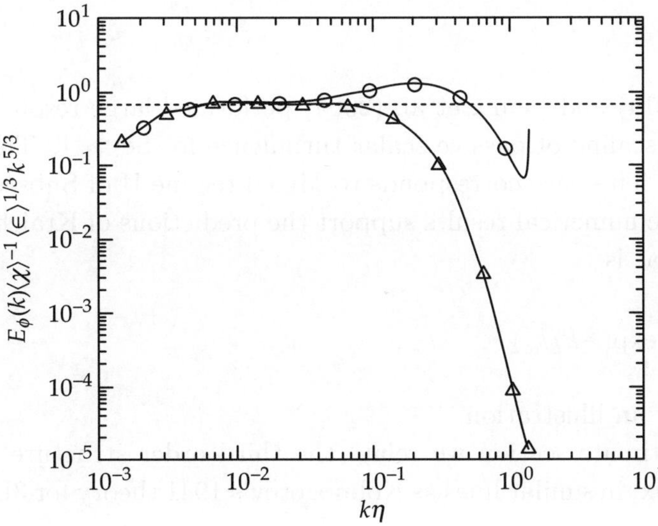

Figure 14.2 The normalized scalar energy spectrum $E_\theta(k)\epsilon_\theta^{-1}\epsilon_u^{1/3}k^{5/3}$ for $\mathrm{Re}_\lambda \approx 700$, and $\mathrm{Sc} = 1$ (circles) and $1/8$ (triangles). The plots verify the spectrum of Eq. (14.11). Read χ, ϵ of the figure as ϵ_θ and ϵ_u respectively. Taken from Yeung et al. (2005). Reprinted with permission from AIP.

We are not aware of any simulation that attempts to verify Eqs. (14.17) for the inertial–dissipation range when $\mathrm{Sc} \sim 1$.

14.4.2 $\mathrm{Sc} \ll 1$

Yeung and Sreenivasan (2013) simulated scalar turbulence with large Re and small Sc ($\mathrm{Re}_\lambda \approx 100$; $\mathrm{Sc} = 1/512$), which corresponds to Mixed regime I of Subsection 14.3.3 and Fig. 14.1(d). They reported that $E_u(k)$ is Kolmogorov-like ($k^{-5/3}$), but that $E_\theta(k)$ follows Eq. (14.25), which is Batchelor et al. (1959)'s scaling.

Batchelor et al. (1959)'s scaling is very different from Eq. (14.21a) derived in the earlier section. Also, refer to Fig. 14.1(d) for an illustration. We believe that the exponential spectrum may provide a better and more reliable fit to $E_\theta(k)$ than the steep power law of Eq. (14.25). We also believe that the variable flux arguments provide a stronger basis for the derivation of $E_\theta(k)$ than the scaling arguments of Batchelor et al. (1959). The proposed exponential spectrum however needs to be tested for $\mathrm{Sc} \ll 1$ limit.

14.4.3 $\mathrm{Sc} \gg 1$

Gotoh et al. (2014) and Yeung et al. (2004) performed large resolution simulations in order to test scaling of passive scalar turbulence for $\mathrm{Sc} \gg 1$. They employed Sc as high as 1000. This case corresponds to Mixed regime II of Subsection 14.3.4 and Fig. 14.1(f). The numerical results support the predictions of Kraichnan (1968) and Eq. (14.21a), that is,

$$E_\theta(k) \sim \frac{1}{k} \exp(-k/\bar{k}_c). \tag{14.32}$$

See Fig. 14.3 for an illustration.

In the penultimate section we relate the third order structure function to the passive scalar flux, in similar lines as Kolmogorov's 1941 theory for 3D hydrodynamic turbulence.

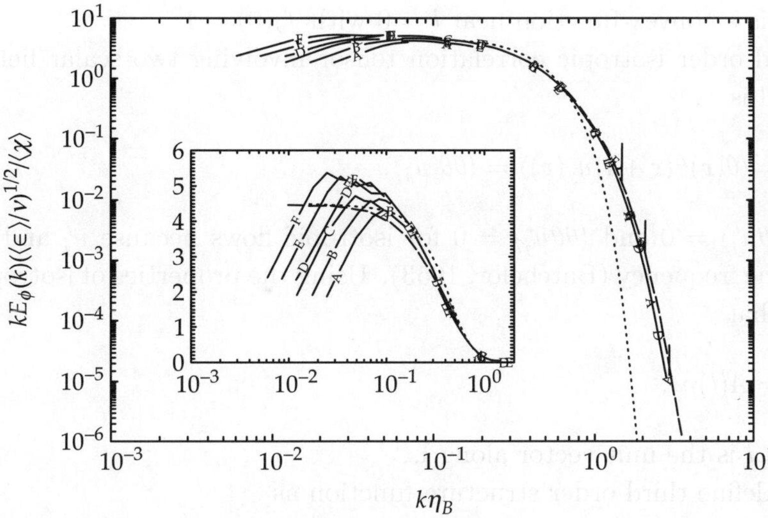

Figure 14.3 The normalized scalar energy spectrum $kE_\theta(k)(\epsilon_u/\nu)^{1/2}/\epsilon_\theta$ vs. k for $\mathrm{Re}_\lambda \approx 8$ and $\mathrm{Sc} = 64, 128, 256 (A - C)$ from a 256^3 grid simulation, and $\mathrm{Sc} = 256, 512, 1024 (D - F)$ from a 512^3 grid simulation. The predictions of Batchelor and Kraichnan models are shown as dotted and dashed lines respectively [Eqs. (14.29, 14.30) respectively]. Numerical results are closer to Kraichnan's predictions, which are same as Eq. (14.21a). Read χ, ϵ of the figure as ϵ_θ and ϵ_u respectively. Taken from Yeung et al. (2004). Reprinted with permission from Springer Nature.

14.5 Third Order Structure Function for Passive Scalar Turbulence: Four-third Law

Kolmogorov's four-fifth law for hydrodynamic turbulence, which was discussed in Chapter 12, was extended to passive scalar turbulence by Yaglom (1949). We present Yaglom's derivation in this section.

We assume the flow to be homogeneous and isotropic. For such flows, the second order correlation function involving two scalar fields is

$$C(\mathbf{l}) = \langle \theta(\mathbf{r})\theta(\mathbf{r}+\mathbf{l}) \rangle = \langle \theta\theta' \rangle. \tag{14.33}$$

Following Chapter 12, the unprimed and primed variables are measured at $\mathbf{r}$ and $\mathbf{r}+\mathbf{l}$ respectively. For isotropic flows,

$$C(\mathbf{l}) = C(l) = \bar{\theta}^2 f_\theta(l), \tag{14.34}$$

where $f_\theta(l)$ is a convex function near $l = 0$ with $f_\theta(0) = 1$.

The third order isotropic correlation tensor involving two scalar fields and the velocity field is

$$C_j(\mathbf{l}) = \langle \theta(\mathbf{r})\theta(\mathbf{r}+\mathbf{l})u_j(\mathbf{r})\rangle = \langle \theta\theta' u_j\rangle. \tag{14.35}$$

Note that $\langle \theta u_j' \rangle = 0$ and $\langle \theta\theta u_j' \rangle = 0$ for isotropic flows because u_j' and $-u_j'$ occur with the same frequency (Batchelor, 1953). Using the properties of isotropic tensors, we deduce that

$$C_j(\mathbf{l}) = A(l) n_j, \tag{14.36}$$

where $\mathbf{n} = \mathbf{l}/l$ is the unit vector along $\mathbf{l}$.

We also define third order structure function as

$$Q_j(\mathbf{l}) \quad = \quad \langle (\theta' - \theta)(\theta' - \theta)(u_j' - u_j)\rangle \tag{14.37}$$

that is expanded as

$$Q_j(\mathbf{l}) \quad = \quad -\langle \theta'\theta' u_j\rangle + \langle \theta\theta u_j'\rangle - 2\langle \theta'\theta u_j'\rangle + 2\langle \theta'\theta u_j\rangle. \tag{14.38}$$

In this expression, $\langle \theta'\theta' u_j \rangle = \langle \theta\theta u_j' \rangle = 0$ due to isotropy. In addition,

$$\langle \theta'\theta' u_j' \rangle - \langle \theta\theta u_j \rangle = 0 \tag{14.39}$$

due to homogeneity of the flow. Now

$$\begin{aligned} \langle \theta'\theta u_j' \rangle &= \langle \theta(\mathbf{r}+\mathbf{l})\theta(\mathbf{r})u_j'(\mathbf{r}+\mathbf{l})\rangle = \langle \theta(\mathbf{r})\theta(\mathbf{r}-\mathbf{l})u_j'(\mathbf{r})\rangle \\ &= C_j(-\mathbf{l}) = -A(l)n_j = -C_j(\mathbf{l}). \end{aligned} \tag{14.40}$$

Therefore,

$$Q_j(\mathbf{l}) \quad = \quad 4\langle \theta'\theta u_j \rangle = 4A(l)n_j, \tag{14.41a}$$

$$\text{Or,} \quad \mathbf{Q}(l) \quad = \quad 4A(l)\mathbf{n}, \tag{14.41b}$$

and

$$S_3^\theta(l) \quad = \quad \langle (\theta' - \theta)^2 (\mathbf{u}' - \mathbf{u})\cdot \mathbf{n}\rangle = Q_1(l) = 4A(l), \tag{14.42}$$

$$\nabla_l \cdot \mathbf{Q}(l) \quad = \quad \frac{1}{l^2}\frac{d}{dl}\left[l^2 4A(l)\right] = \frac{1}{l^2}\frac{d}{dl}\left[l^2 S_3^\theta(l)\right]. \tag{14.43}$$

Now we derive a dynamical equation for the second order correlation function:

$$\begin{aligned}
\frac{\partial}{\partial t}\frac{1}{2}\langle\theta\theta'\rangle &= \frac{1}{2}\langle\theta'\frac{\partial}{\partial t}\theta\rangle + \frac{1}{2}\langle\theta\frac{\partial}{\partial t}\theta'\rangle \\
&= \frac{1}{2}\left[-\partial_j\langle\theta'(u_j\theta)\rangle - \partial'_j\langle\theta(u'_j\theta')\rangle\right. \\
&\quad \left. +\langle\theta' F_\theta\rangle + \langle\theta F'_\theta\rangle + \kappa\langle\theta'\nabla^2\theta\rangle + \kappa\langle\theta\nabla'^2\theta'\rangle\right] \\
&= \frac{1}{2}\left[\partial'_j\langle\theta'(u_j\theta)\rangle - \partial'_j\langle\theta(u'_j\theta')\rangle + 2\langle\theta' F_\theta\rangle + 2\kappa\nabla'^2\langle\theta\theta'\rangle\right] \\
&= \frac{1}{4}\nabla_l\cdot\langle(\theta'-\theta)^2(\mathbf{u}'-\mathbf{u})\rangle + \langle\theta' F_\theta\rangle + \kappa\nabla'^2\langle\theta\theta'\rangle \\
&= T_\theta(\mathbf{l}) + \mathcal{F}_\theta(\mathbf{l}) - D_\theta(\mathbf{l}),
\end{aligned} \tag{14.44}$$

where $T_\theta(\mathbf{l})$ corresponds to the spectral energy transfer term $T_\theta(\mathbf{k})$, whereas $\mathcal{F}_\theta(\mathbf{l})$ and $D_\theta(\mathbf{l})$ are respectively the correlations corresponding to the scalar energy feed by the external force and the diffusion rate.

We make similar assumptions as in Section 12.3—steady state, $\kappa \to 0$, and forcing at large scales. Under these assumptions, in the inertial range, $D_\theta(l) \to 0$, and hence,

$$\mathcal{F}_\theta(l) = \epsilon_\theta = -T_\theta(l) = -\frac{1}{4}\nabla_l\cdot\mathbf{Q}(l) \tag{14.45}$$

or

$$\epsilon_\theta = -\frac{1}{4}\frac{1}{l^2}\frac{d}{dl}\left[l^2 S_3^\theta(l)\right]. \tag{14.46}$$

An integration of this equation yields

$$S_3^\theta(l) = \langle(\theta'-\theta)^2\{(\mathbf{u}'-\mathbf{u})\cdot\mathbf{n}\}\rangle = -\frac{4}{3}\epsilon_\theta l, \tag{14.47}$$

which is the four-third law for a passive scalar, first derived by Yaglom (1949).[1] Yeung et al. (2005) performed numerical simulation of turbulent passive scalar and verified the four-third law.

The diffusion term is

[1]Note that $S_3^\theta(l)$ is a mixed product of $(\delta\theta)^2$ and $\delta u_\parallel$. This combination yields scalar energy flux. From the equation for the scalar energy, Eq. (13.9b), we deduce that

$$k\theta_k^2 u_k = \Pi_\theta, \tag{14.48}$$

which corresponds to Eq. (14.47).

$$\begin{aligned} D_\theta(l) &= -\kappa\nabla'^2\langle\theta\theta'\rangle = -\kappa\bar{\theta}^2\nabla^2 C(l) \\ &= -\kappa\bar{\theta}^2\frac{1}{l^2}\frac{d}{dl}\left[l^2\frac{d}{dl}f_\theta\right]. \end{aligned} \tag{14.49}$$

Now, we combine the nonlinear and diffusion terms as

$$-\frac{1}{l^2}\frac{d}{dl}\left[l^2\left(\kappa\bar{\theta}^2 f_\theta' + \frac{S_3^\theta(l)}{4}\right)\right] = \epsilon_\theta \tag{14.50}$$

that yields

$$\kappa\bar{\theta}^2 f_\theta' + \frac{S_3^\theta(l)}{4} = -\frac{1}{3}\epsilon_\theta l. \tag{14.51}$$

This equation yields $S_3^\theta(l) = -(4/3)\epsilon_\theta l$ in the inertial range, and

$$f_\theta(l) = 1 - \frac{\epsilon_\theta}{6\kappa\bar{\theta}^2}l^2 \tag{14.52}$$

in the diffusive range.

Before we close this session, we contrast the the four-third law described above and the scalar energy flux discussed earlier. In Fig. 14.4 we relate the terms of $-T_\theta(l)$ with $\Pi_\theta(K)$. Here $\delta\mathbf{u}$, which corresponds to $\mathbf{u}(\mathbf{q})$ in spectra space, mediates the energy transfers among $\delta\theta$ fields that appear in the product of $-T_\theta(l)$. Note that the giver mode and the receiver mode cannot be contrasted in real space because $-T_\theta(l)$ is a sum of many mode-to-mode energy transfers.

In the next section we briefly describe the field-theoretic results of passive scalar turbulence.

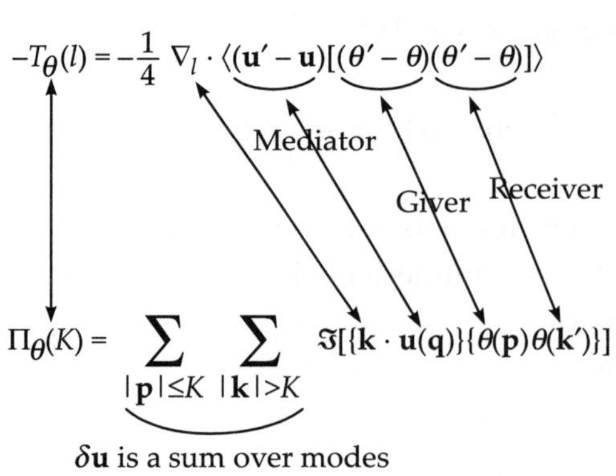

Figure 14.4 Connection between the formulas $-T_\theta(l)$ and the energy flux $\Pi_\theta(K)$. $\delta\mathbf{u}$ mediates energy transfer between $\delta\theta$ fields in the product term of $-T_\theta(l)$.

14.6 Field-theoretic Treatment of Passive Scalar Turbulence

In chapter 10 we described how to compute the renormalized viscosity and Kolmogorov's constant for hydrodynamic turbulence using McComb's procedure (McComb, 1990, 2014). These computations involve energy flux computations. For passive scalar turbulence, Zhou and Vahala (1993), Lin et al. (2000), and Verma (2001b) employed a similar procedure to compute renormalized diffusivity and scalar energy flux. Yakhot and Orszag (1986) employed ϵ-expansion for these computations, and Adzhemyan et al. (1999) employed an alternate scheme of renormalization.

We do not detail the above calculations due to lack of space. We state that Verma (2001b)'s computations yields the renormalized or scale-dependent scalar diffusivity as

$$\kappa(k) = \kappa_* \sqrt{K_{\mathrm{Ko}}} \Pi_u^{1/3} k^{-4/3}, \tag{14.53}$$

where $\kappa_* = 0.85$ for 3D turbulence. He also reported that the Obukhov–Corrsin constant $K_{\mathrm{OC}} = 1.25$. Other field-theoretic computations yields similar results.

There is a vast literature on the computation of intermittency exponents for passive scalar, but these discussions are beyond the scope of the book. The reader can refer to Falkovich et al. (2001).

In summary, passive scalar turbulence is useful for modeling diffusion of light particles and nonreactive fields. In addition, it acts as a starting point for understanding more complex flows such as buoyancy-driven flows, diffusion of pollution, binary fluid, etc. We will discuss buoyancy-driven flows and binary fluids in the subsequent chapters.

Further Reading

There is an extensive literature on passive scalar turbulence. In addition to the references cited in the chapter, we refer the reader to books by Leslie (1973) and Lesieur (2008). For a detailed discussion on earlier works on passive scalar spectrum and flux, refer to Batchelor (1959); Kraichnan (1968); Lesieur (2008); Gotoh and Yeung (2013), and the original papers cited in this chapter.

In this book, we have avoided discussions on intermittency and higher order structure functions of a passive scalar. Refer to Falkovich et al. (2001) for details and references.

Exercises

1. The discussion in this chapter was focused on 3D passive scalar turbulence. How will the energy spectrum and flux change in 2D?

Chapter 15

Stably Stratified Turbulence

In this chapter we will describe physics of stably stratified turbulence, which is a turbulent flow with an active scalar. The omnipresent gravity affects all kinds of flows. For example, most of Earth's atmosphere is strongly stratified with fluid density decreasing with height[1]; this configuration leads to a strong horizontal wind velocity, and a weak vertical wind velocity. Gravity affects the flow via buoyancy. Hence, the material density acts as an *active scalar.*

In the present chapter we will describe the properties of the density and velocity fields of stably stratified turbulence. We start with the description of the governing equations in real space.

15.1 Governing Equations in Real Space

We consider a fluid under a vertical and stable density stratification, as shown in Fig. 15.1. The local density $\varrho(x, y, z)$ is a sum of mean density $\bar{\rho}(z)$ and density fluctuation $\rho(x, y, z)$:

$$\varrho(x, y, z) = \bar{\rho}(z) + \rho(x, y, z). \tag{15.1}$$

[1]The density stratification near the surface of the Earth is more complex, and it could become unstable at some time of the day.

We assume a linear density profile for $\bar{\rho}(z)$:

$$\bar{\rho}(z) = \rho_b + \frac{d\bar{\rho}}{dz}z = \rho_b + \frac{\rho_t - \rho_b}{d}z, \tag{15.2}$$

and that the gravity is along $-\hat{z}$. Under stable stratification, $\rho_t < \rho_b$.

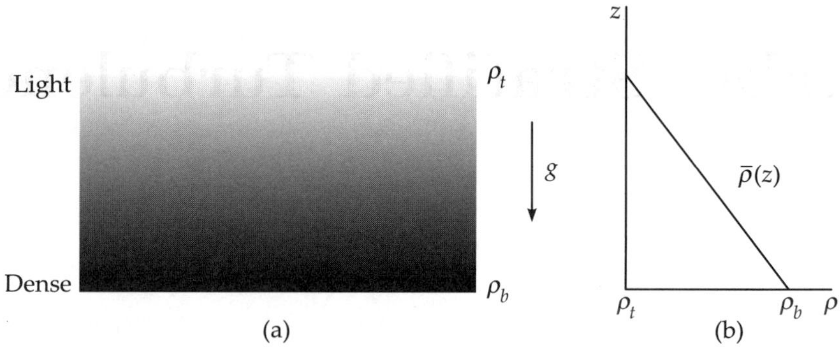

Figure 15.1 (a) Schematic diagram of a stably stratified flow with lighter fluid above the heavier fluid. The densities at the bottom and top are ρ_b and ρ_t respectively. (b) The mean density $\bar{\rho}(z)$ falls linearly with z.

We assume Oberbeck–Boussinesq (OB) approximation under which (a) the density is assumed to be constant except for the buoyancy term; (b) the fluid parameters like kinematic viscosity and thermal diffusivity are constant in space and time. Under this approximation, the governing equations for the flow are:

$$\frac{\partial \mathbf{u}}{\partial t} + (\mathbf{u}\cdot\nabla)\mathbf{u} = -\frac{1}{\rho_m}\nabla\sigma - \frac{\rho}{\rho_m}g\hat{z} + \nu\nabla^2\mathbf{u} + \mathbf{F}_u, \tag{15.3a}$$

$$\frac{\partial \rho}{\partial t} + (\mathbf{u}\cdot\nabla)\rho = -\frac{d\bar{\rho}}{dz}u_z + \kappa\nabla^2\rho, \tag{15.3b}$$

$$\nabla\cdot\mathbf{u} = 0, \tag{15.3c}$$

where σ is the pressure field that includes contributions from gravity, ρ_m is the mean density, $\mathbf{F}_u$ is the external force in addition to gravity, g is the acceleration due to gravity, and ν, κ are respectively the kinematic viscosity and density diffusivity of the fluid. Note that the density field affects the flow via the buoyancy term, $-(\rho/\rho_m)g\hat{z}$. Also, in relation to Eq. (14.1b), Eq. (15.3b) has $F_\rho = -(d\bar{\rho}/dz)u_z$. For a detailed derivation of Eqs. (15.3) and discussion on Oberbeck–Boussinesq (OB) approximation, refer to Verma (2018). Note that in stably stratified turbulence, buoyancy and F_ρ acts at all scales, in contrast to the situation in passive scalar turbulence where the forces are assumed to be at large scales. In this chapter we will study the consequences arising due to these differences.

A linearized version of the aforementioned equations with $\nu = \kappa = 0$ and $\mathbf{F}_u = 0$ are

$$\frac{\partial \mathbf{u}}{\partial t} = -\frac{1}{\rho_m}\nabla\sigma - \frac{\rho}{\rho_m} g\hat{z}, \tag{15.4a}$$

$$\frac{\partial \rho}{\partial t} = -\frac{d\bar{\rho}}{dz} u_z, \tag{15.4b}$$

which are the equations of internal gravity waves. These waves have a frequency of

$$N = \sqrt{\frac{g}{\rho_m}\left|\frac{d\bar{\rho}}{dz}\right|}, \tag{15.5}$$

which is called the *Brunt–Väisälä frequency*. For a detailed solution of Eq. (15.4), refer to Verma (2018). Here we present an internal gravity wave traveling along the x-axis:

$$\mathbf{k} = k_x \hat{x}, \tag{15.6a}$$

$$\mathbf{u} = \hat{z} A \cos(k_x x - Nt), \tag{15.6b}$$

$$\rho = A\sqrt{\frac{\rho_m}{g}\left|\frac{d\bar{\rho}}{dz}\right|}\sin(k_x x - Nt). \tag{15.6c}$$

Note that $\langle \rho u_z \rangle = 0$ (volume averaged quantity) for internal gravity waves.

We employ N, ρ_m, and g to construct a new variable ρ that has the dimensions of velocity:

$$\rho \to \frac{g}{N}\frac{\rho}{\rho_m}. \tag{15.7}$$

In terms of the new variable, the equations for the stably stratified flows are as follows.

$$\frac{\partial \mathbf{u}}{\partial t} + (\mathbf{u}\cdot\nabla)\mathbf{u} = -\frac{1}{\rho_m}\nabla\sigma - N\rho\hat{z} + \nu\nabla^2\mathbf{u} + \mathbf{F}_u, \tag{15.8a}$$

$$\frac{\partial \rho}{\partial t} + (\mathbf{u}\cdot\nabla)\rho = N u_z + \kappa\nabla^2\rho. \tag{15.8b}$$

The above equations indicate the following quadratic quantities for stably stratified flows:

$$\text{Kinetic energy density } E_u(\mathbf{r}) = \frac{1}{2}u^2, \tag{15.9a}$$

$$\text{Potential energy density} \quad E_\rho(\mathbf{r}) = \frac{1}{2}\rho^2. \tag{15.9b}$$

The volume average of these quantities are the total kinetic and potential energies, and they are denoted by E_u and E_ρ respectively. The evolution equations for $E_u(\mathbf{r})$ and $E_\rho(\mathbf{r})$ are

$$\frac{\partial}{\partial t}\frac{u^2}{2} + \nabla \cdot \left[\frac{u^2}{2}\mathbf{u}\right] = -\nabla \cdot (\sigma \mathbf{u}) - N\rho u_z + \nu \mathbf{u} \cdot \nabla^2 \mathbf{u} + \mathbf{F}_u \cdot \mathbf{u}, \tag{15.10a}$$

$$\frac{\partial}{\partial t}\frac{\rho^2}{2} + \nabla \cdot \left[\frac{\rho^2}{2}\mathbf{u} - \kappa\rho\nabla\rho\right] = N\rho u_z - \kappa(\nabla\rho)^2. \tag{15.10b}$$

Under the approximation that $\nu = \kappa = 0$ and $\mathbf{F}_u = 0$, a sum of the volume integrals of the aforementioned equations for periodic or vanishing boundary conditions yields

$$\frac{d}{dt}(E_u + E_\rho) = 0, \tag{15.11}$$

which is a statement of the *conservation of total energy.* Note that E_u and E_ρ are not conserved individually.

In the absence of the nonlinear, viscous, and diffusive terms, the system supports gravity waves for which E_u and E_ρ oscillate around their mean values while keeping $E_u + E_\rho =$ const. Here, $\langle \rho u_z \rangle = 0$. This is the neutral equilibrium of the system. Note that a wave solution yields neither a growth nor a decay of E_u and E_ρ. Also, the energy exchange between E_u and E_ρ is due to the linear interactions of Eqs. (15.4), and is different from the nonlinear energy transfers.

In the presence of nonlinearity, $\langle \rho u_z \rangle > 0$ for the following reasons. The volume average of nondissipative and force-free ($\mathbf{F}_u = 0$) version of Eq. (15.10a) yields

$$\frac{1}{2}\frac{\partial}{\partial t}\langle u^2 \rangle = -N\langle \rho u_z \rangle. \tag{15.12}$$

This is because the volume average of the nonlinear and pressure gradients vanish for periodic or vanishing boundary conditions. From Eq. (15.12), we deduce that the stability of a stably stratified flow is possible only when the average energy feed by buoyancy, $-N\langle \rho u_z \rangle < 0$ or $\langle \rho u_z \rangle > 0$. If $-N\langle \rho u_z \rangle$ were to be positive, the kinetic energy would grow in time and make the flow unstable, which is a contradiction for a stably stratified flow. Thus, we prove that $\langle \rho u_z \rangle > 0$ for stably stratified flows.

Due to the above reasons, E_u is transferred to E_ρ. Therefore, E_u decays when $\mathbf{F}_u = 0$. This is the reason why we need $\mathbf{F}_u$ to generate a steady state. In the next section we will describe the equations of stably stratified flows in Fourier space.

15.2 Governing Equations in Fourier Space

The aforementioned equations in real space can be written in Fourier space as follows:

$$\frac{d}{dt}\mathbf{u}(\mathbf{k}) + \mathbf{N}_u(\mathbf{k}) = -i\mathbf{k}\sigma(\mathbf{k}) - N\rho(\mathbf{k})\hat{z} - \nu k^2\mathbf{u}(\mathbf{k}) + \mathbf{F}_u(\mathbf{k}), \tag{15.13a}$$

$$\frac{d}{dt}\rho(\mathbf{k}) + N_\rho(\mathbf{k}) = Nu_z(\mathbf{k}) - \kappa k^2\rho(\mathbf{k}), \tag{15.13b}$$

$$\mathbf{k}\cdot\mathbf{u}(\mathbf{k}) = 0, \tag{15.13c}$$

where

$$\mathbf{N}_u(\mathbf{k}) = i\sum_{\mathbf{p}}[\mathbf{k}\cdot\mathbf{u}(\mathbf{k}-\mathbf{p})]\mathbf{u}(\mathbf{p}), \tag{15.14a}$$

$$N_\rho(\mathbf{k}) = i\sum_{\mathbf{p}}[\mathbf{k}\cdot\mathbf{u}(\mathbf{k}-\mathbf{p})]\rho(\mathbf{p}), \tag{15.14b}$$

and

$$\sigma(\mathbf{k}) = \frac{i}{k^2}[\mathbf{k}\cdot\mathbf{N}_u(\mathbf{k}) + Nk_z\rho(\mathbf{k})]. \tag{15.15}$$

The modal potential energy is defined as

$$E_\rho(\mathbf{k}) = \frac{1}{2}|\rho(\mathbf{k}')|^2. \tag{15.16}$$

The dynamical equations for $E_u(\mathbf{k})$ and $E_\rho(\mathbf{k})$ are

$$\frac{d}{dt}E_u(\mathbf{k}) = \Im\left[\sum_{\mathbf{p}}[\mathbf{k}\cdot\mathbf{u}(\mathbf{q})]\,[\mathbf{u}(\mathbf{p})\cdot\mathbf{u}^*(\mathbf{k})]\right] - N\Re[\rho(\mathbf{k})u_z^*(\mathbf{k})] + \Re\,[\mathbf{F}_u(\mathbf{k})\cdot\mathbf{u}^*(\mathbf{k})] - 2\nu k^2E_u(\mathbf{k}), \tag{15.17a}$$

$$\frac{d}{dt}E_\rho(\mathbf{k}) = \Im\left[\sum_{\mathbf{p}}[\mathbf{k}\cdot\mathbf{u}(\mathbf{q})]\,[\rho(\mathbf{p})\rho^*(\mathbf{k})]\right] + N\Re[\rho(\mathbf{k})u_z^*(\mathbf{k})] - 2\kappa k^2E_\rho(\mathbf{k}). \tag{15.17b}$$

When we compare the above equations with the corresponding equations for hydrodynamics (Eq. (3.19)) and that for passive scalar, we observe the following:

1. Term 1 of $\dot{E}_u(\mathbf{k})$: The nonlinear kinetic energy transfers to $\mathbf{u}(\mathbf{k})$ from other $\mathbf{u}$ modes. This is exactly the same as the nonlinear term of Eq. (3.19).
2. Term 2 of $\dot{E}_u(\mathbf{k})$: Depletion rate of $E_u(\mathbf{k})$ by gravitational force. The depleted energy is transferred to the potential energy.

3. Term 3 of $\dot{E}_u(\mathbf{k})$: The kinetic energy supply rate to $E_u(\mathbf{k})$ by the external force $\mathbf{F}_u$, which is in addition to the buoyancy. Typically, $\mathbf{F}_u$ is employed at large length scales to maintain a steady flow.
4. Term 4 of $\dot{E}_u(\mathbf{k})$: The viscous dissipation rate of $E_u(\mathbf{k})$.
5. Term 1 of $\dot{E}_\rho(\mathbf{k})$: The nonlinear potential energy transfers to $\rho(\mathbf{k})$ from other ρ modes.
6. Term 2 of $\dot{E}_\rho(\mathbf{k})$: Potential energy supply rate to $E_\rho(\mathbf{k})$ by gravitational force. This energy comes from the velocity field.
7. Term 3 of $\dot{E}_\rho(\mathbf{k})$: The diffusion rate of $E_\rho(\mathbf{k})$.

In the absence of an external force $\mathbf{F}_u$, the evolution equation for the total energy, $E_u + E_\rho$, is

$$\frac{d}{dt}(E_u + E_\rho) = -2\nu \sum_{\mathbf{k}} k^2 E_u(\mathbf{k}) - 2\kappa \sum_{\mathbf{k}} k^2 E_\rho(\mathbf{k}). \tag{15.18}$$

Thus, the total energy decays in the presence of ν and κ. Clearly, the total energy is conserved when $\nu = \kappa = 0$.

As argued in the previous section, the stability condition yields $\langle \rho u_z \rangle > 0$. Using Parseval's theorem, we deduce that

$$\sum_{\mathbf{k}} \Re[\rho(\mathbf{k}) u_z^*(\mathbf{k})] > 0. \tag{15.19}$$

This equation does not imply that $\Re[\rho(\mathbf{k}) u_z^*(\mathbf{k})] > 0$ for each $\mathbf{k}$; yet numerical and experimental results indicate that

$$\Re[\rho(\mathbf{k}) u_z^*(\mathbf{k})] > 0 \tag{15.20}$$

for almost all the Fourier modes. Hence, the shell spectrum for the buoyancy feed is negative, that is,

$$\mathcal{F}_B(k) = \sum_{k-1<k'\leq k} -N\langle \rho(\mathbf{k}') u_z^*(\mathbf{k}') \rangle < 0. \tag{15.21}$$

The buoyancy $\mathbf{F}_B(\mathbf{k}) = -N\rho(\mathbf{k})\hat{z}$ feeds kinetic helicity to the mode $\mathbf{u}(\mathbf{k})$. Using Eq. (3.48) we show that

$$\begin{aligned}
\mathcal{F}_{H_K}(\mathbf{k}) &= \Re[\boldsymbol{\omega}^*(\mathbf{k}) \cdot \mathbf{F}_B(\mathbf{k})] \\
&= -N\Re[(-i\mathbf{k} \times \mathbf{u}^*(\mathbf{k})) \cdot \hat{z}\rho(\mathbf{k})]
\end{aligned}$$

$$
\begin{aligned}
&= -N\Re[(i\mathbf{k}\times\hat{z}\cdot\mathbf{u}^*(\mathbf{k}))\rho(\mathbf{k})] \\
&= Nk(\sin\zeta)\Im[u_1^*(\mathbf{k})\rho(\mathbf{k})], \qquad (15.22)
\end{aligned}
$$

where u_1 is the first component of the velocity field in the Craya–Herring basis (with $\hat{z}=\hat{n}$), and ζ is the angle between the wave vector $\mathbf{k}$ and $\hat{z}$. Refer to Section 9.1 for details on Craya–Herring basis. Following Eq. (3.50) and similar lines of arguments as above, we derive the enstrophy feed to the mode by buoyancy as

$$
\begin{aligned}
\mathcal{F}_\omega(\mathbf{k}) &= k^2\Re[\mathbf{u}^*(\mathbf{k})\cdot\mathbf{F}_B(\mathbf{k})] \\
&= -k^2N\Re[\rho(\mathbf{k})u_z^*(\mathbf{k})]. \qquad (15.23)
\end{aligned}
$$

In the next section we describe energy transfers and fluxes in stably stratified turbulence.

15.3 Energy Transfers and Fluxes for Stably Stratified Turbulence

As described in Section 13.2, the mode-to-mode kinetic and potential energy transfers are

$$S^{uu}(\mathbf{k}|\mathbf{p}|\mathbf{q}) = -\Im\left[\{\mathbf{k}'\cdot\mathbf{u}(\mathbf{q})\}\{\mathbf{u}(\mathbf{p})\cdot\mathbf{u}(\mathbf{k}')\}\right], \qquad (15.24a)$$

$$S^{\rho\rho}(\mathbf{k}|\mathbf{p}|\mathbf{q}) = -\Im\left[\{\mathbf{k}'\cdot\mathbf{u}(\mathbf{q})\}\{\rho(\mathbf{p})\rho(\mathbf{k}')\}\right]. \qquad (15.24b)$$

Similarly, from Section 13.3, the flux formulas for the kinetic and potential energies are as follows.

$$\Pi_u(k_0) = \sum_{|\mathbf{k}'|>k_0}\sum_{|\mathbf{p}|\le k_0} S^{uu}(\mathbf{k}'|\mathbf{p}|\mathbf{q}), \qquad (15.25a)$$

$$\Pi_\rho(k_0) = \sum_{|\mathbf{k}'|>k_0}\sum_{|\mathbf{p}|\le k_0} S^{\rho\rho}(\mathbf{k}'|\mathbf{p}|\mathbf{q}). \qquad (15.25b)$$

The shell-to-shell kinetic and potential energy transfers from shell m to shell n are defined as follows:

$$T^{u,m}_{u,n} = \sum_{\mathbf{k}'\in n}\sum_{\mathbf{p}\in m} S^{uu}(\mathbf{k}'|\mathbf{p}|\mathbf{q}), \qquad (15.26a)$$

$$T^{\rho,m}_{\rho,n} = \sum_{\mathbf{k}'\in n}\sum_{\mathbf{p}\in m} S^{\rho\rho}(\mathbf{k}'|\mathbf{p}|\mathbf{q}). \qquad (15.26b)$$

Following the arguments of variable energy flux as in Sections 4.4 and 13.3, we obtain the following set of equations for the inertial–dissipative range:

$$\frac{\partial}{\partial t}E_u(k,t) = -\frac{\partial}{\partial k}\Pi_u(k,t) + \mathcal{F}_B(k,t) - 2\nu k^2 E_u(k,t), \qquad (15.27a)$$

$$\frac{\partial}{\partial t}E_\rho(k,t) = -\frac{\partial}{\partial k}\Pi_\rho(k,t) - \mathcal{F}_B(k,t) - 2\kappa k^2 E_\rho(k,t), \tag{15.27b}$$

where

$$\mathcal{F}_B(k)dk = -\sum_{k<k'\leq k+dk} N\Re[\rho(\mathbf{k}')u_z^*(\mathbf{k}')]. \tag{15.28}$$

We assume that the external force $\mathbf{F}_u$ is active only at small wavenumbers. Hence, $\mathbf{F}_u(\mathbf{k}) = 0$ in the inertial range.

We limit our attention to the inertial range where the dissipative effects are negligible. Therefore, under a steady state ($\partial_t = 0$), we obtain

$$\frac{d}{dk}\Pi_u(k) = \mathcal{F}_B(k), \tag{15.29a}$$

$$\frac{d}{dk}\Pi_\rho(k) = -\mathcal{F}_B(k). \tag{15.29b}$$

Hence, the total energy flux

$$\Pi_u(k) + \Pi_\rho(k) = \text{const.} \tag{15.30}$$

We will discuss the properties of these fluxes in the subsequent sections.

15.4 Various Regimes of Stably Stratified Turbulence

Before getting into a discussion of phenomenological theories of stably stratified turbulence, we define a quantity called *Richardson number* Ri, which is the ratio of buoyancy and the nonlinear term:

$$\mathrm{Ri} = \frac{N|\rho|_{\mathrm{rms}}L}{u_{\mathrm{rms}}^2}, \tag{15.31}$$

where L is the length scale of the system. The other important parameter is the Prandtl number $\mathrm{Pr} = \nu/\kappa$.

Using Re and Ri we can broadly classify the stably stratified turbulence (SST) into the following three categories:

1. $\mathrm{Re} \gg 1$ and $\mathrm{Ri} \approx 1$ (turbulent SST with moderate buoyancy): In this regime, the flow is nearly isotropic because buoyancy and nonlinearity are of the same order. For this parameter regime, Bolgiano (1959) and Obukhov (1959) showed that $E_u(k) \sim k^{-11/5}$ and $E_\rho(k) \sim k^{-7/5}$ in the wavenumber range affected by buoyancy, and $E_u(k) \sim k^{-5/3}$ and $E_\rho(k) \sim k^{-5/3}$ beyond this range. In Section 15.5.2, we also present a modified version of Bolgiano–Obukhov scaling.

2. Re $\gg$ 1 and Ri $\ll$ 1 (turbulent SST with weak buoyancy): In this regime, strong nonlinearity yields behavior similar to passive scalar turbulence, that is, $E_u(k) \sim k^{-5/3}$ and $E_\rho(k) \sim k^{-5/3}$. Since passive scalar turbulence has already been discussed in Chapter 14, we will not describe this regime in the present chapter.

3. Re $\gg$ 1 and Ri $\gg$ 1 (turbulent SST with strong buoyancy): In this regime, due to strong gravity, the flow becomes quasi-2D with strong $u_\perp$ and weak $u_\parallel$. Such flows are quite complex with many unsolved issues; hence, they are not discussed in the present book.

Note that Re $\gg$ 1 for all the aforementioned cases. In the following discussion, we provide further details for regime 1.

15.5 Stably Stratified Turbulence with Moderate Buoyancy

For moderate stratification in which buoyancy is of the same order as the nonlinear term, Bolgiano (1959) and Obukhov (1959) constructed a phenomenology that provides the spectra and fluxes for the kinetic and potential energies. We describe this phenomenology in the following subsection.

15.5.1 Bolgiano–Obukhov phenomenology

In this section we describe the Bolgiano–Obukhov phenomenology that is applicable to moderately buoyant SST. A key ingredient of this phenomenology is that

$$\mathcal{F}_B(k) < 0 \tag{15.32}$$

due to the stable nature of the flow. We proved the negativity of $\mathcal{F}_B(k)$ at the end of Section 15.1. We showed that for an inviscid, nondiffusive and linear stratification, a negative $\mathcal{F}_B(k)$ enables a transfer from kinetic energy to potential energy that makes the flow stable.

The condition that $\mathcal{F}_B(k) < 0$ yields

$$\frac{d}{dk}\Pi_u(k) = \mathcal{F}_B(k) < 0 \tag{15.33}$$

in the inertial range. Hence, the kinetic energy flux $\Pi_u(k)$ decreases with k in this range. See Fig. 15.2(a) for an illustration. A decreasing $\Pi_u(k)$ is in contrast to the constant $\Pi_u(k)$ in 3D hydrodynamics. A consequence of a decreasing $\Pi_u(k)$ is that $E_u(k)$ is steeper than Kolmogorov's $k^{-5/3}$ spectrum.

Since $\mathcal{F}_B(k) < 0$, following Eq. (15.29b), we obtain

$$\frac{d}{dk}\Pi_\rho(k) = -\mathcal{F}_B(k) > 0. \tag{15.34}$$

Hence, as illustrated in Fig. 15.2(b), $\Pi_\rho(k)$ increases with k. Note, however, that in the inertial range,

$$\frac{d}{dk}[\Pi_u(k) + \Pi_\rho(k)] = 0. \tag{15.35}$$

Hence, the sum, the total energy flux,

$$\Pi_u(k) + \Pi_\rho(k) = \Pi \tag{15.36}$$

is a constant, and equals the total energy dissipation rate. In Section 15.5.2, we will show that this constancy leads to a rapid decrease of $\Pi_u(k)$, but a very slow increase of $\Pi_\rho(k)$.

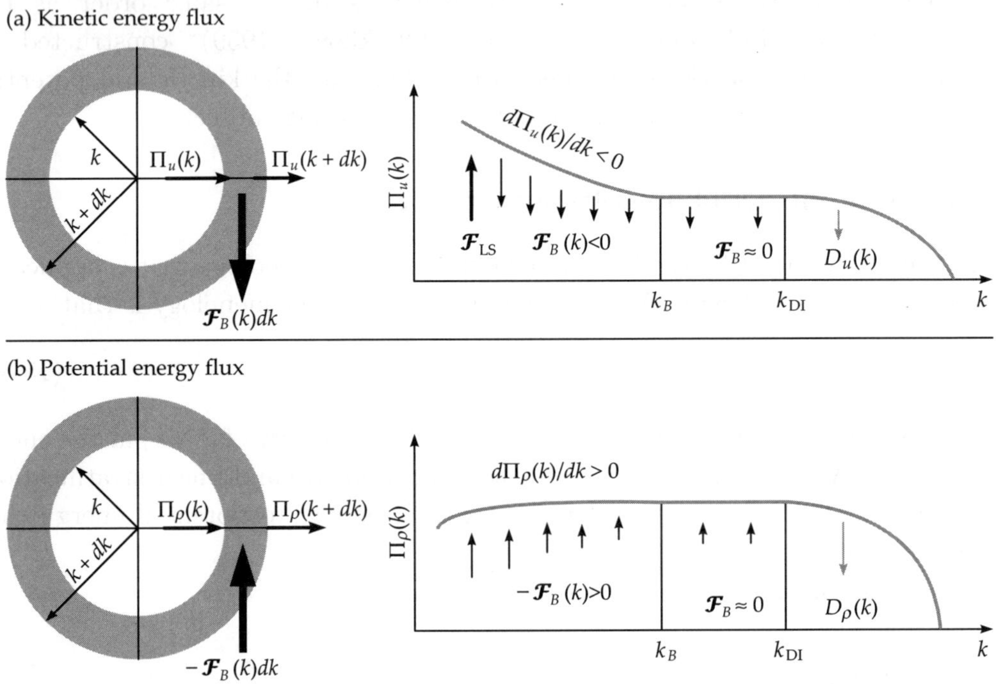

Figure 15.2 Bolgiano (1959) and Obukhov (1959) model for SST: Schematic diagrams of (a) kinetic energy flux $\Pi_u(k)$; (b) potential energy flux $\Pi_\rho(k)$. In the band $k_f \ll k \ll k_B$, $\Pi_u(k)$ decreases with k. However, for $k_B \ll k \ll k_{DI}$, $\mathcal{F}_B(k) \approx 0$, hence $\Pi_u(k) = \text{const.}$ $\Pi_\rho(k)$ first increases with k and then becomes constant.

After this background, in the following discussion, we describe the model of Bolgiano (1959) and Obukhov (1959) for SST. We denote the velocity and density fluctuations at wavenumber k using symbols u_k and ρ_k respectively. Since Ri ~ 1, the balance of nonlinear term and buoyancy yields

$$ku_k^2 = N\rho_k. \tag{15.37}$$

Bolgiano (1959) and Obukhov (1959) argued that the potential energy flux is approximately constant in the inertial range, and that it equals the dissipation rate of the potential energy (ϵ_ρ). As argued above, Π_ρ increases with k, yet constancy of Π_ρ is a reasonable assumption for the inertial range, as will be shown in Section 15.5.2. Hence, using dimensional analysis we deduce that

$$\Pi_\rho = k\rho_k^2 u_k = \epsilon_\rho. \tag{15.38}$$

Using Eqs. (15.37, 15.38) we can solve for u_k and ρ_k that yields

$$E_u(k) = \frac{u_k^2}{k} = \epsilon_\rho^{2/5} N^{4/5} k^{-11/5}, \tag{15.39a}$$

$$E_\rho(k) = \frac{\rho_k^2}{k} = \epsilon_\rho^{4/5} N^{-2/5} k^{-7/5}, \tag{15.39b}$$

$$\Pi_u(k) = ku_k^3 = \epsilon_\rho^{3/5} N^{6/5} k^{-4/5}, \tag{15.39c}$$

$$\Pi_\rho(k) = \epsilon_\rho. \tag{15.39d}$$

This is how Bolgiano (1959) and Obukhov (1959) showed that for stably stratified turbulence with moderate stratification, $\Pi_u(k) \sim k^{-4/5}$ and $E_u(k) \sim k^{-11/5}$. Clearly, $\Pi_u(k)$ decreases with k, consistent with Eq. (15.33).

Bolgiano and Obukhov went further and argued that buoyancy weakens at large k. For negligible buoyancy, the spectra and fluxes of SST would be same as those of passive scalar turbulence:

$$E_u(k) = K_{\mathrm{Ko}} \epsilon_u^{2/3} k^{-5/3}, \tag{15.40a}$$

$$E_\rho(k) = K_{\mathrm{OC}} \epsilon_u^{-1/3} \epsilon_\rho k^{-5/3}, \tag{15.40b}$$

$$\Pi_u(k) = \epsilon_u, \tag{15.40c}$$

$$\Pi_\rho(k) = \epsilon_\rho, \tag{15.40d}$$

where ϵ_u is the viscous dissipation rate, and $K_{\mathrm{Ko}}, K_{\mathrm{OC}}$ are respectively Kolmogorov's and Obukhov–Corrsin's constants. The behavioral transition from

Eq. (15.39) to Eq. (15.40) occurs near Bolgiano wavenumber k_B, which is obtained by matching $\Pi_u(k)$ in the two regimes. This operation yields

$$k_B \approx N^{3/2} \epsilon_u^{-5/4} \epsilon_\rho^{3/4}. \tag{15.41}$$

In the following discussion, we however show that with the increase of k, u_k gets weaker compared to ρ_k, and hence the second regime is impossible. Therefore, we expect only one scaling, that of Eq. (15.39) for SST. This is the topic of the next subsection.

15.5.2 Modified Bolgiano–Obukhov scaling

The aforementioned derivation of Bolgiano and Obukhov is based on the following two assumptions:

1. $\Pi_\rho(k) \approx$ const.

2. Buoyancy weakens at large wavenumbers (beyond Bolgiano wavenumber), and hence the scaling of passive scalar turbulence for this regime.

In the following discussion, we do not make either of the aforementioned assumptions. Here we make use of Eqs. (15.37), and conservation of total energy flux [Eq. (15.36)].

The conservation of total energy flux yields

$$\Pi_u(k) + \Pi_\rho(k) = k u_k^3 + k \rho_k^2 u_k = \epsilon_u + \epsilon_\rho = \epsilon, \tag{15.42}$$

where ϵ is the total energy dissipation rate. Using Eq. (15.37) we eliminate ρ_k in Eq. (15.42) that yields the following fifth-order polynomial in u_k:

$$k u_k^3 \left[1 + \frac{k^2 u_k^2}{N^2} \right] = \epsilon. \tag{15.43}$$

Alam et al. (2018) solved Eq. (15.43) for u_k and then computed ρ_k using Eq. (15.37). After this, they computed $\Pi_u(k)$, $\Pi_\rho(k)$, $E_u(k)$, and $E_\rho(k)$, which are plotted in Fig. 15.3. For a wide range of wavenumbers (from $k = 1$ to 10^{10}), they observed a scaling consistent with Eqs. (15.39) with no transition to Eqs. (15.40). Hence, the transition proposed by Bolgiano (1959) and Obukhov (1959) is incorrect. In the following paragraph we show why only a single scaling of Eqs. (15.39) is expected for SST with moderate stratification.

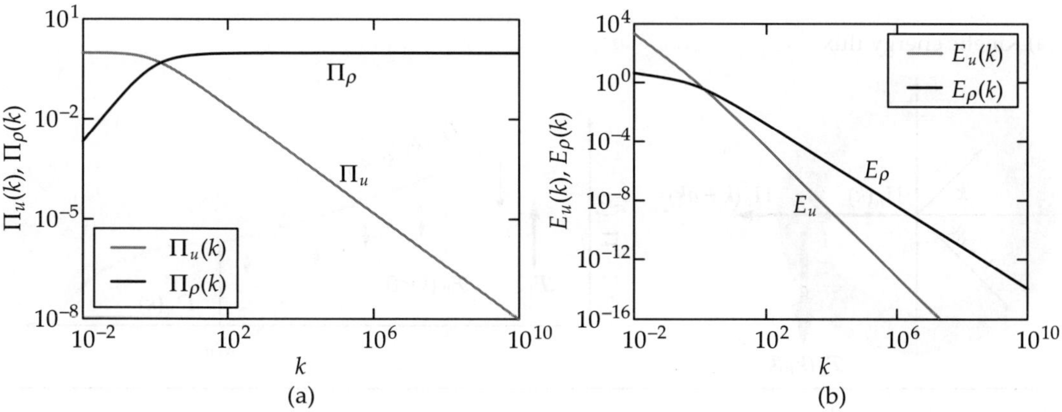

Figure 15.3 Solution of Eqs. (15.37, 15.43) yields: (a) Fluxes $\Pi_u(k)$ and $\Pi_\rho(k)$. (b) Kinetic and potential energy spectra $E_u(k)$ and $E_\rho(k)$. These plots exhibit scaling of Eqs. (15.39), and no transition from Eqs. (15.39) to Eqs. (15.40). Adopted from figures of Alam et al. (2018).

Using Eqs. (15.39), which is the solution of Eqs. (15.37, 15.43), we deduce that

$$\begin{aligned} u_k &\approx \epsilon_\rho^{1/5} N^{2/5} k^{-3/5}, \\ \rho_k &\approx \epsilon_\rho^{2/5} N^{-1/5} k^{-1/5}. \end{aligned} \tag{15.44}$$

Hence, with the increase of k, u_k decreases faster than ρ_k, and

$$\frac{\Pi_u(k)}{\Pi_\rho(k)} \approx \frac{k u_k^3}{k \rho_k^2 u_k} \approx \epsilon_\rho^{-2/5} N^{6/5} k^{-4/5}. \tag{15.45}$$

Therefore, with the increase of k, $\Pi_u(k)$ decreases faster than $\Pi_\rho(k)$ [see Fig. 15.3(a)]. Since u_k is weaker than ρ_k in this regime, there is no possibility that $k u_k^2$ will ever overcome the buoyancy term at large k. Hence, the second scaling of Eqs. (15.40) is not expected.

In Fig. 15.4, we sketch $\Pi_u(k)$ and $\Pi_\rho(k)$ obtained using the aforementioned arguments. The kinetic energy flux $\Pi_u(k)$ decreases quite rapidly as $k^{-4/5}$. Therefore,

$$\Pi_u(k) \approx A k^{-4/5} \implies \frac{d}{dk}\Pi_u(k) = \mathcal{F}_B(k) = -\frac{4}{5} A k^{-9/5}. \tag{15.46}$$

Thus, the energy feed by buoyancy becomes weaker with increasing k. Using

$$\frac{d}{dk}\Pi_\rho(k) = -\mathcal{F}_B(k) = \frac{4}{5} A k^{-9/5}, \tag{15.47}$$

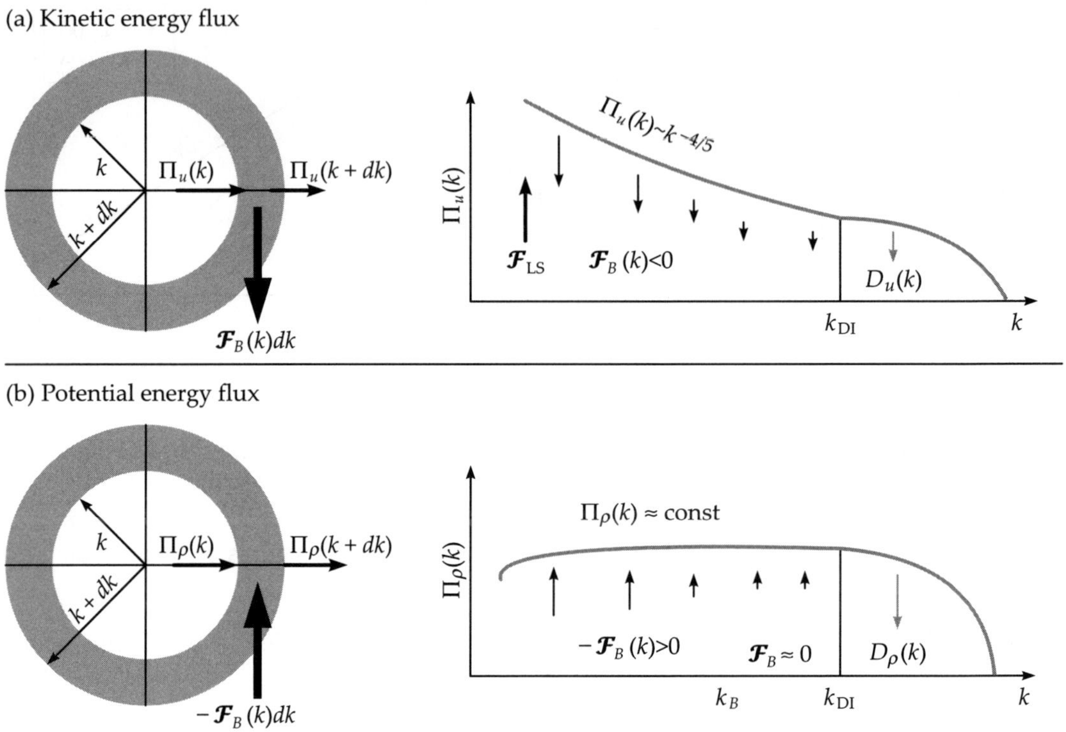

Figure 15.4 Modified Bolgoano–Obhukhv model: A schematic diagram of (a) kinetic energy flux; (b) potential energy flux. $\Pi_u(k) \sim k^{-4/5}$, while $\Pi_\rho(k) \approx$ const. These plots differ from the flux predictions of the Bolgiano–Obukhov model, which is exhibited in Fig. 15.2.

we obtain

$$\Pi_\rho(k) = \Pi_\rho(k_0) + \int_{k_0}^{k} dk' \frac{4}{5} A k'^{-9/5} \approx \Pi_\rho(k_0). \tag{15.48}$$

Due to power law physics, $\Pi_u(k)$ decreases rapidly, but $\Pi_\rho(k)$ does not change significantly (still keeping the sum to be a constant).

The above scaling is valid for $k > 1$. But for $k < 1$, the role of the two energy fluxes switch. From Eq. (15.45), we deduce that $\Pi_u(k) \gg \Pi_\rho(k)$ for $k \ll 1$. A simple analysis shows that for $k \ll 1$, the scaling relations are

$$\Pi_u(k) = \epsilon_u, \tag{15.49a}$$

$$\Pi_\rho(k) = N^{-2} \epsilon_u^{5/3} k^{4/3}, \tag{15.49b}$$

$$E_u(k) = K_{\mathrm{Ko}}\epsilon_u^{2/3}k^{-5/3}, \tag{15.49c}$$

$$E_\rho(k) = N^{-2}\epsilon_u^{4/3}k^{-1/3}. \tag{15.49d}$$

However, it is not certain if $k \ll 1$ regime is indeed realizable in experiments or in numerical simulations.

Before we end this subsection, we state that the Bolgiano–Obuknov scaling can be derived by substituting the variable kinetic energy flux in formulas for passive scalar turbulence. When we substitute $\Pi_u(k)$ of Eq. (15.39c) in Kolmogorov's formula for $E_u(k)$, we obtain

$$E_u(k) = K_{\mathrm{Ko}}[\Pi_u(k)]^{2/3}k^{-5/3} \sim K_{\mathrm{Ko}}\epsilon_\rho^{2/5}N^{4/5}k^{-11/5}. \tag{15.50}$$

Similar substitution in the formula for the potential energy spectrum [Eq. (14.11)] yields

$$E_\theta(k) = K_{\mathrm{OC}}\Pi_\theta[\Pi_u(k)]^{-1/3}k^{-5/3} \sim K_{\mathrm{OC}}\epsilon_\rho^{4/5}N^{-2/5}k^{-7/5}. \tag{15.51}$$

In the next subsection we will describe a numerical simulation that shows partial agreement with this phenomenology.

15.5.3 Numerical results on moderately stratified turbulence

To verify the Bolgiano–Obukhov scaling, Kumar et al. (2014a) performed a numerical simulation of stably stratified turbulence on a 1024^3 periodic grid. They chose $\mathrm{Pr} = 1$ and $\mathrm{Ri} = 0.01$; this choice of modest Ri leads to a moderate stratification of the flow (nonlinearity $\approx$ buoyancy). Kumar et al. forced the flow at large length scales and obtained a steady state with Reynolds number $\mathrm{Re} \approx 649$. Using the steady-state data, they computed the spectra and fluxes of the kinetic and potential energies, which are exhibited in Figs. 15.5 and 15.6.

Figure 15.5 shows that $E_u(k)k^{11/5} \approx$ const. for a relatively narrow wavenumber range, hence $E_u(k) \sim k^{-11/5}$ for this regime. Similarly, we observe that $E_\rho(k)k^{7/5} \approx$ const., hence $E_\rho(k) \sim k^{-7/5}$. Note that the plots of $E_u(k)k^{5/3}$ and $E_\rho(k)k^{5/3}$ are not flat. Thus, Kumar et al. (2014a)'s simulations provide evidences for Bolgiano and Obukhov's predictions of Eqs. (15.39).

In adidition, Kumar et al. (2014a) did not observe any transition from the spectra of Eqs. (15.39) to those of Eqs. (15.40), consistent with the modified Bolgiano and Obukhov scaling. This statement however needs verification from numerical simulations with more grid points (say 2048^3).

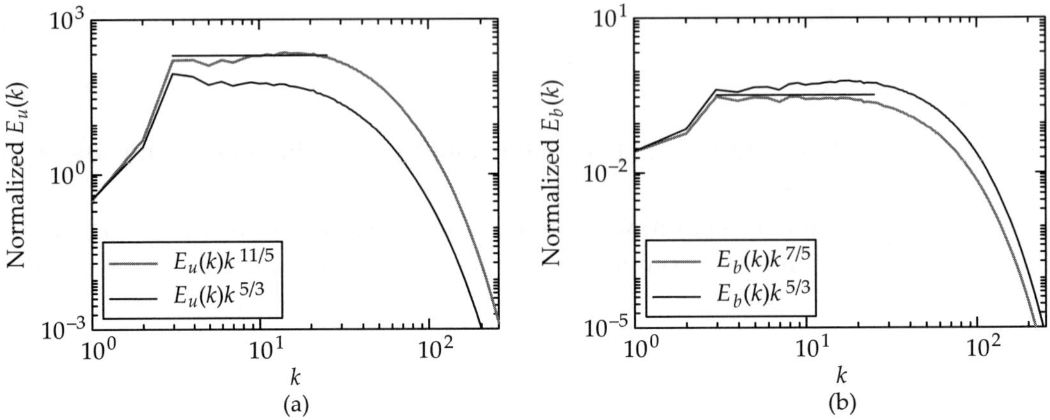

Figure 15.5 For stably-stratified turbulence with Pr = 1, Re = 649, and Ri = 0.01, plots of (a) normalized kinetic energy spectrum, (b) normalized potential energy spectrum. The Bolgiano–Obukhov scaling fits better with the data than Kolmogorov scaling. Read E_b as E_ρ. From Kumar et al. (2014a). Reprinted with permission from APS.

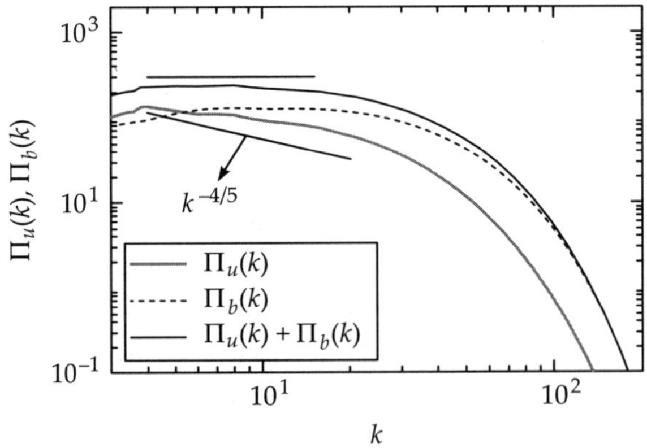

Figure 15.6 For stably-stratified turbulence with Pr = 1, Re = 649, and Ri = 0.01, plots of $\Pi_u(k)$, $\Pi_\rho(k)$, and $\Pi_u(k) + \Pi_\rho(k)$. Read $\Pi_b(k)$ as $\Pi_\rho(k)$. The energy fluxes are somewhat consistent with the predictions of Bolgiano–Obukhov phenomenology. From Verma (2018). Reprinted with permission from World Scientific.

The kinetic energy flux $\Pi_u(k)$ exhibited in Fig. 15.6 does not fall as steeply as $k^{-4/5}$, which is a prediction of Bolgiano–Obukhov scaling. However, there is a certain decrease in $\Pi_u(k)$ with k. The potential energy flux plotted in Fig. 15.6 show that $\Pi_\rho(k) \sim$ const. As described in the previous subsection, we expect that the sum $\Pi_u(k) + \Pi_\rho(k) \approx$ const.; but the numerical data show a minor decrease with k. Thus, the plots of the fluxes of kinetic and potential energies agree with the Bolgiano–Obukhov scaling only qualitatively.

15.6 Stably Stratified Turbulence with Strong Buoyancy

When Re $\gg 1$ and Ri $\gg 1$, the strong buoyancy tends to make the flow quasi-2D with $|\mathbf{u}_\perp| \gg u_\parallel$. Lindborg (2006) and Davidson (2013) provide scaling analysis for this regime, and argue that

$$E_\perp(k_\perp) \sim \epsilon_u^{2/3} k_\perp^{-5/3}, \tag{15.52a}$$

$$E_\perp(k_\parallel) \sim N^2 k_\parallel^{-3}. \tag{15.52b}$$

The tools described in Chapter 11 would be very useful for quantifying anisotropy in this regime. However, such discussions are skipped for brevity.

Though there are interesting numerical and phenomenological work in this area, there are a number of unresolved issues. Refer to Davidson (2013), Lindborg (2006), and references therein for more details.

With this, we end our discussion on stably stratified turbulence. In the next chapter we will describe scaling of turbulent thermal convection.

Further Reading

For more details on stably stratified flows, refer to books by Davidson (2013), Sagaut and Cambon (2008), and Verma (2018). The original sources for the Bolgiano–Obukhov scaling are Bolgiano (1959) and Obukhov (1959). For numerical verification of the Bolgiano–Obukhov scaling, refer to Kumar et al. (2014a).

For SST with strong stratification, refer to Davidson (2013), Lindborg (2006), and references therein.

Chapter 16

Thermal Convection

When we heat a fluid from the bottom and cool it from the top, hot fluid elements ascend and cold ones descend. This phenomena, called *thermal convection*, is due to the density variations induced by heating. Since the temperature affects the fluid flow, it is an *active scalar*. In this chapter we will study the properties of thermal convection, especially from energy transfer perspectives.

The equations for thermal convection are identical to those for stably stratified flows, except for the sign of density stratification: $d\rho/dz < 0$ for stably stratified flows, but $d\rho/dz > 0$ for thermal convection. This sign difference has strong ramifications on the stability and dynamics of the flow, as will be shown below.

We start with the governing equation of thermal convection.

16.1 Governing Equations

We assume that the fluid is confined between two conducting horizontal plates that are kept at $z = 0$ and d. See Fig. 16.1 for an illustration. The temperature of the corresponding plates are T_b and T_t respectively, with $T_b > T_t$. The local temperature is a combination of the imposed linear temperature $\bar{T}(z)$ and temperature fluctuation θ:

$$T(x, y, z) = \bar{T}(z) + \theta(x, y, z), \tag{16.1}$$

where

$$\bar{T}(z) = T_b + \frac{d\bar{T}}{dz}z = T_b - \frac{T_b - T_t}{d}z. \tag{16.2}$$

This particular configuration of thermal convection is the *Rayleigh–Bénard convection*, denoted in short by RBC.

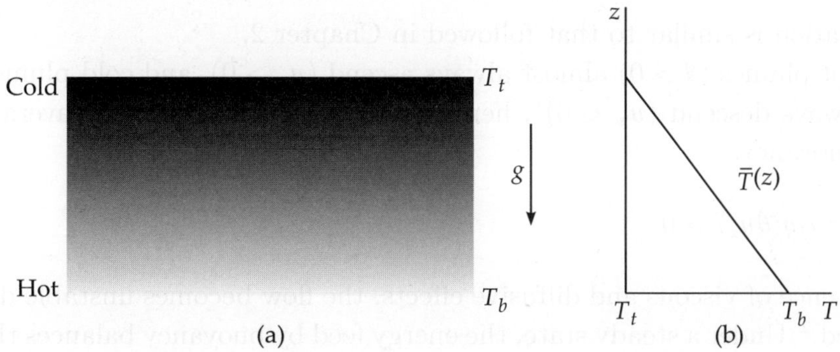

Figure 16.1 (a) A schematic diagram of Rayleigh–Bénard convection. The temperature of the bottom and top thermal plates are T_b and T_t respectively. (b) The mean temperature $\bar{T}(z)$ falls linearly with z.

For such flows, the equations of motion for the velocity and temperature fluctuations are

$$\frac{\partial \mathbf{u}}{\partial t} + (\mathbf{u}\cdot\nabla)\mathbf{u} = -\frac{1}{\rho_m}\nabla\sigma + \alpha g\theta\hat{z} + \nu\nabla^2\mathbf{u}, \tag{16.3a}$$

$$\frac{\partial \theta}{\partial t} + (\mathbf{u}\cdot\nabla)\theta = \frac{\Delta}{d}u_z + \kappa\nabla^2\theta, \tag{16.3b}$$

$$\nabla\cdot\mathbf{u} = 0, \tag{16.3c}$$

where $\Delta = T_b - T_t$, and ν, κ are respectively the kinematic viscosity and *thermal diffusivity* of the fluid. The temperature field is an active scalar since it affects the velocity field. For details refer to Verma (2018).

The equations for the kinetic energy and scalar energy ($\theta^2/2$) are

$$\frac{\partial}{\partial t}\frac{u^2}{2} + \nabla\cdot\left[\frac{u^2}{2}\mathbf{u}\right] = -\nabla\cdot(\sigma\mathbf{u} - \nu\mathbf{u}\times\boldsymbol{\omega}) + \alpha g\theta u_z - \nu\omega^2, \tag{16.4a}$$

$$\frac{\partial}{\partial t}\frac{\theta^2}{2} + \nabla\cdot\left[\frac{\theta^2}{2}\mathbf{u} - \kappa\theta\nabla\theta\right] = \frac{\Delta}{d}\theta u_z - \kappa(\nabla\theta)^2. \tag{16.4b}$$

In literature, $\theta^2/2$ is also called *entropy*. However, in this book, $\theta^2/2$ is referred to as scalar energy for consistency with the convention followed in the earlier chapters. It is easy to show that for $\nu = \kappa = 0$, and for periodic or vanishing boundary condition, the conserved quantity for RBC is

$$\frac{1}{2}\int d\mathbf{r}\left(u^2 - \frac{\alpha g d}{\Delta}\theta^2\right). \tag{16.5}$$

The derivation is similar to that followed in Chapter 2.

The hot plumes ($\theta > 0$) almost always ascend ($u_z > 0$), and cold plumes ($\theta < 0$) almost always descend ($u_z < 0$)[1], hence $\langle u_z\theta\rangle > 0$. Therefore, the average energy feed by buoyancy,

$$\mathcal{F}_B = \alpha g\langle \theta u_z\rangle > 0. \tag{16.6}$$

In the absence of viscous and diffusive effects, the flow becomes unstable due to this energy feed.[2] Under a steady state, the energy feed by buoyancy balances the viscous dissipation, and $\langle u^2\rangle$ is constant statistically. Note that θ^2 too gets a positive feed from $(\Delta/d)\langle u_z\theta\rangle$, which is balanced by the diffusion term. A key difference between thermal convection and stably stratified flows is the sign of $\mathcal{F}_B$: $\mathcal{F}_B > 0$ in thermal convection, but it is negative for stably stratified flows.

Equations (16.3) can be nondimensionalized using d as the length scale, free-fall velocity $\sqrt{\alpha g\Delta d}$ as the velocity scale, and temperature difference Δ as the temperature scale. Hence,

$$\mathbf{u} \to \mathbf{u}\sqrt{\alpha g\Delta d}; \qquad \theta \to \theta\Delta. \tag{16.7}$$

In terms of these variables, Eqs. (16.3) translate to the following nondimensional counterparts:

$$\frac{\partial \mathbf{u}}{\partial t} + (\mathbf{u}\cdot\nabla)\mathbf{u} = -\nabla\sigma + \theta\hat{z} + \sqrt{\frac{\mathrm{Pr}}{\mathrm{Ra}}}\nabla^2\mathbf{u}, \tag{16.8a}$$

$$\frac{\partial \theta}{\partial t} + (\mathbf{u}\cdot\nabla)\theta = u_z + \frac{1}{\sqrt{\mathrm{RaPr}}}\nabla^2\theta, \tag{16.8b}$$

$$\nabla\cdot\mathbf{u} = 0, \tag{16.8c}$$

[1] This pattern may be altered briefly during a flow reversal, as shown by Chandra and Verma (2013).

[2] This is a generic feature of all unstably stratified flows—bubbly turbulence, Rayleigh–Taylor turbulence, etc.

where

$$\text{Ra} = \frac{\alpha g d^3 \Delta}{\nu\kappa}, \tag{16.9}$$

$$\text{Pr} = \frac{\nu}{\kappa} \tag{16.10}$$

are the *Rayleigh number* and *Prandtl number* respectively.

In the next section, we will derive the above equations in Fourier space.

16.2 Governing Equations in Fourier Space, Energy Transfers, and Fluxes

The RBC equations in Fourier space are

$$\frac{d}{dt}\mathbf{u}(\mathbf{k}) + \mathbf{N}_u(\mathbf{k}) = -i\mathbf{k}\sigma(\mathbf{k}) + \alpha g\theta(\mathbf{k})\hat{z} - \nu k^2\mathbf{u}(\mathbf{k}), \tag{16.11a}$$

$$\frac{d}{dt}\theta(\mathbf{k}) + N_\theta(\mathbf{k}) = \frac{\Delta}{d}u_z(\mathbf{k}) - \kappa k^2\theta(\mathbf{k}), \tag{16.11b}$$

$$\mathbf{k}\cdot\mathbf{u}(\mathbf{k}) = 0, \tag{16.11c}$$

where

$$\mathbf{N}_u(\mathbf{k}) = i\sum_{\mathbf{p}}[\mathbf{k}\cdot\mathbf{u}(\mathbf{k}-\mathbf{p})]\mathbf{u}(\mathbf{p}), \tag{16.12a}$$

$$N_\theta(\mathbf{k}) = i\sum_{\mathbf{p}}[\mathbf{k}\cdot\mathbf{u}(\mathbf{k}-\mathbf{p})]\theta(\mathbf{p}) \tag{16.12b}$$

are the nonlinear terms, which were derived in Chapters 4 and 13, and the pressure is

$$\sigma(\mathbf{k}) = \frac{i}{k^2}[\mathbf{k}\cdot\mathbf{N}_u(\mathbf{k}) - \alpha g k_z\rho(\mathbf{k})]. \tag{16.13}$$

The governing equations for the modal kinetic energy $|\mathbf{u}(\mathbf{k})|^2/2$ and the modal scalar energy $|\theta(k)|^2/2$ are as follows:

$$\frac{d}{dt}E_u(\mathbf{k}) = \Im\left[\sum_{\mathbf{p}}[\mathbf{k}\cdot\mathbf{u}(\mathbf{q})]\,[\mathbf{u}(\mathbf{p})\cdot\mathbf{u}^*(\mathbf{k})]\right] + \alpha g\Re[\theta(\mathbf{k})u_z^*(\mathbf{k})] - 2\nu k^2E_u(\mathbf{k}), \tag{16.14a}$$

$$\frac{d}{dt}E_\theta(\mathbf{k}) = \Im\left[\sum_{\mathbf{p}}[\mathbf{k}\cdot\mathbf{u}(\mathbf{q})]\,[\theta(\mathbf{p})\theta^*(\mathbf{k})]\right] + \frac{\Delta}{d}\Re[\theta(\mathbf{k})u_z^*(\mathbf{k})] - 2\kappa k^2E_\theta(\mathbf{k}), \tag{16.14b}$$

where $\mathbf{q} = \mathbf{k} - \mathbf{p}$, and $\Re, \Im$ stand respectively for the real and imaginary parts of the argument. In the aforementioned equation, $E_u(\mathbf{k})$ changes with time due to (a) the nonlinear energy transfers from other Fourier modes, (b) the kinetic energy feed by buoyancy,

$$\mathcal{F}_B(\mathbf{k}) = \alpha g \Re[\theta(\mathbf{k}) u_z^*(\mathbf{k})], \tag{16.15}$$

and (c) the viscous dissipation rate

$$D_u(\mathbf{k}) = -2\nu k^2 E_u(\mathbf{k}). \tag{16.16}$$

Similar interpretations hold for $E_\theta(\mathbf{k})$. When we contrast the above equations with those for passive scalar, we observe that in RBC, both kinetic and scalar energies receive additional contributions from buoyancy.

In the previous section we showed that the volume averaged kinetic energy feed by buoyancy, $\alpha g \langle u_z \theta \rangle > 0$. Hence, using Parseval's theorem, we deduce that

$$\sum_{\mathbf{k}} \alpha g \Re[\theta(\mathbf{k}) u_z^*(\mathbf{k})] > 0. \tag{16.17}$$

Note however that each term of this sum is not necessarily positive. Yet, numerical and experimental results indicate that

$$\mathcal{F}_B(\mathbf{k}) = \alpha g \Re[\theta(\mathbf{k}) u_z^*(\mathbf{k})] > 0 \tag{16.18}$$

for all the Fourier modes. This relation implies that the shell spectrum

$$\mathcal{F}_B(k) = \sum_{k-1 < k' \le k} \alpha g \langle \theta(\mathbf{k}') u_z^*(\mathbf{k}') \rangle > 0. \tag{16.19}$$

The formulas for the mode-to-mode kinetic energy and scalar energy transfers, as well as their associated energy fluxes, would be the same as those discussed in earlier chapters because of the identical forms of the nonlinear term. Hence,

$$S^{uu}(\mathbf{k}|\mathbf{p}|\mathbf{q}) = -\Im\left[\{\mathbf{k}' \cdot \mathbf{u}(\mathbf{q})\}\{\mathbf{u}(\mathbf{p}) \cdot \mathbf{u}(\mathbf{k}')\}\right], \tag{16.20a}$$

$$S^{\theta\theta}(\mathbf{k}|\mathbf{p}|\mathbf{q}) = -\Im\left[\{\mathbf{k}' \cdot \mathbf{u}(\mathbf{q})\}\{\theta(\mathbf{p})\theta(\mathbf{k}')\}\right], \tag{16.20b}$$

and

$$\Pi_u(k_0) = \sum_{|\mathbf{k}'| > k_0} \sum_{|\mathbf{p}| \le k_0} S^{uu}(\mathbf{k}'|\mathbf{p}|\mathbf{q}), \tag{16.21a}$$

$$\Pi_\theta(k_0) = \sum_{|\mathbf{k}'| > k_0} \sum_{|\mathbf{p}| \le k_0} S^{\theta\theta}(\mathbf{k}'|\mathbf{p}|\mathbf{q}). \tag{16.21b}$$

Following the same set of energetics arguments as in Sections 4.4 and 13.3, we derive the following set of equations for E_u and E_θ in the inertial–dissipative range:

$$\frac{\partial}{\partial t}E_u(k,t) = -\frac{\partial}{\partial k}\Pi_u(k,t) + \mathcal{F}_B(k,t) - 2\nu k^2 E_u(k,t), \tag{16.22a}$$

$$\frac{\partial}{\partial t}E_\theta(k,t) = -\frac{\partial}{\partial k}\Pi_\theta(k,t) + \frac{\Delta}{\alpha g d}\mathcal{F}_B(k,t) - 2\kappa k^2 E_\theta(k,t). \tag{16.22b}$$

Under a steady state ($\partial/\partial t = 0$), in the inertial range where the dissipation and diffusion terms are negligible, the aforementioned equations yield

$$\frac{d}{dk}\Pi_u(k) = \mathcal{F}_B(k), \tag{16.23a}$$

$$\frac{d}{dk}\Pi_\theta(k) = \frac{\Delta}{\alpha g d}\mathcal{F}_B(k). \tag{16.23b}$$

Hence, under the above assumptions, we obtain

$$\Pi_u(k) - \frac{\alpha g d}{\Delta}\Pi_\theta(k) = \text{const.} = C_1. \tag{16.24}$$

To compare the two fluxes, we construct their nondimensional counterparts—$\Pi'_u(k), \Pi'_\theta(k)$—as:

$$\Pi_u(k) = \frac{U^3}{d}\Pi'_u(k), \tag{16.25a}$$

$$\Pi_\theta(k) = \frac{U\Delta^2}{d}\Pi'_\theta(k), \tag{16.25b}$$

where U is the rms velocity of the flow. Substitution of these terms in Eq. (16.24) yields

$$\frac{U^3}{d}\left[\Pi'_u(k) - \frac{\alpha g \Delta d}{U^2}\Pi'_\theta(k)\right] = \text{const.} \tag{16.26}$$

Since $U \approx \sqrt{\alpha g \Delta d}$, we deduce that

$$\Pi'_u(k) - \Pi'_\theta(k) \approx C_2. \tag{16.27}$$

Numerical simulations however reveal that $C_2 \approx 0$.

The buoyancy $\alpha g \theta(\mathbf{k})\hat{z}$ feeds kinetic helicity to the Fourier mode $\mathbf{u}(\mathbf{k})$. Using Eq. (3.48) we obtain

$$\begin{aligned}\mathcal{F}_{H_K}(\mathbf{k}) &= \Re[\boldsymbol{\omega}^*(\mathbf{k}) \cdot \mathbf{F}_u(\mathbf{k})] \\ &= \alpha g \Re[(-i\mathbf{k} \times \mathbf{u}^*(\mathbf{k})) \cdot \hat{z}\theta(\mathbf{k})]\end{aligned}$$

$$
\begin{aligned}
&= \alpha g \Re[(i\mathbf{k} \times \hat{z} \cdot \mathbf{u}^*(\mathbf{k}))\theta(\mathbf{k})] \\
&= -\alpha g k (\sin\zeta) \Im[u_1^*(\mathbf{k})\theta(\mathbf{k})], \qquad (16.28)
\end{aligned}
$$

where u_1 is the first component of the velocity field in the Craya–Herring basis (see Section 9.1), and ζ is the angle between the wave vector $\mathbf{k}$ and $\hat{z}$. Similarly, we can derive the enstrophy feed to $\mathbf{u}(\mathbf{k})$ by buoyancy as (see Eq. (3.50))

$$
\begin{aligned}
\mathcal{F}_\omega(\mathbf{k}) &= k^2 \Re[\mathbf{u}^*(\mathbf{k}) \cdot \mathbf{F}_u(\mathbf{k})] \\
&= k^2 \alpha g \Re[\theta(\mathbf{k}) u_z^*(\mathbf{k})]. \qquad (16.29)
\end{aligned}
$$

In the next section we briefly describe the structure of the temperature field in RBC.

16.3 Structure of Temperature Field in Thermal Convection

In RBC, the temperature field has an interesting large-scale pattern, as will be described in this section. For convenience, we work with nondimensional variables in which the temperature of the bottom and top plates located at $z = 0, 1$ are $T = 1$ and 0 respectively. We take planar average of Eq. (16.1) over the xy plane that yields

$$
T_m(z) = \bar{T}(z) + \theta_m(z) = 1 - z + \theta_m(z), \qquad (16.30)
$$

where the subscript m stands for mean.

For small thermal diffusivity, $T_m(z)$ varies sharply in the thermal boundary layers, but is nearly 1/2 in the bulk. Hence,

$$
\theta_m(z) \approx z - 1 + 1/2 \approx z - 1/2 \qquad (16.31)
$$

in the bulk (which is nearly all of the flow when the boundary layers are thin). See Fig. 16.2 for an illustration.

Pandey and Verma (2016) and Verma et al. (2017) performed Fourier transform of Eq. (16.31) that yields

$$
\theta_m(0,0,k_z) \approx \begin{cases} -\frac{1}{\pi k_z} & \text{for even } k_z \\ 0 & \text{otherwise} \end{cases} \qquad (16.32)
$$

Hence, the scalar energy spectrum of θ_m is

$$
E_{\theta_m}(k) = \frac{1}{2}|\theta_m(0,0,k_z)|^2 \sim k^{-2}. \qquad (16.33)
$$

We will revisit this result in Section 16.6.2.

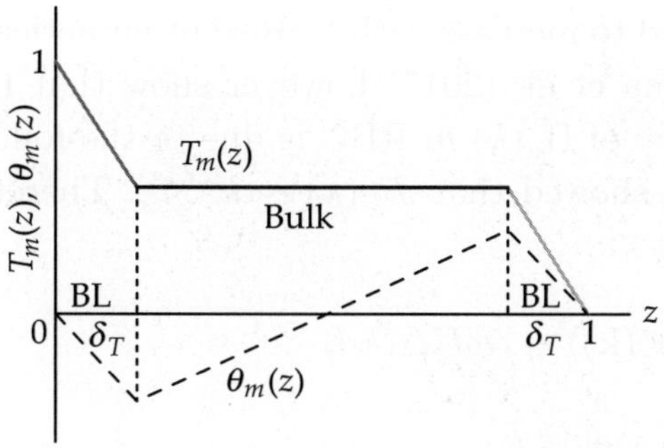

Figure 16.2 A schematic diagram of the planar-averaged temperature $T_m(z)$ and $\theta_m(z)$. From Pandey and Verma (2016). Reprinted with permission from AIP.

In the next section, we will describe a phenomenology of turbulent thermal convection.

16.4 Phenomenology of Turbulent Thermal Convection

The equations for RBC are very similar to those of stably stratified flows. Motivated by these similarities, many researchers argued that the Bolgiano–Obukhov (BO) phenomenology for stably stratified turbulence would also apply to turbulent thermal convection (TTC), that is, $E_u(k) \sim k^{-11/5}$ and $E_\theta(k) \sim k^{-7/5}$ (L'vov, 1991; L'vov and Falkovich, 1992). We will show in this section that the energetics of stably stratified turbulence and TTC are very different. Based on new energetics arguments and some subtle properties of TTC, we deduce that TTC exhibits Kolmogorov-like kinetic energy spectrum, that is, $E_u(k) \sim k^{-5/3}$.

For the inertial range of TTC, we start with Eq. (16.23a). Since hot plumes ascend and cold plumes descend, θ and u_z are positively correlated, or

$$\langle \theta(\mathbf{r}) u_z(\mathbf{r}) \rangle > 0. \tag{16.34}$$

As argued in Section 16.2, we claim that

$$\mathcal{F}_B(k)dk = \sum_{k<k' \le k+dk} \alpha g \langle \theta(\mathbf{k}') u_z^*(\mathbf{k}') \rangle > 0. \tag{16.35}$$

Therefore,

$$\frac{d}{dk}\Pi_u(k) = \mathcal{F}_B(k) > 0. \tag{16.36}$$

Hence, $\Pi_u(k)$ is expected to increase with k. Recent numerical simulations of Kumar et al. (2014a) and Verma et al. (2017) however show that $\Pi_u(k)$ of RBC is nearly constant. The constancy of $\Pi_u(k)$ in RBC is due to the following reasons.

Verma et al. (2017) showed that $\mathcal{F}_B(k) \sim k^{-5/3}$. Therefore, using dimensional analysis we deduce that

$$\mathcal{F}_B(k) = \alpha g \langle \theta(\mathbf{k}) u_z^*(\mathbf{k}) \rangle \sim \alpha g U \Delta (kd)^{-5/3}. \tag{16.37}$$

Therefore, Eq. (16.23a) yields

$$\begin{aligned} \frac{1}{\Pi_u(k)} \frac{d}{dk} \Pi_u(k) &= \frac{1}{\Pi_u(k)} \mathcal{F}_B(k) \\ &\sim \frac{d}{U^3} \alpha g U \Delta (kd)^{-5/3} \\ &\sim (kd)^{-5/3}. \end{aligned} \tag{16.38}$$

In this derivation, we employ $\alpha g \Delta d \sim U^2$. Using Eq. (16.38) we infer that the buoyancy weakens with the increase of k. Hence, in the inertial range, it is not strong enough to be able to increase $\Pi_u(k)$ significantly. See Fig. 16.3 for an illustration.

There is however a subtle difference between the hydrodynamic turbulence (Kolmogorov's picture) and turbulent thermal convection. As described earlier, thermal plumes drive the flow at all scales, though prominently at large scales. Under a steady state,

$$\frac{d}{dk} \Pi_u(k) = \mathcal{F}_B(k) - D_u(k). \tag{16.39}$$

An integration of this equation from k_0 to ∞ yields

$$\cancel{\Pi_u(\infty)} - \Pi_u(k_0) = \int_{k_0}^{\infty} \mathcal{F}_B(k) dk - \int_{k_0}^{\infty} D_u(k) dk = \int_{k_0}^{\infty} \mathcal{F}_B(k) dk - \epsilon_u. \tag{16.40}$$

Note that

$$\int_{k_0}^{\infty} \mathcal{F}_B(k) dk > 0 \tag{16.41}$$

unlike in Kolmogorov's model of hydrodynamics turbulence for which the external force in the inertial range is zero. Hence,

$$\Pi_u(k_0) = \epsilon_u - \int_{k_0}^{\infty} \mathcal{F}_B(k) dk < \epsilon_u. \tag{16.42}$$

Thus, $\Pi_u(k_0)$ in the inertial range is less than the energy dissipation rate. To illustrate, Bhattacharya et al. (2018b) show that for very large Ra and Pr ≈ 1, the inertial range Π_u is around half of ϵ_u. We however remark that the constant kinetic energy flux is still a good assumption. The sum of weak buoyancy feed over the whole range of wavenumbers is only half of ϵ_u.

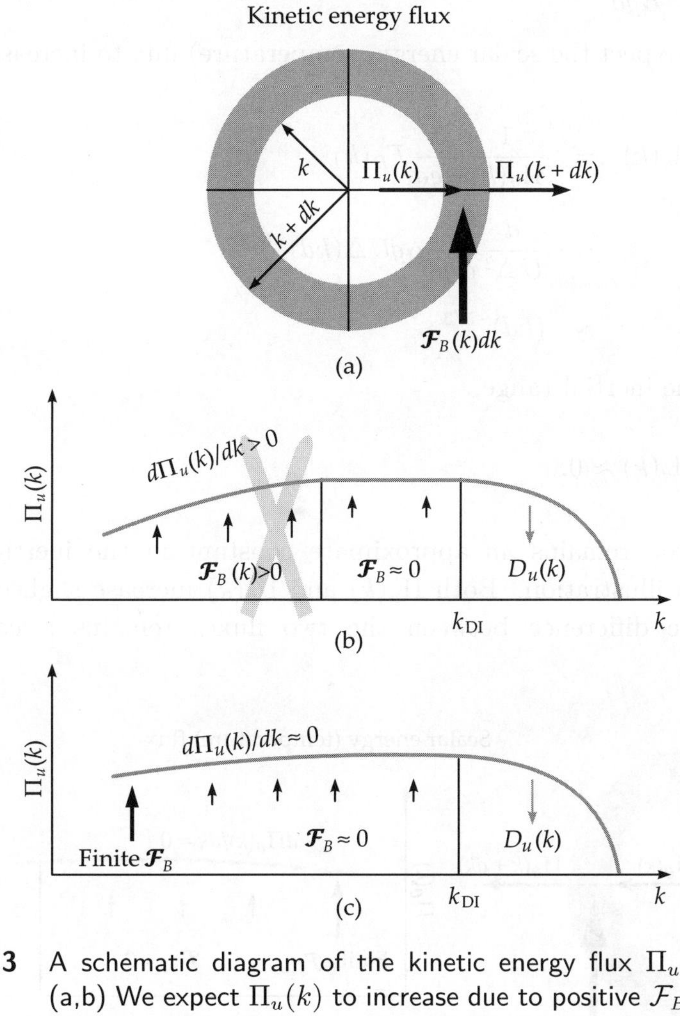

Figure 16.3 A schematic diagram of the kinetic energy flux $\Pi_u(k)$ of TTC. (a,b) We expect $\Pi_u(k)$ to increase due to positive $\mathcal{F}_B(k)$, but this is not so. (c) $\mathcal{F}_B(k)$ is much smaller than the pressure gradient. Therefore, $d\Pi_u(k)/dk \approx 0$, or $\Pi_u(k) \approx$ const.

This result is consistent with the numerical results of Pandey and Verma (2016) who showed that in the momentum equation, buoyancy is much smaller than the pressure gradient.[3] Hence, turbulent thermal convection is driven primarily by the

[3]In a related work, Verma (2018) also showed that Ri ≈ 0.1 for turbulent thermal convection (see Sec. 11.3

pressure gradient, as in 3D hydrodynamics. Buoyancy essentially supplies energy at large scales. Therefore, the statement that plumes drive RBC is incorrect.

Similar constancy is also observed for the scalar energy or temperature flux. From Eq. (16.23b) we observe that

$$\frac{d}{dk}\Pi_\theta(k) = \frac{\Delta}{\alpha g d}\mathcal{F}_B > 0. \tag{16.43}$$

Hence, we may expect the scalar energy (temperature) flux to increase with k. Note however that

$$\begin{aligned} \frac{1}{\Pi_\theta(k)}\frac{d}{dk}\Pi_\theta(k) &= \frac{1}{\Pi_\theta(k)}\frac{\Delta}{\alpha g d}\mathcal{F}_B(k) \\ &\sim \frac{d}{U\Delta^2}\frac{\Delta}{\alpha g d}\alpha g U \Delta (kd)^{-5/3} \\ &\sim (kd)^{-5/3}. \end{aligned} \tag{16.44}$$

Therefore, in the inertial range,

$$\frac{1}{\Pi_\theta(k)}\frac{d}{dk}\Pi_\theta(k) \approx 0. \tag{16.45}$$

Hence, $\Pi_\theta(k)$ too remains an approximate constant in the inertial range. See Fig. 16.4 for an illustration. Both $\Pi_u(k)$ and $\Pi_\theta(k)$ increase slightly due to small $\mathcal{F}_B(k)$, but the difference between the two fluxes remains a constant, as in Eq. (16.24).

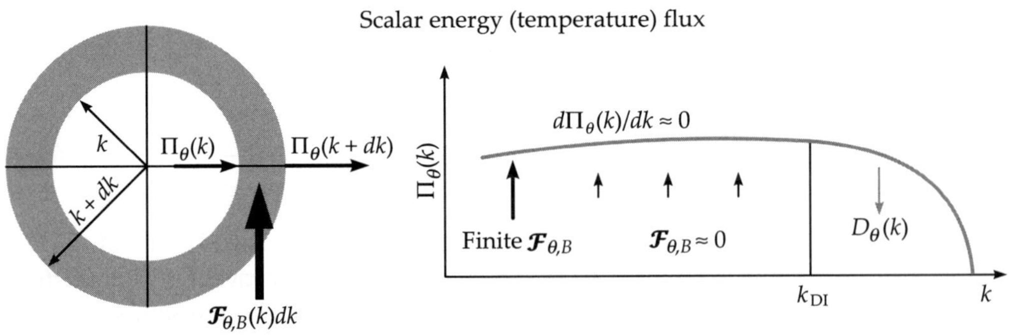

Figure 16.4 A schematic diagram of the scalar energy or temperature flux, $\Pi_\theta(k)$. The scalar energy flux $d\Pi_\theta(k)/dk \approx 0$, or $\Pi_\theta(k) \approx \text{const}$.

of Verma (2018)).

Constant values for Π_u and Π_θ, and similarities between TTC and hydrodynamic turbulence lead us to conclude that the kinetic energy spectrum $E_u(k)$ of TTC follows Kolmogorov's spectrum:

$$E_u(k) = K_{\mathrm{Ko}}(\Pi_u)^{2/3}k^{-5/3}. \tag{16.46}$$

Refer to Chapter 5 for the derivation of Kolmogorov's spectrum for hydrodynamic turbulence. Note that the literature is divided on the suitability of Bologiano–Obukhov phenomenology or Kolmogorov's phenomenology for TTC. The aforementioned arguments however conclusively show that TTC should have constant kinetic energy flux and $k^{-5/3}$ kinetic energy spectrum. For clarity, in Table 16.1, we contrast some of the important properties of stably stratified turbulence and TTC.

Table 16.1 Turbulent thermal convection vs. stably stratified turbulence with moderate stratification.

Property	Thermal convection	Stably stratified turbulence
Nature of stability	Unstable	Stable
Phenomenology	Kolmogorov	Bolgiano–Obukhov
$\mathcal{F}_B(k)$	Positive	Negative
$\Pi_u(k)$	$\Pi_u(k) \approx$ const.	$\Pi_u(k) \sim k^{-4/5}$
$E_u(k)$	$k^{-5/3}$	$k^{-11/5}$
Density/temperature flux	constant in k	constant in k
Density/temperature spec.	bi-spectrum	$k^{-7/5}$

In the next section we describe the structure function of turbulent thermal convection that is similar to that of hydrodynamic turbulence (apart from some subtle differences).

16.5 Structure Functions of Turbulent Thermal Convection

In this section we describe the structure functions for the velocity field of TTC. We show that these structure functions are very similar to those of hydrodynamic turbulence. The derivation for TTC is very similar to that by Kolmogorov (1941c,a). Earlier, Nath et al. (2016) had shown that the distribution of the velocity field in turbulent convection is nearly isotropic. Hence, we employ the assumptions of homogeneity and isotropy to TTC.

The qth order structure function for the velocity field is defined as

$$S_q^u(l) = \langle [\{\mathbf{u}(\mathbf{r}+\mathbf{l}) - \mathbf{u}(\mathbf{r})\} \cdot \mathbf{n}]^q \rangle, \tag{16.47}$$

where $\mathbf{n} = \mathbf{l}/l$. See Fig. 12.1 for an illustration. As described in the previous section, in the inertial range of turbulent thermal convection, the buoyancy is much weaker than the pressure gradient. Hence, its flow behavior is similar to that of hydrodynamic turbulence. Hence, we expect (see Sec. 12.3)

$$S_3(l) = -\frac{4}{5}\Pi_u l. \tag{16.48}$$

The only difference is that $\Pi_u < \epsilon_u$, and the difference is the energy supply rate by buoyancy in the inertial range.

For a more rigorous derivation, we start with Eq. (12.47):

$$\frac{\partial}{\partial t}\frac{1}{2}\langle u_i(\mathbf{r})u_i(\mathbf{r}+\mathbf{l})\rangle = T_u(\mathbf{l}) + \mathcal{F}_B(\mathbf{l}) - D_u(\mathbf{l}), \tag{16.49}$$

where $T_u(\mathbf{l})$ is the nonlinear energy transfer at length scale l, and $\mathcal{F}_B(\mathbf{l})$ and $D_u(\mathbf{l})$ are respectively the correlations associated with the energy feed by buoyancy and viscous dissipation. Under steady state, $\partial\langle u_i(\mathbf{r})u_i(\mathbf{r}+\mathbf{l})\rangle/\partial t = 0$. In addition we focus on the inertial range where $D_u(\mathbf{l}) \to 0$. Therefore,

$$-T_u(\mathbf{l}) = \mathcal{F}_B(\mathbf{l}). \tag{16.50}$$

Numerical simulations of Verma et al. (2017) show that $\mathcal{F}_B(\mathbf{l})$ of TTC is dominant at small wavenumbers, but has a power law tail in the inertial range ($k^{-5/3}$). For simplification of the calculation, we model $\mathcal{F}_B(\mathbf{l})$ as

$$\mathcal{F}_B(\mathbf{k}) = \frac{A}{2}(\delta_{\mathbf{k},\mathbf{k}_0} + \delta_{\mathbf{k},-\mathbf{k}_0}) + Bk^{-\alpha}, \tag{16.51}$$

where A, B are constants but $B \ll A$, and $\alpha \approx 5/3$. Inverse Fourier transform of this equation yields

$$\begin{aligned}
\mathcal{F}_B(\mathbf{l}) &= \langle \mathbf{F}(\mathbf{r}) \cdot \mathbf{u}(\mathbf{r}+\mathbf{l})\rangle \\
&= \sum_{\mathbf{k}} \mathcal{F}(\mathbf{k}) \exp(i\mathbf{k}\cdot\mathbf{l}) \\
&= A\cos(\mathbf{k}_0\cdot\mathbf{l}) + \int d\mathbf{k} Bk^{-\alpha}\exp(i\mathbf{k}\cdot\mathbf{l}) \\
&\approx A + DBl^{\alpha-1} \approx A + DBl^{2/3},
\end{aligned} \tag{16.52}$$

where D is a nondimensional constant coming out of the integral.

In Eq. (16.50), we substitute this $\mathcal{F}_B(\mathbf{l})$, and $T_u(\mathbf{l})$ of Eq. (12.53). These operations yield

$$-\frac{1}{12}\frac{1}{l^2}\frac{d}{dl}\left[\frac{1}{l}\frac{d}{dl}(l^4 S_3(l))\right] = A + DBl^{2/3}. \tag{16.53}$$

Following similar procedure as in Section 12.3, we obtain

$$S_3(l) = -\frac{4}{5}(Al + DBl^{5/3}). \tag{16.54}$$

The large-scale forcing at $k = k_0$ equals the energy flux Π_u. We ignore the second term under the assumption that B is small, and hence,

$$S_3(l) = -\frac{4}{5}\Pi_u l, \tag{16.55}$$

which is of the similar form as Kolmogorov (1941a). Note that $\Pi_u \approx \epsilon_u/2$ for $\mathrm{Pr} \approx 1$ (Bhattacharya et al., 2018b).

In Section 12.6 we discussed multiscaling exponents for hydrodynamic turbulence. We show in Section 16.6.3 that the multiscaling exponents for $\mathbf{u}$ of turbulent convection are quite similar to those of hydrodynamic turbulence. However, the temperature field of TTC is inhomogeneous as shown in Fig. 16.2. Hence, it is difficult to extend the derivation of Section 14.5 to thermal turbulence. Therefore, we do not discuss the structure function of θ in this book.

In the next section we will test the aforementioned phenomenology of TTC using numerical simulations.

16.6 Numerical Verification of the Phenomenology of Turbulent Thermal Convection

There have been a large number of numerical simulations of turbulent thermal convection. Some of these simulations report agreement with Kolmogorov-like spectrum for TTC, while some show similarities with Bolgiano–Obukhov scaling. In the present section we present recent numerical results that conclusively show that TTC and hydrodynamic turbulence have similar properties.

Verma et al. (2017) simulated RBC on a 4096^3 grid for $\mathrm{Pr} = 1$ and $\mathrm{Ra} = 1.1 \times 10^{11}$. At the thermal plates, they employed free-slip boundary for the velocity field and conducting boundary condition for the temperature field. Periodic boundary condition was employed along the horizontal directions.

Though most results presented here are for free-slip and conducting boundary conditions, we also report kinetic energy spectrum and flux of a no-slip simulation by Kumar and Verma (2018) with $\mathrm{Pr} = 1$ and $\mathrm{Ra} = 10^8$.

16.6.1 Kinetic energy spectrum and flux; Scalar energy flux

Using the steady state data, Verma et al. (2017) computed the spectrum and flux of the kinetic energy. They also computed the scalar energy flux. We exhibit these quantities in Fig. 16.5, and describe their properties in this section.

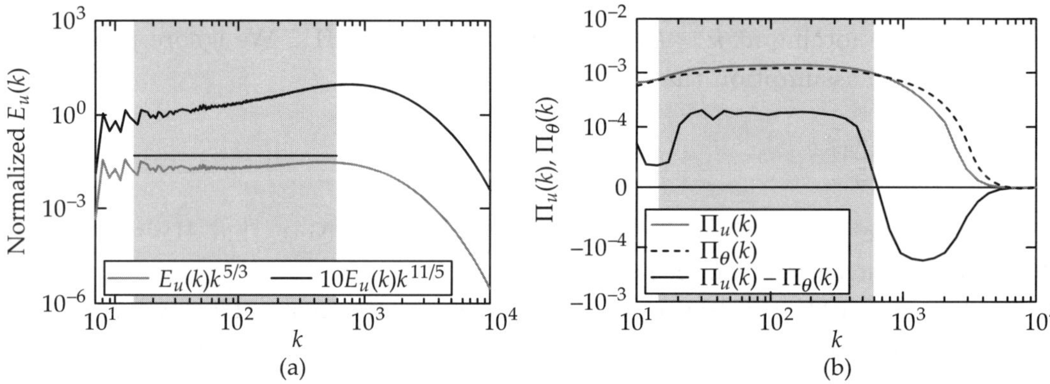

Figure 16.5 For the RBC simulation with $\mathrm{Pr} = 1$ and $\mathrm{Ra} = 1.1 \times 10^{11}$ on a 4096^3 grid with free-slip and isothermal boundary conditions: (a) plots of normalized kinetic energy spectra $E_u(k)k^{5/3}$ and $E_u(k)k^{11/5}$ indicate that $E_u(k) \sim k^{-5/3}$. (b) The fluxes $\Pi_u(k), \Pi_\theta(k)$ are approximate constants in the inertial range (shaded region in the plot). The flux difference, $\Pi_u(k) - \Pi_\theta(k)$, is flat. From Verma (2018). Reprinted with permission from World Scientific.

As shown in Fig. 16.5(a), $E_u(k)k^{5/3} \sim$ const., hence $E_u(k) \sim k^{-5/3}$. Note that $E_u(k)k^{11/5}$ is not flat, thus ruling out the Bolgiano–Obukhov spectrum for kinetic energy. We also observe that both $\Pi_u(k)$ and $\Pi_\theta(k)$ are nearly constant in the inertial range. These fluxes however show a slight increase with k due to buoyancy, but the difference, $\Pi_u(k) - \Pi_\theta(k)$, is a constant, as predicted by Eq. (16.24). Here $\alpha g d/\Delta = 1$ due to normalization.

RBC simulations with the no-slip boundary condition too exhibit similar spectrum and flux for the kinetic energy. For example, Kumar and Verma (2018) simulated RBC under no-slip boundary condition for $\mathrm{Pr} = 1$ and $\mathrm{Ra} = 10^8$. Figure 16.6 illustrates $E_u(k)$ and $\Pi_u(k)$ computed using the steady-state data. Clearly, $E_u(k) \sim k^{-5/3}$, and $\Pi_u(k) \approx$ const., consistent with Kolmogorov's phenomenology.

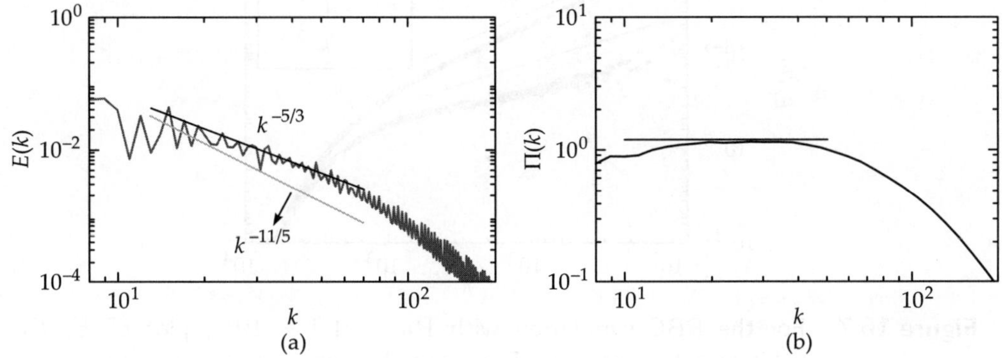

Figure 16.6 For a RBC simulation with $\mathrm{Pr} = 1$ and $\mathrm{Ra} = 10^8$ under no-slip boundary condition along all the walls. The kinetic energy spectrum $E_u(k) \sim k^{-5/3}$, and kinetic energy flux $\Pi_u(k) \approx$ constant in the inertial range. From Kumar and Verma (2018). Reprinted under Creative Commons Attribution License.

16.6.2 Scalar energy or temperature spectrum

Considering the similarities between the TTC and hydrodynamic turbulence, we may expect the scalar energy spectrum to be Kolmogorov-like, as in Eq. (14.11). However, the scalar energy or temperature spectrum, $E_\theta(k)$, is surprisingly quite different for that of passive scalar. As shown in Fig. 16.7, $E_\theta(k)$ exhibits a bi-spectrum with the upper branch as k^{-2}. The lower branch however follows neither $k^{-5/3}$ nor $k^{-7/5}$ spectrum.

In Section 16.3 we showed that the planar average of the temperature fluctuation, $\theta_m(z)$, yields (Pandey and Verma, 2016; Verma et al., 2017)

$$E_{\theta_m}(k) \sim k^{-2}. \tag{16.56}$$

The upper branch of Fig. 16.7 corresponds to $E_{\theta_m}(k)$. The remaining modes (not of the form $\theta(0, 0, k_z)$) make up the lower branch of $E_\theta(k)$. Equation (16.45) demands that the scalar energy (temperature) flux is constant in the inertial range of RBC. Here, both the branches of $E_\theta(k)$ contribute to a constant $\Pi_\theta(k)$.

For the no-slip turbulent thermal convection, the temperature spectrum and flux, not shown here, too have similar behavior.

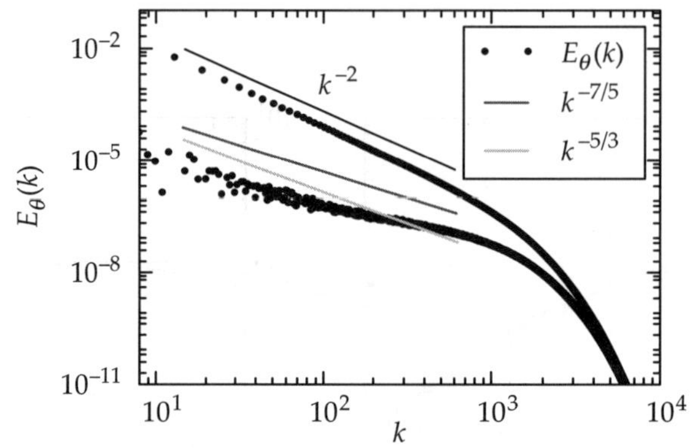

Figure 16.7 For the RBC simulation with $\mathrm{Ra} = 1.1 \times 10^{11}$, plot of $E_\theta(k)$ exhibits a bispectrum. The upper branch $\sim k^{-2}$, while the lower branch scales neither as $k^{-7/5}$ nor as $k^{-5/3}$. From Verma et al. (2017). Reprinted under Creative Commons Attribution 3.0 license.

16.6.3 Structure functions

Bhattacharya et al. (2018b) took a filtered data from the 4096^3 grid simulation described earlier, and computed the structure functions $S_q^u(l)$ for the velocity field. Figure 16.8(a) illustrates the $S_q^u(l)$ vs. l plot along with ζ_q values in the inertial range, while Fig. 16.8(b) exhibits the ζ_q vs. q plot. As shown in the figure, the exponents ζ_q of thermal convection are in good agreement with She–Leveque's predictions (shown as dashed line) that describes ζ_q of hydrodynamic turbulence

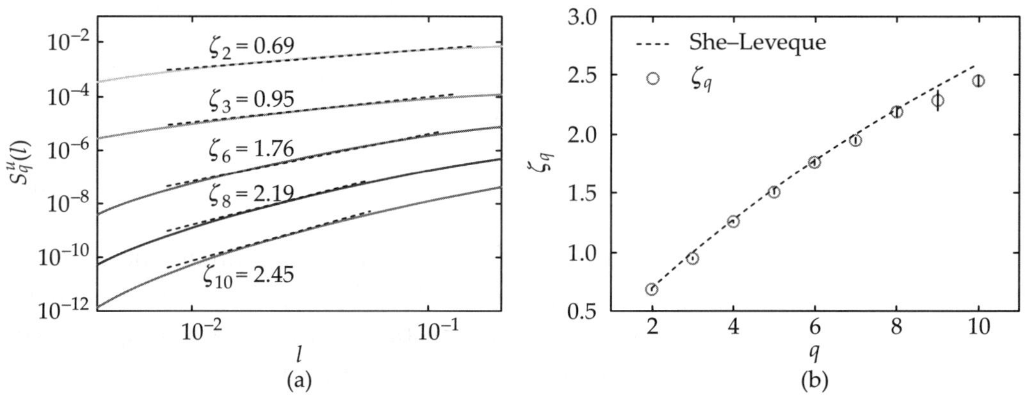

Figure 16.8 For RBC simulation with $\mathrm{Ra} = 1.1 \times 10^{11}$: (a) Plots of velocity structure functions $S_q^u(l)$. (b) Plot of multiscaling exponents ζ_q vs. q. ζ_q's are in good agreement with She–Leveque's predictions (dashed line) for hydrodynamic turbulence. Adopted from a figure of Bhattacharya et al. (2018b).

quite well (see Section 12.6 for details on ζ_q for hydrodynamic turbulence). Thus, the structure function exponents ζ_q of TTC are quite close to those of hydrodynamic turbulence, hence reinforcing our viewpoint that TTC has a similar behavior as hydrodynamic turbulence.

We remark that the temperature structure functions of TTC are not similar to that of passive scalar turbulence. The differences between the two are due to the walls. We do not present these results here.

16.6.4 Shell-to-shell energy transfers

In this subsection we report the shell-to-shell kinetic energy transfers for TTC. Verma et al. (2017) computed these transfers for the velocity field using the 4096^3 data described earlier. Figure 16.9 illustrates the shell-to-shell transfers with x and y axes representing the receiver and giver shells respectively. As shown in the figure, shell n gives energy primarily to shell $n+1$, and receives maximum energy from shell $n-1$. Thus, the shell-to-shell energy transfers of TTC are local and forward as in hydrodynamic turbulence.

Thus, there are many evidences to claim that the turbulent thermal convection has behavior very similar to hydrodynamic turbulence. Vashishtha et al. (2018) exploited this property to perform large-eddy simulation (LES) of turbulent RBC using a subgrid model of hydrodynamics with turbulent Prandtl number as unity. Their LES results provide satisfactory results.

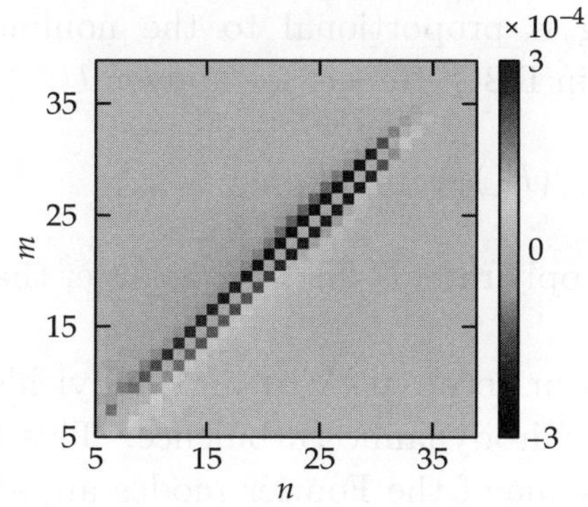

Figure 16.9 For RBC simulation with $\mathrm{Ra} = 1.1 \times 10^{11}$: Density plot of shell-to-shell energy transfers $T_{u,n}^{u,m}$ reveals that the energy transfers are local and forward as in 3D hydrodynamic turbulence. From Verma et al. (2017). Reprinted under Creative Commons Attribution 3.0 license (see Color Plates).

16.7 Forcing, Energy Dissipation, and Drag Reduction in Turbulent Convection

In 3D hydrodynamic turbulence, the kinetic energy dissipation rate and turbulent drag (F_L) are respectively

$$\epsilon_u = \frac{U^3}{L}, \tag{16.57a}$$

$$F_L = \frac{U^2}{L}, \tag{16.57b}$$

apart from prefactors that are of the order of unity. Note that $\epsilon_u \approx f_L U$. In this section we will explore if the kinetic energy dissipation rate and turbulent drag are altered in turbulent thermal convection.

Using numerical simulations, Pandey and Verma (2016), Bhattacharya et al. (2018a), and Verma (2018) showed that the kinetic energy dissipation rate in turbulent thermal convection is smaller than its hydrodynamic counterpart:

$$\epsilon_u = \frac{U^3}{L}\mathrm{Ra}^{-0.18}. \tag{16.58}$$

They also reported that the ratio of the nonlinear term and the viscous term is lower than Re:

$$\frac{\text{Nonlinear term}}{\text{Viscous term}} = \frac{\langle|(\mathbf{u}\cdot\nabla)\mathbf{u}|\rangle}{\langle|\nu\nabla^2\mathbf{u}|\rangle} = \frac{UL}{\nu}\mathrm{Ra}^{-0.14} \sim \mathrm{Re}\ \mathrm{Ra}^{-0.14}. \tag{16.59}$$

Since the turbulent drag is proportional to the nonlinear term, we can claim turbulent drag reduction in RBC. Hence, for a given U,

$$(\epsilon_u)_{\mathrm{TTC}} < (\epsilon_u)_{\mathrm{Hydro}}; \quad (F_L)_{\mathrm{TTC}} < (F_L)_{\mathrm{Hydro}}. \tag{16.60}$$

Or, for the same energy supply rate, U for TTC is larger than that for hydrodynamic turbulence.

The suppressed nonlinear term $(\mathbf{u}\cdot\nabla)\mathbf{u}$ in TTC yields a lower energy cascade rate $\Pi_u(k)$ compared to hydrodynamic turbulence. This is consistent with Verma (2018)'s arguments that some of the Fourier modes are absent in RBC due to the walls. For example, 2D hydrodynamic turbulence with periodic boundary condition contains modes $\{(0,1),(1,0),(1,1)\}$, but the Fourier modes $\{(0,1),\ (1,0)\}$ are absent in RBC with walls on all sides. In fact, RBC excludes modes of the form $\{(0,n),\ (n,0)\}$, where n is an integer. As a result, some of the important triads of hydrodynamics are absent in RBC, leading to a suppression of nonlinearity compared to hydrodynamic turbulence. We believe this to be one of the key reasons for the reduction in ϵ_u and f_L in TTC.

A related issue is the nature of forcing, which is buoyancy for RBC. Compared to the random forcing hypothesized in Kolmogorov's model, buoyancy is more orderly. This is because large-scale velocity and temperature are proportional to each other in RBC. Such ordered flow suppresses nonlinearity. The truncation of some of the Fourier modes mentioned above is related to such ordered structures. A lesson to be learned from the aforementioned observations is that the turbulent drag and dissipation rates not only depend on the magnitude of the velocity, but also on the phases of the large-scale Fourier modes [Eqs. (16.20)]. That is,

$$\epsilon_u = C\frac{U^3}{L}, \tag{16.61}$$

where C is a function of the phase relations between the large-scale Fourier modes. Note that random phases of the Fourier modes would yield zero flux irrespective of the magnitude of U.

The third factor is the presence of $\mathcal{F}_B$ at all scales of TTC. Consequently, as indicated by Eq. (16.42), $\Pi_u < \epsilon_u$. This factor too plays a role in the reduction of the kinetic energy flux and turbulent drag in thermal convection.

Spandan et al. (2018) showed that bubbly turbulence too exhibits turbulent drag reduction. Since bubbly flows are also driven by buoyancy, we can extrapolate that the drag reduction mechanisms described earlier would also be applicable to bubbly turbulence. It will be interesting to study bubbly turbulence from the energy transfer perspectives. In Chapters 23 and 27 we will revisit turbulent drag reduction in magnetohydrodynamic (MHD) turbulence and in polymeric flows from the viewpoint of energy transfers.

In the next section, we will describe anisotropy in TTC.

16.8 Anisotropy in Turbulent Thermal Convection

In this section we describe the anisotropic properties of TTC using the diagnostics tools described in Chapter 11. Nath et al. (2016) simulated RBC for a set of Pr and Ra, and computed various anisotropy diagnostics—$A(k)$ of Eq. (11.12), ring spectrum, energy fluxes $\Pi_\perp$ and $\Pi_\parallel$, and energy transfers from $u_\parallel$ to $\mathbf{u}_\perp$. Here, we present their main results. We focus on Pr = 1 and Ra = 10^8.

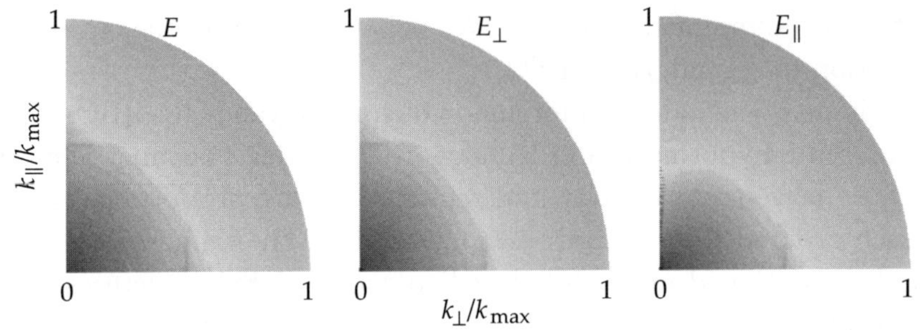

Figure 16.10 For turbulent thermal convection with $\mathrm{Pr} = 1$ and $\mathrm{Ra} = 10^8$, the density plots of the ring spectra $E(k, \beta)$, $E_{u,\perp}(k, \beta)$, and $E_{u,\parallel}(k, \beta)$. The magnitude of kinetic energy is proportional to the darkness. These plots indicate near isotropy of the flow. From Nath et al. (2016). Reprinted with permission from APS

Figure 16.10 exhibits the density plot of the spherical ring spectra: $E(k, \beta)$, $E_{u,\perp}(k, \beta)$, and $E_{u,\parallel}(k, \beta)$ for spherical rings. See Section 11.1 for definition. Nath et al. (2016) employed $\mathrm{Norm}_2(k, \beta)$ [Eq. (11.3b)] for their computation. The ring spectra are nearly independent of the polar angle ζ apart from minor deviations near the $k_\perp$ and $k_\parallel$ axes (Nath et al., 2016). Near isotropy of the ring spectra is yet another similarity between TTC and hydrodynamic turbulence.

Figure 16.11(a) exhibits the energy fluxes $\Pi_u(k)$, $\Pi_{u,\perp}(k)$, and $\Pi_{u,\parallel}(k)$. See Section 11.4 for definitions. Figure 16.11 exhibits the following:

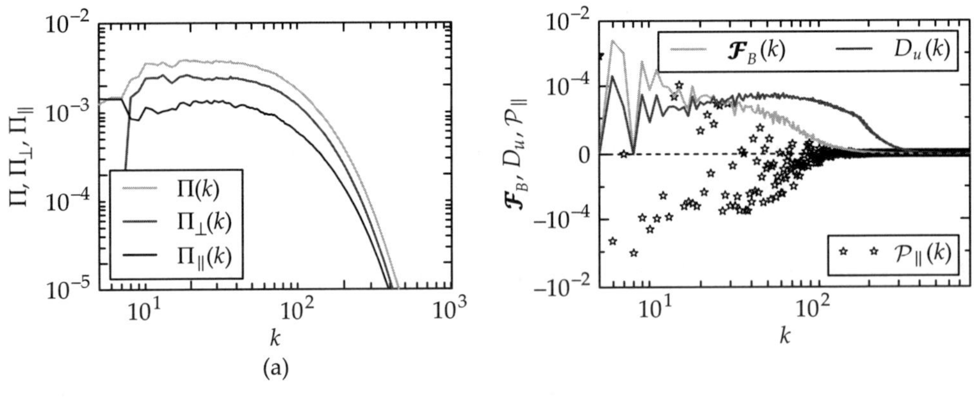

Figure 16.11 For turbulent thermal convection with $\mathrm{Pr} = 1$ and $\mathrm{Ra} = 10^8$: (a) Plots of $\Pi_u(k)$, $\Pi_{u,\perp}(k)$, and $\Pi_{u,\parallel}(k)$; (b) Plots of $\mathcal{F}_B(k)$, $D_u(k)$, and $\mathcal{P}_\parallel(k)$. From Verma (2018). Reprinted with permission from World Scientific.

1. $\Pi_{u,\perp}(k), \Pi_{u,\parallel}(k)$: Both these fluxes are positive with $\Pi_{u,\perp}(k) > \Pi_{u,\parallel}(k)$, except for small wavenumbers.

2. $\mathcal{F}_B(k)$: The kinetic energy feed by buoyancy is positive as argued in Section 16.4.

3. $D_u(k)$: The viscous dissipation rate is naturally positive.

4. $\mathcal{P}_\parallel(k)$: This is the energy transfer rate from $\mathbf{u}_\perp$ to $u_\parallel$ via pressure. $\mathcal{P}_\parallel(k) < 0$ for many k's indicting that $u_\parallel$ loses kinetic energy to $\mathbf{u}_\perp$. Since $\mathcal{P}_\parallel(k) = -\mathcal{P}_\perp(k)$, we conclude that $\mathbf{u}_\perp$ gains energy from $u_\parallel$. Without this transfer, $\mathbf{u}_\perp$, which is without any external force, would decay to zero.

16.9 Various Regimes of Thermal Convection

In Section 14.3 we described the spectral properties of passive scalar turbulence for various regimes of Schmidt number (small, moderate, and large). We may expect RBC to have a similar behavior as that of passive scalar turbulence. However, this is not so because temperature in RBC is an active scalar.

When Pr is either too large or too small, the Kolmogorov wavenumber k_c and the diffusion wavenumber k_d could be quite far apart. Following Eq. (14.19), in the turbulent regime, we expect that

$$\frac{k_c}{k_d} = \mathrm{Pr}^{3/4}. \tag{16.62}$$

Now we examine the spectral properties of TTC in the inertial–dissipation range for various parameter regimes.

16.9.1 Re ≫ 1; Pe ≫ 1; Pr ≈ 1

For this case, the kinetic energy spectrum and flux are the same as those for hydrodynamic or passive scalar turbulence, and they are given by

$$\Pi_u(k) = \epsilon_u \exp\left(-\frac{3}{2}K_{\mathrm{Ko}}(k/k_d)^{4/3}\right), \tag{16.63a}$$

$$E_u(k) = K_{\mathrm{Ko}}\epsilon_u^{2/3}k^{-5/3} \exp\left(-\frac{3}{2}K_{\mathrm{Ko}}(k/k_d)^{4/3}\right), \tag{16.63b}$$

See Section 5.5.1 for a derivation.

As described in Section 16.6.2, the scalar energy has a bispectrum. The dissipative range may have behavior similar to those of Eqs. (14.17). This conjecture needs to be verified.

16.9.2 $\mathrm{Re} \gg 1$; $\mathrm{Pr} = 0$

In this book, the zero or very small Prandtl numbers limit corresponds to the thermal diffusivity $\kappa \to \infty$. Using $\mathrm{Pe} = \mathrm{Re}\,\mathrm{Pr}$, we deduce that $\mathrm{Pe} = 0$ for such flows. To derive the spectra for such flows, we start with the following equation (Verma, 2018):

$$u_z + \nabla^2 \theta = 0 \implies u_z(\mathbf{k}) = \theta(\mathbf{k}) k^2. \tag{16.64}$$

Therefore, the energy feed by buoyancy,

$$\mathcal{F}_B(\mathbf{k}) \sim \mathrm{Ra} \langle \theta(\mathbf{k}) u_z^*(\mathbf{k}) \rangle \sim \mathrm{Ra} \langle |u_z(\mathbf{k})|^2 \rangle / k^2, \tag{16.65}$$

has a very steep spectrum. Hence, buoyancy is active at small wavenumbers only, as assumed in Kolmogorov's phenomenology. Therefore, the velocity field has the following spectrum:

$$E_u(k) = K_{\mathrm{Ko}} \Pi_u^{2/3} k^{-5/3}. \tag{16.66}$$

Using Eq. (16.64) we deduce that

$$E_\theta(k) = \frac{E_z(k)}{k^4} \approx \Pi_u^{2/3} k^{-17/3}. \tag{16.67}$$

Note that for $\mathrm{Pr} \to 0$, the thermal boundary layer spans the whole volume. Hence, $\theta_m(z) \to 0$, and the k^{-2} branch is absent. See Mishra and Verma (2010) for a numerical demonstration of this scaling.

When we contrast this scaling with those of passive scalar, we notice that in RBC, the diffusion term matches with F_θ that leads to a very steep power law ($k^{-17/3}$) for $E_\theta(k)$. In passive scalar turbulence, an absence of F_θ yields the following equation:

$$\frac{\partial \theta}{\partial t} = \kappa \nabla^2 \theta, \tag{16.68}$$

which is the diffusion equation.

16.9.3 $\mathrm{Re} \gg 1$; Small Pr

For small Pr, we expect Eq. (16.64) to hold approximately. Hence, the kinetic energy feed by buoyancy would still be at small wavenumbers. Thus, we expect $E_u(k) \sim k^{-5/3}$, as in Section 16.9.2.

For this case, the mean profile $\theta_m(z)$ starts to emerge and generate a weak k^{-2} branch for $E_\theta(k)$, albeit for a small range. See Fig. 16.12 for an illustration for

Pr = 0.02 (Mishra and Verma, 2010). To study the other branch, we start with Eq. (16.22a) and set $\partial/\partial t = 0$ that yields

$$\frac{d}{dk}\Pi_\theta(k) + 2\kappa k^2 E_\theta(k) = \frac{\Delta}{\alpha g d}\mathcal{F}_B(k). \tag{16.69}$$

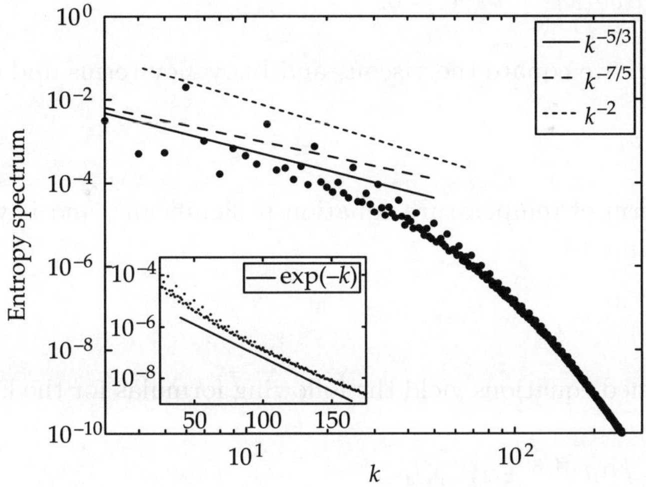

Figure 16.12 For RBC simulation with Pr = 0.02, plot of $E_\theta(k)$ exhibits a bispectrum with an exponential lower branch ($E_\theta(k) \sim \exp(-k/k_c)$) and a weak k^{-2} upper branch. From Mishra and Verma (2010). Reprinted with permission from APS.

For small Pr, $\mathcal{F}_B(k)$ is active at small wavenumbers. Hence, for the inertial–dissipation range,

$$\frac{d}{dk}\Pi_\theta(k) = -2\kappa k^2 E_\theta(k). \tag{16.70}$$

Therefore, using the smallness of nonlinear terms, we derive the following equations for the temperature spectrum and flux (see Section 14.3.2):

$$E_\theta(k) = B\frac{1}{k}\exp(-k/\bar{k}_c), \tag{16.71a}$$

$$\Pi_\theta(k) = 2B\kappa\bar{k}_c^2(1 + k/\bar{k}_c)\exp(-k/\bar{k}_c), \tag{16.71b}$$

Interestingly, the lower branch of $E_\theta(k)$ is proportional to $\exp(-k)$, which is borne out in Fig. 16.12. See subfigure.

16.9.4 Pe $\gg$ 1; Pr $= \infty$

In the limit of infinite Prandtl number ($\nu \to \infty$), the momentum equation is linear (Pandey et al., 2014; Verma, 2018):

$$-i\mathbf{k}\sigma(\mathbf{k}) + \alpha g\theta(\mathbf{k})\hat{z} - \nu k^2\mathbf{u} = 0. \tag{16.72}$$

For simplification, we equate the viscous and buoyancy terms and obtain

$$\alpha g\theta_k \approx \nu k^2 u_k. \tag{16.73}$$

The nonlinear term of temperature equation is significant, and it yields a constant entropy flux leading to

$$\Pi_\theta \approx k\theta_k^2 u_k. \tag{16.74}$$

The aforementioned equations yield the following formulas for the kinetic and scalar energy spectra:

$$E_u(k) = \left(\frac{\alpha g}{\nu}\right)^{4/3} \epsilon_\theta^{2/3} k^{-13/3}, \tag{16.75a}$$

$$E_\theta(k) = \left(\frac{\alpha g}{\nu}\right)^{-2/3} \epsilon_\theta^{2/3} k^{-1/3}. \tag{16.75b}$$

Pandey et al. (2014) performed numerical simulations for Pr $= \infty$ and 100. Figure 16.13 illustrates $E_u(k)$ and $E_\theta(k)$ computed by them. From Fig. 16.13(a) we deduce that $E_u(k) \sim k^{-13/3}$. In the scalar energy spectrum exhibited in Fig. 16.13(b), the upper branch k^{-2} is due to $\theta_m(z)$, and the lower branch is

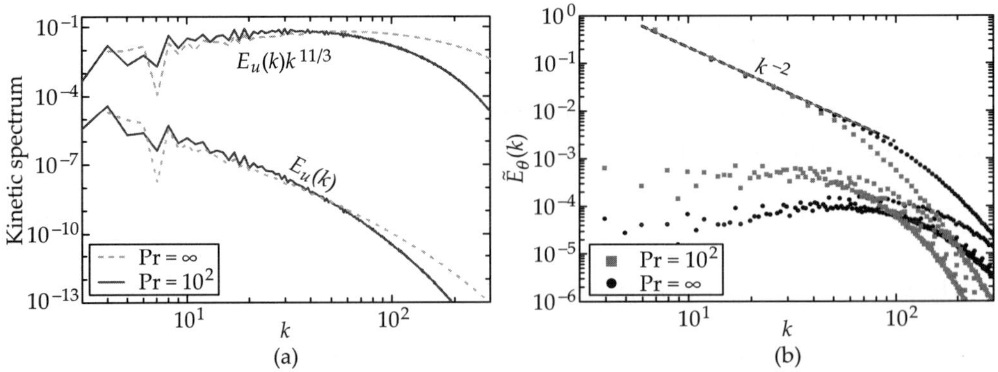

Figure 16.13 For RBC simulation with Pr $= \infty$ and 100: (a) Plots of $E_u(k)$ and normalized $E_u(k)k^{13/3}$ indicating that $E_u(k) \sim k^{-13/3}$. (b) Plots of $E_\theta(k)$ indicating bispectrum. From Pandey et al. (2014). Reprinted with permission from APS.

somewhat flat. Thus, we do not observe $E_\theta(k) \sim k^{-1/3}$, as predicted by the aforementioned scaling. This discrepancy is due to the presence of walls.

Pandey et al. (2014) showed that the spectra for very large Prandtl numbers are very similar to those for $\mathrm{Pr} = \infty$. See Fig. 16.13 for an illustration. There is a minor difference between the spectra for $\mathrm{Pr} = \infty$ and $\mathrm{Pr} = 100$. In the dissipation range, the spectra for $\mathrm{Pr} = 100$ are steeper than those for infinite Prandtl number. This is because $k_c \propto \mathrm{Pr}^{4/3}$ [see Eq. (16.62)]; larger the Prandtl number, larger the k_c, and hence a larger range of power law.

Note that the passive scalar turbulence with very large Schmidt number has $E_u(k) \sim k^{-1} \exp(-k/\bar{k}_c)$ [see Eqs. (14.21)], which is very different from the kinetic energy spectrum described earlier. This difference arises due to the active nature of temperature. In Eq. (16.73), the buoyancy matches with the viscous term; this is the origin for the $k^{-13/3}$ power law for $E_u(k)$. In passive scalar turbulence, $E_u(k) \sim \exp(-k)$ due to lack of such scale-dependent force. Thus, the nature of energy spectra of active and passive scalars are very different.

16.10 Two-dimensional Turbulent Thermal Convection

The scaling arguments described so far are for 3D TTC. Since 2D turbulence tends to promote an inverse cascade of kinetic energy (see Chapter 7), the phenomenological arguments described in Section 16.4 needs modification. This topic has not been explored in detail. Here we make some conjectures that require verification.

A possible kinetic energy flux scenario for 2D TTC is illustrated in Fig. 16.14. We start with the kinetic energy flux equation:

$$\frac{d}{dk}\Pi_u(k) \approx \mathcal{F}_B(k) > 0. \tag{16.76}$$

Since $\Pi_u(k) < 0$ for 2D flows, we obtain

$$\frac{d}{dk}|\Pi_u(k)| \approx -\mathcal{F}_B(k) < 0. \tag{16.77}$$

Hence $|\Pi_u(k)|$ is expected to decrease with k, as in stably stratified turbulence (see Section 15.5). The aforementioned flux scenario is illustrated in Fig. 16.14. Motivated by this simple picture, it may be tempting to adopt the derivation of Section 15.5 to 2D TTC and conjecture that

$$E_u(k) \;=\; \frac{u_k^2}{k} = \epsilon_\theta^{2/5}(\alpha g)^{4/5}k^{-11/5}, \tag{16.78a}$$

$$E_\theta(k) = \frac{\rho_k^2}{k} = \epsilon_\theta^{4/5}(\alpha g)^{-2/5}k^{-7/5}, \tag{16.78b}$$

$$\Pi_u(k) = ku_k^3 = \epsilon_\theta^{3/5}(\alpha g)^{6/5}k^{-4/5}, \tag{16.78c}$$

$$\Pi_\theta(k) = \epsilon_\theta. \tag{16.78d}$$

For the conversion of Eqs. (15.39) to Eqs. (16.78), we employ $N \to \alpha g$ and $\epsilon_\rho \to \epsilon_\theta$. This proposed phenomenology however requires verification using numerical simulations and/or experiments. This scaling may also apply to the boundary layers of TTC.

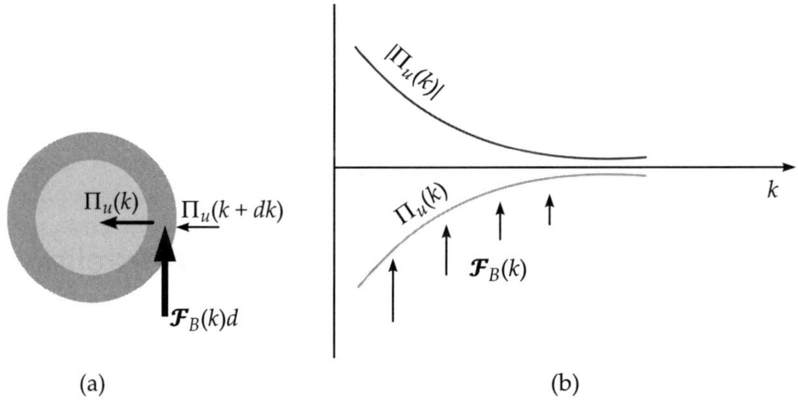

Figure 16.14 A possible schematic diagram of the kinetic energy flux $\Pi_u(k)$ for two-dimensional TTC, and in the boundary layers of TTC.

Before we close this chapter, we remark that many unstably buoyant systems have similar behavior as turbulent thermal convection. Leading examples in this category are Rayleigh–Taylor turbulence, Taylor–Couette turbulence, and bubbly turbulence. These similarities are due to the energetics arguments discussed in Section 16.4. Refer to Verma (2018) for a detailed discussion on these topics.

With this we end our discussion on turbulent thermal convection. In the next chapter we will describe another active scalar—binary fluid mixture.

Further Reading

This chapter discusses only one aspect of turbulent thermal convection—the energy spectrum and flux. The phenomenology discussed in Section 16.4 was first described by Kumar et al. (2014a), who also performed a numerical simulation to verify it. Verma et al. (2017) found similar results for a higher resolution RBC simulation. See Verma (2018) for a comprehensive review on this topic.

For a more comprehensive discussion on thermal convection, refer to books by Chandrasekhar (2013), Getling (1998), and Verma (2018); and review articles by Ahlers et al. (2009), Lohse and Xia (2010), Chillà and Schumacher (2012), Bodenschatz et al. (2000), and Verma et al. (2017).

Chapter 17

A More Complex Example of an Active Scalar: Binary Fluid Mixture

In Chapters 15 and 16 we described how the material density acts as an active scalar and affects a buoyant flow. In these flows, the forces $\mathbf{F}_u(\mathbf{k})$ and $F_\theta(\mathbf{k})$ are linear functions of $\theta(\mathbf{k})$ and $u_z(\mathbf{k})$ respectively. Hence, the energy feed by buoyancy at wavenumber $\mathbf{k}$ are functions of the same wavenumber:

$$\mathcal{F}_u(\mathbf{k}) \propto \mathcal{F}_\theta(\mathbf{k}) \sim \Re[u_z(\mathbf{k})\theta^*(\mathbf{k})]. \tag{17.1}$$

From the stability criteria, we could infer the sign of $\mathcal{F}_u(\mathbf{k})$ that determines the nature of kinetic and scalar energy fluxes.

In this short chapter, we will introduce a more complex example of an active scalar for which the determination of the energy flux is much more complex. We will show how the concepts of energy transfers help us understand these systems better.

17.1 Dynamics of a Binary Fluid Mixture

In this section, we present the dynamics of a binary fluid mixture, as in Ruiz and Nelson (1981). We consider a binary fluid mixture that contains two fluids in approximately equal proportions. The difference in the mass densities of the two species is the following scalar field:

$$\theta(\mathbf{r}) = \rho_A(\mathbf{r}) - \rho_B(\mathbf{r}), \tag{17.2}$$

whose equation of motion is (Ruiz and Nelson, 1981)

$$\frac{\partial \theta}{\partial t} + (\mathbf{u} \cdot \nabla)\theta = \kappa \nabla^2 \theta. \tag{17.3}$$

This equation is same as that of a passive scalar with $F_\theta = 0$ [see Eq. (14.1b)]. Ruiz and Nelson (1981) argued that the velocity field is affected by the scalar field, and its evolution equation is

$$\frac{\partial \mathbf{u}}{\partial t} + (\mathbf{u} \cdot \nabla)\mathbf{u} \;\; = \;\; -\nabla p/\rho_0 - \alpha(\nabla\theta)(\nabla^2\theta) + \nu\nabla^2\mathbf{u}, \tag{17.4}$$

where ρ_0 is the mean density of the fluid, ν is the kinematic velocity, and α is a positive constant.

Ruiz and Nelson (1981) studied various aspects of the aforementioned fluid. In this chapter we will briefly describe the spectral properties of the system. From Eq. (17.4), we deduce that $\mathbf{F}_u = -\alpha(\nabla\theta)(\nabla^2\theta)$; hence the kinetic energy feed by the external force is

$$\mathcal{F}_u(\mathbf{r}) = \mathbf{F}_u(\mathbf{r}) \cdot \mathbf{u}(\mathbf{r}). \tag{17.5}$$

In Fourier space, $\mathcal{F}_u(\mathbf{r})$ transforms to

$$\begin{aligned} \mathcal{F}_u(\mathbf{k}) &= -\alpha \sum_{\mathbf{p}} \Re[iq^2\theta(\mathbf{q})\theta(\mathbf{p})\{\mathbf{p} \cdot \mathbf{u}^*(\mathbf{k})\}] \\ &= \alpha \sum_{\mathbf{p}} \Im[q^2\theta(\mathbf{q})\theta(\mathbf{p})\{\mathbf{p} \cdot \mathbf{u}^*(\mathbf{k})\}], \end{aligned} \tag{17.6}$$

where $\mathbf{q} = \mathbf{k} - \mathbf{p}$.

Compared to $\mathcal{F}_u(\mathbf{k})$ of Rayleigh–Bénard convection (RBC) and stably stratified turbulence, $\mathcal{F}_u(\mathbf{k})$ of the binary fluid is a convolution involving many triads whose analysis is much more complex. For example, we cannot easily determine the sign of $\mathcal{F}_u(\mathbf{k})$. Determination of the effects of $\mathcal{F}_u(\mathbf{k})$ on the kinetic energy flux $\Pi_u(k)$ needs further inputs from numerical simulation and/or from field-theoretic computations (see Chapter 10 for illustration).

In the turbulent regime, let us assume that

$$\Pi_u(k) \sim \epsilon_u k^\beta, \tag{17.7}$$

where k is nondimensionalized wavenumber (using the box size), and β is a constant. Substitution of this form in Kolmogorov's model yields the following kinetic energy spectrum:

$$E_u(k) = K_{\mathrm{Ko}}[\Pi_u(k)]^{2/3}k^{-5/3} \sim K_{\mathrm{Ko}}\epsilon^{2/3}k^{-5/3+(2\beta)/3}. \tag{17.8}$$

This derivation is similar to that used for Bolgiano–Obukhov scaling where $\beta = -4/5$.

For the scalar field, we can derive the scalar energy flux following the arguments of Section 14.2. Note that the scalar field is force-free, hence

$$\mathcal{F}_\theta(k) = 0. \tag{17.9}$$

Therefore, in the inertial range, where the diffusion process can be ignored, we obtain

$$\Pi_\theta(k) = \text{const.} = \epsilon_\theta. \tag{17.10}$$

After this, we can employ Eq. (14.11) of Section 14.2:

$$E_\theta(k) = K_{\mathrm{OC}}\Pi_\theta[\Pi_u(k)]^{-1/3}k^{-5/3}. \tag{17.11}$$

Now using Eq. (17.7), we deduce that

$$E_\theta(k) = \epsilon_\theta(\epsilon_u)^{-1/3}k^{-5/3-\beta/3}. \tag{17.12}$$

We need to determine β using numerical simulations or by other means such as field theory.

The aforementioned discussion is for the inertial range of 3D turbulent flows with Schmidt number of the order of unity. Appropriate modifications are required for 2D flows, and for flows with very small or very large Schmidt numbers.

The present discussion shows that the dynamics of active scalars can become quite complex. We also remark that spinodal decomposition and domain growth are described using coupled Cahn–Hilliard and Navier–Stokes equations (Perlekar et al., 2017) for which we could employ some of the aforementioned scaling ideas.

In the next set of chapters we will describe flows with vectors.

Part III

FLOWS WITH VECTORS

Chapter 18

Energy Transfers in Flows with Vectors

A fluid flow often advects vector fields. A prime example of such fields is the magnetic field in magnetohydrodynamics (MHD). Other examples are flow of nematic liquid crystal and dipoles, motion of a flock, etc. In similar lines as for scalars, we classify such vector fields as *active vectors* or *passive vectors* depending on whether they affect the velocity field or not. The magnetic field in MHD is an example of an active vector field because it affects the velocity field via Lorentz force. However, a dilute concentration of dipoles could act as a passive vector.

In this chapter we introduce a general framework for flows with vectors. In Chapter 19 we will describe the spectral properties of passive vector turbulence, and in Chapters 20–24 we will detail properties of MHD turbulence with a focus on energy transfers. As we show in the following sections, the equations for a vector flow are similar to those of a scalar flow discussed in Chapter 13. Hence, we state the final equations and results without derivation, and we refer the reader to Chapter 13 for details.

In the next section we describe the governing equations for a flow field with vectors embedded in it.

18.1 Governing Equations

The equations for a flow with a vector field (denoted by $\mathbf{w}$) are similar to those of a scalar flow, which was discussed in Chapter 13:

$$\frac{\partial \mathbf{u}}{\partial t} + (\mathbf{u}\cdot\nabla)\mathbf{u} = -\nabla(p/\rho) + \nu\nabla^2\mathbf{u} + \mathbf{F}_u, \tag{18.1a}$$

$$\frac{\partial \mathbf{w}}{\partial t} + (\mathbf{u}\cdot\nabla)\mathbf{w} = \eta\nabla^2\mathbf{w} + \mathbf{F}_w, \tag{18.1b}$$

$$\nabla\cdot\mathbf{u} = 0, \tag{18.1c}$$

where $\mathbf{u}$ is the velocity field, ρ is density which is assumed to be unity, ν is the kinematic viscosity for the velocity field, η is the diffusion coefficient for the vector field, and $\mathbf{F}_u, \mathbf{F}_w$ are respectively the force fields for the velocity and vector fields. The ratio

$$\mathrm{Pr}_w = \frac{\nu}{\eta} \tag{18.2}$$

is called *Prandtl number* of the flow. For MHD, it is referred to as magnetic Prandtl number. In Eq. (18.1b), the ratio of the nonlinear term and the diffusion term is *Reynolds number* based on w, that is,

$$\mathrm{Re}_w = \frac{UL}{\eta}, \tag{18.3}$$

where L is the system size, and U is the large-scale velocity field. Note that the vector field $\mathbf{w}$ may or may not be div-free.

In this book we will cover the following vector flows:

1. Passive vector flow: $\mathbf{F}_u$ is independent of $\mathbf{w}$.
2. Magnetohydrodynamics: $\mathbf{F}_u \propto (\nabla\times\mathbf{B})\times\mathbf{B}$ and $\mathbf{F}_b \propto (\mathbf{B}\cdot\nabla)\mathbf{u}$, where $\mathbf{B}$ is the magnetic field. Clearly, $\mathbf{B}$ is an active vector.
3. Quasi-static magnetohydrodynamics: $\mathbf{F}_u \propto (\mathbf{B}_0\cdot\nabla)^2\mathbf{u}$ and $\mathbf{F}_b \propto (\mathbf{B_0}\cdot\nabla)\mathbf{u}$; here too the magnetic field $\mathbf{B}$ acts as an active vector.

Due to lack of space, in this book we do not detail the following vector flows:

1. Nematic liquid crystal: Liquid crystals often flow with molecules oriented along a particular direction. We can treat the orientation of the molecules as a vector field.
2. Flock of animals: Animals like birds and fish travel in flocks. Their flow can be treated as vector field $\mathbf{w}$ (Ramaswamy, 2010).

For the vector field $\mathbf{w}$, we define the vector energy density as

$$E_w(\mathbf{r}) = \frac{1}{2}w^2, \tag{18.4}$$

whose evolution equation is

$$\frac{\partial}{\partial t}\frac{w^2}{2} + \nabla \cdot \left(\frac{1}{2}w^2\mathbf{u}\right) = \mathbf{F}_w \cdot \mathbf{w} + \eta\mathbf{w} \cdot (\nabla^2\mathbf{w}). \tag{18.5}$$

Clearly, in the absence of $\mathbf{F}_w$ and η, for periodic or vanishing boundary condition,

$$\int \frac{1}{2}w^2 d\mathbf{r} = \text{const.} \tag{18.6}$$

This is a statement of *conservation of vector energy* in a nondiffusive and force-free vector flow.

In Fourier space, the equations of motion for the velocity field and the vector field $\mathbf{w}$ are

$$\frac{d}{dt}\mathbf{u}(\mathbf{k}) + \mathbf{N}_u(\mathbf{k}) = -i\mathbf{k}p(\mathbf{k}) + \mathbf{F}_u(\mathbf{k}) - \nu k^2\mathbf{u}(\mathbf{k}), \tag{18.7a}$$

$$\frac{d}{dt}\mathbf{w}(\mathbf{k}) + \mathbf{N}_w(\mathbf{k}) = \mathbf{F}_w(\mathbf{k}) - \eta k^2\mathbf{w}(\mathbf{k}), \tag{18.7b}$$

$$\mathbf{k} \cdot \mathbf{u}(\mathbf{k}) = 0, \tag{18.7c}$$

where the nonlinear terms are

$$\mathbf{N}_u(\mathbf{k}) = i\sum_{\mathbf{p}} \{\mathbf{k} \cdot \mathbf{u}(\mathbf{q})\}\mathbf{u}(\mathbf{p}), \tag{18.8}$$

$$\mathbf{N}_w(\mathbf{k}) = i\sum_{\mathbf{p}} \{\mathbf{k} \cdot \mathbf{u}(\mathbf{q})\}\mathbf{w}(\mathbf{p}), \tag{18.9}$$

with $\mathbf{q} = \mathbf{k} - \mathbf{p}$. Using the above equations, we derive the following evolution equations for the kinetic energy and vector energy:

$$\frac{d}{dt}E_u(\mathbf{k}) = \sum_{\mathbf{p}} \Im\left[\{\mathbf{k} \cdot \mathbf{u}(\mathbf{q})\}\{\mathbf{u}(\mathbf{p}) \cdot \mathbf{u}^*(\mathbf{k})\}\right] + \Re[\mathbf{F}_u(\mathbf{k}) \cdot \mathbf{u}^*(\mathbf{k})] - 2\nu k^2 E_u(\mathbf{k}), \tag{18.10}$$

$$\frac{d}{dt}E_w(\mathbf{k}) = \sum_{\mathbf{p}} \Im\left[\{\mathbf{k} \cdot \mathbf{u}(\mathbf{q})\}\{\mathbf{w}(\mathbf{p}) \cdot \mathbf{w}^*(\mathbf{k})\}\right] + \Re[\mathbf{F}_w(\mathbf{k}) \cdot \mathbf{w}^*(\mathbf{k})] - 2\eta k^2 E_w(\mathbf{k}). \tag{18.11}$$

In the next section we will state the formulas for the energy transfers in a vector flow.

18.2 Mode-to-mode Vector Energy Transfers and Energy Fluxes

Following similar arguments as that for scalars (Chapter 13), we derive the formulas for the mode-to-mode energy transfers and related quantities. For kinetic energy, these transfers take place via the nonlinear term $(\mathbf{u}\cdot\nabla)\mathbf{u}$. Hence, the formula for the mode-to-mode kinetic energy transfer from Fourier mode $\mathbf{u}(\mathbf{p})$ to mode $\mathbf{u}(\mathbf{k}')$ with $\mathbf{u}(\mathbf{q})$ acting as a mediator is same as that derived in Chapter 4:

$$S^{uu}(\mathbf{k}'|\mathbf{p}|\mathbf{q}) = -\Im\left[\{\mathbf{k}'\cdot\mathbf{u}(\mathbf{q})\}\{\mathbf{u}(\mathbf{p})\cdot\mathbf{u}(\mathbf{k}')\}\right]. \tag{18.12}$$

The energy flux and shell-to-shell energy transfer are also the same as those derived in the same chapter.

However, for the vector field, the nonlinear term $(\mathbf{u}\cdot\nabla)\mathbf{w}$ facilitates the energy transfers among the $\mathbf{w}$ modes with $\mathbf{u}$ acting as a mediator. For a triad $(\mathbf{k}',\mathbf{p},\mathbf{q})$ satisfying $\mathbf{k}'+\mathbf{p}+\mathbf{q}=0$, using the force-free and diffusionless version of Eq. (18.11), we derive that

$$\begin{aligned}\frac{d}{dt}E_w(\mathbf{k}') &= S^{ww}(\mathbf{k}'|\mathbf{p},\mathbf{q})\\ &= -\Im\left[\{\mathbf{k}'\cdot\mathbf{u}(\mathbf{q})\}\{\mathbf{w}(\mathbf{p})\cdot\mathbf{w}(\mathbf{k}')\}\right]+\Im\left[\{\mathbf{k}'\cdot\mathbf{u}(\mathbf{p})\}\{\mathbf{w}(\mathbf{q})\cdot\mathbf{w}(\mathbf{k}')\}\right],\end{aligned} \tag{18.13}$$

where $S^{ww}(\mathbf{k}'|\mathbf{p},\mathbf{q})$ is the *combined vector energy transfer* to mode $\mathbf{w}(\mathbf{k}')$ from modes $\mathbf{w}(\mathbf{p})$ and $\mathbf{w}(\mathbf{q})$. It is easy to show that

$$S^{ww}(\mathbf{k}'|\mathbf{p},\mathbf{q}) + S^{ww}(\mathbf{p}|\mathbf{q},\mathbf{k}') + S^{ww}(\mathbf{q}|\mathbf{k}',\mathbf{p}) = 0, \tag{18.14}$$

which is a statement of the *detailed energy conservation via S^{ww} channel.*

Following arguments similar to those in Section 13.2, we derive the mode-to-mode vector energy transfer from vector mode $\mathbf{w}(\mathbf{p})$ to vector mode $\mathbf{w}(\mathbf{k}')$ with the mediation of velocity mode $\mathbf{u}(\mathbf{q})$ as

$$S^{ww}(\mathbf{k}'|\mathbf{p}|\mathbf{q}) = -\Im\left[\{\mathbf{k}'\cdot\mathbf{u}(\mathbf{q})\}\{\mathbf{w}(\mathbf{p})\cdot\mathbf{w}(\mathbf{k}')\}\right]. \tag{18.15}$$

In the derivation, the only difference is that we replace the scalar field θ by the vector field $\mathbf{w}$, and the normal product between two scalar modes with a dot product between two vector modes.

In Eq. (18.1b), the nonlinear term $(\mathbf{u}\cdot\nabla)\mathbf{w}$ represents advection of the vector field $\mathbf{w}$ by the velocity field $\mathbf{u}$. Therefore, $\mathbf{u}$ mediates the vector energy transfer between $\mathbf{w}$ fields of $[(\mathbf{u}\cdot\nabla)\mathbf{w}]\cdot\mathbf{w}$. In Fourier space, this transfer translates to the vector energy transfer from mode $\mathbf{w}(\mathbf{p})$ to mode $\mathbf{w}(\mathbf{k}')$ with mode $\mathbf{u}(\mathbf{q})$ acting as a mediator. See Sections 4.1.1 and 13.2.1 for detailed arguments. In Fig. 18.1 we illustrate the aforementioned energy transfer. Note that we can also derive Eq. (18.15) using mathematical arguments similar to those of Section 13.2.2.

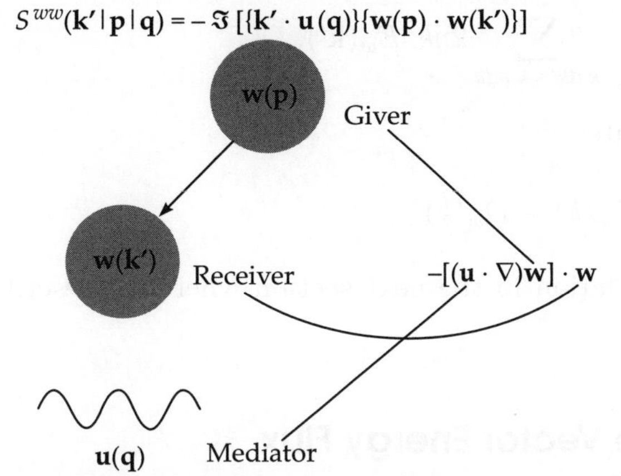

Figure 18.1 A schematic diagram exhibiting mode-to-mode vector energy transfer from mode $\mathbf{w}(\mathbf{p})$ to mode $\mathbf{w}(\mathbf{k}')$ with mode $\mathbf{u}(\mathbf{q})$ acting as a mediator. For this energy transfer, $\mathbf{u}(\mathbf{q})$ advects the vector modes $\mathbf{w}(\mathbf{p})$ and $\mathbf{w}(\mathbf{k}')$ who exchange energy among themselves.

Now we derive the flux and shell-to-shell transfers for a vector flow. For a wavenumber sphere of radius k_0, the vector energy flux $\Pi_w(k_0)$ is the net vector energy transferred from the modes inside the sphere to the modes outside the sphere. Using the definition of the mode-to-mode vector energy transfer, we obtain

$$\Pi_w(k_0) = \sum_{|\mathbf{k}'|>k_0} \sum_{|\mathbf{p}|\le k_0} S^{ww}(\mathbf{k}'|\mathbf{p}|\mathbf{q}). \tag{18.16}$$

Similarly, we can derive the shell-to-shell vector energy transfer from shell m to shell n as

$$T^{w,m}_{w,n} = \sum_{\mathbf{k}'\in n} \sum_{\mathbf{p}\in m} S^{ww}(\mathbf{k}'|\mathbf{p}|\mathbf{q}). \tag{18.17}$$

Note that the above formulas are applicable to both active and passive vectors.

Following the arguments of Section 13.4, we deduce the evolution equation for the vector energy spectrum $E_w(k)$ as

$$\frac{\partial}{\partial t}E_w(k,t) = -\frac{\partial}{\partial k}\Pi_w(k,t) + \mathcal{F}_w(k,t) - D_w(k,t), \tag{18.18}$$

where $\mathcal{F}_w(k)$ is the vector energy supply rate by $\mathbf{F}_w$, and $D_w(k)$ is the diffusion rate of vector energy in shell k. These quantities are defined as follows:

$$\mathcal{F}_w(k)dk = \sum_{k<k'\leq k+dk} \Re[\mathbf{F}_w(\mathbf{k}')\cdot \mathbf{w}^*(\mathbf{k}')], \tag{18.19}$$

$$D_w(k)dk = \sum_{k<k'\leq k+dk} 2\eta k^2 E_w(\mathbf{k}'). \tag{18.20}$$

Under a steady state,

$$\frac{d}{dk}\Pi_w(k) = \mathcal{F}_w(k) - D_w(k). \tag{18.21}$$

We will use this relation in the next section where we describe the variable vector energy flux.

18.3 Variable Vector Energy Flux

In this section, we briefly describe the behavior of vector energy flux that depends on the nature of $\mathcal{F}_w(k)$. The arguments are similar to those of Section 13.4; hence, we only state the results, and skip the intermediate steps.

In the inertial range, $D_w(k) \approx 0$. Hence, the vector energy flux is governed by the following equation:

$$\frac{d}{dk}\Pi_w(k) = \mathcal{F}_{w,\mathrm{in}}(k), \tag{18.22}$$

where $\mathcal{F}_{w,\mathrm{in}}(k)$ is the vector energy supply rate in the inertial range. The behavior of vector energy flux depends crucially on $\mathcal{F}_{w,\mathrm{in}}(k)$, as was the case for scalar energy flux. In Table 18.1 we list the four possibilities. In MHD turbulence, we expect the magnetic energy flux to vary with k due to $\mathcal{F}_{w,\mathrm{in}}(k)$. We will describe these variations in Chapters 21 and 22.

Table 18.1 Behavior of vector energy flux for different kinds of $\mathcal{F}_{w,\mathrm{in}}(k)$.

$\mathcal{F}_{w,\mathrm{in}}(k)$	$\Pi_w(k)$	Figure reference
0	0	13.3(a)
0	Positive const.	13.3(b)
<0	$d\Pi_w(k)/dk < 0$	13.3(c)
>0	$d\Pi_w(k)/dk > 0$	13.3(d)

In the next section we describe how equations of motion and energy transfers of vector turbulence are represented in Craya–Herring basis.

18.4 Vector Flow in Craya–Herring Basis

In Chapter 9 we employed Craya–Herring basis to derive equations of motion for the velocity modes of hydrodynamic flows. In this section we will repeat this exercise for vector flows without forces $\mathbf{F}_u, \mathbf{F}_w$ and dissipative terms. Under this condition, the momentum equations remain the same as those of hydrodynamics. Hence, the equations for the velocity field would be exactly the same as those derived in Section 9.2.

To derive equations of motion for the vector field $\mathbf{w}$, we follow the same steps as in Section 9.2. For a triad with wavenumbers $(\mathbf{k}', \mathbf{p}, \mathbf{q})$, following the condition $\mathbf{k}' + \mathbf{p} + \mathbf{q} = 0$, the equation of motion for mode $\mathbf{w}(\mathbf{k}')$ is

$$\frac{d}{dt}\mathbf{w}(\mathbf{k}') = -i[\mathbf{k}' \cdot \mathbf{u}(-\mathbf{q})]\mathbf{w}(-\mathbf{p}) - i[\mathbf{k}' \cdot \mathbf{u}(-\mathbf{p})]\mathbf{w}(-\mathbf{q}). \tag{18.23}$$

A dot product of this equation with $\hat{e}_1(\mathbf{k}')$ yields

$$\dot{w}_1(\mathbf{k}') = ik'[\sin\beta\cos\gamma u_1^*(\mathbf{q})w_1^*(\mathbf{p}) - \sin\gamma\cos\beta u_1^*(\mathbf{p})w_1^*(\mathbf{q})], \tag{18.24}$$

where α, β, γ are internal angles of the triangle formed by the wavenumbers of the triad. See Fig. 9.5 for an illustration. Similarly, a dot product of Eq. (18.23) with $\hat{e}_2(\mathbf{k}')$ yields

$$\dot{w}_2(\mathbf{k}') = ik'\{\sin\gamma u_1^*(\mathbf{p})w_2^*(\mathbf{q}) - \sin\beta u_1^*(\mathbf{q})w_2^*(\mathbf{p})\}. \tag{18.25}$$

Following similar steps, we derive equations of motion for the other two modes:

$$\dot{w}_1(\mathbf{p}) = ip[\sin\gamma\cos\alpha u_1^*(\mathbf{k}')w_1^*(\mathbf{q}) - \sin\alpha\cos\gamma u_1^*(\mathbf{q})w_1^*(\mathbf{k}')], \tag{18.26a}$$

$$\dot{w}_1(\mathbf{q}) = iq[\sin\alpha\cos\beta u_1^*(\mathbf{p})w_1^*(\mathbf{k}') - \sin\beta\cos\alpha u_1^*(\mathbf{k}')w_1^*(\mathbf{p})]. \tag{18.26b}$$

and

$$\dot{w}_2(\mathbf{p}) = ip\{\sin\alpha u_1^*(\mathbf{q})w_2^*(\mathbf{k}') - \sin\gamma u_1^*(\mathbf{k}')w_2^*(\mathbf{q})\}, \tag{18.27a}$$

$$\dot{w}_2(\mathbf{q}) = iq\{\sin\beta u_1^*(\mathbf{k}')w_2^*(\mathbf{p}) - \sin\alpha u_1^*(\mathbf{p})w_2^*(\mathbf{k}')\}. \tag{18.27b}$$

18.5 Energy Transfers in Craya–Herring and Helical Basis

Following the same procedure as in Chapter 9, we compute the mode-to-mode vector energy transfers in Craya–Herring basis. The mode-to-mode vector energy transfer from $\mathbf{w}(\mathbf{p})$ to $\mathbf{w}(\mathbf{k}')$ with $\mathbf{u}(\mathbf{q})$ acting as a mediator is

$$S^{ww}(\mathbf{k}'|\mathbf{p}|\mathbf{q}) = S^{w_1 w_1}(\mathbf{k}'|\mathbf{p}|\mathbf{q}) + S^{w_2 w_2}(\mathbf{k}'|\mathbf{p}|\mathbf{q}) \tag{18.28}$$

where

$$S^{w_1w_1}(\mathbf{k}'|\mathbf{p}|\mathbf{q}) = k'\sin\beta\cos\gamma\Im\{u_1(\mathbf{q})w_1(\mathbf{p})w_1(\mathbf{k}')\} \tag{18.29a}$$

$$S^{w_2w_2}(\mathbf{k}'|\mathbf{p}|\mathbf{q}) = -k'\sin\beta\Im\{u_1(\mathbf{q})w_2(\mathbf{p})w_2(\mathbf{k}')\}. \tag{18.29b}$$

As in Section 9.3, $S^{w_1w_1}(\mathbf{k}'|\mathbf{p}|\mathbf{q})$ represents vector energy transfer from $w_1(\mathbf{p})$ to $w_1(\mathbf{k}')$ with $\mathbf{u}(\mathbf{q})$ acting as a mediator. Similarly, $S^{w_2w_2}(\mathbf{k}'|\mathbf{p}|\mathbf{q})$ represents the transfer from $w_2(\mathbf{p})$ to $w_2(\mathbf{k}')$ with $\mathbf{u}(\mathbf{q})$ acting as a mediator. Therefore, for w_1 and w_2 fields, the vector energy fluxes crossing a wavenumber sphere of radius k_0 are

$$\Pi^{w_1<}_{w_1>}(k_0) = \sum_{|\mathbf{p}|\le k_0}\sum_{|\mathbf{k}|>k_0} S^{w_1w_1}(\mathbf{k}|\mathbf{p}|\mathbf{q}), \tag{18.30a}$$

$$\Pi^{w_2<}_{w_2>}(k_0) = \sum_{|\mathbf{p}|\le k_0}\sum_{|\mathbf{k}|>k_0} S^{w_2w_2}(\mathbf{k}|\mathbf{p}|\mathbf{q}). \tag{18.30b}$$

The other mode-to-mode vector energy transfers for the triad are

$$S^{ww}(\mathbf{p}|\mathbf{q}|\mathbf{k}') = p\sin\gamma\cos\alpha\Im\{w_1(\mathbf{k}')Bw_1(\mathbf{q})u_1(\mathbf{p})\} -p\sin\gamma\Im\{u_1(\mathbf{k}')w_2(\mathbf{q})w_2(\mathbf{p})\}, \tag{18.31a}$$

$$S^{ww}(\mathbf{q}|\mathbf{k}'|\mathbf{p}) = q\sin\alpha\cos\beta\Im\{u_1(\mathbf{p})w_1(\mathbf{k}')w_1(\mathbf{q})\} -q\sin\alpha\{u_1(\mathbf{p})w_2(\mathbf{k}')w_2(\mathbf{q})\}. \tag{18.31b}$$

In helical basis, the mode-to-mode energy transfer from $\mathbf{w}(\mathbf{p})$ to $\mathbf{w}(\mathbf{k}')$ is given by

$$\begin{aligned} S^{ww}(\mathbf{k}'|\mathbf{p}|\mathbf{q}) &= -\Im\left[\{\mathbf{k}'\cdot\mathbf{u}(\mathbf{q})\}\{\mathbf{w}(\mathbf{p})\cdot\mathbf{w}(\mathbf{k}')\}\right] \\ &= -\sum_{s_p,s_{k'}}\Im\left[\{\mathbf{k}'\cdot\mathbf{u}(\mathbf{q})\}w_{s_p}(\mathbf{p})w_{s_{k'}}(\mathbf{k}')\{\hat{e}_{s_p}(\mathbf{p})\cdot\hat{e}_{s_{k'}}(\mathbf{k}')\}\right] \\ &= \sum_{s_p,s_{k'}} S^{ww}_{s_{k'}s_p}(\mathbf{k}'|\mathbf{p}|\mathbf{q}). \end{aligned} \tag{18.32}$$

The above expression is simplified further into

$$S^{ww}_{s_{k'}s_p}(\mathbf{k}'|\mathbf{p}|\mathbf{q}) = -\frac{k'}{2}\sin\beta(1+s_ps_{k'}\cos\gamma)\Im\{u_1(\mathbf{q})w_{s_p}(\mathbf{p})w_{s_{k'}}(\mathbf{k}')\}. \tag{18.33}$$

Note that $S^{ww}_{s_{k'}s_p}(\mathbf{k}'|\mathbf{p}|\mathbf{q})$ represents the vector energy transfer from $w_{s_p}(\mathbf{p})$ to $w_{s_{k'}}(\mathbf{k}')$ with $\mathbf{u}(\mathbf{q})$ acting as a mediator. Hence, the vector energy flux from helical modes w_{s_g} to w_{s_r}, where s_g and s_r are the signs of the giver and receiver modes respectively, is

$$\Pi^{w_{s_g}<}_{w_{s_r}>}(k_0) = \sum_{|\mathbf{p}|\le k_0}\sum_{|\mathbf{k}|>k_0} S^{ww}_{s_rs_g}(\mathbf{k}'|\mathbf{p}|\mathbf{q}). \tag{18.34}$$

The shell-to-shell energy transfer from w_{s_g} of shell m to w_{s_r} of shell n is

$$T^{u_{s_g},m}_{u_{s_r},n} = \sum_{\mathbf{p}\in m}\sum_{\mathbf{k}\in n} S^{uu}_{s_r s_g}(\mathbf{k}'|\mathbf{p}|\mathbf{q}). \tag{18.35}$$

Example 18.1: In a periodic box $[2\pi, 2\pi]$, consider the following force-free vector flow with $\nu = \eta = 0$:

$$\begin{aligned}\mathbf{u} &= \hat{x}2D_u \cos y + \hat{y}2C_u \cos x + (\hat{x}-\hat{y})2A_u \sin(x+y),\\ \mathbf{w} &= \hat{x}2D_w \cos y + \hat{y}2C_w \cos x + (\hat{x}-\hat{y})2A_w \sin(x+y).\end{aligned}$$

Derive equations for A_u, C_u, D_u, A_w, C_w, and D_w as in Example 3.4.

Solution: These fields are constructed with wavenumbers (1,0), (0,1), (1,1), $(-1,0)$, $(0,-1)$, and $(-1,-1)$. We choose $\hat{n} = \hat{z}$. For the interacting triad $\mathbf{k}' = (-1,-1)$, $\mathbf{p} = (0,1)$, and $\mathbf{q} = (1,0)$, the amplitudes of the Fourier modes in the Craya–Herring basis are listed in Table 18.2. Note that $u_2 = w_2 = 0$ since the field is two-dimensional.

Table 18.2 Example 18.1: The amplitudes of the velocity and vector Fourier modes.

Mode	u_1	w_1
$\mathbf{k}' = (-1,-1)$	$-\frac{A_u\sqrt{2}}{i}$	$-\frac{A_w\sqrt{2}}{i}$
$\mathbf{p} = (0,1)$	D_u	D_w
$\mathbf{q} = (1,0)$	$-C_u$	$-C_w$

Without $\mathbf{F}_u$, the equations of motion of A, B, and C are the same as those in Example 3.4 or Example 9.2. The solution of these equations are

$$\begin{aligned}A &= \text{constant},\\ D &= c\cos(At),\\ C &= c\sin(At),\end{aligned}$$

where c is a constant.

For the vector field $\mathbf{w}$, following similar steps as earlier, and as in Example 13.1, we derive the following equations of motion for the vector modes:

$$\begin{aligned}\dot{w}_1(-1,-1) &= \frac{-\dot{A}_w\sqrt{2}}{i} = \frac{i}{\sqrt{2}}(-C_u D_w + D_u C_w),\\ \dot{w}_1(0,1) &= \dot{D}_w = i(0 + C_u(A_w/i)),\end{aligned}$$

$$\dot{w}_1(1,0) = -\dot{C}_w = i(D_u(A/i)+0).$$

Or

$$\begin{aligned}
\dot{A}_w &= \frac{1}{2}(-C_u D_w + D_u C_w), \\
\dot{D}_w &= C_u A_w, \\
\dot{C}_w &= -D_u A_w.
\end{aligned}$$

Given A_u, D_u, C_u, we can solve for A_w, D_w, and C_w. Note

$$2A_w^2 + D_w^2 + C_w^2 = \text{const.},$$

that follows from the conservation of vector energy for the diffusionless case.

Example 18.2: Compute the mode-to-mode vector energy transfers for the flow field of Example 18.1.

Solution: We employ Eqs. (18.28, 18.31a, 18.31b) to compute the mode-to- mode energy transfers:

$$\begin{aligned}
S^{ww}(\mathbf{k}'|\mathbf{p}|\mathbf{q}) &= k' \sin\beta\cos\gamma \Im\{u_1(\mathbf{q})w_1(\mathbf{p})w_1(\mathbf{k}')\} \\
&= \frac{\sqrt{2}}{2}\Im\{(-C_u)D_w\left(-\frac{A_w\sqrt{2}}{i}\right)\} = -C_u A_w D_w \\
S^{ww}(\mathbf{p}|\mathbf{q}|\mathbf{k}') &= p\sin\gamma\cos\alpha\Im\{u_1(\mathbf{k}')w_1(\mathbf{p})w_1(\mathbf{q})\} = 0, \\
S^{ww}(\mathbf{q}|\mathbf{k}'|\mathbf{p}) &= q\sin\alpha\cos\beta\Im\{u_1(\mathbf{p})w_1(\mathbf{q})w_1(\mathbf{k}')\} \\
&= -\Im\{D_u(-C_w)\left(-\frac{A_w}{i}\right)\} = D_u A_w C_w.
\end{aligned}$$

Note that the pattern of the energy transfer is similar to that of Example 4.3 (see Fig. 4.4), except that $S^{ww}(\mathbf{k}'|\mathbf{p}|\mathbf{q})$ and $S^{ww}(\mathbf{q}|\mathbf{k}'|\mathbf{p})$ are not equal, unlike Example 4.3 where these quantities are equal.

With this, we end our discussion on flows with vectors.

Chapter 19

Flow with a Passive Vector

In the previous chapter we introduced governing equations and energy transfers for a fluid flow with a vector field. In the present chapter we will describe the spectral properties of a passive vector flow. In the next section we describe the governing equations of the flow.

19.1 Governing Equations

To derive equations of motion for an incompressible passive vector flow, we assume that $\mathbf{F}_u$ is independent of $\mathbf{w}$, and that both $\mathbf{F}_u, \mathbf{F}_w$ act at large scales. Under these assumptions, the equations of motion are

$$\frac{\partial \mathbf{u}}{\partial t} + (\mathbf{u} \cdot \nabla)\mathbf{u} = -\nabla(p/\rho) + \nu \nabla^2 \mathbf{u} + \mathbf{F}_u, \tag{19.1a}$$

$$\frac{\partial \mathbf{w}}{\partial t} + (\mathbf{u} \cdot \nabla)\mathbf{w} = \eta \nabla^2 \theta + \mathbf{F}_w, \tag{19.1b}$$

$$\nabla \cdot \mathbf{u} = 0, \tag{19.1c}$$

where $\mathbf{u}, \mathbf{w}, p$ are the velocity, passive vector, and pressure fields respectively; ρ is density which is assumed to be unity; $\mathbf{F}_u, \mathbf{F}_w$ are the respective force field for the velocity and vector fields; ν is the kinematic viscosity; and η is the diffusion coefficient of $\mathbf{w}$. The ratio

$$\mathrm{Pm} = \frac{\nu}{\eta}. \tag{19.2}$$

is the *Prandtl number*.

Following the same set of arguments as those for the passive scalar (see Chapter 14), we can derive the following results for the passive vector $\mathbf{w}$:

1. When $\eta = 0$ and $\mathbf{F}_w = 0$, the total vector energy is conserved, that is,

$$\int \frac{1}{2} w^2 d\mathbf{r} = \text{const.} \tag{19.3}$$

2. In Fourier space the equations for $\mathbf{u}$ and $\mathbf{w}$ are same as Eqs. (18.7).
3. The formulas for the mode-to-mode vector energy transfer and vector energy flux are the same as those derived in Chapter 18
4. The evolution equation for the vector energy spectrum $E_w(k)$ is

$$\frac{\partial}{\partial t} E_w(k,t) = -\frac{\partial}{\partial k} \Pi_w(k,t) + \mathcal{F}_w(k,t) - D_w(k,t). \tag{19.4}$$

Under a steady state, $\partial E_w(k)/\partial t \approx 0$, the aforementioned equation for $\mathbf{w}$ transforms to

$$\frac{d}{dk} \Pi_w(k) = \mathcal{F}_w(k) - D_w(k) = \mathcal{F}_w(k) - 2\eta k^2 E_w(k). \tag{19.5}$$

We will use this relation in the next section.

19.2 Phenomenology of a Passive Vector Turbulence

We can derive the vector energy flux and spectrum for the passive vector turbulence using the following arguments. The derivation is very similar to that of Section 14.2. Note that the kinetic energy spectra for the 3D and 2D flows would be the same as those in Chapters 5 and 7 because the velocity field is not affected by the passive vector $\mathbf{w}$.

In the inertial range, $\mathcal{F}_w(k,t) = 0$, and the diffusion term is quite weak. Hence, Eq. (19.5) yields

$$\Pi_w(k) = \text{const.} = \epsilon_w. \tag{19.6}$$

From scaling analysis, we expect that $E_w(k)$ depends on Π_u, Π_w, and k, whose dimensions are

$$[\Pi_u] = [L^2/T^3]; \quad [\Pi_w] = [w^2/T]; \quad [E_w(k)] = [w^2 L]. \tag{19.7}$$

Using dimensional analysis, we postulate that

$$E_w(k) = (\Pi_w)^\alpha (\Pi_u)^\beta k^\gamma. \tag{19.8}$$

Now matching the dimensions of $[w]$, $[L]$, and $[T]$ yields

$$\alpha = 1; \quad \beta = -1/3; \quad \gamma = -5/3. \tag{19.9}$$

Therefore,

$$E_\theta(k) = K\Pi_w (\Pi_u)^{-1/3} k^{-5/3}, \tag{19.10}$$

where K is a nondimensional constant. Note that this derivation and the final formula are very similar to those for passive scalar.

In the next section, we will describe the spectral properties of a passive vector flow in various regimes.

19.3 Various Regimes of a Passive Vector Flow

A passive vector flow has two nonlinear and two dissipative terms. The energy spectra and fluxes for $\mathbf{u}$ and $\mathbf{w}$ depend on the relative strengths of the nonlinear and dissipative terms. A careful analysis of the relevant equations shows that the energy spectra and fluxes of passive scalar and passive vector flows are very similar. Refer to Section 14.3 for a detailed derivation. To go from a passive scalar to a passive vector, we employ the following translation:

$$\theta \;\to\; \mathbf{w}. \tag{19.11a}$$

$$k_c \;\to\; k_w = \left(\frac{\epsilon_w}{\eta^3}\right)^{1/4}. \tag{19.11b}$$

$$\frac{k_c}{k_d} \;\to\; \frac{k_w}{k_d} = \left(\frac{\nu}{\eta}\right)^{3/4} = (\mathrm{Pr}_w)^{3/4}. \tag{19.11c}$$

$$\bar{k}_c \;\to\; \bar{k}_w = \frac{1}{L}\sqrt{\frac{UL}{\eta}} = \frac{\sqrt{\mathrm{Re}_w}}{L}. \tag{19.11d}$$

With this, we close our discussion on passive vector flows. In the next set of chapters we will describe the properties of magnetohydrodynamics, which is an example of an active vector flow.

Chapter 20

Magnetohydrodynamics: Formalism

Charged fluids support electric current and electromagnetic field. A special class of quasi-neutral and highly conducting charged fluid is called *magnetofluid*, and its dynamics is described by *magnetohydrodynamics (MHD)*. A significant part of the universe is magnetofluid, for example, stellar convection zone, interstellar medium, etc. Hence, a good understanding of MHD is important for understanding these systems.

It is also important to note that in MHD, ions and electrons carry linear momentum and electric current respectively. It is the low frequency limit of the flow. Later, in a chapter on electron MHD (EMHD), we will describe the high frequency regime of the flow in which ions are stationary, and electrons carry both linear momentum and electric current. Note that we follow CGS system of units for these discussions.

In the following six chapters of the book, we will cover MHD and EMHD as examples of active vector flows, mostly focusing on energy transfer issues. For a broader discussion on MHD turbulence, refer to Biskamp (2003) and Verma (2004).

We start with the governing equations for MHD turbulence.

20.1 Governing Equations in Real Space

A magnetofluid experiences Lorentz force:

$$\mathbf{F}_u = \frac{1}{c}\mathbf{J} \times \mathbf{B}, \tag{20.1}$$

where $\mathbf{B}$ is the magnetic field, and $\mathbf{J}$ is the electric current density.[1] Under MHD approximation (Verma, 2004),

$$\mathbf{J} = \frac{c}{4\pi}\nabla \times \mathbf{B}, \tag{20.2}$$

substitution of which in Eq. (20.1) yields

$$\mathbf{F}_u = \frac{1}{4\pi}[(\nabla \times \mathbf{B}) \times \mathbf{B}] = -\frac{1}{4\pi}\nabla B^2 + \frac{1}{4\pi}(\mathbf{B} \cdot \nabla)\mathbf{B}. \tag{20.3}$$

Since $\mathbf{F}_u$ is a function of $\mathbf{B}$, the magnetic field is an active vector. Note that $\mathbf{F}_u$ is a nonlinear function of the field variable $\mathbf{B}$. The nonlinear form of $\mathbf{F}_u$ induces nonlinear energy transfers among modes (to be covered in the next chapter). This is unlike $\mathbf{F}_u$ of a stably stratified flow and of RBC that induces energy exchange between $\mathbf{u}(\mathbf{k})$ and $\theta(\mathbf{k})$ of the same wavenumber (linear interaction).

With the Lorentz force, the Navier–Stokes equations are

$$\frac{\partial \mathbf{u}}{\partial t} + \mathbf{u} \cdot \nabla \mathbf{u} = -\frac{1}{\rho}\nabla p + \frac{1}{4\pi\rho}(\mathbf{B} \cdot \nabla)\mathbf{B} + \nu\nabla^2\mathbf{u}, \tag{20.4a}$$

$$\nabla \cdot \mathbf{u} = 0, \tag{20.4b}$$

where ρ, ν are respectively the density and kinematic viscosity of the fluid, and p is sum of the hydrodynamic and magnetic pressure:

$$p = p_{\text{hydro}} + \frac{B^2}{8\pi}. \tag{20.5}$$

In the present and subsequent chapters on MHD, we set $\rho = 1$.

The evolution of the magnetic field is described by the following Maxwell's equations:

$$\frac{\partial \mathbf{B}}{\partial t} = \nabla \times (\mathbf{u} \times \mathbf{B}) + \eta\nabla^2\mathbf{B}, \tag{20.6a}$$

$$\nabla \cdot \mathbf{B} = 0, \tag{20.6b}$$

where η is the magnetic diffusivity of the fluid. Equation (20.6a) can be simplified further to

$$\frac{\partial \mathbf{B}}{\partial t} + (\mathbf{u} \cdot \nabla)\mathbf{B} = (\mathbf{B} \cdot \nabla)\mathbf{u} + \eta\nabla^2\mathbf{B}. \tag{20.7}$$

[1]Under MHD approximation, the electric field is much smaller than the magnetic field; hence, the electric force is ignored.

When we compare this equation with Eq. (18.1b), we observe that $\mathbf{B}$ is a vector field with

$$\mathbf{F}_B = (\mathbf{B} \cdot \nabla)\mathbf{u}. \tag{20.8}$$

Note that $\mathbf{F}_B$ is a nonlinear function of the fields $\mathbf{B}$ and $\mathbf{u}$, similar to $\mathbf{F}_u$. Hence, $\mathbf{F}_B$ induces nonlinear energy transfers among the Fourier modes (to be covered in the next chapter). Also note that $\mathbf{B}$ is advected by $\mathbf{u}$.

Equation (20.6a) has a very similar form as the vorticity equation (see Section 2.2). In analogy with the vorticity equation, the term $(\mathbf{u} \cdot \nabla)\mathbf{B}$ yields magnetic energy transfers within the magnetic field, while the term $(\mathbf{B} \cdot \nabla)\mathbf{u}$ represents the stretching of the magnetic field by the velocity field that causes a transfer of kinetic energy to magnetic energy, and hence a growth of the magnetic field.

It is convenient to express $\mathbf{B}$ in velocity units using a transformation:

$$\frac{\mathbf{B}_{\mathrm{CGS}}}{\sqrt{4\pi\rho}} \to \mathbf{B} \tag{20.9}$$

that converts Eqs. (20.4, 20.6) to

$$\frac{\partial \mathbf{u}}{\partial t} + (\mathbf{u} \cdot \nabla)\mathbf{u} = -\nabla p + (\mathbf{B} \cdot \nabla)\mathbf{B} + \nu \nabla^2 \mathbf{u}, \tag{20.10a}$$

$$\frac{\partial \mathbf{B}}{\partial t} + (\mathbf{u} \cdot \nabla)\mathbf{B} = (\mathbf{B} \cdot \nabla)\mathbf{u} + \eta \nabla^2 \mathbf{B}, \tag{20.10b}$$

$$\nabla \cdot \mathbf{u} = 0, \tag{20.10c}$$

$$\nabla \cdot \mathbf{B} = 0. \tag{20.10d}$$

In this chapter we will work with this set of equations. These equations can also be expressed in tensorial form as

$$\frac{\partial u_i}{\partial t} + \partial_j(u_j u_i - B_j B_i) = -\partial_i p + \nu \nabla^2 u_i, \tag{20.11a}$$

$$\frac{\partial B_i}{\partial t} + \partial_j(u_j B_i - B_j u_i) = \eta \nabla^2 B_i, \tag{20.11b}$$

$$\partial_i u_i = 0, \tag{20.11c}$$

$$\partial_i B_i = 0, \tag{20.11d}$$

where $\partial_i f$ is shorthand for $\partial f / \partial x_i$.

In MHD the equation for the vorticity is

$$\frac{\partial \boldsymbol{\omega}}{\partial t} = \nabla \times (\mathbf{u} \times \boldsymbol{\omega} + \mathbf{J} \times \mathbf{B}) + \nu \nabla^2 \boldsymbol{\omega}. \tag{20.12}$$

There is another interesting quantity called vector potential $\mathbf{A}$, which is defined using

$$\mathbf{B} = \nabla \times \mathbf{A}. \tag{20.13}$$

Using Eq. (20.6a) we derive the following dynamical equation for $\mathbf{A}$:

$$\frac{\partial \mathbf{A}}{\partial t} = \mathbf{u} \times \mathbf{B} + \eta \nabla^2 \mathbf{A} + \nabla \Psi, \tag{20.14}$$

where Ψ is a scalar, which is determined using a gauge condition (Biskamp, 2003). Using $\mathbf{B} = \nabla \times \mathbf{A}$, Eq. (20.14) is rewritten as

$$\frac{\partial A_i}{\partial t} + u_j \partial_j A_i = u_j \partial_i A_j + \eta \nabla^2 A_i + \partial_i \Psi. \tag{20.15}$$

In 2D,

$$\mathbf{A} = A(x, y)\hat{z}. \tag{20.16}$$

Consequently, Eq. (20.15) reduces to

$$\frac{\partial A}{\partial t} + u_j \partial_j A = \eta \nabla^2 A. \tag{20.17}$$

This form of equation indicates that in 2D, A can be treated as a passive scalar (see Chapter 14).

Often magnetofluids are subjected to a constant magnetic field, denoted by $\mathbf{B}_0$, that is,

$$\mathbf{B} = \mathbf{B}_0 + \mathbf{b}, \tag{20.18}$$

where $\mathbf{b}$ is the magnetic fluctuation. In terms of these variables, the MHD equations are

$$\frac{\partial \mathbf{u}}{\partial t} - (\mathbf{B}_0 \cdot \nabla)\mathbf{b} + (\mathbf{u} \cdot \nabla)\mathbf{u} - (\mathbf{b} \cdot \nabla)\mathbf{b} = -\nabla p + \nu \nabla^2 \mathbf{u}, \tag{20.19a}$$

$$\frac{\partial \mathbf{b}}{\partial t} - (\mathbf{B}_0 \cdot \nabla)\mathbf{u} + (\mathbf{u} \cdot \nabla)\mathbf{b} - (\mathbf{b} \cdot \nabla)\mathbf{u} = \eta \nabla^2 \mathbf{b}, \tag{20.19b}$$

$$\nabla \cdot \mathbf{u} = 0, \tag{20.19c}$$

$$\nabla \cdot \mathbf{b} = 0. \tag{20.19d}$$

When $u \ll B_0$ and $b \ll B_0$, the nonlinear terms can be dropped. Linearized incompressible and inviscid MHD generates *Alfvén waves* (Verma, 2004). These waves, which will be discussed in Section 20.4, are conveniently described using *Elsässer variables*:

$$\mathbf{z}^{\pm} = \mathbf{u} \pm \mathbf{b}. \tag{20.20}$$

The dynamical equations for these variables are

$$\begin{aligned} \frac{\partial \mathbf{z}^{\pm}}{\partial t} \mp (\mathbf{B}_0 \cdot \nabla)\mathbf{z}^{\pm} + (\mathbf{z}^{\mp} \cdot \nabla)\mathbf{z}^{\pm} &= -\nabla p + \nu_+ \nabla^2 \mathbf{z}^{\pm} + \nu_- \nabla^2 \mathbf{z}^{\mp}, && (20.21a) \\ \nabla \cdot \mathbf{z}^{\pm} &= = 0, && (20.21b) \end{aligned}$$

where

$$\nu_{\pm} = \frac{1}{2}(\nu \pm \eta). \tag{20.22}$$

In tensorial form, the MHD equations for $\mathbf{z}^{\pm}$ are

$$\begin{aligned} \frac{\partial}{\partial t} z_i^{\pm} \mp B_{0j}\partial_j z_i^{\pm} + \partial_j (z_j^{\mp} z_i^{\pm}) &= -\partial_i p + \nu_+ \partial_{jj} z_i^{\pm} + \nu_- \partial_{jj} z_i^{\mp}, && (20.23a) \\ \partial_j z_j^{\pm} &= 0. && (20.23b) \end{aligned}$$

In the next section we will describe the conservation laws of MHD.

20.2 Conservation Laws

In MHD, in addition to kinetic energy and kinetic helicity, we have the following quadratic quantities:

$$\begin{aligned} \text{Magnetic energy density} \quad E_b(\mathbf{r}) &= \frac{1}{2}B^2, && (20.24a) \\ \text{Energy density of } \mathbf{z}^{\pm} \quad E_{z^{\pm}}(\mathbf{r}) &= \frac{1}{2}(z^{\pm})^2, && (20.24b) \\ \text{Cross helicity density} \quad H_c(\mathbf{r}) &= \frac{1}{2}\mathbf{u} \cdot \mathbf{B}, && (20.24c) \\ \text{Magnetic helicity density} \quad H_M(\mathbf{r}) &= \frac{1}{2}\mathbf{A} \cdot \mathbf{B}, && (20.24d) \\ \text{Enstrophy density} \quad E_{\omega}(\mathbf{r}) &= \frac{1}{2}\omega^2, && (20.24e) \\ \text{Mean square vector potential density} \quad E_A(\mathbf{r}) &= \frac{1}{2}A^2. && (20.24f) \end{aligned}$$

The evolution equations for $E_u(\mathbf{r})$ and $E_b(\mathbf{r})$ are as follows:

$$\begin{aligned}\frac{\partial}{\partial t}E_u(\mathbf{r}) + \partial_j\left(\frac{1}{2}u^2 u_j\right) &= -\partial_j(u_j p) + u_i\partial_j\left(B_j B_i\right) - \nu\omega^2, && (20.25a)\\ \frac{\partial}{\partial t}E_b(\mathbf{r}) + \partial_j\left(\frac{1}{2}B^2 u_j\right) &= B_i\partial_j\left(B_j u_i\right) - \eta J^2. && (20.25b)\end{aligned}$$

Adding these two equations yields

$$\frac{\partial}{\partial t}(E_u + E_b)(\mathbf{r}) + \partial_j\left[(E_u + E_b)u_j\right] = -\partial_j(u_j p) + \partial_j\left(2H_c B_j\right) - \nu\omega^2 - \eta J^2. \quad (20.26)$$

Taking a dot product of Eqs. (20.11a, 20.11b) with $\mathbf{B}$ and $\mathbf{u}$ respectively, and then adding them yields

$$\frac{\partial}{\partial t}2H_c(\mathbf{r}) + \partial_j\left(2H_c u_j\right) = -\partial_j(B_j p) + \partial_j\left[(E_u + E_b)B_j\right] - \nu B_i\nabla^2 u_i - \eta u_i\nabla^2 B_i. \quad (20.27)$$

We can define the total of the aforementioned quantities by integrating them over the whole volume. Also note that in terms of Alfvén variables,

$$\begin{aligned}E_{z^\pm} &= E_u + E_b \pm 2H_c, && (20.28a)\\ E &= \frac{1}{2}(E_{z^+} + E_{z^-}), && (20.28b)\\ H_c &= \frac{1}{4}(E_{z^+} - E_{z^-}). && (20.28c)\end{aligned}$$

Two commonly used quantities of MHD are Alfvén ratio r_A and normalized cross helicity σ_c, which are defined as

$$\begin{aligned}r_A &= \frac{E_u}{E_b}, && (20.29a)\\ \sigma_c &= \frac{2H_c}{E}. && (20.29b)\end{aligned}$$

From Eq. (20.28a) equations, it follows that

$$\epsilon_{z^\pm} = \epsilon_u + \epsilon_b \pm 2\epsilon_c, \quad (20.30)$$

where $\epsilon_\pm, \epsilon_u, \epsilon_b, \epsilon_c$ are respectively the dissipation rates of $E_\pm, E_u, E_b$ and H_c.

In dissipationless and diffusionless MHD, the quadratic invariants are the total energy, total cross helicity, and total magnetic helicity. The first two invariants are quite easy to prove using Eqs. (20.26, 20.27). When we integrate these equations for periodic or vanishing boundary conditions, the terms of the form $\partial_j(fu_j)$ and $\partial_j(fB_j)$ vanish (by Gauss theorem) leading to the conservation of the total E and

H_c. Given these conservation laws, using Eq. (20.28a), we conclude that the total E_+ and E_- are also conserved.

It is important to note that neither the kinetic energy nor the magnetic energy is conserved in dissipationless MHD. This is because of the energy exchange between the velocity and magnetic fields. We will derive these transfers in Chapter 21. The conservation of magnetic helicity is proven in the following manner.

Proof: Using the notation $\boldsymbol{\varepsilon} = \mathbf{u} \times \mathbf{B}$ and Eq. (2.34), we obtain

$$\begin{aligned} \frac{d}{dt}\int H_M(\mathbf{r})d\mathbf{r} &= \int d\mathbf{r}\left[\mathbf{B}\cdot\frac{D\mathbf{A}}{Dt} + \mathbf{A}\cdot\frac{D\mathbf{B}}{Dt}\right] \\ &= \int d\mathbf{r}\,[\mathbf{B}\cdot(\boldsymbol{\varepsilon} + \nabla\Psi) + \mathbf{A}\cdot(\nabla\times\boldsymbol{\varepsilon})] \\ &= \int d\mathbf{r}\,[\mathbf{B}\cdot\boldsymbol{\varepsilon} + \boldsymbol{\varepsilon}\cdot(\nabla\times\mathbf{A})] + \int d\mathbf{S}\cdot(\Psi\mathbf{B}) \\ &= \int d\mathbf{r}[2\mathbf{B}\cdot\boldsymbol{\varepsilon}] + \int d\mathbf{S}\cdot(\Psi\mathbf{B}) \\ &= 0, \end{aligned} \tag{20.31}$$

provided the above surface term vanishes.[2] Note that $\mathbf{B}\cdot\boldsymbol{\varepsilon} = 0$. In the third step of the above derivation we employ integration by parts to obtain

$$\int d\mathbf{r}\mathbf{A}\cdot(\nabla\times\boldsymbol{\varepsilon}) = (\mathbf{A}\times\boldsymbol{\varepsilon})_i|_{S_i} + \int d\mathbf{r}\boldsymbol{\varepsilon}\cdot(\nabla\times\mathbf{A}). \tag{20.32}$$

The surface term vanishes for periodic boundary condition. □

There is an important nonquadratic invariant of MHD, which is the magnetic flux crossing a surface enclosed by a comoving closed contour:

$$\Phi = \int \mathbf{B}\cdot d\mathbf{S}. \tag{20.33}$$

This conservation law is called *Alfvén frozen-in theorem*. A physical interpretation of this theorem is that the magnetic field lines are frozen in a magnetofluid, and they move along with the fluid. This theorem is analogous to Kelvin's circulation theorem of hydrodynamics.

Proof: The time derivative of Φ is

$$\frac{d}{dt}\int \mathbf{B}\cdot d\mathbf{S} = \frac{d}{dt}\oint \mathbf{A}\cdot d\mathbf{l}$$

[2]This method can also be used to prove conservation of kinetic helicity.

$$
\begin{aligned}
&= \oint \frac{D\mathbf{A}}{Dt} \cdot d\mathbf{l} + \mathbf{A} \cdot \frac{Dd\mathbf{l}}{Dt} \\
&= \oint dl_i u_j \partial_i A_j + \oint (\nabla\Psi) \cdot d\mathbf{l} + \mathbf{A} \cdot [d\mathbf{l} \cdot \nabla\mathbf{u}] \\
&= \oint [dl_i u_j \partial_i A_j + A_j dl_i \partial_i u_j] + \oint (\nabla\Psi) \cdot d\mathbf{l} \\
&= \oint d\mathbf{l} \cdot \nabla(\mathbf{u} \cdot \mathbf{A} + \Psi) = 0. \qquad (20.34)
\end{aligned}
$$

In this derivation we have employed Eq. (20.15) and an identity—$Dd\mathbf{l}/Dt = d\mathbf{l} \cdot \nabla\mathbf{u}$ (see Section 2.3).

We can also prove this theorem using the following arguments:

$$
\frac{d}{dt}\int \mathbf{B} \cdot d\mathbf{S} = \int \frac{D\mathbf{B}}{Dt} \cdot d\mathbf{S} + \mathbf{B} \cdot \frac{Dd\mathbf{S}}{Dt}. \qquad (20.35)
$$

During a short interval, a movement of the closed contour forms a cylinder. The change in the surface area enclosed by the closed contours, $\Delta(d\mathbf{S})$, is the surface area of the cylinder. Note that $\Delta(d\mathbf{S}) = (\mathbf{u}dt) \times d\mathbf{l}$, where $\mathbf{u}$ is the velocity of the line segment $d\mathbf{l}$. Hence,

$$
\begin{aligned}
\frac{d}{dt}\int \mathbf{B} \cdot d\mathbf{S} &= \int \frac{D\mathbf{B}}{Dt} \cdot d\mathbf{S} + \mathbf{B} \cdot \frac{Dd\mathbf{S}}{Dt} \\
&= \int \nabla \times (\mathbf{u} \times \mathbf{B}) \cdot d\mathbf{S} + \mathbf{B} \cdot (\mathbf{u} \times d\mathbf{l}) \\
&= \int \nabla \times (\mathbf{u} \times \mathbf{B}) \cdot d\mathbf{S} + (\mathbf{B} \times \mathbf{u}) \cdot d\mathbf{l} \\
&= \int \nabla \times (\mathbf{u} \times \mathbf{B}) \cdot d\mathbf{S} + \nabla \times (\mathbf{B} \times \mathbf{u}) \cdot d\mathbf{S} \\
&= 0. \qquad (20.36)
\end{aligned}
$$

□

In 2D MHD, total magnetic helicity is zero; hence, it is trivially conserved. The nontrivial conserved quantity in diffusionless 2D MHD is the total of the mean square vector potential, $E_A = A^2/2$. This is conversed because A acts as a passive scalar (see Chapter 14).

Note that Alfvén's frozen-in theorem is very similar to Kelvin's circulation theorem, which was discussed in Section 2.4. In addition, the evolution equations of the magnetic and vorticity fields are very similar. Hence, there are certain similarities between the behavior of these two fields. For example, the velocity field stretches the vorticity field as well as the magnetic field. Note however that the magnetic field is an independent field, while vorticity, $\nabla \times \mathbf{u}$, is a derivative of the velocity field. As a result, the analogy is not perfect. We will revisit these issues in subsequent chapters.

To summarize, the conserved quantities for 3D MHD are the total energy, total cross helicity, and total magnetic helicity. However, 2D MHD has conservation of total energy, total cross helicity, and total of mean square vector potential. It is easy to show that the total kinetic energy, total magnetic energy, total kinetic helicity, total enstrophy, and circulation are not conserved in MHD.

In the next section we describe the MHD equations in Fourier space.

20.3 Governing Equations in Fourier Space

In Fourier space, the MHD equations are

$$\frac{d}{dt}\mathbf{u}(\mathbf{k}) + \mathbf{N}_u(\mathbf{k}) = -i\mathbf{k}p(\mathbf{k}) + \mathbf{F}_u(\mathbf{k}) - \nu k^2\mathbf{u}(\mathbf{k}), \tag{20.37a}$$

$$\frac{d}{dt}\mathbf{B}(\mathbf{k}) + \mathbf{N}_B(\mathbf{k}) = \mathbf{F}_B(\mathbf{k}) - \eta k^2\mathbf{u}(\mathbf{k}), \tag{20.37b}$$

$$\mathbf{k}\cdot\mathbf{u}(\mathbf{k}) = 0, \tag{20.37c}$$

$$\mathbf{k}\cdot\mathbf{B}(\mathbf{k}) = 0, \tag{20.37d}$$

where the nonlinear terms are

$$\mathbf{N}_u(\mathbf{k}) = i\sum_{\mathbf{p}}\{\mathbf{k}\cdot\mathbf{u}(\mathbf{q})\}\mathbf{u}(\mathbf{p}), \tag{20.38a}$$

$$\mathbf{N}_B(\mathbf{k}) = i\sum_{\mathbf{p}}\{\mathbf{k}\cdot\mathbf{u}(\mathbf{q})\}\mathbf{B}(\mathbf{p}), \tag{20.38b}$$

$$\mathbf{F}_u(\mathbf{k}) = i\sum_{\mathbf{p}}\{\mathbf{k}\cdot\mathbf{B}(\mathbf{q})\}\mathbf{B}(\mathbf{p}), \tag{20.38c}$$

$$\mathbf{F}_B(\mathbf{k}) = i\sum_{\mathbf{p}}\{\mathbf{k}\cdot\mathbf{B}(\mathbf{q})\}\mathbf{u}(\mathbf{p}), \tag{20.38d}$$

with $\mathbf{k} = \mathbf{p} + \mathbf{q}$. Using Eq. (20.6), we obtain

$$-\mathbf{N}_B(\mathbf{k}) + \mathbf{F}_B(\mathbf{k}) = i\sum_{\mathbf{p}}\mathbf{k}\times[\mathbf{u}(\mathbf{q})\times\mathbf{B}(\mathbf{p})]. \tag{20.39}$$

When the magnetic field is decomposed into the mean field $\mathbf{B}_0$ and fluctuation $\mathbf{b}$, the dynamical equations for the fluctuations $\mathbf{u}, \mathbf{b}$ are given by

$$\frac{d}{dt}\mathbf{u}(\mathbf{k}) - i(\mathbf{B}_0\cdot\mathbf{k})\mathbf{b}(\mathbf{k}) + \mathbf{N}_u(\mathbf{k}) = -i\mathbf{k}p(\mathbf{k}) + \mathbf{f}_u(\mathbf{k}) - \nu k^2\mathbf{u}(\mathbf{k}), \tag{20.40a}$$

$$\frac{d}{dt}\mathbf{b}(\mathbf{k}) - i(\mathbf{B}_0 \cdot \mathbf{k})\mathbf{u}(\mathbf{k}) + \mathbf{N}_b(\mathbf{k}) = \mathbf{f}_b(\mathbf{k}) - \eta k^2 \mathbf{u}(\mathbf{k}), \tag{20.40b}$$

$$\mathbf{k} \cdot \mathbf{u}(\mathbf{k}) = 0, \tag{20.40c}$$

$$\mathbf{k} \cdot \mathbf{b}(\mathbf{k}) = 0, \tag{20.40d}$$

where the nonlinear terms are

$$\mathbf{N}_u(\mathbf{k}) = i\sum_{\mathbf{p}} \{\mathbf{k} \cdot \mathbf{u}(\mathbf{q})\}\mathbf{u}(\mathbf{p}), \tag{20.41a}$$

$$\mathbf{N}_b(\mathbf{k}) = i\sum_{\mathbf{p}} \{\mathbf{k} \cdot \mathbf{u}(\mathbf{q})\}\mathbf{b}(\mathbf{p}). \tag{20.41b}$$

$$\mathbf{f}_u(\mathbf{k}) = i\sum_{\mathbf{p}} \{\mathbf{k} \cdot \mathbf{b}(\mathbf{q})\}\mathbf{b}(\mathbf{p}), \tag{20.41c}$$

$$\mathbf{f}_b(\mathbf{k}) = i\sum_{\mathbf{p}} \{\mathbf{k} \cdot \mathbf{b}(\mathbf{q})\}\mathbf{u}(\mathbf{p}). \tag{20.41d}$$

We can derive Eqs. (20.40, 20.41) from Eqs. (20.37, 20.38) by setting $\mathbf{B}(\mathbf{k}=0) = \mathbf{B}_0$.

The equations for $\mathbf{z}^{\pm}$ in Fourier space are

$$\frac{d}{dt}\mathbf{z}^{\pm}(\mathbf{k}) \mp i(\mathbf{B}_0 \cdot \mathbf{k})\mathbf{z}^{\pm}(\mathbf{k}) + \mathbf{N}_{z^{\pm}}(\mathbf{k})(\mathbf{k}) = -i\mathbf{k}p(\mathbf{k}) - \nu_+ k^2 \mathbf{z}^{\pm}(\mathbf{k}) - \nu_- k^2 \mathbf{z}^{\mp}(\mathbf{k}), \tag{20.42a}$$

$$\mathbf{k} \cdot \mathbf{z}^{\pm}(\mathbf{k}) = 0, \tag{20.42b}$$

where the nonlinear terms are

$$\mathbf{N}_{z^{\pm}}(\mathbf{k}) = i\sum_{\mathbf{p}} \{\mathbf{k} \cdot \mathbf{z}^{\mp}(\mathbf{q})\}\mathbf{z}^{\pm}(\mathbf{p}). \tag{20.43}$$

The equation for the vorticity is

$$\frac{d}{dt}\boldsymbol{\omega}(\mathbf{k}) + \mathbf{N}_{\omega}(\mathbf{k}) = \mathbf{F}_{\omega}(\mathbf{k}) - \nu k^2 \boldsymbol{\omega}(\mathbf{k}), \tag{20.44}$$

where

$$\mathbf{N}_{\omega}(\mathbf{k}) = -i\mathbf{k} \times \sum_{\mathbf{p}} \{\mathbf{u}(\mathbf{q}) \times \boldsymbol{\omega}(\mathbf{p})\}, \tag{20.45a}$$

$$\mathbf{F}_{\omega}(\mathbf{k}) = i\mathbf{k} \times \sum_{\mathbf{p}} \{\mathbf{J}(\mathbf{q}) \times \mathbf{B}(\mathbf{p})\}. \tag{20.45b}$$

Using Eq. (20.14), we deduce that

$$\frac{d}{dt}\mathbf{A}(\mathbf{k}) = i\sum_{\mathbf{p}}[\mathbf{u}(\mathbf{k}-\mathbf{p})\times\mathbf{B}(\mathbf{p})] - \eta k^2\mathbf{u}(\mathbf{k}) - i\mathbf{k}\Psi(\mathbf{k}). \tag{20.46}$$

In MHD, in addition to the modal kinetic energy, modal enstrophy, and modal kinetic helicity (see Section 2.3), we have modal magnetic energy and modal magnetic helicity:

$$E_B(\mathbf{k}) = \frac{1}{2}|\mathbf{B}(\mathbf{k})|^2, \tag{20.47a}$$

$$H_M(\mathbf{k}) = \frac{1}{2}\Re[\mathbf{A}(\mathbf{k})\cdot\mathbf{B}^*(\mathbf{k})]. \tag{20.47b}$$

The spectral equations for the modal kinetic energy ($|\mathbf{u}(\mathbf{k})|^2/2$) and magnetic energy ($|\mathbf{B}(\mathbf{k})|^2/2$) are as follows:

$$\begin{aligned}\frac{d}{dt}E_u(\mathbf{k}) &= -\Re[\mathbf{N}_u(\mathbf{k})\cdot\mathbf{u}^*(\mathbf{k})] + \Re[\mathbf{F}_u(\mathbf{k})\cdot\mathbf{u}^*(\mathbf{k})] - 2\nu k^2 E_u(\mathbf{k}) \\ &= \sum_{\mathbf{p}}\Im\left[\{\mathbf{k}\cdot\mathbf{u}(\mathbf{q})\}\{\mathbf{u}(\mathbf{p})\cdot\mathbf{u}^*(\mathbf{k})\} - \{\mathbf{k}\cdot\mathbf{B}(\mathbf{q})\}\{\mathbf{B}(\mathbf{p})\cdot\mathbf{u}^*(\mathbf{k})\}\right] \\ &\quad -2\nu k^2 E_u(\mathbf{k}),\end{aligned} \tag{20.48a}$$

$$\begin{aligned}\frac{d}{dt}E_B(\mathbf{k}) &= -\Re[\mathbf{N}_B(\mathbf{k})\cdot\mathbf{B}^*(\mathbf{k})] + \Re[\mathbf{F}_B(\mathbf{k})\cdot\mathbf{B}^*(\mathbf{k})] - 2\eta k^2 E_B(\mathbf{k}) \\ &= \sum_{\mathbf{p}}\Im\left[\{\mathbf{k}\cdot\mathbf{u}(\mathbf{q})\}\{\mathbf{B}(\mathbf{p})\cdot\mathbf{B}^*(\mathbf{k})\} - \{\mathbf{k}\cdot\mathbf{B}(\mathbf{q})\}\{\mathbf{u}(\mathbf{p})\cdot\mathbf{B}^*(\mathbf{k})\}\right] \\ &\quad -2\eta k^2 E_B(\mathbf{k}).\end{aligned} \tag{20.48b}$$

Similarly,

$$\begin{aligned}\frac{d}{dt}E_{z^\pm}(\mathbf{k}) &= \sum_{\mathbf{p}}\Im\left[\{\mathbf{k}\cdot\mathbf{z}^\mp(\mathbf{q})\}\{\mathbf{z}^\pm(\mathbf{p})\cdot\mathbf{z}^{\pm*}(\mathbf{k})\}\right] - 2\nu_\pm k^2 E_{z^\pm}(\mathbf{k}) \\ &\quad -2\nu_- k^2[E_u(\mathbf{k}) - E_b(\mathbf{k})].\end{aligned} \tag{20.49}$$

Now let us write down the spectral equation for the kinetic helicity. We start from Eq. (3.45), which is

$$\frac{d}{dt}H_K(\mathbf{k}) = \frac{1}{k^2}\Re[i\mathbf{k}\cdot\{\dot{\boldsymbol{\omega}}(\mathbf{k})\times\boldsymbol{\omega}^*(\mathbf{k})\}]. \tag{20.50}$$

By following the same algebra as in the derivation of $\dot{H}_K$ in Section 3.3, we obtain

$$
\begin{aligned}
i\mathbf{k}\cdot\{\dot{\boldsymbol{\omega}}(\mathbf{k})\times\boldsymbol{\omega}^*(\mathbf{k})\} &= i\mathbf{k}\cdot\sum_{\mathbf{p}}[i\{\mathbf{k}\times(\mathbf{u}(\mathbf{q})\times\boldsymbol{\omega}(\mathbf{p})+\mathbf{J}(\mathbf{q})\times\mathbf{B}(\mathbf{p}))\}\times\boldsymbol{\omega}^*(\mathbf{k})]\\
&\quad -2\nu k^2 H_K(\mathbf{k})\\
&= \mathbf{k}\cdot\sum_{\mathbf{p}}[\mathbf{k}\{\mathbf{u}(\mathbf{q})\times\boldsymbol{\omega}(\mathbf{p})+\mathbf{J}(\mathbf{q})\times\mathbf{B}(\mathbf{p})\}\cdot\boldsymbol{\omega}^*(\mathbf{k})]\\
&\quad -2\nu k^2 H_K(\mathbf{k})\\
&= k^2\sum_{\mathbf{p}}\mathbf{u}(\mathbf{q})\cdot\{\boldsymbol{\omega}(\mathbf{p})\times\boldsymbol{\omega}^*(\mathbf{k})\}\\
&\quad +k^2\sum_{\mathbf{p}}\mathbf{J}(\mathbf{q})\cdot\{\mathbf{B}(\mathbf{p})\times\boldsymbol{\omega}^*(\mathbf{k})\}-2\nu k^2 H_K(\mathbf{k}), \qquad (20.51)
\end{aligned}
$$

where $\mathbf{q}=\mathbf{k}-\mathbf{p}$. Hence,

$$\frac{d}{dt}H_K(\mathbf{k})=\sum_{\mathbf{p}}\Re[\mathbf{u}(\mathbf{q})\cdot\{\boldsymbol{\omega}(\mathbf{p})\times\boldsymbol{\omega}^*(\mathbf{k})\}+\mathbf{J}(\mathbf{q})\cdot\{\mathbf{B}(\mathbf{p})\times\boldsymbol{\omega}^*(\mathbf{k})\}]-2\nu k^2 H_K(\mathbf{k}). \quad (20.52)$$

The equation for the magnetic helicity is derived similarly. Using Coulomb gauge $\mathbf{k}\cdot\mathbf{A}(\mathbf{k})=0$, we derive that

$$\mathbf{A}(\mathbf{k})=\frac{i}{k^2}\mathbf{k}\times\mathbf{B}(\mathbf{k}). \quad (20.53)$$

Substitution of this form of $\mathbf{A}(\mathbf{k})$ in Eq. (20.47b) yields

$$H_M(\mathbf{k})=\frac{1}{2k^2}\Re[i\mathbf{k}\cdot\{\mathbf{B}(\mathbf{k})\times\mathbf{B}^*(\mathbf{k})\}]. \quad (20.54)$$

Following the same algebra as that for H_K, we obtain

$$
\begin{aligned}
\frac{d}{dt}H_M(\mathbf{k}) &= \frac{1}{k^2}\Re[i\mathbf{k}\cdot\{\dot{\mathbf{B}}(\mathbf{k})\times\mathbf{B}^*(\mathbf{k})\}]\\
&= -\frac{1}{k^2}\Re\sum_{\mathbf{p}}\mathbf{k}\cdot[\{\mathbf{k}\times(\mathbf{u}(\mathbf{q})\times\mathbf{B}(\mathbf{p}))\}\times\mathbf{B}^*(\mathbf{k})]-2\eta k^2 H_M(\mathbf{k})\\
&= \Re\sum_{\mathbf{p}}\mathbf{u}(\mathbf{q})\cdot\{\mathbf{B}(\mathbf{p})\times\mathbf{B}^*(\mathbf{k})\}-2\eta k^2 H_M(\mathbf{k}). \qquad (20.55)
\end{aligned}
$$

The equation for enstrophy is derived similarly. In 2D, the equation for mean square vector potential E_A is

$$\frac{d}{dt}E_A(\mathbf{k})=\sum_{\mathbf{p}}\Im\left[\{\mathbf{k}\cdot\mathbf{u}(\mathbf{q})\}A(\mathbf{p})A^*(\mathbf{k})\right]-2\eta k^2 E_A(\mathbf{k}). \quad (20.56)$$

In the next section, we describe the properties of Alfvén waves.

20.4 Alfvén Waves

When $u \ll B_0$ and $b \ll B_0$, the nonlinear terms can be dropped. The resulting linear equations support waves that will be described in this section.

The linearized, incompressible, dissipationless, and diffusionless MHD equations are

$$\frac{d}{dt}\mathbf{u}(\mathbf{k}) = -i\mathbf{k}p(\mathbf{k}) + i[\mathbf{B}_0 \cdot \mathbf{k}]\mathbf{b}(\mathbf{k}), \quad (20.57a)$$

$$\frac{d}{dt}\mathbf{b}(\mathbf{k}) = i[\mathbf{B}_0 \cdot \mathbf{k}]\mathbf{u}(\mathbf{k}), \quad (20.57b)$$

$$\mathbf{k} \cdot \mathbf{u}(\mathbf{k}) = \mathbf{k} \cdot \mathbf{b}(\mathbf{k}) = 0. \quad (20.57c)$$

We work in Craya–Herring basis that eliminates pressure and yields relatively simpler set of equations:

$$\frac{d}{dt}\begin{pmatrix} u_1(\mathbf{k}) \\ u_2(\mathbf{k}) \\ b_1(\mathbf{k}) \\ b_2(\mathbf{k}) \end{pmatrix} = \begin{pmatrix} 0 & 0 & i\omega & 0 \\ 0 & 0 & 0 & i\omega \\ i\omega & 0 & 0 & 0 \\ 0 & i\omega & 0 & 0 \end{pmatrix} \begin{pmatrix} u_1(\mathbf{k}) \\ u_2(\mathbf{k}) \\ b_1(\mathbf{k}) \\ b_2(\mathbf{k}) \end{pmatrix}, \quad (20.58)$$

where

$$\omega = \mathbf{B}_0 \cdot \mathbf{k} \quad (20.59)$$

is the dispersion relation. The above matrix has eigenvalues $-i\omega, -i\omega, i\omega, i\omega$, and the corresponding eigenvectors are

$$\begin{pmatrix} -1 \\ 0 \\ 1 \\ 0 \end{pmatrix}; \begin{pmatrix} 0 \\ -1 \\ 0 \\ 1 \end{pmatrix}; \begin{pmatrix} 1 \\ 0 \\ 1 \\ 0 \end{pmatrix}; \begin{pmatrix} 0 \\ 1 \\ 0 \\ 1 \end{pmatrix}. \quad (20.60)$$

In real space, these eigenvectors correspond to the following wave solutions:

$$\begin{pmatrix} -1 \\ 0 \\ 1 \\ 0 \end{pmatrix} \cos(kz' - \omega t + \Psi_{k1}); \quad \begin{pmatrix} 0 \\ -1 \\ 0 \\ 1 \end{pmatrix} \cos(kz' - \omega t + \Psi_{k2});$$

$$\begin{pmatrix} 1 \\ 0 \\ 1 \\ 0 \end{pmatrix} \cos(kz' + \omega t + \Psi_{k3}); \quad \begin{pmatrix} 0 \\ 1 \\ 0 \\ 1 \end{pmatrix} \cos(kz' + \omega t + \Psi_{k4}), \quad (20.61)$$

where z' is the coordinate along $\mathbf{k}$, and Ψ_k's are the phases of the wave modes. From the phases of the cosine function we deduce that the first two solutions represent the waves propagating along $\mathbf{k}$, while the next two are the waves propagating along $-\mathbf{k}$. These waves are called *Alfvén waves*, and they are depicted in Fig. 20.1. Note that the first two waves have $\mathbf{u} = -\mathbf{b}$, but the next two have $\mathbf{u} = \mathbf{b}$.

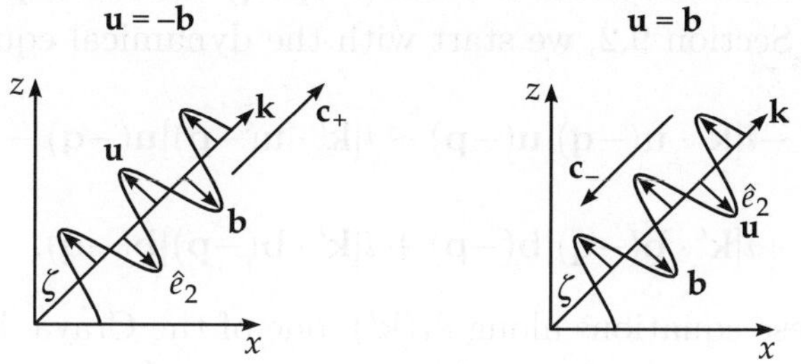

Figure 20.1 (a) An Alfvén wave traveling along $\mathbf{k}$ with a phase velocity of $c_+ = B_0 \cos\zeta$. It has $\mathbf{u} = -\mathbf{b}$ (b) An Alfvén wave traveling along $-\mathbf{k}$ with a phase velocity of $c_- = -B_0 \cos\zeta$. It has $\mathbf{u} = \mathbf{b}$. From Verma (2018). Reprinted with permission from World Scientific.

The phase velocities of the waves are

$$\mathbf{c}_\pm = \pm \frac{\omega}{k} \hat{k} = \pm B_0 \cos\zeta \hat{k}. \tag{20.62}$$

Note that for $\mathbf{k} = \mathbf{k}_\perp = k_x \hat{x} + k_y \hat{y}$, we obtain $\omega = 0$ (because $\zeta = \pi/2$). Hence, the fluctuations in the $k_z = 0$ plane are not waves; they are two-dimensional fluctuations.

20.5 MHD Equations in Craya–Herring Basis

As described in Chapter 9, Craya–Herring basis helps present the hydrodynamic Fourier modes (their equations, energies, and helicities) in a compact manner. Here we describe how the magnetic field is represented in Craya–Herring basis. Since $\mathbf{k} \cdot \mathbf{b}(\mathbf{k}) = 0$, we can employ Craya–Herring basis vectors to represent the magnetic Fourier mode ($\mathbf{k} \neq 0$) as

$$\mathbf{b}(\mathbf{k}) \quad = \quad b_1(\mathbf{k})\hat{e}_1 + b_2(\mathbf{k})\hat{e}_2. \tag{20.63}$$

In Craya–Herring basis, the magnetic energy and magnetic helicity are

$$E_b(\mathbf{k}) \quad = \quad \frac{1}{2}\mathbf{b}^*(\mathbf{k}) \cdot \mathbf{b}(\mathbf{k}) = \frac{1}{2}[b_1(\mathbf{k})|^2 + b_2(\mathbf{k})|^2], \tag{20.64a}$$

$$H_M(\mathbf{k}) \quad = \quad \frac{1}{2}\Re[\mathbf{b}^*(\mathbf{k}) \cdot \mathbf{a}(\mathbf{k})] = \frac{1}{k}\Im[b_1^*(\mathbf{k}) b_2(\mathbf{k})]. \tag{20.64b}$$

In Chapters 9, 13, and 18 we presented derivations of the equations of motion for hydrodynamics, and for flows with scalar and vector fields. In the present section, we extend these derivations to MHD equations. These equations are very useful for constructing low-dimensional models for dynamos (see Chapter 23).

As in Section 9.2, we consider a triad $(\mathbf{k}', \mathbf{p}, \mathbf{q})$ with $\mathbf{k}' + \mathbf{p} + \mathbf{q} = 0$. Following similar steps as in Section 9.2, we start with the dynamical equations for $\mathbf{u}(\mathbf{k}')$:

$$\begin{aligned}\frac{d}{dt}\mathbf{u}(\mathbf{k}') &= -i[\mathbf{k}' \cdot \mathbf{u}(-\mathbf{q})]\mathbf{u}(-\mathbf{p}) - i[\mathbf{k}' \cdot \mathbf{u}(-\mathbf{p})]\mathbf{u}(-\mathbf{q}) - i\mathbf{k}p(\mathbf{k}') \\ &\quad + i[\mathbf{k}' \cdot \mathbf{b}(-\mathbf{q})]\mathbf{b}(-\mathbf{p}) + i[\mathbf{k}' \cdot \mathbf{b}(-\mathbf{p})]\mathbf{b}(-\mathbf{q}). \qquad (20.65)\end{aligned}$$

A projection of these equations along $\hat{e}_1(\mathbf{k}')$, one of the Craya–Herring basis vector, yields the following equation for $u_1((\mathbf{k}')$:

$$\begin{aligned}\dot{u}_1(\mathbf{k}') &= [-ik' \sin\beta \hat{e}_1(\mathbf{p}) \cdot \hat{e}_1(\mathbf{k}') + ik' \sin\gamma \hat{e}_1(\mathbf{q}) \cdot \hat{e}_1(\mathbf{k}')] \\ &\quad \times [u_1^*(\mathbf{p})u_1^*(\mathbf{q}) - b_1^*(\mathbf{p})b_1^*(\mathbf{q})] \\ &= ik' \sin(\beta - \gamma)[u_1^*(\mathbf{p})u_1^*(\mathbf{q}) - b_1^*(\mathbf{p})b_1^*(\mathbf{q})]. \qquad (20.66)\end{aligned}$$

Similar computation yields the following equations for $u_1(\mathbf{p})$ and $u_1(\mathbf{q})$:

$$\dot{u}_1(\mathbf{p}) = ip\sin(\gamma - \alpha)[u_1^*(\mathbf{q})u_1^*(\mathbf{k}') - b_1^*(\mathbf{q})b_1^*(\mathbf{k}')], \qquad (20.67a)$$

$$\dot{u}_1(\mathbf{q}) = iq\sin(\alpha - \beta)[u_1^*(\mathbf{k}')u_1^*(\mathbf{p}) - b_1^*(\mathbf{k}')b_1^*(\mathbf{p})]. \qquad (20.67b)$$

A projection of Eqs. (20.65) along $\hat{e}_2$ yields the following equations:

$$\begin{aligned}\dot{u}_2(\mathbf{k}') &= ik'\{\sin\gamma u_1^*(\mathbf{p})u_2^*(\mathbf{q}) - \sin\beta u_1^*(\mathbf{q})u_2^*(\mathbf{p})\} \\ &\quad - ik'\{\sin\gamma b_1^*(\mathbf{p})b_2^*(\mathbf{q}) - \sin\beta b_1^*(\mathbf{q})b_2^*(\mathbf{p})\}, \qquad (20.68a)\end{aligned}$$

$$\begin{aligned}\dot{u}_2(\mathbf{p}) &= ip\{\sin\alpha u_1^*(\mathbf{q})u_2^*(\mathbf{k}') - \sin\gamma u_1^*(\mathbf{k}')u_2^*(\mathbf{q})\} \\ &\quad - ip\{\sin\alpha b_1^*(\mathbf{q})b_2^*(\mathbf{k}') - \sin\gamma b_1^*(\mathbf{k}')b_2^*(\mathbf{q})\}, \qquad (20.68b)\end{aligned}$$

$$\begin{aligned}\dot{u}_2(\mathbf{q}) &= iq\{\sin\beta u_1^*(\mathbf{k}')u_2^*(\mathbf{p}) - \sin\alpha u_1^*(\mathbf{p})u_2^*(\mathbf{k}')\} \\ &\quad - iq\{\sin\beta b_1^*(\mathbf{k}')b_2^*(\mathbf{p}) - \sin\alpha b_1^*(\mathbf{p})b_2^*(\mathbf{k}')\}. \qquad (20.68c)\end{aligned}$$

The equations for the the magnetic mode $\mathbf{b}(\mathbf{k}')$ are

$$\begin{aligned}\frac{d}{dt}\mathbf{b}(\mathbf{k}') &= -i[\mathbf{k}'\cdot\mathbf{u}(-\mathbf{q})]\mathbf{b}(-\mathbf{p}) - i[\mathbf{k}'\cdot\mathbf{u}(-\mathbf{p})]\mathbf{b}(-\mathbf{q}) \\ &\quad +i[\mathbf{k}'\cdot\mathbf{b}(-\mathbf{q})]\mathbf{u}(-\mathbf{p}) + i[\mathbf{k}'\cdot\mathbf{b}(-\mathbf{p})]\mathbf{u}(-\mathbf{q}). \qquad (20.69)\end{aligned}$$

Projections of these equations and those for $\mathbf{b}(\mathbf{p})$ and $\mathbf{b}(\mathbf{q})$ along the Craya–Herring basis vectors yield the following equations:

$$\begin{aligned}\dot{b}_1(\mathbf{k}') &= ik'\sin(\beta+\gamma)[b_1^*(\mathbf{p})u_1^*(\mathbf{q}) - u_1^*(\mathbf{p})b_1^*(\mathbf{q})], \\ &= ik'\sin\alpha[b_1^*(\mathbf{p})u_1^*(\mathbf{q}) - u_1^*(\mathbf{p})b_1^*(\mathbf{q})], \qquad (20.70a)\\ \dot{b}_1(\mathbf{p}) &= ip\sin(\gamma+\alpha)[b_1^*(\mathbf{q})u_1^*(\mathbf{k}') - u_1^*(\mathbf{q})b_1^*(\mathbf{k}')], \\ &= ip\sin\beta[b_1^*(\mathbf{q})u_1^*(\mathbf{k}') - u_1^*(\mathbf{q})b_1^*(\mathbf{k}')], \qquad (20.70b)\\ \dot{b}_1(\mathbf{q}) &= iq\sin(\alpha+\beta)[b_1^*(\mathbf{k}')u_1^*(\mathbf{p}) - u_1^*(\mathbf{k}')b_1^*(\mathbf{p})] \\ &= iq\sin\gamma[b_1^*(\mathbf{k}')u_1^*(\mathbf{p}) - u_1^*(\mathbf{k}')b_1^*(\mathbf{p})]. \qquad (20.70c)\end{aligned}$$

Similarly,

$$\begin{aligned}\dot{b}_2(\mathbf{k}') &= ik'\{\sin\gamma u_1^*(\mathbf{p})b_2^*(\mathbf{q}) - \sin\beta u_1^*(\mathbf{q})b_2^*(\mathbf{p})\} \\ &\quad -ik'\{\sin\gamma b_1^*(\mathbf{p})u_2^*(\mathbf{q}) - \sin\beta b_1^*(\mathbf{q})u_2^*(\mathbf{p})\}, \qquad (20.71a)\\ \dot{b}_2(\mathbf{p}) &= ip\{\sin\alpha u_1^*(\mathbf{q})b_2^*(\mathbf{k}') - \sin\gamma u_1^*(\mathbf{k}')b_2^*(\mathbf{q})\} \\ &\quad -ip\{\sin\alpha b_1^*(\mathbf{q})u_2^*(\mathbf{k}') - \sin\gamma b_1^*(\mathbf{k}')u_2^*(\mathbf{q})\}, \qquad (20.71b)\\ \dot{b}_2(\mathbf{q}) &= iq\{\sin\beta u_1^*(\mathbf{k}')b_2^*(\mathbf{p}) - \sin\alpha u_1^*(\mathbf{p})b_2^*(\mathbf{k}')\} \\ &\quad -iq\{\sin\beta b_1^*(\mathbf{k}')u_2^*(\mathbf{p}) - \sin\alpha b_1^*(\mathbf{p})u_2^*(\mathbf{k}')\}. \qquad (20.71c)\end{aligned}$$

Example 20.1: In a periodic box $[2\pi, 2\pi]$, consider the following velocity and magnetic fields in MHD:

$$\begin{aligned}\mathbf{u} &= \hat{x}2D_u\cos y + \hat{y}2C_u\cos x + (\hat{x}-\hat{y})2A_u\sin(x+y), \\ \mathbf{b} &= \hat{x}2D_b\cos y + \hat{y}2C_b\cos x + (\hat{x}-\hat{y})2A_b\sin(x+y).\end{aligned}$$

Assume $\nu = \eta = 0$. Derive equations for A_u, C_u, D_u, A_b, C_b, and D_b as in Example 3.4.

Solution: These vector fields are the same as those in Example 18.1 (with $\mathbf{w}$ replaced by $\mathbf{b}$). As in Example 18.1, we decompose the fields in Craya–Herring basis, and then write down the equations for the components. The amplitudes of the components are same as those in Table 18.2.

The difference between the present example and Example 18.1 is the following. In MHD, the velocity and magnetic fields are forced by $(\mathbf{b} \cdot \nabla)\mathbf{b}$ and $(\mathbf{b} \cdot \nabla)\mathbf{u}$ respectively, in contrast to the passive vector of Example 18.1 having $\mathbf{F}_u = \mathbf{F}_B = 0$. Consequently, the dynamical equations for the modes have additional terms compared to Example 18.1. The aforementioned two-dimensional field configurations have $u_2 = b_2 = 0$; hence, we write down the evolution equations for u_1 and b_1 components only. Following Eqs. (20.66, 20.67), we obtain

$$\begin{aligned}
\dot{u}_1(-1,-1) &= \dot{A}_u(-\sqrt{2}/i) = 0, \\
\dot{u}_1(1,0) &= \dot{D}_u = -i\sin(\gamma-\alpha)[-C_u(A_u\sqrt{2}/i) + C_b(A_b\sqrt{2}/i)]. \\
\dot{u}_1(0,1) &= -\dot{C}_u = i\sin(\alpha-\beta)(D_u(A_u\sqrt{2}/i) - D_b(A_b\sqrt{2}/i)),
\end{aligned}$$

or

$$\begin{aligned}
\dot{A}_u &= 0, \\
\dot{D}_u &= C_u A_u - A_b C_b, \\
\dot{C}_u &= -D_u A_u + A_b D_b.
\end{aligned}$$

For the magnetic modes we employ Eqs. (20.70) that yields

$$\begin{aligned}
\dot{b}_1(-1,-1) &= \dot{A}_b(-\sqrt{2}/i) = i\sqrt{2}\sin\alpha[-C_u D_b + D_u C_b], \\
\dot{b}_1(1,0) &= \dot{D}_b = i\sin\beta[-C_b(A_u\sqrt{2}/i) + C_u(A_b\sqrt{2}/i)], \\
\dot{b}_1(0,1) &= -\dot{C}_b = i\sin\gamma((A_b\sqrt{2}/i)D_u - (A_u\sqrt{2}/i)D_b),
\end{aligned}$$

or

$$\begin{aligned}
\dot{A}_b &= -C_u D_b + D_u C_b, \\
\dot{D}_b &= -A_u C_b + C_u A_b, \\
\dot{C}_b &= A_u D_b - D_u A_b.
\end{aligned}$$

This set of differential equations could be solved numerically.

It is interesting to note that the total energy

$$(A_u^2 + A_b^2) + \frac{1}{2}(D_u^2 + D_b^2) + \frac{1}{2}(C_u^2 + D_u^2) = \text{const.}$$

which follows from the conservation of total energy for inviscid MHD. Note that the total kinetic energy and total magnetic energy are not conserved.

In the next section, we present the equations of MHD in helical basis.

20.6 MHD Equations in Helical Basis

Helical basis, discussed in Chapter 9, is very useful for dynamo studies. Magnetic energy and magnetic helicity can be written in terms of helical basis vectors as follows:

$$E_b(\mathbf{k}) = \frac{1}{2}\mathbf{b}^*(\mathbf{k})\cdot\mathbf{b}(\mathbf{k}) = \frac{1}{2}[|b_+(\mathbf{k})|^2 + |b_-(\mathbf{k})|^2], \tag{20.72a}$$

$$H_M(\mathbf{k}) = \frac{1}{2}\Re[\mathbf{u}^*(\mathbf{k})\cdot\mathbf{a}(\mathbf{k})] = \frac{1}{2k}[|b_+(\mathbf{k})|^2 - |b_-(\mathbf{k})|^2]. \tag{20.72b}$$

To derive the equations of motion for $\mathbf{u}(\mathbf{k})$ in helical basis, we start with the Navier–Stokes equations with $\nu = 0$:

$$\begin{aligned}\frac{d}{dt}\mathbf{u}(\mathbf{k}') &= -i\mathbf{k}'p(\mathbf{k}') + \mathbf{u}(-\mathbf{q})\times\boldsymbol{\omega}(-\mathbf{p}) + \mathbf{u}(-\mathbf{p})\times\boldsymbol{\omega}(-\mathbf{q})\\ &\quad -\mathbf{b}(-\mathbf{q})\times\mathbf{J}(-\mathbf{p}) - \mathbf{b}(-\mathbf{p})\times\mathbf{j}(-\mathbf{q}),\end{aligned} \tag{20.73}$$

where

$$\mathbf{j}(\mathbf{k}) = i\mathbf{k}\times\mathbf{b}(\mathbf{k}). \tag{20.74}$$

Following the same procedure as in Section 9.7, we expand the fields $\mathbf{u}, \mathbf{j}, \mathbf{b}$ and $\boldsymbol{\omega}$ in helical basis, and take a dot product of the equation with helical basis vectors $\hat{e}^*_{s_{k'}}(\mathbf{k}')$. These operations yield the following equations (when $\nu = 0$):

$$\begin{aligned}\dot{u}_{s_k}(\mathbf{k}') &= [\mathbf{u}^*(\mathbf{q})\times\boldsymbol{\omega}^*(\mathbf{p}) + \mathbf{u}^*(\mathbf{p})\times\boldsymbol{\omega}^*(\mathbf{q})\\ &\quad -\mathbf{b}^*(\mathbf{q})\times\mathbf{j}^*(\mathbf{p}) - \mathbf{b}^*(\mathbf{p})\times\mathbf{j}^*(\mathbf{q})]\cdot\hat{e}^*_{s_{k'}}(\mathbf{k}')\\ &= \sum_{s_p,s_q}[ps_p\{u^*_{s_q}(\mathbf{q})u^*_{s_p}(\mathbf{p}) - b^*_{s_q}(\mathbf{q})b^*_{s_p}(\mathbf{p})\}\hat{e}^*_{s_q}(\mathbf{q})\times\hat{e}^*_{s_p}(\mathbf{p})\cdot\hat{e}^*_{s_{k'}}(\mathbf{k}')]\\ &\quad + \sum_{s_p,s_q}[qs_q\{u^*_{s_q}(\mathbf{q})u^*_{s_p}(\mathbf{p}) - b^*_{s_q}(\mathbf{q})b^*_{s_p}(\mathbf{p})\}\hat{e}^*_{s_p}(\mathbf{p})\times\hat{e}^*_{s_q}(\mathbf{q})\cdot\hat{e}^*_{s_{k'}}(\mathbf{k}')]\\ &= \sum_{s_p,s_q}(ps_p - qs_q)[u^*_{s_q}(\mathbf{q})u^*_{s_p}(\mathbf{p}) - b^*_{s_q}(\mathbf{q})b^*_{s_p}(\mathbf{p})]\hat{e}^*_{s_q}(\mathbf{q})\times\hat{e}^*_{s_p}(\mathbf{p})\cdot\hat{e}^*_{s_{k'}}(\mathbf{k}').\end{aligned} \tag{20.75}$$

Hence,

$$\dot{u}_{s_k}(\mathbf{k}') = \sum_{s_p,s_q} g(ps_p - qs_q)[u^*_{s_q}(\mathbf{q})u^*_{s_p}(\mathbf{p}) - b^*_{s_q}(\mathbf{q})b^*_{s_p}(\mathbf{p})], \tag{20.76}$$

where

$$\begin{aligned} g &= \hat{e}^*_{s_q}(\mathbf{q}) \times \hat{e}^*_{s_p}(\mathbf{p}) \cdot \hat{e}^*_{s_{k'}}(\mathbf{k}') \\ &= \frac{1}{2\sqrt{2}} s_{k'} s_p s_q \{ s_{k'} \sin\alpha + s_p \sin\beta + s_q \sin\gamma \}, \end{aligned} \tag{20.77}$$

with α, β, γ as the angles of the triangle formed by the wavenumbers $\mathbf{k}', \mathbf{p}, \mathbf{q}$. See Fig. 9.5 for an illustration.

To derive the equations for the magnetic modes, we start with the induction equation:

$$\begin{aligned} \frac{d}{dt}\mathbf{b}(\mathbf{k}') &= i\mathbf{k}' \times \boldsymbol{\varepsilon}(\mathbf{k}') \\ &= \sum_{s'_k} \varepsilon_{s_k}(k') i\mathbf{k}' \times \hat{\mathbf{e}}_{\mathbf{s_{k'}}}(\mathbf{k}') \\ &= \sum_{s'_k} k' s_{k'} \varepsilon_{s_k}(k') \hat{e}_{s_{k'}}(\mathbf{k}'), \end{aligned} \tag{20.78}$$

where $\boldsymbol{\varepsilon} = \mathbf{u} \times \mathbf{b}$ is the induction term. From this equation, we deduce that

$$\dot{b}_{s'_k}(\mathbf{k}') = k' s_{k'} \varepsilon_{s_k}(\mathbf{k}'). \tag{20.79}$$

Using $\varepsilon = \mathbf{u} \times \mathbf{b}$, we obtain

$$\begin{aligned} \varepsilon_{s_{k'}}(\mathbf{k}') &= \sum_{s_p,s_q} [\mathbf{u}^*(\mathbf{q}) \times \mathbf{b}^*(\mathbf{p}) + \mathbf{u}^*(\mathbf{p}) \times \mathbf{b}^*(\mathbf{q})] \cdot \hat{e}^*_{s_{k'}}(\mathbf{k}') \\ &= \sum_{s_p,s_q} g[u^*_{s_q}(\mathbf{q})b^*_{s_p}(\mathbf{p}) - u^*_{s_p}(\mathbf{p})b^*_{s_q}(\mathbf{q})], \end{aligned} \tag{20.80}$$

where g is given by Eq. (20.77). Therefore,

$$\dot{b}_{s_{k'}}(\mathbf{k}') = k' s_{k'} \sum_{s_p,s_q} g[u^*_{s_q}(\mathbf{q})b^*_{s_p}(\mathbf{p}) - u^*_{s_p}(\mathbf{p})b^*_{s_q}(\mathbf{q})]. \tag{20.81}$$

It is interesting to note that

$$\frac{1}{2}[|u_{s_{k'}}(\mathbf{k}')|^2 + |b_{s'_k}(\mathbf{k}')|^2 + |u_{s_p}(\mathbf{p})|^2 + |b_{s_p}(\mathbf{p})|^2 + |u_{s_q}(\mathbf{q})|^2 + |b_{s_q}(\mathbf{q})|^2] = \text{const.} \tag{20.82}$$

This is the statement of conservation of total energy under a triadic interaction.

We will revisit these equations in Chapter 23 when we discuss dynamos. In the next section we will describe the nondimensionalized version of MHD equations.

20.7 Nondimensionalized MHD Equations

We start with the MHD equations with a mean magnetic field $\mathbf{B}_0$:

$$\frac{\partial \mathbf{u}}{\partial t} - \mathbf{B}_0 \cdot \nabla \mathbf{b} + \mathbf{u} \cdot \nabla \mathbf{u} - \mathbf{b} \cdot \nabla \mathbf{b} = -\nabla p + \nu \nabla^2 \mathbf{u}, \tag{20.83a}$$

$$\frac{\partial \mathbf{b}}{\partial t} - \mathbf{B}_0 \cdot \nabla \mathbf{u} + \mathbf{u} \cdot \nabla \mathbf{b} - \mathbf{b} \cdot \nabla \mathbf{u} = \eta \nabla^2 \mathbf{B}. \tag{20.83b}$$

We nondimensionalize these equations using the system's large-scale velocity U_0 and length L:

$$\mathbf{r} \to L\mathbf{r}; \quad \mathbf{u} \to U_0 \mathbf{u}; \quad \mathbf{b} \to U_0 \mathbf{b}; \quad p \to (U_0^2)p. \tag{20.84}$$

Substitution of these terms in Eqs. (20.83) yields

$$\frac{\partial \mathbf{u}}{\partial t} - \frac{1}{U_0}\mathbf{B}_0 \cdot \nabla \mathbf{b} + \mathbf{u} \cdot \nabla \mathbf{u} - \mathbf{b} \cdot \nabla \mathbf{b} = -\nabla p + \frac{1}{\mathrm{Re}} \nabla^2 \mathbf{u}, \tag{20.85a}$$

$$\mathrm{Rm}\left[\frac{\partial \mathbf{b}}{\partial t} + \mathbf{u} \cdot \nabla \mathbf{b} - \mathbf{b} \cdot \nabla \mathbf{u}\right] = \frac{\mathbf{B}_0 L}{\eta} \cdot \nabla \mathbf{u} + \nabla^2 \mathbf{b}, \tag{20.85b}$$

where

$$\mathrm{Re} = \frac{\mathbf{u} \cdot \nabla \mathbf{u}}{\nu \nabla^2 \mathbf{u}} = \frac{U_0 L}{\nu}, \tag{20.86}$$

$$\mathrm{Rm} = \frac{\mathbf{u} \cdot \nabla \mathbf{b}}{\eta \nabla^2 \mathbf{b}} = \frac{U_0 L}{\eta}, \tag{20.87}$$

are *Reynolds number* and *magnetic Reynolds number* respectively.

In this book we consider the following two broad categories of flows:

1. $\mathbf{B}_0 = 0$: For this case, the equations are

$$\frac{\partial \mathbf{u}}{\partial t} + \mathbf{u} \cdot \nabla \mathbf{u} = -\nabla p + \mathbf{b} \cdot \nabla \mathbf{b} + \frac{1}{\mathrm{Re}} \nabla^2 \mathbf{u}, \tag{20.88a}$$

$$\frac{\partial \mathbf{b}}{\partial t} + \mathbf{u} \cdot \nabla \mathbf{b} = \mathbf{b} \cdot \nabla \mathbf{u} + \frac{1}{\mathrm{Rm}} \nabla^2 \mathbf{b}, \tag{20.88b}$$

2. $\mathrm{Rm} = 0$: This case, called *quasi-static MHD*, has $b \ll B_0$, and the resulting equations are

$$\frac{\partial \mathbf{u}}{\partial t} + \mathbf{u} \cdot \nabla \mathbf{u} = -\nabla p + \frac{1}{U_0}\mathbf{B}_0 \cdot \nabla \mathbf{b} + \frac{1}{\mathrm{Re}} \nabla^2 \mathbf{u}, \tag{20.89a}$$

$$\frac{\mathbf{B}_0 L}{\eta} \cdot \nabla \mathbf{u} + \nabla^2 \mathbf{b} = 0. \tag{20.89b}$$

We will analyze these equations in Chapter 24.

Based on Re and Rm, we can classify MHD flows into four categories:

1. Re $\gg$ 1, Rm $\gg$ 1: Turbulent MHD in which all the nonlinear terms are significant. See Eqs. (20.88)
2. Re $\gg$ 1, Rm $\to$ 0: For such flows, the nonlinear terms of the induction equation can be ignored. This is quasi-static MHD limit. See Eqs. (20.89).
3. Re $\lessapprox$ 1, Rm $\gg$ 1: The nonlinear terms of the induction equation is significant, but those of the Navier–Stokes are not.
4. Re $\lessapprox$ 1, Rm $\lessapprox$ 1: Laminar MHD in which all the nonlinear terms are negligible.

We will discuss these cases in the subsequent chapters.

Further Reading

There are many books and review articles on MHD formalism. For the topics discussed in this book, we refer the reader to Biskamp (2003) and Verma (2004). More articles and books will be referred to in the subsequent chapters.

Exercises

1. Derive evolution equations for the kinetic helicity and enstrophy. Show that these quantities are not conserved.
2. Derive the conservation laws of MHD using the equations in Fourier space.

Chapter 21

Energy Transfers in MHD

In the present chapter we describe various mode-to-mode energy transfers and energy fluxes of MHD. These quantities are very useful for the construction of turbulence phenomenologies, for understanding experimental and observational data, and for understanding dynamo mechanism. These applications will be described in later chapters.

We start with the combined energy transfers in MHD.

21.1 Combined Energy Transfers in MHD

We start with the equations for kinetic and magnetic energies without dissipation (see Eqs. (20.48)):

$$
\begin{aligned}
\frac{d}{dt}E_u(\mathbf{k}) &= -\Re[\mathbf{N}_u(\mathbf{k})\cdot\mathbf{u}^*(\mathbf{k})] + \Re[\mathbf{F}_u(\mathbf{k})\cdot\mathbf{u}^*(\mathbf{k})] \\
&= \sum_{\mathbf{p}} \Im\left[\{\mathbf{k}\cdot\mathbf{u}(\mathbf{q})\}\{\mathbf{u}(\mathbf{p})\cdot\mathbf{u}^*(\mathbf{k})\} - \{\mathbf{k}\cdot\mathbf{B}(\mathbf{q})\}\{\mathbf{B}(\mathbf{p})\cdot\mathbf{u}^*(\mathbf{k})\}\right], \quad (21.1a) \\
\frac{d}{dt}E_B(\mathbf{k}) &= -\Re[\mathbf{N}_B(\mathbf{k})\cdot\mathbf{B}^*(\mathbf{k})] + \Re[\mathbf{F}_B(\mathbf{k})\cdot\mathbf{B}^*(\mathbf{k})] \\
&= \sum_{\mathbf{p}} \Im\left[\{\mathbf{k}\cdot\mathbf{u}(\mathbf{q})\}\{\mathbf{B}(\mathbf{p})\cdot\mathbf{B}^*(\mathbf{k})\} - \{\mathbf{k}\cdot\mathbf{B}(\mathbf{q})\}\{\mathbf{u}(\mathbf{p})\cdot\mathbf{B}^*(\mathbf{k})\}\right], (21.1b)
\end{aligned}
$$

where $\mathbf{k} = \mathbf{p} + \mathbf{q}$. Here, $\nu = \eta = 0$.

Now we focus on a pair of triads, $(\mathbf{k}', \mathbf{p}, \mathbf{q})$ and $(-\mathbf{k}', -\mathbf{p}, -\mathbf{q})$, with $\mathbf{k}' = -\mathbf{k}$. See Fig. 4.1 for an illustration. For these modes, the aforementioned equations yield

$$\frac{d}{dt}E_u(\mathbf{k}') = S^{uu}(\mathbf{k}'|\mathbf{p},\mathbf{q}) + S^{ub}(\mathbf{k}'|\mathbf{p},\mathbf{q}), \tag{21.2a}$$

$$\frac{d}{dt}E_b(\mathbf{k}') = S^{bb}(\mathbf{k}'|\mathbf{p},\mathbf{q}) + S^{bu}(\mathbf{k}'|\mathbf{p},\mathbf{q}), \tag{21.2b}$$

where

$$S^{uu}(\mathbf{k}'|\mathbf{p},\mathbf{q}) = -\Im\left[\{\mathbf{k}'\cdot\mathbf{u}(\mathbf{q})\}\{\mathbf{u}(\mathbf{p})\cdot\mathbf{u}(\mathbf{k}')\}\right] - \Im\left[\{\mathbf{k}'\cdot\mathbf{u}(\mathbf{p})\}\{\mathbf{u}(\mathbf{q})\cdot\mathbf{u}(\mathbf{k}')\}\right], \tag{21.3a}$$

$$S^{bb}(\mathbf{k}'|\mathbf{p},\mathbf{q}) = -\Im\left[\{\mathbf{k}'\cdot\mathbf{u}(\mathbf{q})\}\{\mathbf{B}(\mathbf{p})\cdot\mathbf{B}(\mathbf{k}')\}\right] - \Im\left[\{\mathbf{k}'\cdot\mathbf{u}(\mathbf{p})\}\{\mathbf{B}(\mathbf{q})\cdot\mathbf{B}(\mathbf{k}')\}\right], \tag{21.3b}$$

$$S^{ub}(\mathbf{k}'|\mathbf{p},\mathbf{q}) = \Im\left[\{\mathbf{k}'\cdot\mathbf{B}(\mathbf{q})\}\{\mathbf{B}(\mathbf{p})\cdot\mathbf{u}(\mathbf{k}')\}\right] + \Im\left[\{\mathbf{k}'\cdot\mathbf{B}(\mathbf{p})\}\{\mathbf{B}(\mathbf{q})\cdot\mathbf{u}(\mathbf{k}')\}\right], \tag{21.3c}$$

$$S^{bu}(\mathbf{k}'|\mathbf{p},\mathbf{q}) = \Im\left[\{\mathbf{k}'\cdot\mathbf{B}(\mathbf{q})\}\{\mathbf{u}(\mathbf{p})\cdot\mathbf{B}(\mathbf{k}')\}\right] + \Im\left[\{\mathbf{k}'\cdot\mathbf{B}(\mathbf{p})\}\{\mathbf{u}(\mathbf{q})\cdot\mathbf{B}(\mathbf{k}')\}\right]. \tag{21.3d}$$

These are combined energy transfers to wavenumber $\mathbf{k}'$ from the other two wavenumbers $\mathbf{p}$ and $\mathbf{q}$, as described in Table 21.1. Similar equations can be derived for wavenumbers $\mathbf{p}$ and $\mathbf{q}$. Note that the energy transfer $S^{uu}(\mathbf{k}'|\mathbf{p},\mathbf{q})$ has been discussed in Chapter 4, while $S^{bb}(\mathbf{k}'|\mathbf{p},\mathbf{q})$ has been discussed in Chapter 18 (with $\mathbf{w} \to \mathbf{b}$); in these chapters we showed the following conservation laws:

$$S^{uu}(\mathbf{k}'|\mathbf{p},\mathbf{q}) + S^{uu}(\mathbf{p}|\mathbf{k}',\mathbf{q}) + S^{uu}(\mathbf{q}|\mathbf{k}',\mathbf{p}) = 0, \tag{21.4a}$$

$$S^{bb}(\mathbf{k}'|\mathbf{p},\mathbf{q}) + S^{bb}(\mathbf{p}|\mathbf{k}',\mathbf{q}) + S^{bb}(\mathbf{q}|\mathbf{k}',\mathbf{p}) = 0. \tag{21.4b}$$

These are the *laws of detailed conservation*; Eq. (21.4a) is related to kinetic energy exchanges among the wavenumbers of the triad via uu channel, while Eq. (21.4b) is related to the corresponding exchanges of the magnetic energy via bb channel.

Table 21.1 Summary of various combined energy transfers in MHD turbulence.

ET	Receiver mode	Giver modes
$S^{uu}(\mathbf{X}\|\mathbf{Y},\mathbf{Z})$	$\mathbf{u}(\mathbf{X})$	$\mathbf{u}(\mathbf{Y}), \mathbf{u}(\mathbf{Z})$
$S^{bb}(\mathbf{X}\|\mathbf{Y},\mathbf{Z})$	$\mathbf{B}(\mathbf{X})$	$\mathbf{B}(\mathbf{Y}), \mathbf{B}(\mathbf{Z})$
$S^{ub}(\mathbf{X}\|\mathbf{Y},\mathbf{Z})$	$\mathbf{u}(\mathbf{X})$	$\mathbf{B}(\mathbf{Y}), \mathbf{B}(\mathbf{Z})$
$S^{bu}(\mathbf{X}\|\mathbf{Y},\mathbf{Z})$	$\mathbf{B}(\mathbf{X})$	$\mathbf{u}(\mathbf{Y}), \mathbf{u}(\mathbf{Z})$

In addition to the aforementioned equations, we also have the following equation for the triad:

$$S^{ub}(\mathbf{k}'|\mathbf{p},\mathbf{q}) + S^{ub}(\mathbf{p}|\mathbf{k}',\mathbf{q}) + S^{ub}(\mathbf{q}|\mathbf{k}',\mathbf{p})$$

$$+S^{bu}(\mathbf{k}'|\mathbf{p},\mathbf{q}) + S^{bu}(\mathbf{p}|\mathbf{k}',\mathbf{q}) + S^{bu}(\mathbf{q}|\mathbf{k}',\mathbf{p}) = 0. \tag{21.5}$$

This statement, the third *detailed energy conservation law*, is related to the energy exchanges among $\mathbf{u}$ and $\mathbf{B}$ modes. Note however that the total magnetic energy in a triad can change via *bu* transfer since

$$S^{bu}(\mathbf{k}'|\mathbf{p},\mathbf{q}) + S^{bu}(\mathbf{p}|\mathbf{k}',\mathbf{q}) + S^{bu}(\mathbf{q}|\mathbf{k}',\mathbf{p}) \neq 0. \tag{21.6}$$

Similarly, the total kinetic energy changes via the following *ub* transfers:

$$S^{ub}(\mathbf{k}'|\mathbf{p},\mathbf{q}) + S^{ub}(\mathbf{p}|\mathbf{k}',\mathbf{q}) + S^{ub}(\mathbf{q}|\mathbf{k}',\mathbf{p}) \neq 0. \tag{21.7}$$

But, the total energy is conserved when we add the S^{bu} and S^{ub} transfers. These results lead to the conservation of the total energy $\sum_{\mathbf{k}}[E_u(\mathbf{k}) + E_B(\mathbf{k})]$ in a triad.

Now we are ready to derive the mode-to-mode energy transfers in MHD.

21.2 Mode-to-mode Energy Transfers in MHD

In MHD, the active modes in a triad are $\mathbf{u}(\mathbf{k}'), \mathbf{u}(\mathbf{p}), \mathbf{u}(\mathbf{q}), \mathbf{B}(\mathbf{k}'), \mathbf{B}(\mathbf{p})$, and $\mathbf{B}(\mathbf{q})$. In this section we will derive mode-to-mode energy transfers from velocity to velocity, magnetic to magnetic, velocity to magnetic, and magnetic to velocity modes, which are illustrated in Fig. 21.1.

In MHD, the kinetic energy transfers among the velocity modes are facilitated by the nonlinear term S^{uu} of Eq. (21.3a), whose structure is the same as those for hydrodynamic turbulence (see Chapter 4). Therefore, the mode-to-mode kinetic energy transfer from $\mathbf{u}(\mathbf{p})$ to $\mathbf{u}(\mathbf{k}')$ with the mediation of $\mathbf{u}(\mathbf{q})$ is given by the following expression (see Section 4.1 for derivation):

$$S^{uu}(\mathbf{k}'|\mathbf{p}|\mathbf{q}) = -\Im\left[\{\mathbf{k}'\cdot\mathbf{u}(\mathbf{q})\}\{\mathbf{u}(\mathbf{p})\cdot\mathbf{u}(\mathbf{k}')\}\right]. \tag{21.8}$$

Similarly, the magnetic energy exchange via the nonlinear term S^{bb} has the same structure as that for vector flow (see Chapter 18). Therefore, as discussed in Section 18.2, the magnetic energy transfer from $\mathbf{B}(\mathbf{p})$ to $\mathbf{B}(\mathbf{k}')$ with the mediation of $\mathbf{u}(\mathbf{q})$ is

$$S^{bb}(\mathbf{k}'|\mathbf{p}|\mathbf{q}) = -\Im\left[\{\mathbf{k}'\cdot\mathbf{u}(\mathbf{q})\}\{\mathbf{B}(\mathbf{p})\cdot\mathbf{B}(\mathbf{k}')\}\right]. \tag{21.9}$$

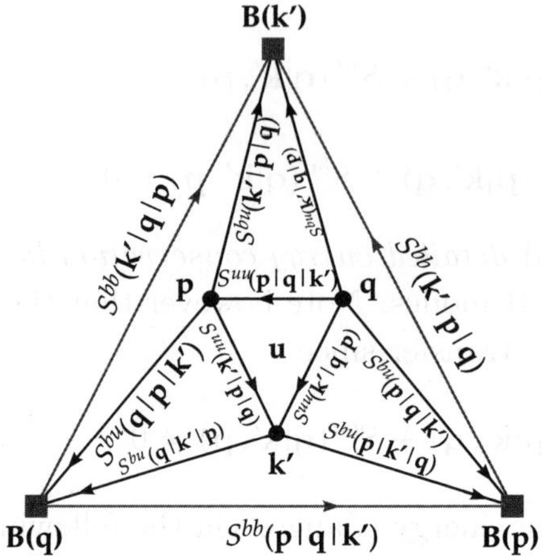

Figure 21.1 In a triad $(\mathbf{k}', \mathbf{p}, \mathbf{q})$ with $\mathbf{k}' + \mathbf{p} + \mathbf{q} = 0$, circles and squares represent the velocity and magnetic modes respectively. In the figure, S^{uu}s are the mode-to-mode kinetic energy transfers between velocity modes, S^{bb}s are the magnetic energy transfers between magnetic modes, and S^{bu}s are the energy transfers from velocity modes to magnetic modes. The magnetic-to-velocity transfers S^{ub} are not shown in the figure. Note that $S^{ub}(\mathbf{k}'|\mathbf{p}|\mathbf{q}) = -S^{bu}(\mathbf{p}|\mathbf{k}'|\mathbf{q})$.

These mode-to-mode energy transfers are exhibited in Fig. 21.2(a,b). In both these kinds of transfers, the velocity mode $\mathbf{u}(\mathbf{q})$ acts a mediator since it advects the modes (see Sections 4.1, 18.2 for detailed arguments). For convenience, the energy transfers S^{uu} and S^{bb} are denoted by $U2U$ and $B2B$ respectively.

As discussed in the previous section, mode $\mathbf{u}(\mathbf{k}')$ receives energy from the magnetic modes via S^{ub} terms. Similarly, mode $\mathbf{B}(\mathbf{k}')$ receives energy from the velocity modes via S^{bu} terms. Let us try to construct formulas for the velocity-to-magnetic and magnetic-to-velocity mode-to-mode energy transfers, which are denoted by S^{bu} and S^{ub} respectively. Figures 21.2(c, d) illustrate these transfers.

For convenience, we denote the wavenumbers in the triad as $\mathbf{X}, \mathbf{Y}, \mathbf{Z}$ with $\mathbf{X} + \mathbf{Y} + \mathbf{Z} = 0$. We denote the mode-to-mode energy transfer from $\mathbf{u}(\mathbf{Y})$ to $\mathbf{B}(\mathbf{X})$ with the mediation of mode $\mathbf{B}(\mathbf{Z})$ as $S^{bu}(\mathbf{X}|\mathbf{Y}|\mathbf{Z})$, which are illustrated in Fig. 21.1. The energy transfer from $\mathbf{B}(\mathbf{Y})$ to $\mathbf{u}(\mathbf{X})$ with the mediation of mode $\mathbf{B}(\mathbf{Z})$ is denoted by $S^{ub}(\mathbf{X}|\mathbf{Y}|\mathbf{Z})$.

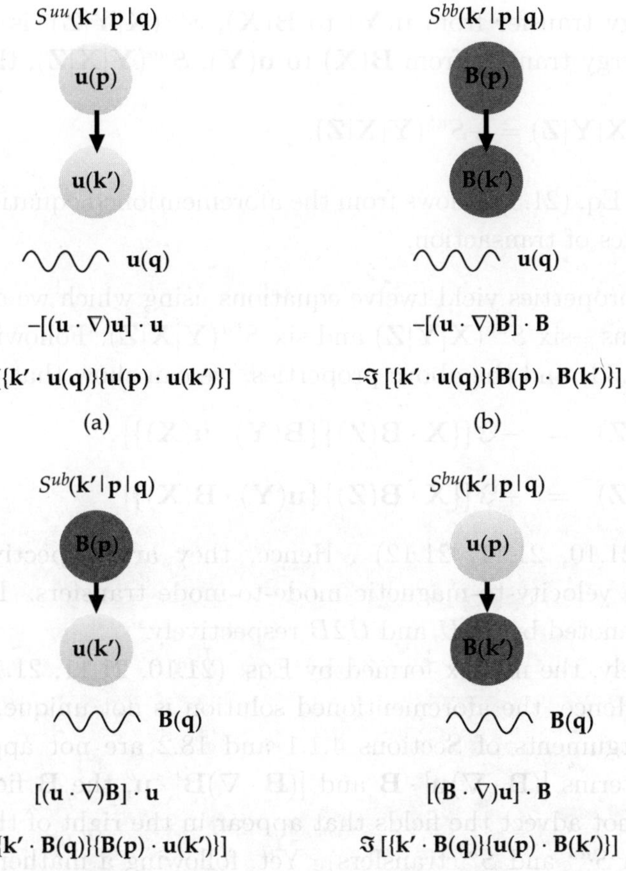

Figure 21.2 A schematic diagram exhibiting mode-to-mode energy transfers in MHD turbulence: (a) $U2U$, (b) $B2B$, (c) $B2U$, (d) $U2B$. Here the wavy lines represent the mediator modes.

The functions $S^{ub}(\mathbf{X}|\mathbf{Y}|\mathbf{Z})$ and $S^{bu}(\mathbf{X}|\mathbf{Y}|\mathbf{Z})$ satisfy the following properties:

1. The sum of $S^{ub}(\mathbf{X}|\mathbf{Y}|\mathbf{Z})$ and $S^{ub}(\mathbf{X}|\mathbf{Z}|\mathbf{Y})$ is the combined energy transfer $S^{ub}(\mathbf{X}|\mathbf{Y},\mathbf{Z})$. That is,

$$S^{ub}(\mathbf{X}|\mathbf{Y}|\mathbf{Z}) + S^{ub}(\mathbf{X}|\mathbf{Z}|\mathbf{Y}) = S^{ub}(\mathbf{X}|\mathbf{Y},\mathbf{Z}). \tag{21.10}$$

Similarly, for S^{bu},

$$S^{bu}(\mathbf{X}|\mathbf{Y}|\mathbf{Z}) + S^{bu}(\mathbf{X}|\mathbf{Z}|\mathbf{Y}) = S^{bu}(\mathbf{X}|\mathbf{Y},\mathbf{Z}). \tag{21.11}$$

2. The energy transfer from $\mathbf{u}(\mathbf{Y})$ to $\mathbf{B}(\mathbf{X})$, $S^{bu}(\mathbf{X}|\mathbf{Y}|\mathbf{Z})$, is equal and opposite to the energy transfer from $\mathbf{B}(\mathbf{X})$ to $\mathbf{u}(\mathbf{Y})$, $S^{ub}(\mathbf{Y}|\mathbf{X}|\mathbf{Z})$; that is,

$$S^{bu}(\mathbf{X}|\mathbf{Y}|\mathbf{Z}) = -S^{ub}(\mathbf{Y}|\mathbf{X}|\mathbf{Z}). \tag{21.12}$$

Note that Eq. (21.5) follows from the aforementioned equation, which are based on the rules of transaction.

The above properties yield twelve equations using which we attempt to solve for twelve unknowns—six $S^{ub}(\mathbf{X}|\mathbf{Y}|\mathbf{Z})$ and six $S^{bu}(\mathbf{Y}|\mathbf{X}|\mathbf{Z})$. Following the structure in Eqs. (21.3c, 21.3d) and the above properties, we can show that

$$\begin{aligned} S^{ub}(\mathbf{X}|\mathbf{Y}|\mathbf{Z}) &= -\Im\left[\{\mathbf{X}\cdot\mathbf{B}(\mathbf{Z})\}\{\mathbf{B}(\mathbf{Y})\cdot\mathbf{u}(\mathbf{X})\}\right], \\ S^{bu}(\mathbf{X}|\mathbf{Y}|\mathbf{Z}) &= -\Im\left[\{\mathbf{X}\cdot\mathbf{B}(\mathbf{Z})\}\{\mathbf{u}(\mathbf{Y})\cdot\mathbf{B}(\mathbf{X})\}\right] \end{aligned} \tag{21.13a}$$

satisfy Eqs. (21.10, 21.11, 21.12) . Hence, they are respectively the magnetic-to-velocity and velocity-to-magnetic mode-to-mode transfers. In shorthand, these transfers are denoted by $B2U$ and $U2B$ respectively.

Unfortunately, the matrix formed by Eqs. (21.10, 21.11, 21.12) has a vanishing determinant. Hence, the aforementioned solution is not unique. Strictly speaking, the physical arguments of Sections 4.1.1 and 18.2 are not applicable because in the nonlinear terms $[(\mathbf{B}\cdot\nabla)\mathbf{u}]\cdot\mathbf{B}$ and $[(\mathbf{B}\cdot\nabla)\mathbf{B}]\cdot\mathbf{u}$, the $\mathbf{B}$ field in the left of ∇ operator does not advect the fields that appear in the right of the ∇ (unlike $\mathbf{u}$ field that advects in S^{uu} and S^{bb} transfers). Yet, following a mathematical analogy, we claim that the $\mathbf{B}$ field in the left of ∇ operator in $[(\mathbf{B}\cdot\nabla)\mathbf{u}]\cdot\mathbf{B}$ and $[(\mathbf{B}\cdot\nabla)\mathbf{B}]\cdot\mathbf{u}$ advects the modes that appear in the right of ∇ operator.

In addition, we provide a convincing mathematical derivation in the lines of Section 4.1.2. From the structure of the nonlinear terms of Eqs. (21.3) we demand that $S^{ub}(\mathbf{k}'|\mathbf{p}|\mathbf{q})$ and $S^{bu}(\mathbf{k}'|\mathbf{p}|\mathbf{q})$ satisfy the following properties:

1. $S^{ub}(\mathbf{k}'|\mathbf{p}|\mathbf{q})$ and $S^{bu}(\mathbf{k}'|\mathbf{p}|\mathbf{q})$ are real.
2. $S^{ub}(\mathbf{k}'|\mathbf{p}|\mathbf{q})$ and $S^{bu}(\mathbf{k}'|\mathbf{p}|\mathbf{q})$ are linear functions of the wave vector $\mathbf{k}'$.
3. $S^{ub}(\mathbf{k}'|\mathbf{p}|\mathbf{q})$ and $S^{bu}(\mathbf{k}'|\mathbf{p}|\mathbf{q})$ are linear functions of the Fourier modes $\mathbf{u}(\mathbf{k}')$, $\mathbf{u}(\mathbf{p})$, $\mathbf{u}(\mathbf{q})$, $\mathbf{B}(\mathbf{k}')$, $\mathbf{B}(\mathbf{p})$, $\mathbf{B}(\mathbf{q})$ with the following conditions:
   (a) One velocity mode and two magnetic modes appear in the function.
   (b) The wavenumbers of the Fourier modes must be distinct.
4. Due to the equivalence of the triads $(\mathbf{k},\mathbf{p},\mathbf{q})$ and $(-\mathbf{k},-\mathbf{p},-\mathbf{q})$,

$$S^{ub}(-\mathbf{k}'|-\mathbf{p}|-\mathbf{q}) = S^{ub}(\mathbf{k}'|\mathbf{p}|\mathbf{q}). \tag{21.14}$$

Using these properties and Eqs. (21.10, 21.11, 21.12), we derive the formulas for $S^{ub}(\mathbf{k}'|\mathbf{p}|\mathbf{q})$ and $S^{bu}(\mathbf{k}'|\mathbf{p}|\mathbf{q})$. The function $S^{ub}(\mathbf{k}'|\mathbf{p}|\mathbf{q})$ involving two magnetic modes and a single velocity mode is of the following form:

$$\begin{aligned} S^{ub}(\mathbf{k}'|\mathbf{p}|\mathbf{q}) &= c_1\Im\left[\{\mathbf{k}'\cdot\mathbf{u}(\mathbf{q})\}\{\mathbf{B}(\mathbf{p})\cdot\mathbf{B}(\mathbf{k}')\}\right] \\ &+ c_2\Im\left[\{\mathbf{k}'\cdot\mathbf{u}(\mathbf{p})\}\{\mathbf{B}(\mathbf{q})\cdot\mathbf{B}(\mathbf{k}')\}\right] \\ &+ c_3\Im\left[\{\mathbf{k}'\cdot\mathbf{B}(\mathbf{q})\}\{\mathbf{B}(\mathbf{k}')\cdot\mathbf{u}(\mathbf{p})\}\right] \\ &+ c_4\Im\left[\{\mathbf{k}'\cdot\mathbf{B}(\mathbf{p})\}\{\mathbf{B}(\mathbf{k}')\cdot\mathbf{u}(\mathbf{q})\}\right] \\ &+ c_5\Im\left[\{\mathbf{k}'\cdot\mathbf{B}(\mathbf{q})\}\{\mathbf{B}(\mathbf{p})\cdot\mathbf{u}(\mathbf{k}')\}\right] \\ &+ c_6\Im\left[\{\mathbf{k}'\cdot\mathbf{B}(\mathbf{p})\}\{\mathbf{B}(\mathbf{q})\cdot\mathbf{u}(\mathbf{k}')\}\right]. \end{aligned} \tag{21.15}$$

Here we have dropped terms involving $\mathbf{k}'\cdot\mathbf{u}(\mathbf{k}')$ and $\mathbf{k}'\cdot\mathbf{b}(\mathbf{k}')$ because they are zeros. Following Eq. (21.9), the first two terms in this expression are $S^{bb}(\mathbf{k}'|\mathbf{p}|\mathbf{q})$ and $S^{bb}(\mathbf{k}'|\mathbf{q}|\mathbf{p})$ respectively. Hence, we drop them from the expression.

For $S^{ub}(\mathbf{k}'|\mathbf{p}|\mathbf{q})$, a closer examination of Eq. (21.3c) reveals that the receiver mode is $\mathbf{u}(\mathbf{k}')$, and it appears in the dot product with one of the magnetic modes. Hence,

$$\begin{aligned} S^{ub}(\mathbf{k}'|\mathbf{p}|\mathbf{q}) &= c_5\Im\left[\{\mathbf{k}'\cdot\mathbf{B}(\mathbf{q})\}\{\mathbf{B}(\mathbf{p})\cdot\mathbf{u}(\mathbf{k}')\}\right] \\ &+ c_6\Im\left[\{\mathbf{k}'\cdot\mathbf{B}(\mathbf{p})\}\{\mathbf{B}(\mathbf{q})\cdot\mathbf{u}(\mathbf{k}')\}\right]. \end{aligned} \tag{21.16}$$

In real space, the nonlinear term corresponding to $S^{ub}(\mathbf{k}'|\mathbf{p}|\mathbf{q})$ is $[(\mathbf{B}\cdot\nabla)\mathbf{B}]\cdot\mathbf{u} = u_i\partial_j(B_jB_i)$. Here, the receiver, giver, and mediator fields are u_i, B_i, and B_j respectively. Hence, in Eq. (21.16), $\mathbf{B}(\mathbf{q})$ must be the mediator mode. Therefore, $c_6 = 0$, and

$$S^{ub}(\mathbf{k}'|\mathbf{p}|\mathbf{q}) = c_5\Im\left[\{\mathbf{k}'\cdot\mathbf{B}(\mathbf{q})\}\{\mathbf{B}(\mathbf{p})\cdot\mathbf{u}(\mathbf{k}')\}\right]. \tag{21.17}$$

Substitution of this term in Eq. (21.10) yields $c_5 = 1$. Therefore,

$$S^{ub}(\mathbf{k}'|\mathbf{p}|\mathbf{q}) = \Im\left[\{\mathbf{k}'\cdot\mathbf{B}(\mathbf{q})\}\{\mathbf{B}(\mathbf{p})\cdot\mathbf{u}(\mathbf{k}')\}\right] \tag{21.18}$$

is the mode-to-mode energy transfer from mode $\mathbf{B}(\mathbf{p})$ to mode $\mathbf{u}(\mathbf{k}')$ with the mediation of $\mathbf{B}(\mathbf{q})$.

Similarly for $S^{bu}(\mathbf{k}'|\mathbf{p}|\mathbf{q})$, using Eq. (21.3d) we deduce that $\mathbf{B}(\mathbf{k}')$ is the receiver mode. Hence,

$$\begin{aligned} S^{bu}(\mathbf{k}'|\mathbf{p}|\mathbf{q}) &= c_3\Im\left[\{\mathbf{k}'\cdot\mathbf{B}(\mathbf{q})\}\{\mathbf{B}(\mathbf{k}')\cdot\mathbf{u}(\mathbf{p})\}\right] \\ &+ c_4\Im\left[\{\mathbf{k}'\cdot\mathbf{B}(\mathbf{p})\}\{\mathbf{B}(\mathbf{k}')\cdot\mathbf{u}(\mathbf{q})\}\right]. \end{aligned} \tag{21.19}$$

Note that $S^{bu}(\mathbf{k}'|\mathbf{p}|\mathbf{q})$ is related to the nonlinear term $[(\mathbf{B}\cdot\nabla)\mathbf{u}]\cdot\mathbf{B} = B_i\partial_j(B_j u_i)$, for which B_i, u_i, and B_j are respectively the receiver, giver, and mediator modes. Hence, $c_4 = 0$, and $c_3 = 1$ [following Eq. (21.11)]. Therefore,

$$S^{bu}(\mathbf{k}'|\mathbf{p}|\mathbf{q}) = \Im\left[\{\mathbf{k}'\cdot\mathbf{B}(\mathbf{q})\}\{\mathbf{u}(\mathbf{p})\cdot\mathbf{B}(\mathbf{k}')\}\right] \tag{21.20}$$

is the mode-to-mode magnetic energy transfer from mode $\mathbf{u}(\mathbf{p})$ to mode $\mathbf{B}(\mathbf{k}')$ with the mediation of $\mathbf{B}(\mathbf{q})$. The mode-to-mode energy transfers $S^{ub}(\mathbf{k}'|\mathbf{p}|\mathbf{q})$ and $S^{bu}(\mathbf{k}'|\mathbf{p}|\mathbf{q})$ are illustrated in Fig. 21.2(c, d).

Following similar lines of arguments, we can also derive mode-to-mode energy transfers among the Elsäser variables $\mathbf{z}^+$ and $\mathbf{z}^-$. This is the topic of the next section.

21.3 Mode-to-mode Transfers for Elsäser Variables

The equations for the modal energy of $\mathbf{z}^{\pm}$ variables for inviscid MHD are [see Eqs. (20.49)]

$$\frac{d}{dt}E_{z^\pm}(\mathbf{k}) = \sum_{\mathbf{p}}\Im\left[\{\mathbf{k}\cdot\mathbf{z}^{\mp}(\mathbf{q})\}\{\mathbf{z}^{\pm}(\mathbf{p})\cdot\mathbf{z}^{\pm*}(\mathbf{k})\}\right]. \tag{21.21}$$

For the triads $(\mathbf{k}',\mathbf{p},\mathbf{q})$ and $(-\mathbf{k}',-\mathbf{p},-\mathbf{q})$ with $\mathbf{k}'=-\mathbf{k}$, Eq. (21.21) yields

$$\frac{d}{dt}E_{z^+}(\mathbf{k}) = S^{z^+z^+}(\mathbf{k}'|\mathbf{p},\mathbf{q}), \tag{21.22a}$$

$$\frac{d}{dt}E_{z^-}(\mathbf{k}) = S^{z^-z^-}(\mathbf{k}'|\mathbf{p},\mathbf{q}), \tag{21.22b}$$

where

$$\begin{aligned} S^{z^+z^+}(\mathbf{k}'|\mathbf{p},\mathbf{q}) &= -\Im\left[\{\mathbf{k}'\cdot\mathbf{z}^-(\mathbf{q})\}\{\mathbf{z}^+(\mathbf{p})\cdot\mathbf{z}^+(\mathbf{k}')\}\right] \\ &\quad -\Im\left[\{\mathbf{k}'\cdot\mathbf{z}^-(\mathbf{p})\}\{\mathbf{z}^+(\mathbf{q})\cdot\mathbf{z}^+(\mathbf{k}')\}\right], \end{aligned} \tag{21.23a}$$

$$\begin{aligned} S^{z^-z^-}(\mathbf{k}'|\mathbf{p},\mathbf{q}) &= -\Im\left[\{\mathbf{k}'\cdot\mathbf{z}^+(\mathbf{q})\}\{\mathbf{z}^-(\mathbf{p})\cdot\mathbf{z}^-(\mathbf{k}')\}\right] \\ &\quad -\Im\left[\{\mathbf{k}'\cdot\mathbf{z}^+(\mathbf{p})\}\{\mathbf{z}^-(\mathbf{q})\cdot\mathbf{z}^-(\mathbf{k}')\}\right] \end{aligned} \tag{21.23b}$$

are the combined energy transfers. Using these equations and the condition that $\mathbf{k}\cdot\mathbf{z}^{\pm}(\mathbf{k}) = 0$, we derive the following conservation laws for the triad:

$$S^{z^+z^+}(\mathbf{k}'|\mathbf{p},\mathbf{q}) + S^{z^+z^+}(\mathbf{p}|\mathbf{q},\mathbf{k}') + S^{z^+z^+}(\mathbf{q}|\mathbf{k}',\mathbf{p}) = 0, \tag{21.24a}$$

$$S^{z^-z^-}(\mathbf{k}'|\mathbf{p},\mathbf{q}) + S^{z^-z^-}(\mathbf{p}|\mathbf{q},\mathbf{k}') + S^{z^-z^-}(\mathbf{q}|\mathbf{k}',\mathbf{p}) = 0. \tag{21.24b}$$

Thus, E_{z^+} and E_{z^-} are individually conserved for MHD interactions in a triad.

In the following discussion, we construct mode-to-mode transfers $S^{z^+z^+}(\mathbf{k}'|\mathbf{p}|\mathbf{q})$ and $S^{z^-z^-}(\mathbf{k}'|\mathbf{p},\mathbf{q})$. Following the arguments of the previous section, we show that

$$S^{z^+z^+}(\mathbf{X}|\mathbf{Y}|\mathbf{Z}) + S^{z^+z^+}(\mathbf{X}|\mathbf{Z}|\mathbf{Y}) = S^{z^+z^+}(\mathbf{X}|\mathbf{Y},\mathbf{Z}), \tag{21.25a}$$

$$S^{z^+z^+}(\mathbf{X}|\mathbf{Y}|\mathbf{Z}) = -S^{z^+z^+}(\mathbf{Y}|\mathbf{X}|\mathbf{Z}). \tag{21.25b}$$

We obtain similar equations for $S^{z^-z^-}$. A solution for Eqs. (21.25) is as follows:

$$S^{z^+z^+}(\mathbf{k}'|\mathbf{p}|\mathbf{q}) = -\Im\left[\{\mathbf{k}'\cdot\mathbf{z}^-(\mathbf{q})\}\{\mathbf{z}^+(\mathbf{p})\cdot\mathbf{z}^+(\mathbf{k}')\}\right], \tag{21.26a}$$

$$S^{z^-z^-}(\mathbf{k}'|\mathbf{p}|\mathbf{q}) = -\Im\left[\{\mathbf{k}'\cdot\mathbf{z}^+(\mathbf{q})\}\{\mathbf{z}^-(\mathbf{p})\cdot\mathbf{z}^-(\mathbf{k}')\}\right]. \tag{21.26b}$$

Unfortunately, the determinant of the matrix formed by Eqs. (21.25a, 21.25b) is zero. Hence, we invoke physical and mathematical arguments to prove uniqueness of this solution.

Physically, in Eq. (21.26a), mode $\mathbf{z}^-(\mathbf{q})$ advects $\mathbf{z}^+(\mathbf{p})$ and $\mathbf{z}^+(\mathbf{k}')$ modes that exchange energy among themselves. Hence, Eq. (21.26a) is the mode-to-mode energy transfer from $\mathbf{z}^+(\mathbf{p})$ to $\mathbf{z}^+(\mathbf{k}')$ with the mediation of $\mathbf{z}^-(\mathbf{q})$. See Fig. 21.3 for an illustration.

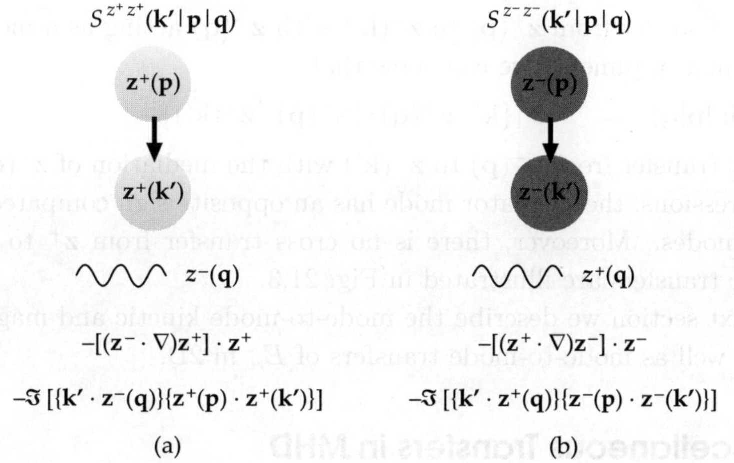

Figure 21.3 Schematic diagram exhibiting mode-to-mode energy transfers among the Elsäser variables $\mathbf{z}^\pm$: (a) z^+ to z^+, (b) z^- to z^-. Here the wavy lines represent the mediator modes. Note that there is no energy transfer from z^- to z^+ and vice versa.

We also provide a mathematical argument for the above formula. In the following discussion we derive $S^{z^+z^+}(\mathbf{k}'|\mathbf{p}|\mathbf{q})$ for which the additional conditions are

1. $S^{z^+z^+}(\mathbf{k}'|\mathbf{p}|\mathbf{q})$ is real.

2. $S^{z^+z^+}(\mathbf{k}'|\mathbf{p}|\mathbf{q})$ is a linear function of $\mathbf{k}'$.
3. $S^{z^+z^+}(\mathbf{k}'|\mathbf{p}|\mathbf{q})$ consists of a scalar product of $\mathbf{z}^+(\mathbf{k}')$ with another $\mathbf{z}^+$ mode. This quantity is multiplied by a scalar product of wavenumber $\mathbf{k}'$ and a $\mathbf{z}^-$ mode. The arguments of the Fourier modes are distinct wavenumbers chosen from $\mathbf{k}', \mathbf{p}, \mathbf{q}$.
4. $S^{z^+z^+}(\mathbf{k}'|\mathbf{p}|\mathbf{q}) = S^{z^+z^+}(-\mathbf{k}'|-\mathbf{p}|-\mathbf{q})$.

Under these constraints, we deduce that

$$\begin{aligned} S^{z^+z^+}(\mathbf{k}'|\mathbf{p}|\mathbf{q}) &= c_1 \Im\left[\{\mathbf{k}' \cdot \mathbf{z}^-(\mathbf{q})\}\{\mathbf{z}^+(\mathbf{p}) \cdot \mathbf{z}^+(\mathbf{k}')\}\right] \\ &\quad + c_2 \Im\left[\{\mathbf{k}' \cdot \mathbf{z}^-(\mathbf{p})\}\{\mathbf{z}^+(\mathbf{q}) \cdot \mathbf{z}^+(\mathbf{k}')\}\right]. \end{aligned} \tag{21.27}$$

Substitution of the above equation and their counterparts in Eqs. (21.25a, 21.25b) yields

$$c_1 = -1; \quad c_2 = 0. \tag{21.28}$$

Therefore,

$$S^{z^+z^+}(\mathbf{k}'|\mathbf{p}|\mathbf{q}) = -\Im\left[\{\mathbf{k}' \cdot \mathbf{z}^-(\mathbf{q})\}\{\mathbf{z}^+(\mathbf{p}) \cdot \mathbf{z}^+(\mathbf{k}')\}\right] \tag{21.29}$$

is the energy transfer from $\mathbf{z}^+(\mathbf{p})$ to $\mathbf{z}^+(\mathbf{k}')$ with $\mathbf{z}^-(\mathbf{q})$ acting as a mediator.

Using similar arguments we can show that

$$S^{z^-z^-}(\mathbf{k}'|\mathbf{p}|\mathbf{q}) = -\Im\left[\{\mathbf{k}' \cdot \mathbf{z}^+(\mathbf{q})\}\{\mathbf{z}^-(\mathbf{p}) \cdot \mathbf{z}^-(\mathbf{k}')\}\right] \tag{21.30}$$

is the energy transfer from $\mathbf{z}^-(\mathbf{p})$ to $\mathbf{z}^-(\mathbf{k}')$ with the mediation of $\mathbf{z}^+(\mathbf{q})$. Note that in these expressions, the mediator mode has an opposite sign compared to the giver or receiver modes. Moreover, there is no cross transfer from $\mathbf{z}^+$ to $\mathbf{z}^-$, and vice versa. These transfers are illustrated in Fig. 21.3.

In the next section we describe the mode-to-mode kinetic and magnetic helicity transfers, as well as mode-to-mode transfers of E_A in 2D.

21.4 Miscellaneous Transfers in MHD

In this section we will describe mode-to-mode transfers of magnetic helicity, kinetic helicity, and E_A.

21.4.1 Mode-to-mode magnetic helicity transfers in MHD

For the derivation of mode-to-mode magnetic helicity transfers, we start with the dissipationless version of the evolution equation for the magnetic helicity [Eq. (20.55)]:

$$\frac{d}{dt}H_M(\mathbf{k}) = \Re\sum_{\mathbf{p}}\mathbf{u}(\mathbf{q})\cdot\{\mathbf{B}(\mathbf{p})\times\mathbf{B}^*(\mathbf{k})\}. \tag{21.31}$$

For the pair of triads $(\mathbf{k}',\mathbf{p},\mathbf{q})$ and $(-\mathbf{k}',-\mathbf{p},-\mathbf{q})$ with $\mathbf{k}'=-\mathbf{k}$, we obtain

$$\begin{aligned}\frac{d}{dt}H_M(\mathbf{k}') &= S^{H_M}(\mathbf{k}'|\mathbf{p},\mathbf{q})\\ &= \Re[\mathbf{u}(\mathbf{q})\cdot\{\mathbf{B}(\mathbf{p})\times\mathbf{B}^*(\mathbf{k})\}+\mathbf{u}(\mathbf{p})\cdot\{\mathbf{B}(\mathbf{q})\times\mathbf{B}^*(\mathbf{k})\}],\end{aligned} \tag{21.32}$$

where $S^{H_M}(\mathbf{k}'|\mathbf{p},\mathbf{q})$ is the combined energy transfer to wavenumber **k'** from wavenumbers **p** and **q**. The above combined helicity transfer and related formulas satisfy the following relation:

$$S^{H_M}(\mathbf{k}'|\mathbf{p},\mathbf{q})+S^{H_K}(\mathbf{p}|\mathbf{k}',\mathbf{q})+S^{H_K}(\mathbf{q}|\mathbf{k}',\mathbf{p})=0. \tag{21.33}$$

Hence,

$$H_M(\mathbf{k}')+H_M(\mathbf{p})+H_M(\mathbf{q})=\text{const}. \tag{21.34}$$

Or, the total magnetic helicity is conserved in a triadic interaction, which is a statement of the *detailed conservation of magnetic helicity* in a triad.

Now we proceed to solve for the mode-to-mode magnetic helicity transfers from wavenumber **Y** to wavenumber **X** with the mediation of wavenumber **Z**. Following arguments as in the previous section, we claim that $S^{H_K}(\mathbf{X}|\mathbf{Y}|\mathbf{Z})$ satisfies the following equations:

$$S^{H_M}(\mathbf{X}|\mathbf{Y}|\mathbf{Z})+S^{H_M}(\mathbf{X}|\mathbf{Z}|\mathbf{Y}) = S^{H_M}(\mathbf{X}|\mathbf{Y},\mathbf{Z}), \tag{21.35a}$$

$$S^{H_M}(\mathbf{X}|\mathbf{Y}|\mathbf{Z}) = -S^{H_M}(\mathbf{Y}|\mathbf{X}|\mathbf{Z}). \tag{21.35b}$$

Clearly,

$$S^{H_M}(\mathbf{X}|\mathbf{Y}|\mathbf{Z})=\Re[\mathbf{u}(\mathbf{Z})\cdot\{\mathbf{B}(\mathbf{Y})\times\mathbf{B}(\mathbf{X})\}] \tag{21.36}$$

is a solution to the above set of equations. However this solution is not unique because the determinant of the solution matrix is zero. Hence, we employ more constraints using tensor analysis. Using the structure of Eq. (21.32), we infer that $S^{H_M}(\mathbf{k}'|\mathbf{p}|\mathbf{q})$ is a function of $\mathbf{u}(\mathbf{k}')$, $\mathbf{u}(\mathbf{p})$, $\mathbf{u}(\mathbf{q})$, $\mathbf{B}(\mathbf{k}')$, $\mathbf{B}(\mathbf{p})$, and $\mathbf{B}(\mathbf{q})$ with the following conditions (in addition to Eqs. (21.35a, 21.35b)):

1. $S^{H_M}(\mathbf{k}'|\mathbf{p}|\mathbf{q})$ is real.
2. $S^{H_M}(\mathbf{k}'|\mathbf{p}|\mathbf{q})$ has two **B**'s and one **u** with distinct wavenumbers $\mathbf{k}',\mathbf{p},\mathbf{q}$ as arguments.

3. $S^{H_M}(\mathbf{k}'|\mathbf{p}|\mathbf{q})$ must have $\mathbf{B}(\mathbf{k}')$ in the expression.

4. $S^{H_M}(\mathbf{k}'|\mathbf{p}|\mathbf{q}) = S^{H_M}(-\mathbf{k}'|-\mathbf{p}|-\mathbf{q})$.

Using these conditions and the structure of Eqs. (21.35a, 21.35b), we deduce that

$$S^{H_M}(\mathbf{k}'|\mathbf{p}|\mathbf{q}) = c_1\Re[\mathbf{u}(\mathbf{q})\cdot\{\mathbf{B}(\mathbf{p})\times\mathbf{B}(\mathbf{k}')\}] + c_2\Re[\mathbf{u}(\mathbf{p})\cdot\{\mathbf{B}(\mathbf{q})\times\mathbf{B}(\mathbf{k}')\}]. \quad (21.37)$$

We do not take the imaginary part due to condition 4. Substitution of Eq. (21.37) in Eqs. (21.35a, 21.35b) yields

$$c_1 = 1; \quad c_2 = 0. \quad (21.38)$$

Hence,

$$S^{H_M}(\mathbf{k}'|\mathbf{p}|\mathbf{q}) = \Re[\mathbf{u}(\mathbf{q})\cdot\{\mathbf{B}(\mathbf{p})\times\mathbf{B}(\mathbf{k}')\}] \quad (21.39)$$

is the mode-to-mode kinetic helicity transfer from wavenumber $\mathbf{p}$ to wavenumber $\mathbf{k}'$ with the mediation of wavenumber $\mathbf{q}$.

21.4.2 Mode-to-mode kinetic helicity transfers in MHD

For the kinetic helicity, we start with Eq. (20.52) with $\nu = 0$:

$$\frac{d}{dt}H_K(\mathbf{k}) = \sum_{\mathbf{p}} \Re[\mathbf{u}(\mathbf{q})\cdot\{\boldsymbol{\omega}(\mathbf{p})\times\boldsymbol{\omega}^*(\mathbf{k})\} + \mathbf{J}(\mathbf{q})\cdot\{\mathbf{B}(\mathbf{p})\times\boldsymbol{\omega}^*(\mathbf{k})\}]. \quad (21.40)$$

Compared with the hydrodynamic counterpart, this equation for MHD gets an addition term $\Re[\mathbf{J}(\mathbf{q})\cdot\{\mathbf{B}(\mathbf{p})\times\boldsymbol{\omega}^*(\mathbf{k})\}]$. In Chapter 8 we showed that the term $\Re[\mathbf{u}(\mathbf{q})\cdot\{\boldsymbol{\omega}(\mathbf{p})\times\boldsymbol{\omega}^*(\mathbf{k})\}]$ conserves total kinetic energy in a triad, and the mode-to-mode kinetic helicity with this term is (see Eq. (8.10))

$$S^{H_K}(\mathbf{k}'|\mathbf{p}|\mathbf{q}) = \Re[\mathbf{u}(\mathbf{q})\cdot\{\boldsymbol{\omega}(\mathbf{p})\times\boldsymbol{\omega}(\mathbf{k}')\}]. \quad (21.41)$$

The term $\Re[\mathbf{J}(\mathbf{q})\cdot\{\mathbf{B}(\mathbf{p})\times\boldsymbol{\omega}^*(\mathbf{k})\}]$ feeds a net kinetic helicity to the triad, and hence breaks the conservation of triadic kinetic helicity. With the inclusion of this term, the mode-to-mode kinetic helicity transfer from wavenumber $\mathbf{p}$ to wavenumber $\mathbf{k}'$ with the mediation of wavenumber $\mathbf{q}$ is

$$S^{H_K}(\mathbf{k}'|\mathbf{p}|\mathbf{q}) = \Re[\mathbf{u}(\mathbf{q})\cdot\{\boldsymbol{\omega}(\mathbf{p})\times\boldsymbol{\omega}(\mathbf{k}')\} + \mathbf{J}(\mathbf{q})\cdot\{\mathbf{B}(\mathbf{p})\times\boldsymbol{\omega}(\mathbf{k}')\}]. \quad (21.42)$$

21.4.3 Mode-to-mode transfers of E_A in 2D

To derive mode-to-mode energy transfer for the mean square vector potential E_A in 2D, we start with the dynamical equation for the vector potential in 2D. From Eq. (20.17), which is

$$\frac{\partial A}{\partial t} + (\mathbf{u} \cdot \nabla)A = \eta \nabla^2 A, \tag{21.43}$$

we deduce that A acts as a passive scalar. Therefore, following the derivation of Section 13.2, we deduce that the mode-to-mode transfer from $A(\mathbf{p})$ to $A(\mathbf{k}')$ with the mediation of $\mathbf{u}(\mathbf{q})$ is

$$S^{AA}(\mathbf{k}'|\mathbf{p}|\mathbf{q}) = -\Im\left[\{\mathbf{k}' \cdot \mathbf{u}(\mathbf{q})\}\{A(p)A(k')\}\right]. \tag{21.44}$$

We will use this relation while discussing the anti-dynamo theorem in 2D MHD turbulence.

In Table 21.2, we summarize all the mode-to-mode energy transfers described here:

Table 21.2 Summary of various mode-to-mode energy transfers in MHD turbulence. Here the wavenumbers of the receiver, giver, and mediator modes are $\mathbf{k}', \mathbf{p}, \mathbf{q}$ respectively.

ET	Receiver	Giver	Mediator	Formula
U2U	$\mathbf{u}(\mathbf{k}')$	$\mathbf{u}(\mathbf{p})$	$\mathbf{u}(\mathbf{q})$	$-\Im\left[\{\mathbf{k}' \cdot \mathbf{u}(\mathbf{q})\}\{\mathbf{u}(\mathbf{p}) \cdot \mathbf{u}(\mathbf{k}')\}\right]$
B2B	$\mathbf{B}(\mathbf{k}')$	$\mathbf{B}(\mathbf{p})$	$\mathbf{u}(\mathbf{q})$	$-\Im\left[\{\mathbf{k}' \cdot \mathbf{u}(\mathbf{q})\}\{\mathbf{B}(\mathbf{p}) \cdot \mathbf{B}(\mathbf{k}')\}\right]$
B2U	$\mathbf{u}(\mathbf{k}')$	$\mathbf{B}(\mathbf{p})$	$\mathbf{B}(\mathbf{q})$	$\Im\left[\{\mathbf{k}' \cdot \mathbf{B}(\mathbf{q})\}\{\mathbf{B}(\mathbf{p}) \cdot \mathbf{u}(\mathbf{k}')\}\right]$
U2B	$\mathbf{B}(\mathbf{k}')$	$\mathbf{u}(\mathbf{p})$	$\mathbf{B}(\mathbf{q})$	$\Im\left[\{\mathbf{k}' \cdot \mathbf{B}(\mathbf{q})\}\{\mathbf{u}(\mathbf{p}) \cdot \mathbf{B}(\mathbf{k}')\}\right]$
z^+ to z^+	$\mathbf{z}^+(\mathbf{k}')$	$\mathbf{z}^+(\mathbf{p})$	$\mathbf{z}^-(\mathbf{q})$	$-\Im\left[\{\mathbf{k}' \cdot \mathbf{z}^-(\mathbf{q})\}\{\mathbf{z}^+(\mathbf{p}) \cdot \mathbf{z}^+(\mathbf{k}')\}\right]$
z^- to z^-	$\mathbf{z}^-(\mathbf{k}')$	$\mathbf{z}^-(\mathbf{p})$	$\mathbf{z}^+(\mathbf{q})$	$-\Im\left[\{\mathbf{k}' \cdot \mathbf{z}^+(\mathbf{q})\}\{\mathbf{z}^-(\mathbf{p}) \cdot \mathbf{z}^-(\mathbf{k}')\}\right]$
A2A	$A(\mathbf{k}')$	$A(\mathbf{p})$	$\mathbf{u}(\mathbf{q})$	$-\Im\left[\{\mathbf{k}' \cdot \mathbf{u}(\mathbf{q})\}\{A(\mathbf{p})A(\mathbf{k}')\}\right]$
H_M	$\mathbf{k}'$	$\mathbf{p}$	$\mathbf{q}$	$\Re[\mathbf{u}(\mathbf{q}) \cdot \{\mathbf{B}(\mathbf{p}) \times \mathbf{B}(\mathbf{k}')\}]$

Example 21.1: In a periodic box $[2\pi, 2\pi]$, consider the following velocity and magnetic fields in MHD:

$$\mathbf{u} = \hat{x}2D_u \cos y + \hat{y}2C_u \cos x + (\hat{x} - \hat{y})2A_u \sin(x + y).$$

$$\mathbf{b} = \hat{x}2D_w \cos y + \hat{y}2C_w \cos x + (\hat{x} - \hat{y})2A_w \sin(x + y)$$

Compute various energy transfers of MHD turbulence.

Solution: The mode amplitudes (in Craya–Herring basis) for these velocity and magnetic fields are listed in Table 18.2 (with $\mathbf{w}$ replaced by $\mathbf{b}$). Note that $u_2 = b_2 = 0$ for these fields.

The $U2U$ and $B2B$ energy transfers have been computed in Examples 4.5 and 18.2 respectively, and they are

$$\begin{aligned}
S^{uu}(\mathbf{k}'|\mathbf{p}|\mathbf{q}) &= -A_u D_u C_u \\
S^{uu}(\mathbf{p}|\mathbf{q}|\mathbf{k}') &= 0, \\
S^{uu}(\mathbf{q}|\mathbf{k}'|\mathbf{p}) &= A_u D_u C_u \\
S^{bb}(\mathbf{k}'|\mathbf{p}|\mathbf{q}) &= -C_u A_b D_b \\
S^{bb}(\mathbf{p}|\mathbf{q}|\mathbf{k}') &= 0, \\
S^{bb}(\mathbf{q}|\mathbf{k}'|\mathbf{p}) &= D_u A_b C_b.
\end{aligned}$$

Using Eqs. (21.96, 21.97, 21.98, 21.99), we obtain

$$\begin{aligned}
S^{ub}(\mathbf{k}'|\mathbf{p}|\mathbf{q}) &= A_u D_b C_b \\
S^{ub}(\mathbf{p}|\mathbf{q}|\mathbf{k}') &= 0, \\
S^{ub}(\mathbf{q}|\mathbf{k}'|\mathbf{p}) &= A_b D_b C_u \\
S^{bu}(\mathbf{k}'|\mathbf{p}|\mathbf{q}) &= A_b C_b D_u \\
S^{bu}(\mathbf{p}|\mathbf{q}|\mathbf{k}') &= 0, \\
S^{bu}(\mathbf{q}|\mathbf{k}'|\mathbf{p}) &= A_u D_b C_b.
\end{aligned}$$

We can determine other transfers using variants of these formulas.

Note that the energy transfer functions satisfy:

$$\sum_{\mathbf{X},\mathbf{Y}} S^{uu}(\mathbf{X}|\mathbf{Y}|\mathbf{Z}) = \sum_{\mathbf{X},\mathbf{Y}} S^{bb}(\mathbf{X}|\mathbf{Y}|\mathbf{Z}) = 0,$$

$$\sum_{\mathbf{X},\mathbf{Y}} S^{ub}(\mathbf{X}|\mathbf{Y}|\mathbf{Z}) + \sum_{\mathbf{X},\mathbf{Y}} S^{bu}(\mathbf{X}|\mathbf{Y}|\mathbf{Z}) = 0.$$

21.5 Transfers for Many Triads and Fluxes

In the presence of a large number of triads, the evolution equations for the modal kinetic and magnetic energies are

$$\begin{aligned}
\frac{d}{dt}E_u(\mathbf{k}) &= \sum_{\mathbf{p}} S^{uu}(\mathbf{k}|\mathbf{p}|\mathbf{q}) + \sum_{\mathbf{p}} S^{ub}(\mathbf{k}|\mathbf{p}|\mathbf{q}) - 2\nu k^2 E_u(\mathbf{k}) \\
&= T_{uu}(\mathbf{k}) + T_{ub}(\mathbf{k}) - D_u(\mathbf{k}),
\end{aligned} \tag{21.45a}$$

$$\frac{d}{dt}E_b(\mathbf{k}) = \sum_{\mathbf{p}} S^{bb}(\mathbf{k}|\mathbf{p}|\mathbf{q}) + \sum_{\mathbf{p}} S^{bu}(\mathbf{k}|\mathbf{p}|\mathbf{q}) - 2\eta k^2 E_b(\mathbf{k})$$
$$= T_{bb}(\mathbf{k}) + T_{bu}(\mathbf{k}) - D_b(\mathbf{k}). \tag{21.45b}$$

Here $T_{uu}(\mathbf{k})$, $T_{ub}(\mathbf{k})$, $T_{bu}(\mathbf{k})$, and $T_{bb}(\mathbf{k})$ are transfer functions, and $D_u(\mathbf{k}), D_b(\mathbf{k})$ are respectively the kinetic and magnetic energy dissipation rates. Using the laws of detailed energy conservation of Section 21.1, we can show that for any wavenumber region R,

$$\sum_{\mathbf{k}\in R}\sum_{\mathbf{p}\in R} S^{uu}(\mathbf{k}|\mathbf{p}|\mathbf{q}) = 0, \tag{21.46a}$$

$$\sum_{\mathbf{k}\in R}\sum_{\mathbf{p}\in R} S^{bb}(\mathbf{k}|\mathbf{p}|\mathbf{q}) = 0, \tag{21.46b}$$

$$\sum_{\mathbf{k}\in R}\sum_{\mathbf{p}\in R} [S^{ub}(\mathbf{k}|\mathbf{p}|\mathbf{q}) + S^{bu}(\mathbf{k}|\mathbf{p}|\mathbf{q})] = 0. \tag{21.46c}$$

However,

$$\sum_{\mathbf{k}\in R}\sum_{\mathbf{p}\in R} S^{bu}(\mathbf{k}|\mathbf{p}|\mathbf{q}) \neq 0, \tag{21.47a}$$

$$\sum_{\mathbf{k}\in R}\sum_{\mathbf{p}\in R} S^{ub}(\mathbf{k}|\mathbf{p}|\mathbf{q}) \neq 0. \tag{21.47b}$$

Hence, the magnetic modes in R can gain or lose energy. When R represents the complete Fourier space, Eqs. (21.45) yield

$$\frac{d}{dt}(E_u + E_b) = \epsilon_u + \epsilon_b, \tag{21.48}$$

where ϵ_u, ϵ_b represent the kinetic and magnetic energy dissipation rates respectively, and they arise due to viscous dissipation and Joule heating respectively. This equation implies that the total energy decays in the absence of an external force.

Now we can define various energy fluxes of MHD for a wavenumber sphere of radius k_0. These fluxes are as follows:

1. $\Pi^{u<}_{u>}(k_0)$: Energy transfers from all the velocity modes inside the sphere to all the velocity modes outside the sphere:

$$\Pi^{u<}_{u>}(k_0) = \sum_{|\mathbf{p}|\leq k_0}\sum_{|\mathbf{k}'|>k_0} S^{uu}(\mathbf{k}'|\mathbf{p}|\mathbf{q}). \tag{21.49}$$

2. $\Pi^{u<}_{b>}(k_0)$: Energy transfers from all the velocity modes inside the sphere to all the magnetic modes outside the sphere:

$$\Pi^{u<}_{b>}(k_0) = \sum_{|\mathbf{p}|\le k_0} \sum_{|\mathbf{k}'|>k_0} S^{bu}(\mathbf{k}'|\mathbf{p}|\mathbf{q}). \tag{21.50}$$

3. $\Pi^{b<}_{b>}(k_0)$: Energy transfers from all the magnetic modes inside the sphere to all the magnetic modes outside the sphere:

$$\Pi^{b<}_{b>}(k_0) = \sum_{|\mathbf{p}|\le k_0} \sum_{|\mathbf{k}'|>k_0} S^{bb}(\mathbf{k}'|\mathbf{p}|\mathbf{q}). \tag{21.51}$$

4. $\Pi^{b<}_{u>}(k_0)$: Energy transfers from all the magnetic modes inside the sphere to all the velocity modes outside the sphere:

$$\Pi^{b<}_{u>}(k_0) = \sum_{|\mathbf{p}|\le k_0} \sum_{|\mathbf{k}'|>k_0} S^{ub}(\mathbf{k}'|\mathbf{p}|\mathbf{q}). \tag{21.52}$$

5. $\Pi^{u<}_{b<}(k_0)$: Energy transfers from all the velocity modes inside the sphere to all the magnetic modes inside the sphere:

$$\Pi^{u<}_{b<}(k_0) = \sum_{|\mathbf{p}|\le k_0} \sum_{|\mathbf{k}'|\le k_0} S^{ub}(\mathbf{k}'|\mathbf{p}|\mathbf{q}). \tag{21.53}$$

6. $\Pi^{u>}_{b>}(k_0)$: Energy transfers from all the velocity modes outside the sphere to all the magnetic modes outside the sphere:

$$\Pi^{u>}_{b>}(k_0) = \sum_{|\mathbf{p}|>k_0} \sum_{|\mathbf{k}'|>k_0} S^{bu}(\mathbf{k}'|\mathbf{p}|\mathbf{q}). \tag{21.54}$$

In the above notation, the superscript represents the giver modes, and subscript represents the receiver modes. Also, $<$ and $>$ represent the modes residing inside and outside the sphere respectively. These fluxes are illustrated in Fig. 21.4. Note that the flux $\Pi^{u<}_{b<}(k_0)$ involves velocity and magnetic modes inside the sphere, while $\Pi^{u>}_{b>}(k_0)$ involves modes outside the sphere. From the above reasoning we can deduce that the flux of the total energy, which is the total energy leaving the sphere of radius k_0, is

$$\Pi_{\text{tot}}(k_0) = \Pi^{u<}_{u>} + \Pi^{u<}_{b>} + \Pi^{b<}_{b>} + \Pi^{b<}_{u>}. \tag{21.55}$$

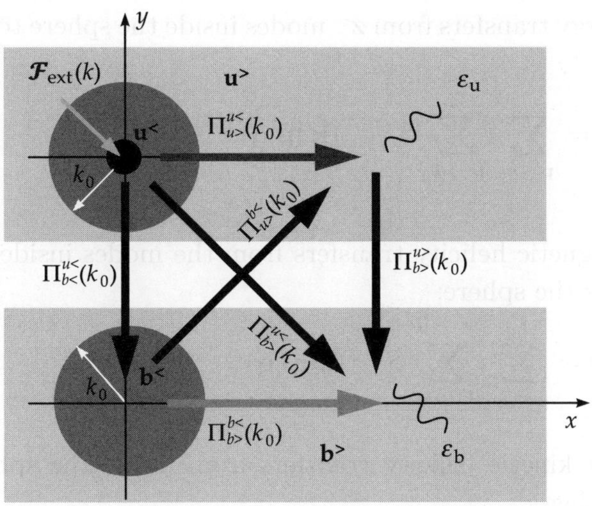

Figure 21.4 Various energy fluxes of MHD turbulence. ϵ_u, ϵ_b are the dissipation rates of the velocity and magnetic fields respectively.

In addition to conservation laws of Eq. (21.46), we also have conservation laws of $E_{z^\pm}$ and H_M for any wavenumber region R:

$$\sum_{\mathbf{k}\in R}\sum_{\mathbf{p}\in R} S^{z^+}(\mathbf{k}|\mathbf{p}|\mathbf{q}) = 0, \tag{21.56a}$$

$$\sum_{\mathbf{k}\in R}\sum_{\mathbf{p}\in R} S^{z^-}(\mathbf{k}|\mathbf{p}|\mathbf{q}) = 0, \tag{21.56b}$$

$$\sum_{\mathbf{k}\in R}\sum_{\mathbf{p}\in R} S^{H_M}(\mathbf{k}|\mathbf{p}|\mathbf{q}) = 0. \tag{21.56c}$$

In 2D MHD, E_A is conserved in place of H_M:

$$\sum_{\mathbf{k}\in R}\sum_{\mathbf{p}\in R} S^{AA}(\mathbf{k}|\mathbf{p}|\mathbf{q}) = 0. \tag{21.57}$$

In the following we define the fluxes for the aforementioned quantities, as well as for kinetic helicity:

1. $\Pi_{z^+}(k_0)$: Energy transfers from $\mathbf{z}^+$ modes inside the sphere to $\mathbf{z}^+$ modes outside the sphere:

$$\Pi_{z^+}(k_0) = \sum_{|\mathbf{p}|\le k_0}\sum_{|\mathbf{k}'|>k_0} S^{z^+z^+}(\mathbf{k}'|\mathbf{p}|\mathbf{q}). \tag{21.58}$$

2. $\Pi_{z^-}(k_0)$: Energy transfers from $\mathbf{z}^-$ modes inside the sphere to $\mathbf{z}^-$ modes outside the sphere:

$$\Pi_{z^-}(k_0) = \sum_{|\mathbf{p}|\le k_0} \sum_{|\mathbf{k}'|>k_0} S^{z^-z^-}(\mathbf{k}'|\mathbf{p}|\mathbf{q}). \tag{21.59}$$

3. $\Pi_{H_M}(k_0)$: Magnetic helicity transfers from the modes inside the sphere to the modes outside the sphere:

$$\Pi_{H_M}(k_0) = \sum_{|\mathbf{p}|\le k_0} \sum_{|\mathbf{k}'|>k_0} S^{H_M}(\mathbf{k}'|\mathbf{p}|\mathbf{q}). \tag{21.60}$$

4. $\Pi_{H_K}(k_0)$: Net kinetic helicity transfers from inside the sphere to the modes outside the sphere:

$$\Pi_{H_K}(k_0) = \sum_{|\mathbf{p}|\le k_0} \sum_{|\mathbf{k}'|>k_0} S^{H_K}(\mathbf{k}'|\mathbf{p}|\mathbf{q}). \tag{21.61}$$

5. $\Pi_A(k_0)$: Net E_A transfer from inside the sphere to the modes outside the sphere:

$$\Pi_A(k_0) = \sum_{|\mathbf{p}|\le k_0} \sum_{|\mathbf{k}'|>k_0} S^{AA}(\mathbf{k}'|\mathbf{p}|\mathbf{q}). \tag{21.62}$$

In Fig. 21.5, we illustrate the energy flux $\Pi_{z^\pm}$.

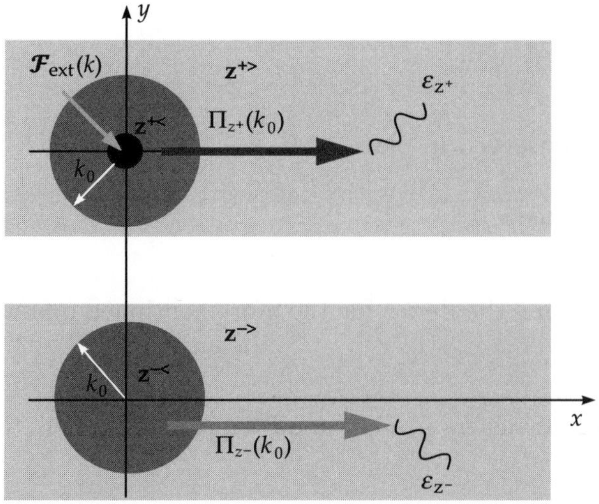

Figure 21.5 Energy fluxes Π^+ and Π^- in MHD turbulence. $\epsilon_{z^+}, \epsilon_{z^-}$ are the dissipation rates of $\mathbf{z}^+$ and $\mathbf{z}$ fields respectively.

Note that

$$\Pi_{\text{tot}} = \frac{1}{2}[\Pi_{z^+} + \Pi_{z^-}], \tag{21.63a}$$

$$\Pi_{\text{Hc}} = \frac{1}{4}[\Pi_{z^+} - \Pi_{z^-}]. \tag{21.63b}$$

We can also derive the above fluxes using transfer functions, which are defined in Eqs. (21.45). Equation (4.41) shows that

$$\sum_{|\mathbf{k}|>k_0} T_{uu}(\mathbf{k}) = \Pi^{u<}_{u>}(k_0). \tag{21.64}$$

Similar derivation can be used to derive

$$\sum_{|\mathbf{k}|>k_0} T_{bb}(\mathbf{k}) = \Pi^{b<}_{b>}(k_0). \tag{21.65}$$

$$\begin{aligned} \sum_{|\mathbf{k}|>k_0} T_{ub}(\mathbf{k}) &= \sum_{|\mathbf{k}'|>k_0} \sum_{\mathbf{p}} S^{ub}(\mathbf{k}'|\mathbf{p}|\mathbf{q}) \\ &= \sum_{|\mathbf{k}'|>k_0} \sum_{|\mathbf{p}|<k_0} S^{ub}(\mathbf{k}'|\mathbf{p}|\mathbf{q}) + \sum_{|\mathbf{k}'|>k_0} \sum_{|\mathbf{p}|>k_0} S^{ub}(\mathbf{k}'|\mathbf{p}|\mathbf{q}) \\ &= \Pi^{b<}_{u>}(k_0) + \Pi^{b>}_{u>}(k_0), \end{aligned} \tag{21.66}$$

and

$$\begin{aligned} \sum_{|\mathbf{k}|>k_0} T_{bu}(\mathbf{k}) &= \sum_{|\mathbf{k}'|>k_0} \sum_{\mathbf{p}} S^{bu}(\mathbf{k}'|\mathbf{p}|\mathbf{q}) \\ &= \sum_{|\mathbf{k}'|>k_0} \sum_{|\mathbf{p}|<k_0} S^{bu}(\mathbf{k}'|\mathbf{p}|\mathbf{q}) + \sum_{|\mathbf{k}'|>k_0} \sum_{|\mathbf{p}|>k_0} S^{bu}(\mathbf{k}'|\mathbf{p}|\mathbf{q}) \\ &= \Pi^{u<}_{b>}(k_0) + \Pi^{u>}_{b>}(k_0). \end{aligned} \tag{21.67}$$

Adding Eqs. (21.64-21.67) yields

$$\sum_{|\mathbf{k}|>k_0} [T_{uu}(\mathbf{k}) + T_{bb}(\mathbf{k}) + T_{bu}(\mathbf{k}) + T_{ub}(\mathbf{k})] = \Pi_{\text{tot}}(k_0). \tag{21.68}$$

Here we employed $\Pi^{u>}_{b>}(k_0) + \Pi^{b>}_{u>}(k_0) = 0$.

In the next section, we relate these transfer functions to variable energy fluxes.

21.6 Variable Energy Fluxes and Conserved Fluxes of MHD Turbulence

In MHD, the velocity field experiences Lorentz force, while the magnetic field is advected and stretched by the velocity field. In the following discussion we show how these forces affect the kinetic and magnetic energy fluxes.

21.6.1 Kinetic and magnetic energy fluxes

In the beginning of Section 21.5, we showed that the total energy decays due to viscous and Joule dissipation. Hence, to obtain a steady state, we need to employ an external force. Typically, it is assumed that the large-scale force $\mathbf{F}_{\text{ext}}$ acts on the velocity field. This force feeds kinetic energy at the rate of

$$\mathcal{F}_{\text{ext}}(\mathbf{k}) = \Re[\mathbf{F}_{\text{ext}}(\mathbf{k}) \cdot \mathbf{u}^*(\mathbf{k})]. \tag{21.69}$$

Note that $\mathcal{F}_{\text{ext}}(\mathbf{k}) = 0$ in the inertial and dissipation range.

With $\mathcal{F}_{\text{ext}}$, the modified equation of the kinetic and magnetic energies are

$$\frac{d}{dt}E_u(\mathbf{k}) = T_{uu}(\mathbf{k}) + \mathcal{F}_u(\mathbf{k}) + \mathcal{F}_{\text{ext}}(\mathbf{k}) - D_u(\mathbf{k}), \tag{21.70a}$$

$$\frac{d}{dt}E_b(\mathbf{k}) = T_{bb}(\mathbf{k}) + \mathcal{F}_B(\mathbf{k}) - D_b(\mathbf{k}), \tag{21.70b}$$

where $\mathcal{F}_u$ is the Lorentz force, while

$$\mathcal{F}_B(\mathbf{k}) = i \sum_{\mathbf{p}} \{\mathbf{k} \cdot \mathbf{B}(\mathbf{q})\}\{\mathbf{u}(\mathbf{p}) \cdot \mathbf{B}^*(\mathbf{k})\}. \tag{21.71}$$

Note that $\mathcal{F}_u(\mathbf{k}) = T_{ub}(\mathbf{k})$, $\mathcal{F}_B(\mathbf{k}) = T_{bu}(\mathbf{k})$. Now, using Eqs. (21.64, 21.65) we deduce that under a steady state,

$$T_{uu}(k,t) = -\frac{d}{dk}\Pi^{u<}_{u>}(k,t), \tag{21.72a}$$

$$T_{bb}(k,t) = -\frac{d}{dk}\Pi^{b<}_{b>}(k,t). \tag{21.72b}$$

An addition of Eq. (21.70a) and Eq. (21.70b) and subsequent summation over all wavenumbers yield (under a steady state)

$$\sum_{\mathbf{k}} \mathcal{F}_{\text{ext}}(\mathbf{k}) = \sum_{\mathbf{k}} [D_u(\mathbf{k}) + D_b(\mathbf{k})] = \epsilon_u + \epsilon_b = \epsilon_{\text{tot}}. \tag{21.73}$$

This equation implies that the kinetic energy feed by the external force balances the total dissipation rate ϵ_{tot}.

To derive the variations of the energy fluxes with k, we start with Eqs. (21.70a, 21.70b), and use the arguments of Section 4.4. We perform two sums of the Fourier modes: first sum over sphere of radus k, and the second sum over the sphere of radius $k + dk$. The differences of the two sums for E_u and E_b yield the following equations:

$$\frac{\partial}{\partial t}E_u(k,t) = -\frac{\partial}{\partial k}\Pi^{u<}_{u>}(k) + \mathcal{F}_u(k,t) + \mathcal{F}_{\text{ext}}(k,t) - D_u(k,t), \tag{21.74a}$$

$$\frac{\partial}{\partial t}E_b(k,t) = -\frac{\partial}{\partial k}\Pi^{b<}_{b>}(k) + \mathcal{F}_B(k,t) - D_b(k,t). \tag{21.74b}$$

Under a steady state $(\partial E_u(k,t)/\partial t = 0; \partial E_b(k,t)/\partial t = 0)$, Eq. (21.74) gets converted to

$$\frac{d}{dk}\Pi^{u<}_{u>}(k) = \mathcal{F}_u(k) + \mathcal{F}_{\text{ext}}(k) - D_u(k), \tag{21.75a}$$

$$\frac{d}{dk}\Pi^{b<}_{b>}(k) = \mathcal{F}_B(k) - D_b(k). \tag{21.75b}$$

In the inertial range, where $\mathcal{F}_{\text{ext}} = 0$ and the dissipation rates are weak, these equations yield

$$\frac{d}{dk}\Pi^{u<}_{u>}(k) = \mathcal{F}_u(k), \tag{21.76a}$$

$$\frac{d}{dk}\Pi^{b<}_{b>}(k) = \mathcal{F}_B(k). \tag{21.76b}$$

A physical interpretation of the aforementioned equations is as follows. The forces $\mathcal{F}_u(k)$ and $\mathcal{F}_B(k)$ are present in the inertial range; hence, they can induce variations in the fluxes $\Pi^{u<}_{u>}(k)$ and $\Pi^{b<}_{b>}(k)$ with k. This is in a similar spirit as in Sections 4.5 and 18.3.

Using Eqs. (21.66, 21.67) we deduce that

$$\mathcal{F}_u(k) = T_{ub}(k) = -\frac{d}{dk}[\Pi^{b<}_{u>}(k) + \Pi^{b>}_{u>}(k)], \tag{21.77a}$$

$$\mathcal{F}_B(k) = T_{bu}(k) = -\frac{d}{dk}[\Pi^{u<}_{b>}(k) + \Pi^{u>}_{b>}(k)], \tag{21.77b}$$

substitution of which in Eqs. (21.76) yields

$$\Pi^{\text{all}}_{u>}(k) = \Pi^{u<}_{u>}(k) + \Pi^{b<}_{u>}(k) + \Pi^{b>}_{u>}(k) = C_u, \tag{21.78a}$$

$$\Pi^{\text{all}}_{b>}(k) = \Pi^{b<}_{b>}(k) + \Pi^{u<}_{b>}(k) + \Pi^{u>}_{b>}(k) = C_b, \tag{21.78b}$$

where C_u and C_b are constants.

We choose a wavenumber k_0 such that $k_0 > k_f$, but it is in the beginning of the inertial range. Integration of Eqs. (21.75) from k_0 to ∞ yields

$$C_u = \Pi^{\text{all}}_{u>}(k_0) = \int_{k_0}^{\infty} D_u(k) \approx \int_0^{\infty} D_u(k) = \epsilon_u, \tag{21.79a}$$

$$C_b = \Pi^{\text{all}}_{b>}(k_0) = \int_{k_0}^{\infty} D_b(k) \approx \int_0^{\infty} D_b(k) = \epsilon_b. \tag{21.79b}$$

Physically, the net energy flux convergent on the velocity channel are dissipated by viscous force, while the total energy flux in the magnetic channel is dissipated by the Joule heating. Also, note that the magnetic energy in each shell reaches a steady state though it is not forced. This is due to the steady energy supply to the magnetic field from the velocity field. The total energy transferred from the velocity field to the magnetic field is dissipated by Joule heating. Hence,

$$\Pi^{u<}_{b<}(k) + \Pi^{u<}_{b>}(k) + \Pi^{u>}_{b<}(k) + \Pi^{u>}_{b>}(k) = \epsilon_b. \tag{21.80}$$

An addition of Eqs. (21.78a, 21.78b) yields the conservation of total energy flux in the inertial range:

$$\Pi_{\mathrm{tot}}(k_0) = \Pi^{u<}_{u>} + \Pi^{u<}_{b>} + \Pi^{b<}_{b>} + \Pi^{b<}_{u>} = C_u + C_b = \epsilon_{\mathrm{tot}}. \tag{21.81}$$

For the derivation of this equation, we employed $\Pi^{b>}_{u>} + \Pi^{u>}_{b>} = 0$. Also, a steady state condition for the magnetic and velocity wavenumber spheres of radius k yields

$$\Pi^{u<}_{b<}(k) + \Pi^{u>}_{b<}(k) = \Pi^{b<}_{b>}(k), \tag{21.82a}$$

$$\Pi^{u<}_{u>}(k) + \Pi^{u<}_{b<}(k) + \Pi^{u<}_{b>}(k) = \sum_{\mathbf{k}} \mathcal{F}_{\mathrm{ext}}(\mathbf{k}) = \epsilon_{\mathrm{tot}}. \tag{21.82b}$$

In summary, Eqs. (21.78, 21.80, 21.81, 21.82) are the identities for the energy fluxes of MHD turbulence. Note that these identities are not independent, and they can be combined to construct new identities.

21.6.2 Fluxes for Elsäser fields and magnetic helicity

For the Elsäser variables,

$$\frac{d}{dt}E_{z^\pm}(\mathbf{k}) = T_{z^\pm z^\pm}(\mathbf{k}) + \mathcal{F}_{\mathrm{ext}}(\mathbf{k}) - D_\pm(\mathbf{k}) - 2\nu_- k^2 E_R(\mathbf{k}), \tag{21.83}$$

where

$$T_{z^\pm z^\pm}(\mathbf{k}) = \sum_{\mathbf{p}} S^{z^\pm z^\pm}(\mathbf{k}|\mathbf{p}|\mathbf{q}), \tag{21.84a}$$

$$D_{z^\pm}(\mathbf{k}) = 2\nu_\pm k^2 E_{z^\pm}(\mathbf{k}), \tag{21.84b}$$

$$E_R(\mathbf{k}) = E_u(\mathbf{k}) - E_b(\mathbf{k}). \tag{21.84c}$$

Using the definition of energy fluxes $\Pi_{z^\pm}$, we obtain

$$\sum_{|\mathbf{k}|>k_0} T_{z^\pm z^\pm}(\mathbf{k}) = \sum_{|\mathbf{k}|>k_0} \sum_{\mathbf{p}} S^{z^\pm z^\pm}(\mathbf{k}|\mathbf{p}|\mathbf{q}) = \Pi_{z^\pm}(k_0). \tag{21.85}$$

Now following similar steps as in the previous subsection, we obtain the following equations for the energy fluxes $\Pi_{z^\pm}(k)$:

$$\frac{\partial}{\partial t}E_{z^+}(k,t) = -\frac{\partial}{\partial k}\Pi_{z^+}(k,t) + \mathcal{F}_{\rm ext}(k) - D_{z^+}(k,t) - 2\nu_- k^2 E_R(k), \quad (21.86a)$$

$$\frac{\partial}{\partial t}E_{z^-}(k,t) = -\frac{\partial}{\partial k}\Pi_{z^-}(k,t) + \mathcal{F}_{\rm ext}(k) - D_{z^-}(k,t) - 2\nu_- k^2 E_R(k). \quad (21.86b)$$

Under a steady state, in the inertial range where $\mathcal{F}_{\rm ext}(k) = 0$ and the dissipation terms are negligible, we obtain

$$\frac{d}{dk}\Pi_{z^\pm}(k) = 0. \quad (21.87)$$

That is, the fluxes $\Pi_{z^+}(k)$ and $\Pi_{z^-}(k)$ are constants in the inertial range. This is unlike $\Pi^{u<}_{u>}$ and $\Pi^{b<}_{b>}$ that vary with k. Therefore, it is recommended that the fluxes $\Pi_{z^+}(k)$ and $\Pi_{z^-}(k)$ be measured in numerical simulations and in experiments.

Similarly, we can derive that

$$\frac{\partial}{\partial t}H_M(k,t) = -\frac{\partial}{\partial k}\Pi_{H_M}(k,t) - D_{H_M}(k,t), \quad (21.88)$$

where

$$D_{H_M}(k) = 2\nu k^2 H_M(k). \quad (21.89)$$

Under a steady state,

$$\frac{d}{dk}\Pi_{H_M}(k) = -D_{H_M}(k). \quad (21.90)$$

Further, in the inertial range where $D_{H_M}(k)$ is negligible, we have constancy of magnetic helicity flux, that is,

$$\Pi_{H_M}(k) = \text{const.} \quad (21.91)$$

21.7 Shell-to-shell Transfers in MHD

For MHD turbulence, we define the following shell-to-shell transfer from shell m to shell n:

$$T^{u,m}_{u,n} = \sum_{\mathbf{p}\in m}\sum_{\mathbf{k}\in n} S^{uu}(\mathbf{k}|\mathbf{p}|\mathbf{q}), \quad (21.92a)$$

$$T^{b,m}_{b,n} = \sum_{\mathbf{p}\in m}\sum_{\mathbf{k}\in n} S^{bb}(\mathbf{k}|\mathbf{p}|\mathbf{q}), \quad (21.92b)$$

$$T^{b,m}_{u,n} = \sum_{\mathbf{p}\in m}\sum_{\mathbf{k}\in n} S^{ub}(\mathbf{k}|\mathbf{p}|\mathbf{q}), \quad (21.92c)$$

$$T^{u,m}_{b,n} = \sum_{\mathbf{p}\in m}\sum_{\mathbf{k}\in n} S^{bu}(\mathbf{k}|\mathbf{p}|\mathbf{q}). \quad (21.92d)$$

These transfers represent $U2U$, $B2B$, $B2U$, and $U2B$ transfers, and are depicted in Fig. 21.6. They provide valuable insights into the physics of turbulence, as well as into dynamo mechanism.

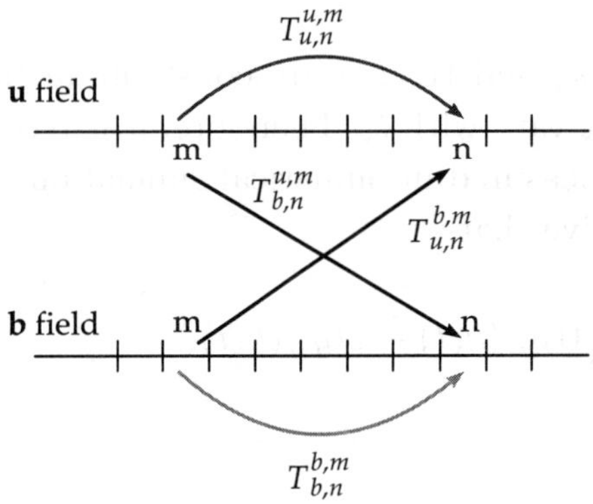

Figure 21.6 Illustration of $U2U$, $B2B$, $U2B$, and $B2U$ shell-to-shell transfers in MHD turbulence.

In addition to the aforementioned shell-to-shell transfers, we define shell-to-shell energy transfers for z^+ and z^- fields as follows:

$$T^{z^+,m}_{z^+,n} = \sum_{\mathbf{p}\in m}\sum_{\mathbf{k}\in n} S^{z^+z^+}(\mathbf{k}|\mathbf{p}|\mathbf{q}), \quad (21.93a)$$

$$T^{z^-,m}_{z^-,n} = \sum_{\mathbf{p}\in m}\sum_{\mathbf{k}\in n} S^{z^-z^-}(\mathbf{k}|\mathbf{p}|\mathbf{q}). \quad (21.93b)$$

These transfers are exhibited in Fig. 21.7. Note that there is no cross transfer from z^+ to z^-. On similar lines, we also define shell-to-shell kinetic and magnetic helicity transfers, as well as shell-to-shell E_A transfers, which are

$$T^{H_M,m}_{H_M,n} = \sum_{\mathbf{p}\in m}\sum_{\mathbf{k}\in n} S^{H_M}(\mathbf{k}|\mathbf{p}|\mathbf{q}), \quad (21.94a)$$

$$T^{H_K,m}_{H_K,n} = \sum_{\mathbf{p}\in m}\sum_{\mathbf{k}\in n} S^{H_K}(\mathbf{k}|\mathbf{p}|\mathbf{q}), \quad (21.94b)$$

$$T^{A,m}_{A,n} = \sum_{\mathbf{p}\in m}\sum_{\mathbf{k}\in n} S^{AA}(\mathbf{k}|\mathbf{p}|\mathbf{q}). \tag{21.94c}$$

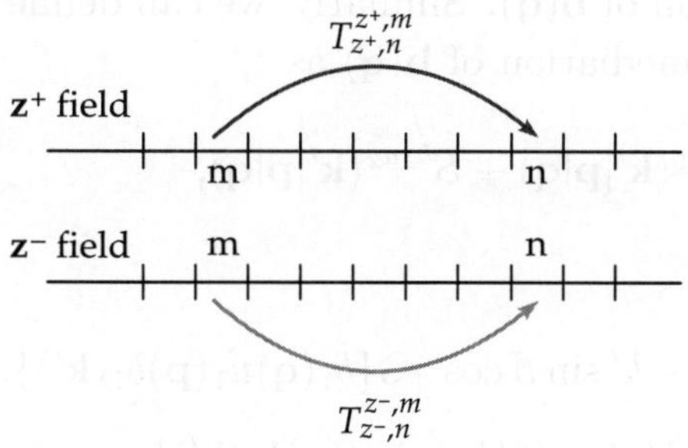

Figure 21.7 Illustration of z^+2z^+ and z^-2z^- shell-to-shell transfers in MHD turbulence.

Note that from energetics considerations, the shell-to-shell transfer from field X of shell m to field Y of shell n is equal and opposite to the shell-to-shell transfer from field Y of shell n to field X of shell m. That is,

$$T^{X,m}_{Y,n} = -T^{Y,n}_{X,m}. \tag{21.95}$$

In the next section, we describe the mode-to-mode energy and helicity transfers in Craya–Herring basis.

21.8 Energy Transfers in Craya–Herring Basis

Following the same procedure as Chapters 9 and 18, we compute the mode-to-mode energy transfers in Craya–Herring basis. The $B2B$ energy transfer is same as the $\mathbf{w}(\mathbf{p})$ to $\mathbf{w}(\mathbf{k}')$ energy transfer discussed in Chapter 18. Since the $U2U$ and $B2B$ transfers have already been discussed in Chapters 4 and 18, we will not repeat them here. In the following, we describe the mode-to-mode energy transfer from $\mathbf{b}(\mathbf{p})$ to $\mathbf{u}(\mathbf{k}')$, and vice versa.

The mode-to-mode energy transfer from $\mathbf{b}(\mathbf{p})$ to $\mathbf{u}(\mathbf{k}')$ with the mediation of $\mathbf{b}(\mathbf{q})$ is

$$S^{ub}(\mathbf{k}'|\mathbf{p}|\mathbf{q}) = S^{u_1b_1}(\mathbf{k}'|\mathbf{p}|\mathbf{q}) + S^{u_2b_2}(\mathbf{k}'|\mathbf{p}|\mathbf{q}), \tag{21.96}$$

where

$$S^{u_1b_1}(\mathbf{k}'|\mathbf{p}|\mathbf{q}) = -k'\sin\beta\cos\gamma\,\Im\{b_1(\mathbf{q})b_1(\mathbf{p})u_1(\mathbf{k}')\}, \tag{21.97a}$$

$$S^{u_2b_2}(\mathbf{k}'|\mathbf{p}|\mathbf{q}) = k'\sin\beta\,\Im\{b_1(\mathbf{q})b_2(\mathbf{p})u_2(\mathbf{k}')\}. \tag{21.97b}$$

In these equations, $S^{u_1 b_1}(\mathbf{k}'|\mathbf{p}|\mathbf{q})$ represents energy transfer from $b_1(\mathbf{p})$ to $u_1(\mathbf{k}')$ with the mediation of $\mathbf{b}(\mathbf{q})$. Similarly, $S^{u_2 b_2}(\mathbf{k}'|\mathbf{p}|\mathbf{q})$ represents the transfer from $b_2(\mathbf{p})$ to $u_2(\mathbf{k}')$ with the mediation of $\mathbf{b}(\mathbf{q})$. Similarly, we can define the energy transfer from $\mathbf{u}(\mathbf{p})$ to $\mathbf{b}(\mathbf{k}')$ with the mediation of $\mathbf{b}(\mathbf{q})$ as

$$S^{bu}(\mathbf{k}'|\mathbf{p}|\mathbf{q}) = S^{b_1 u_1}(\mathbf{k}'|\mathbf{p}|\mathbf{q}) + S^{b_2 u_2}(\mathbf{k}'|\mathbf{p}|\mathbf{q}), \tag{21.98}$$

where

$$S^{b_1 u_1}(\mathbf{k}'|\mathbf{p}|\mathbf{q}) = -k' \sin\beta \cos\gamma \Im\{b_1(\mathbf{q}) u_1(\mathbf{p}) b_1(\mathbf{k}')\}, \tag{21.99a}$$

$$S^{b_2 u_2}(\mathbf{k}'|\mathbf{p}|\mathbf{q}) = k' \sin\beta \Im\{b_1(\mathbf{q}) u_2(\mathbf{p}) b_2(\mathbf{k}')\}. \tag{21.99b}$$

Here, $S^{b_1 u_1}, S^{b_2 u_2}$ represent respective energy transfers from u_1 to b_1, and u_2 to b_2.

Using these mode-to-mode energy transfers, we can define energy fluxes among u_1, b_1, u_2, and b_2 fields. These fluxes are

$$\Pi^{b_1<}_{u_1>}(k_0) = \sum_{|\mathbf{p}|\le k_0} \sum_{|\mathbf{k}|>k_0} S^{u_1 b_1}(\mathbf{k}|\mathbf{p}|\mathbf{q}), \tag{21.100a}$$

$$\Pi^{b_2<}_{u_2>}(k_0) = \sum_{|\mathbf{p}|\le k_0} \sum_{|\mathbf{k}|>k_0} S^{u_2 b_2}(\mathbf{k}|\mathbf{p}|\mathbf{q}), \tag{21.100b}$$

$$\Pi^{u_1<}_{b_1>}(k_0) = \sum_{|\mathbf{p}|\le k_0} \sum_{|\mathbf{k}|>k_0} S^{b_1 u_1}(\mathbf{k}|\mathbf{p}|\mathbf{q}), \tag{21.100c}$$

$$\Pi^{u_2<}_{b_2>}(k_0) = \sum_{|\mathbf{p}|\le k_0} \sum_{|\mathbf{k}|>k_0} S^{b_2 u_2}(\mathbf{k}|\mathbf{p}|\mathbf{q}). \tag{21.100d}$$

The mode-to-mode magnetic helicity transfer in Craya–Herring basis is

$$\begin{aligned} S^{H_M}(\mathbf{k}'|\mathbf{p}|\mathbf{q}) &= \Re\left[\mathbf{u}(\mathbf{q}) \cdot \mathbf{b}(\mathbf{p}) \times \mathbf{b}(\mathbf{k}')\}\right] \\ &= -\sin\gamma \Im\{b_1(\mathbf{k}') b_1(\mathbf{p}) u_2(\mathbf{q})\} - \sin\beta \Im\{b_1(\mathbf{k}') b_2(\mathbf{p}) u_1(\mathbf{q})\} \\ &\quad - \sin\alpha \Im\{b_2(\mathbf{k}') b_1(\mathbf{p}) u_1(\mathbf{q})\}\}. \end{aligned} \tag{21.101}$$

The formulas for the enstrophy and kinetic helicity transfers can be derived in a similar fashion.

In the next section we describe energy transfers in helical basis.

21.9 Energy Transfers in Helical Basis

The energy transfers among the helical velocity modes have been discussed in Section 9.8; and those among the vector modes, here **B** field, have been covered in

Section 18.5. In this section, we will derive formulas for the energy transfers between the helical velocity and magnetic modes.

In helical basis, the mode-to-mode energy transfer from $\mathbf{b(p)}$ to $\mathbf{u(k')}$ with the mediation of $\mathbf{b(q)}$ is given by

$$\begin{aligned} S^{ub}(\mathbf{k'|p|q}) &= \Im\left[\{\mathbf{k'}\cdot\mathbf{b(q)}\}\{\mathbf{b(p)}\cdot\mathbf{u(k')}\}\right] \\ &= \sum_{s_p,s_{k'}} \Im\left[\{\mathbf{k'}\cdot\mathbf{b(q)}\}b_{s_p}(\mathbf{p})u_{s_{k'}}(\mathbf{k'})\{\hat{e}_{s_p}(\mathbf{p})\cdot\hat{e}_{s_{k'}}(\mathbf{k'})\}\right] \\ &= \sum_{s_p,s_{k'}} S^{ub}_{s_{k'}s_p}(\mathbf{k'|p|q}), \end{aligned} \tag{21.102}$$

where $S^{ub}_{s_{k'}s_p}(\mathbf{k'|p|q})$ represents the energy transfer from mode $b_{s_p}(\mathbf{p})$ to mode $u_{s_{k'}}(\mathbf{k'})$ with the mediation of mode $\mathbf{b(q)}$. Note that $S^{ub}_{s_{k'}s_p}(\mathbf{k'|p|q})$ can be simplified further into [see Eq. (9.80)]

$$S^{ub}_{s_{k'}s_p}(\mathbf{k'|p|q}) = \frac{k'}{2}\sin\beta(1+s_p s_{k'}\cos\gamma)\Im\{b_1(\mathbf{q})b_{s_p}(\mathbf{p})u_{s_{k'}}(\mathbf{k'})\}. \tag{21.103}$$

Similarly,

$$\begin{aligned} S^{bu}(\mathbf{k'|p|q}) &= \Im\left[\{\mathbf{k'}\cdot\mathbf{b(q)}\}\{\mathbf{u(p)}\cdot\mathbf{b(k')}\}\right] \\ &= \sum_{s_p,s_{k'}} S^{bu}_{s_{k'}s_p}(\mathbf{k'|p|q}) \end{aligned} \tag{21.104}$$

with

$$S^{bu}_{s_{k'}s_p}(\mathbf{k'|p|q}) = \frac{k'}{2}\sin\beta(1+s_p s_{k'}\cos\gamma)\Im\{b_1(\mathbf{q})u_{s_p}(\mathbf{p})b_{s_{k'}}(\mathbf{k'})\} \tag{21.105}$$

representing the energy transfer from mode $u_{s_p}(\mathbf{p})$ to mode $b_{s_{k'}}(\mathbf{k'})$ with the mediation of mode $\mathbf{b(q)}$.

The mode-to-mode magnetic helicity transfer is

$$\begin{aligned} S^{H_M}(\mathbf{k'|p|q}) &= \Re\left[\mathbf{u(q)}\cdot\mathbf{b(p)}\times\mathbf{b(k')}\}\right] \\ &= \sum_{s_p,s_{k'}} \Re[\{\mathbf{u(q)}\cdot\hat{e}_{s_p}(\mathbf{p})\times\hat{e}_{s_p}(\mathbf{k'})\}b_{s_p}(\mathbf{p})b_{s_{k'}}(\mathbf{k'})] \\ &= \sum_{s_p,s_{k'}} S^{H_M}_{s_{k'}s_p}(\mathbf{k'|p,q}). \end{aligned} \tag{21.106}$$

Using

$$\hat{e}_1(\mathbf{q})\cdot\hat{e}_{s_p}(\mathbf{p})\times\hat{e}_{s_{k'}}(\mathbf{k'}) = \frac{i}{2}[s_{k'}\sin\beta + s_p\sin\alpha], \tag{21.107a}$$

$$\hat{e}_2(\mathbf{q}) \cdot \hat{e}_{s_p}(\mathbf{p}) \times \hat{e}_{s_{k'}}(\mathbf{k}') = \frac{1}{2} s_{k'} s_p \sin\gamma, \tag{21.107b}$$

we derive that

$$S^{H_M}_{s_{k'} s_p}(\mathbf{k}'|\mathbf{p},\mathbf{q}) = -\frac{1}{2}[s_{k'} \sin\beta + s_p \sin\alpha]\Im\{u_1(\mathbf{q}) b_{s_p}(\mathbf{p}) b_{s_{k'}}(\mathbf{k}')\} + \frac{1}{2}\sin\gamma \Re\{u_2(\mathbf{q}) b_{s_p}(\mathbf{p}) b_{s_{k'}}(\mathbf{k}')\}. \tag{21.108}$$

The formulas for the mode-to-mode kinetic helicity and enstrophy transfers can be computed following the procedure outlined in Section 9.8.

Using these energy transfers we can define associated energy fluxes as

$$\Pi^{b_{s_g}<}_{u_{s_r}>}(k_0) = \sum_{|\mathbf{p}|\le k_0} \sum_{|\mathbf{k}|>k_0} S^{ub}_{s_r s_g}(\mathbf{k}'|\mathbf{p}|\mathbf{q}), \tag{21.109a}$$

$$\Pi^{u_{s_g}<}_{b_{s_r}>}(k_0) = \sum_{|\mathbf{p}|\le k_0} \sum_{|\mathbf{k}|>k_0} S^{bu}_{s_r s_g}(\mathbf{k}'|\mathbf{p}|\mathbf{q}), \tag{21.109b}$$

$$\Pi^{H_M s_g}_{H_M s_r}(k_0) = \sum_{|\mathbf{p}|\le k_0} \sum_{|\mathbf{k}|>k_0} S^{H_M}_{s_r s_g}(\mathbf{k}'|\mathbf{p}|\mathbf{q}), \tag{21.109c}$$

where s_g and s_r are the signs of the giver and receiver modes respectively. The corresponding shell-to-shell energy transfers are as follows.

$$T^{b_{s_g},m}_{u_{s_r},n} = \sum_{\mathbf{p}\in m} \sum_{\mathbf{k}\in n} S^{bu}_{s_r s_g}(\mathbf{k}'|\mathbf{p}|\mathbf{q}). \tag{21.110}$$

$$T^{u_{s_g},m}_{b_{s_r},n} = \sum_{\mathbf{p}\in m} \sum_{\mathbf{k}\in n} S^{ub}_{s_r s_g}(\mathbf{k}'|\mathbf{p}|\mathbf{q}). \tag{21.111}$$

In the next four chapters, we will employ the formulation developed in the previous and present chapters to discuss MHD turbulence phenomenologies, dynamo mechanism, etc.

Further Reading

The present chapter presents the most detailed descriptions of energy transfers in MHD turbulence. Historically, Dar et al. (2001) were the first to formulate the mode-to-mode energy transfer for MHD turbulence. Verma (2004) provided more details on this formalism. Alexakis et al. (2007) too discussed shell-to-shell energy transfers in MHD turbulence.

Exercises

1. Show that in MHD turbulence, the kinetic energy feed by the Lorentz force equals T^{ub}.
2. Derive formulas for the cross helicity and enstrophy fluxes in MHD turbulence.

Chapter 22

Models of MHD Turbulence

In the present chapter we present some of the important models of Magnetohydrodynamic (MHD) turbulence. In addition, we also present several solar wind observations and numerical results that provide validation for these models.

22.1 Models of MHD Turbulence

In this section, we briefly describe a number of MHD turbulence phenomenologies. These theories make the following assumptions:

1. As in Kolmogorov's theory for hydrodynamic turbulence, it is assumed that the velocity field is forced by an external force $\mathbf{F}_{\mathrm{ext}}$ that acts at large scales. Note that this force is in addition to the Lorentz force.
2. Most MHD models assume that the flow is three dimensional.
3. For MHD turbulence, both Re and Re_M are large. Hence, both, the velocity and magnetic fields exhibit forward cascade.

Now, we start our discussion on the models.

22.1.1 Kraichnan and Iroshnikov's model—$E(k) \propto k^{-3/2}$

Kraichnan (1965) and Iroshnikov (1964) constructed the first phenomenological model of MHD turbulence. They considered a magnetofluid in an external field

$\mathbf{B}_0$. Here, $\mathbf{z}^+$ and $\mathbf{z}^-$ modes travel in opposite directions with a phase velocity of B_0. Therefore, for $z^\pm \ll B_0$, the oppositely traveling waves of wavenumber k interact weakly during a short interval of Alfvén time $\tau_A(k) = (B_0 k)^{-1}$. Using dimensional arguments, they concluded that the total energy flux

$$\Pi = A^2 \tau_A(k) \left(E_b(k)\right)^2 k^4 = A^2 B_0^{-1} \left(E_b(k)\right)^2 k^3. \tag{22.1}$$

Hence,

$$E_b(k) = A \left(\Pi B_0\right)^{1/2} k^{-3/2}, \tag{22.2}$$

where A is a nondimensional constant of order unity. Note that above arguments assume $z^\pm \ll B_0$. In the absence of a mean magnetic field, Kraichnan (1965) and Iroshnikov (1964) assumed that the magnetic field of the large eddies would act as B_0.

The aforementioned phenomenology belongs to a broad category of *weak turbulence models* because $z^\pm \ll B_0$. Due to weak interactions, Kraichnan and Iroshnikov argued that the kinetic and magnetic energies are equipartitioned. Note that for an Alfvén wave, the average kinetic and magnetic energy are equal. See Section 20.4.

Dobrowolny et al. (1980) generalized this model in the following manner.

22.1.2 Dobrowonly et al.'s model

Dobrowolny et al. (1980) assumed local interactions among the Fourier modes. That is, the wavelengths of the interacting modes z^+ and z^- are approximately equal. We denote the interaction time scales for the fluctuations $z_k^\pm$ as $\tau_k^\pm$. In a unit interaction, the variations in these fluctuations are

$$\delta z_k^\pm \approx \tau_k^\pm z_k^+ z_k^- k. \tag{22.3}$$

Assuming the interactions to be stochastic, in N such interactions, the amplitude variations would be

$$\Delta z_k^\pm \approx (\delta z_k^\pm) \sqrt{N_k^\pm}. \tag{22.4}$$

Therefore, numbers of interactions $N_k^\pm$ required to obtain variations equal to the initial amplitude $z_k^\pm$ are

$$N_k^\pm \approx \frac{1}{k^2 \left(z_k^\mp\right)^2 \left(\tau_k^\pm\right)^2}, \tag{22.5}$$

and the corresponding times scales are

$$T_k^\pm \approx N_k^\pm \tau_k^\pm \approx \frac{1}{k^2 \left(z_k^\mp\right)^2 \tau_k^\pm}. \tag{22.6}$$

Therefore, the fluxes $\Pi_{z^\pm}$ of the fluctuations $z_k^\pm$ would be

$$\Pi_{z^\pm} \approx \frac{\left(z_k^\mp\right)^2}{T_k^\pm} \approx \tau_k^\pm \left(z_k^\pm\right)^2 \left(z_k^\mp\right)^2 k^2. \tag{22.7}$$

In this expression, when we substitute

$$\tau_k^\pm \approx \frac{1}{kB_0}, \tag{22.8}$$

as prescribed by Kraichnan (1965) and Iroshnikov (1964), we obtain

$$\Pi_{z^+} \approx \Pi_{z^-} \approx \frac{1}{B_0} E_{z^+}(k) E_{z^-}(k) k^3 = \Pi. \tag{22.9}$$

For $E_{z^+}(k) \approx E_{z^-}(k)$, we obtain

$$E_{z^+}(k) \approx E_{z^-}(k) \approx \left(B_0 \Pi\right)^{1/2} k^{-3/2}, \tag{22.10}$$

which is the energy spectrum proposed by Kraichnan (1965) and Iroshnikov (1964).

The other choice of the time scale is the nonlinear time scales:

$$\tau_k^\pm \approx \tau_{NL}^\pm \approx \frac{1}{k z_k^\mp}. \tag{22.11}$$

Substitution of these time scales in Eqs. (22.7) yields

$$\begin{aligned} \Pi_{z^\pm} &\approx \left(z_k^\pm\right)^2 \left(z_k^\mp\right) k \\ &\approx E_{z^\pm}(k) \sqrt{E_{z^\mp}(k)} k^{5/2} \end{aligned} \tag{22.12}$$

that in turn leads to

$$E_{z^\pm}(k) = K_\pm (\Pi_{z^\pm})^{4/3} (\Pi_{z^\mp})^{-2/3} k^{-5/3}, \tag{22.13}$$

where $K_\pm$ are constants, referred to as *Kolmogorov's constants for MHD turbulence* (Marsch, 1991; Verma, 2003b, 2004). Due to its similarity with Kolmogorov's fluid turbulence phenomenology, we refer to this phenomenology as

Kolmogorov-like MHD turbulence phenomenology. Equation (22.13) also yields

$$\frac{E_{z^+}(k)}{E_{z^-}(k)} = \frac{K_+}{K_-}\left(\frac{\Pi_{z^+}}{\Pi_{z^-}}\right)^2. \tag{22.14}$$

Marsch (1991) derived Eq. (22.12) by matching the nonlinear term of $\partial E_{z^\pm}/\partial t$ with $\epsilon_{z^\pm}$, which is

$$k z_k^{\mp}\left(z_k^{\pm}\right)^2 \sim \epsilon_{z^\pm}. \tag{22.15}$$

This process too yields Eqs. (22.13).

22.1.3 Model based on energy fluxes

In this subsection, we construct MHD turbulence phenomenology based on energy fluxes. The arguments are in similar lines as those in Secstion 14.2 and 19.2 [also see (Marsch, 1991)].

In the inertial range, where $\mathcal{F}_{\text{ext}} = 0$ and the dissipative terms are weak, Eqs. (21.86) yield constant fluxes for $z^\pm$ variables:

$$\Pi_{z^+} = \epsilon_{z^+}, \tag{22.16a}$$

$$\Pi_{z^-} = \epsilon_{z^-}, \tag{22.16b}$$

where $\epsilon_{z^\pm}$ are the dissipation rates of $z^\pm$ fields. This constancy of the fluxes is valid even in the presence of $\mathbf{B}_0$. Thus, following the arguments of Section 19.2, we expect that the energy spectra $E_{z^\pm}(k)$ depend on B_0, Π_{z^+}, Π_{z^-}, and k. Since $z^\pm$ are in units of velocity, the dimensions of these quantities are as follows:

$$[\Pi_{z^\pm}] = [L^2/T^3]; \quad [E_{z^\pm}(k)] = [L^3/T^2], [B_0] = [L/T]; [k] = [L^{-1}]. \tag{22.17}$$

Since the dynamical equations for $\mathbf{z}^\pm$ are symmetric, we argue that

$$E_{z^\pm}(k) = (\Pi_{z^\pm})^\alpha (\Pi_{z^\mp})^\beta k^\gamma B_0^\delta. \tag{22.18}$$

There are four unknowns and two equations (one each for L and T); hence, it is not possible to determine the four unknowns without additional constraints. For the special case when B_0 effects are ignored ($\delta = 0$), using the dynamical form of the MHD equations [Eq. (22.12)], we obtain

$$\alpha = 4/3; \quad \beta = -2/3; \quad \gamma = -5/3, \tag{22.19}$$

which is the same as Eq. (22.13). This formulation requires further refinement.

We remark that the fluxes $\Pi_{z^\pm}$ are constant in the inertial range, but the energy fluxes such as $\Pi^{u<}_{u>}$, $\Pi^{b<}_{u>}$, $\Pi^{u<}_{b>}$, $\Pi^{b<}_{b>}$ are not constant.

22.1.4 Goldreich and Sridhar—$E(k_\perp) \sim k_\perp^{-5/3}$

Goldreich and Sridhar (1995) constructed a model for strong MHD turbulence. They argued that a *critical balance* is established between the Alfvén and nonlinear time scales, that is,

$$k_\| B_0 \approx k_\perp z^\pm_{k_\perp}. \tag{22.20}$$

An assumption of constant energy flux Π yields

$$\Pi = \left(z^\pm_{k_\perp}\right)^3 k_\perp = \text{const.}, \tag{22.21}$$

or

$$z^\pm_{k_\perp} \sim k_\perp^{-1/3}. \tag{22.22}$$

Hence,

$$E(k_\perp) \approx \frac{\left(z^\pm_{k_\perp}\right)^2}{k_\perp} \propto k_\perp^{-5/3}. \tag{22.23}$$

The condition for critical balance [Eq. (22.20)] and Eq. (22.22) yields

$$k_\| \propto k_\perp^{2/3}. \tag{22.24}$$

Using the above, Goldreich and Sridhar (1995) deduced that

$$E(k_\perp, k_{||}) = \frac{\Pi^{2/3} k_\perp^{-5/3}}{k_\perp k_{||}} = \Pi^{2/3} k_\perp^{-10/3} g\left(k_{||}/k_\perp^{2/3}\right). \tag{22.25}$$

From this, they further deduced that

$$E(k_\perp) \sim \int E(k_\perp, k_{||}) dk_{||} \sim k_\perp^{-8/3}, \tag{22.26a}$$

$$E(k_{||}) \sim \int E(k_\perp, k_{||}) k_\perp dk_\perp \sim k_{||}^{-2}. \tag{22.26b}$$

The aforementioned phenomenologies of Kraichnan (1965), Iroshnikov (1964), Dobrowolny et al. (1980), and Goldreich and Sridhar (1995) assume a forward energy cascade in the presence of a mean magnetic field $\mathbf{B}_0$. Later in this chapter we describe how large B_0 makes the flow quasi-two-dimensional with an inverse cascade of energy. Thus, the ideas discussed in this section need closer examination when the external magnetic field is strong.

22.1.5 Verma—Effective mean magnetic field and $E(k) \propto k^{-5/3}$

As described in the previous section, Kraichnan (1965) and Iroshnikov (1964) assumed that the Alfvén waves propagate with Alfvén velocity, which is of the order of B_0. However, in a strong turbulent regime, magnetic fluctuations are present at all scales. For the fluctuations of Fig. 22.1, we expect that the Alfvénic fluctuation at scale λ_1 would be affected by magnetic field fluctuations of all scales (e.g., of scales λ_2 and λ_3), as well as by the mean magnetic field $\mathbf{B}_0$. Hence, we need to compute the "effective mean magnetic field" that affects the Alfvénic fluctuation at any scale.

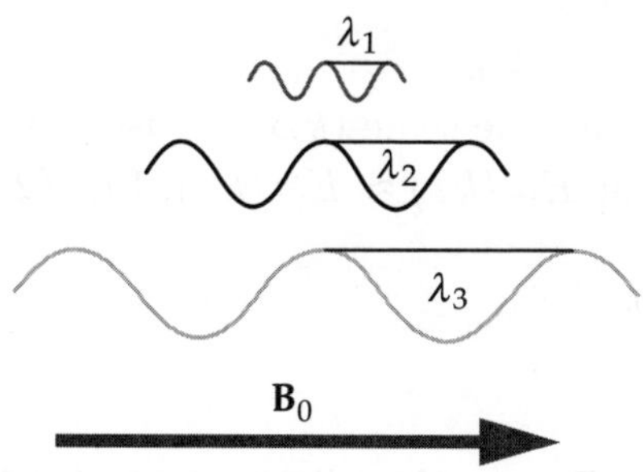

Figure 22.1 A schematic diagram exhibiting magnetic fluctuations at various scales in MHD turbulence. $\mathbf{B}_0$ is the mean magnetic field. A fluctuation with wavelength λ_1 is scattered by the combined effects of all the fluctuations existing at various scales, not by $\mathbf{B}_0$ alone.

Using renormalization group calculations, Verma (1999, 2004) showed that the Alfvén wave of wavenumber k is scattered primarily by the *local* fluctuations with wavenumbers ak, where $a < 1$. This is in a similar spirit as the local interactions of Kolmogorov's theory in which the maximal energy transfer takes places among neighboring wavenumber shells (see Section 5.2). For an Alfvén wave of wavenumber k, Verma (1999, 2004) showed that the *effective mean magnetic field* for the nonlinear interactions is

$$B_0(k) \propto \Pi^{1/3} k^{-1/3}. \tag{22.27}$$

Interestingly, when we substitute this $B_0(k)$ in the Kraichnan and Iroshnikov's model (Eq. (22.2)), we obtain Kolmogorov-like spectrum for MHD turbulence:

$$E_u(k) \approx E_b(k) \approx [\Pi B_0(k)]^{1/2} k^{-3/2} \approx \Pi^{2/3} k^{-5/3} \tag{22.28}$$

Thus, Verma shows consistency between Kraichnan and Iroshnikov's model with Kolmogorov-like scaling, which is observed in the solar wind and in numerical simulations (to be discussed subsequently).

22.1.6 Galtier et al.—Weak turbulence and $E(k_\perp) \propto k_\perp^{-2}$

Galtier et al. (2000) employed weak turbulence technique to MHD turbulence and showed that

$$\Pi \sim \frac{1}{k_{||} B_0} E_{z^+}(k_\perp) E_{z^-}(k_\perp) k_\perp^4 . \tag{22.29}$$

In this expression, the Alfvén time scale $(kB_0)^{-1}$ used by Kraichnan and Iroshnikov is replaced by $(k_{||}B_0)^{-1}$. For $E_{z^+}(k_\perp) \approx E_{z^-}(k_\perp)$, Eq. (22.29) yields

$$E(k_\perp, k_{||}) \propto B_0^{1/2} k_{||}^{1/2} k_\perp^{-2}. \tag{22.30}$$

22.1.7 Boldyrev et al.—Dynamic alignment yields $k^{-3/2}$ spectrum

Boldyrev (2006) hypothesized that in MHD turbulence, the fluctuations in the inertial range have certain alignment. Hence, the interaction time scale is

$$T_k \sim \frac{1}{k u_k \sin\theta_k} \sim \frac{1}{k u_k \theta_k}, \tag{22.31}$$

where θ_k is a measure of the alignment between the velocity and magnetic fluctuations at scale $1/k$. Boldyrev (2006) postulates that $\theta_k \sim k^{-1/4}$ that yields the following energy flux:

$$\Pi \sim \frac{u_k^2}{T_k} \sim k u_k^3 k^{-1/4}. \tag{22.32}$$

Hence,

$$u_k \sim \Pi^{1/3} k^{-1/4} \tag{22.33}$$

that yields the energy spectrum as

$$E_u(k) \sim E_b(k) \sim \frac{u_k^2}{k} \sim \Pi^{2/3} k^{-3/2}. \tag{22.34}$$

Boldyrev and coworkers (Boldyrev, 2006) performed numerical simulations and observed consistency between the numerical results and the aforementioned scaling arguments. A word of caution however is in order. In these simulations, the wavenumber range exhibiting $\theta_k \sim k^{-1/4}$ and $E(k) \sim k^{-3/2}$ is quite narrow. Beresnyak (2011) has contested this scaling.

In addition, Verma (2001a), Verma (2003b), Verma (2003a), and Verma (2004) employed renormalization groups (RG) to MHD turbulence and showed that for zero cross helicity, the following forms of renormalized magnetic diffusivity $\nu(k)$, renormalized viscosity $\eta(k)$, and energy spectra form a self-consistent solution of the RG equations:

$$\nu(k) = \sqrt{K_u} \nu_* \Pi_{\text{tot}}^{1/3} k^{-4/3}, \tag{22.35a}$$

$$\eta(k) = \sqrt{K_u} \eta_* \Pi_{\text{tot}}^{1/3} k^{-4/3}, \tag{22.35b}$$

$$E(k) = K \Pi_{\text{tot}}^{2/3} k^{-5/3}, \tag{22.35c}$$

where K_u, K, ν_*, η_* are constants, and $E(k)$ is the total energy spectrum. These solutions are modified in the presence of cross helicity (Verma, 2003a). These results reinforce Kolmogorov-like phenomenology for MHD turbulence.

The aforementioned MHD turbulence models clearly indicate varying predictions on the energy spectra and fluxes. We will show below that solar wind observations and numerical simulations support 5/3 exponent more strongly than 3/2 exponent. Yet, in this field there are more questions than answers. Moreover, we need more refined work to solve this problem conclusively.

In the next section we will take a detour to real space and extend Kolmogorov's four-fifth law for hydrodynamic turbulence to MHD turbulence.

22.2 Third Order Structure Function: Four-third Law

In this section we derive a relationship between third order structure function and the dissipation rate of MHD turbulence under the assumption of homogeneity and isotropy. The derivation, first performed by Politano and Pouquet (1998), is simpler for $z^{\pm}$ fields because $z^{\pm}$ have constant energy fluxes in the inertial range. The method is very similar to that for passive turbulence (see Section 14.5).

We assume the flow to be homogeneous and isotropic for which the second order correlation functions for $z^{\pm}$ fields are

$$\begin{aligned} C_{z^+}(l) &= \langle \mathbf{z}^+(\mathbf{r}) \cdot \mathbf{z}^+(\mathbf{r}+\mathbf{l}) \rangle = \langle \mathbf{z}^+ \cdot \mathbf{z}^{+'} \rangle, \\ C_{z^-}(l) &= \langle \mathbf{z}^-(\mathbf{r}) \cdot \mathbf{z}^-(\mathbf{r}+\mathbf{l}) \rangle = \langle \mathbf{z}^- \cdot \mathbf{z}^{-'} \rangle, \end{aligned} \tag{22.36}$$

where the unprimed and primed variables are measured at $\mathbf{r}$ and $\mathbf{r}+\mathbf{l}$ respectively. We can model $C_{z^{\pm}}(l)$ as

$$C_{z^{\pm}}(l) = (\overline{z^{\pm}})^2 f_{z^{\pm}}(l), \tag{22.37}$$

where $f_{z^{\pm}}(l)$ are convex functions in l with their maxima at $l = 0$, and $f_{z^{\pm}}(0) = 1$.

The third order correlation functions that appear in the dynamical equations are

$$C_{z^+,j}(\mathbf{l}) = \langle \{\mathbf{z}^+ \cdot \mathbf{z}^{+'}\} z_j^- \rangle, \tag{22.38a}$$

$$C_{z^-,j}(\mathbf{l}) = \langle \{\mathbf{z}^- \cdot \mathbf{z}^{-'}\} z_j^+ \rangle. \tag{22.38b}$$

Using the properties of isotropic tensors we deduce that $C_{z^{\pm},j}$ should transform as vectors that depend only on $\mathbf{l}$. Hence,

$$C_{z^+,j}(\mathbf{l}) = A_{z^+}(l) n_j, \tag{22.39a}$$

$$C_{z^-,j}(\mathbf{l}) = A_{z^-}(l) n_j, \tag{22.39b}$$

where $\mathbf{n} = \mathbf{l}/l$ is the unit vector along $\mathbf{l}$.

The third order structure function for $\mathbf{z}^+$ field is defined as

$$Q_{z^+,j}(\mathbf{l}) = \langle \{(\mathbf{z}^{+'} - \mathbf{z}^+) \cdot (\mathbf{z}^{+'} - \mathbf{z}^+)\} (z_j^{-'} - z_j^-) \rangle, \tag{22.40}$$

whose expansion is

$$\begin{aligned} Q_{z^+,j}(\mathbf{l}) &= -\cancel{\langle \{\mathbf{z}^{+'} \cdot \mathbf{z}^{+'}\} z_j^- \rangle} + \cancel{\langle \{\mathbf{z}^+ \cdot \mathbf{z}^+ z_j^{-'}\} \rangle} \\ &\quad -2\langle \{\mathbf{z}^{+'} \cdot \mathbf{z}^+\} z_j^{-'} \rangle + 2\langle \{\mathbf{z}^{+'} \cdot \mathbf{z}^+\} z_j^- \rangle. \end{aligned} \tag{22.41}$$

In these expressions, the cancelations are due to isotropy. In addition,

$$\langle \{\mathbf{z}^{+'} \cdot \mathbf{z}^{+'}\} z_j^{-'} \rangle - \langle \{\mathbf{z}^+ \cdot \mathbf{z}^+\} z_j^- \rangle = 0 \tag{22.42}$$

due to homogeneity of the flow, and

$$\langle \{\mathbf{z}^{+'} \cdot \mathbf{z}^+\} z_j^{-'} \rangle = C_{z^+,j}(-\mathbf{l}) = -C_{z^+,j}(\mathbf{l}). \tag{22.43}$$

Therefore,

$$Q_{z^+,j}(\mathbf{l}) = 4A_{z^+}(l) n_j, \tag{22.44}$$

and

$$S_3^{z^+}(l) = \langle\{(\mathbf{z}^{+'} - \mathbf{z}^+)\cdot(\mathbf{z}^{+'} - \mathbf{z}^+)\}\{(\mathbf{z}^{-'} - \mathbf{z}^-)\cdot\mathbf{n}\}\rangle = Q_{z^+,1}(l) = 4A_{z^+}(l). \quad (22.45)$$

Hence,

$$\mathbf{Q}_{z^+} = 4A_{z^+}(l)\mathbf{n} = S_3(l)\mathbf{n}. \quad (22.46)$$

We will employ these tensors in the following dynamical equation and derive relationships between the structure function and the energy dissipation rate.

The evolution equation for $C_{z^+}(l)$ is

$$\begin{aligned}
\frac{\partial}{\partial t}\frac{1}{2}\langle\mathbf{z}^+\cdot\mathbf{z}^{+'}\rangle &= \frac{1}{2}\langle\mathbf{z}^{+'}\cdot\frac{\partial}{\partial t}\mathbf{z}^+\rangle + \frac{1}{2}\langle\mathbf{z}^+\cdot\frac{\partial}{\partial t}\mathbf{z}^{+'}\rangle \\
&= \frac{1}{2}\Big[-\partial_j\langle\{\mathbf{z}^+\cdot\mathbf{z}^{+'}\}z_j^-\rangle - \partial_j'\langle\{\mathbf{z}^{+'}\cdot\mathbf{z}^+\}z_j^{-'}\rangle \\
&\quad + \langle\mathbf{z}^{+'}\cdot\mathbf{F}_{z^+}\rangle + \langle\mathbf{z}^+\cdot\mathbf{F}'_{z^+}\rangle + \nu\langle\mathbf{z}^{+'}\cdot\nabla^2\mathbf{z}^+\rangle + \nu\langle\mathbf{z}^+\cdot\nabla'^2\mathbf{z}^{+'}\rangle\Big] \\
&= \frac{1}{2}\Big[\partial_j'\langle\{\mathbf{z}^+\cdot\mathbf{z}^{+'}\}z_j^-\rangle - \partial_j'\langle\{\mathbf{z}^{+'}\cdot\mathbf{z}^+\}z_j^{-'}\rangle\Big] \\
&\quad + 2\langle\mathbf{z}^{+'}\cdot\mathbf{F}_{z^+}\rangle + 2\nu\nabla'^2\langle\mathbf{z}^+\cdot\nabla^2\mathbf{z}^{+'}\rangle\Big] \\
&= \frac{1}{4}\nabla_l\cdot\langle\{(\mathbf{z}^{+'} - \mathbf{z}^+)\cdot(\mathbf{z}^{+'} - \mathbf{z}^+)\}(\mathbf{z}^{-'} - \mathbf{z}^-)\rangle \\
&\quad + \langle\mathbf{z}^{+'}\cdot\mathbf{F}_{z^+}\rangle + \nu\nabla'^2\langle\mathbf{z}^+\cdot\nabla^2\mathbf{z}^{+'}\rangle \\
&= T_{z^+}(\mathbf{l}) + \mathcal{F}_{z^+}(\mathbf{l}) - D_{z^+}(\mathbf{l}),
\end{aligned} \quad (22.47)$$

where $T_{z^+}(\mathbf{l})$ is the nonlinear energy transfer term, and $\mathcal{F}_{z^+}(\mathbf{l})$ and $D_{z^+}(\mathbf{l})$ are respectively the correlations associated with the energy feed by the external force $\mathbf{F}_{z^+}$ and the dissipation rate of $\mathbf{z}^+$ (See Section 12.3).

To derive an expression for the third order structure function, we make similar assumptions as in Section 12.3—steady state, $\nu \to 0$, and the forcing employed at the large length scales. Under these assumptions, in the inertial range, $D_{z^+}(l) \to 0$, and hence,

$$\mathcal{F}_{z^+}(l) = \epsilon_{z^+} = -T_{z^+}(l), \quad (22.48)$$

or

$$\epsilon_{z^+} = -\frac{1}{4}\frac{1}{l^2}\frac{d}{dl}\left[l^2 S_3^{z^+}(l)\right]. \tag{22.49}$$

We perform an integration of this equation that yields

$$S_3^{z^+}(l) = -\frac{4}{3}\epsilon_{z^+} l. \tag{22.50}$$

This derivation could be easily extended to z^- variable too using $\mathbf{z}^+ \leftrightarrow \mathbf{z}^-$. Hence,

$$S_3^{z^\pm}(l) = -\frac{4}{3}\epsilon_{z^\pm} l, \tag{22.51}$$

which is the four-third law for MHD turbulence, first derived by Politano and Pouquet (1998).

The dissipation term with $\nu_- = 0$ is

$$\begin{aligned} D_{z^\pm}(l) &= -\nu\nabla'^2\langle \mathbf{z}^\pm \cdot \mathbf{z}^{\pm'}\rangle = -\nu(\overline{z^\pm})^2\nabla^2 C_{z^\pm}(l) \\ &= -\nu(\overline{z^\pm})^2\frac{d}{l^2}\frac{d}{dl}\left[l^2\frac{d}{dl}f_{z^\pm}\right]. \end{aligned} \tag{22.52}$$

A combination of the nonlinear and diffusion terms yields

$$-\frac{1}{l^2}\frac{d}{dl}\left[l^2\left(\nu(\overline{z^\pm})^2 f'_{z^\pm} + \frac{S_3^{z^\pm}(l)}{4}\right)\right] = \epsilon_{z^\pm}, \tag{22.53}$$

or

$$\nu(\overline{z^\pm})^2 f'_{z^\pm} + \frac{S_3^{z^\pm}(l)}{4} = -\frac{1}{3}\epsilon_{z^\pm} l. \tag{22.54}$$

Equation (22.54) yields $S_3^{z^\pm}(l) = -(4/3)\epsilon_{z^\pm} l$ in the inertial range, and

$$f_{z^\pm}(l) = 1 - \frac{\epsilon_{z^+}}{6\nu(\overline{z^\pm})^2}l^2 = 1 - \frac{1}{2}\frac{l^2}{\lambda_{z^\pm}^2} \tag{22.55}$$

in the dissipation range. Here,

$$\lambda_{z^\pm} = \left(\frac{3\nu(\overline{z^\pm})^2}{\epsilon_{z^\pm}}\right)^{1/2} \tag{22.56}$$

are Taylor's microscales for $z^\pm$ fields.

We make two remarks regarding the connections between the structure functions, energy transfers and spectra discussed earlier.

1. Following the discussion of Section 12.5, we can make connections between the four-third law described earlier and the energy fluxes $\Pi_{z^\pm}(K)$ discussed in Chapter 21. In Fig. 22.2, we relate the terms of $-T_{z^+}(l)$ with $\Pi_{z^+}(K)$. Clearly, $\delta\mathbf{z}^- \leftrightarrow \mathbf{z}^-(\mathbf{q})$ mediates the energy transfers among $\delta\mathbf{z}^+$ fields that appear in the product of $-T_{z^+}(l)$. Note that in Fourier space, the giver mode and the receiver mode can be contrasted, but that is not the case in real space where $-T_{z^+}(l)$ is a sum of all the mode-to-mode energy transfers.

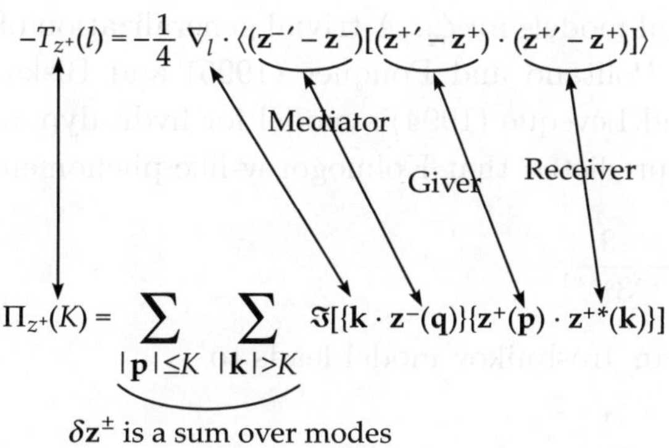

Figure 22.2 Connection between the formulas $-T_{z^+}(l)$ and the energy flux $\Pi_{z^+}(K)$. $\delta\mathbf{z}^-$ mediates energy transfer between $\delta\mathbf{z}^+$ fields in the product term of $-T_{z^+}(l)$.

2. From the third order structure functions derived here, we can estimate that

$$\langle(\delta z^\pm)_l^2(\delta z^\mp)_l\rangle \sim \epsilon_{z^\pm} l. \tag{22.57}$$

In Fourier space, this relation transforms to

$$\left(z_k^\pm\right)^2\left(z_k^\mp\right) k \sim \epsilon_{z^\pm} \tag{22.58}$$

that corresponds to Eq. (22.12) and Kolmogorov-like energy spectrum for MHD turbulence. Thus, the structure function computations are consistent with $k^{-5/3}$ spectrum.

A word of caution—this derivation assumes isotropy that breaks down in the presence of a mean magnetic field.

22.3 Higher Order Structure Functions of MHD Turbulence

Unlike hydrodynamic turbulence, MHD turbulence involves several variables. In this book, for brevity, we present structure function only of $z^{\pm}$. The four-third law discussed in the previous section involves mixed products of δz^{+} and δz^{-}. Hence, $S_q(l)$ may involve various combinations of δz^{+} and δz^{-}. Yet, for simplicity, researchers often define $S_q^{\pm}(l)$ for MHD as follows:

$$S_q(l) = \left\langle (\triangle z^{\pm})_{\parallel}^{q} \right\rangle \sim l^{\zeta_q}. \tag{22.59}$$

There are several models for ζ_q. A trivial generalization of Kolmgorov-like model yields $\zeta_q = q/3$. Politano and Pouquet (1995) and Biskamp and Müller (2000) generalized She and Leveque (1994)'s model for hydrodynamic turbulence to MHD turbulence. They predicted that Kolmogorov-like phenomenology yields

$$\zeta_q^{\mathrm{Ko}} = \frac{q}{9} + 1 - \frac{1}{3^{q/3}}, \tag{22.60}$$

while the Kraichnan–Iroshnikov model leads to

$$\zeta_q^{\mathrm{KI}} = \frac{q}{8} + 1 - \frac{1}{2^{q/4}}. \tag{22.61}$$

These ζ_q's are plotted in Fig. 12.4. Note that $\zeta_3^{\mathrm{Ko}} = 1$ and $\zeta_4^{\mathrm{KI}} = 1$, as expected. Biskamp and Müller (2000) performed numerical simulations and showed that numerical ζ_q's are in closer agreement with ζ_q^{Ko} than with ζ_q^{KI}. Thus, structure functions too provide stronger support to Kolmogorov-like scaling than to Kraichnan–Iroshnikov's scaling. Basu et al. (1998) reported multiscaling exponents for the velocity and magnetic fields of MHD turbulence.

In the following discussions, we go back to Fourier space and describe scaling of various quantities of MHD turbulence.

22.4 Scaling of Cross Helicity and Magnetic Helicity

In MHD turbulence, cross helicity, kinetic helicity, and magnetic helicity play a very important role. In this section, we briefly describe the scaling of cross helicity and magnetic helicity. Note that these are inviscid invariants of MHD turbulence.

22.4.1 Scaling of cross helicity

The easiest description of cross helicity in MHD turbulence is via $E_{z\pm}$ and $\Pi_{z\pm}$. Since $H_c = (E_{z^+} - E_{z^-})/2$, we obtain

$$\Pi_{\mathrm{Hc}} = (\Pi_{z^+} - \Pi_{z^-})/2. \tag{22.62}$$

As described in the previous section, $\Pi_{z^+} = \Pi_{z^-}$ in the Kraichnan–Iroshnikov model, while they are unequal in Kolmogorov-like models of MHD turbulence [See Eqs. (22.9) and (22.14)].

Following a similar procedure as in Section 5.5.1, we can extend the Kolmogorov-like model of MHD turbulence to the inertial–dissipative range. We start with Eqs. (21.86), and search for a steady state solution in the inertial–dissipative range. For simplicity, we also set $\nu_- = 0$, which corresponds to unit magnetic Prandtl number ($\nu_+ = \nu = \eta$).[1] The resulting equations are

$$\frac{d}{dk}\Pi_{z\pm}(k) = -2\nu k^2 E_{z\pm}(k). \tag{22.63}$$

Now we employ Pao's ansatz to $E_{z\pm}(k)$ of Eqs. (22.13) that yields

$$\frac{E_{z\pm}(k}{\Pi_{z\pm}(k)} = K_\pm(\epsilon_{z\pm})^{1/3}(\epsilon_{z\mp})^{-2/3}k^{-5/3}. \tag{22.64}$$

Substitution of this term in Eq. (22.63) yields

$$\frac{d}{dk}\Pi_{z\pm}(k) = -2\nu k^{1/3}K_\pm(\epsilon_{z\pm})^{1/3}(\epsilon_{z\mp})^{-2/3}\Pi_{z\pm}(k) \tag{22.65}$$

whose solution is

$$\Pi_{z\pm}(k) = \epsilon_{z\pm}\exp\left[-\frac{3}{2}K_\pm(k/k_\pm)^{4/3}\right], \tag{22.66a}$$

$$E_{z\pm}(k) = K_\pm(\epsilon_{z\pm})^{4/3}(\epsilon_{z\mp})^{-2/3}k^{-5/3}\exp\left[-\frac{3}{2}K_\pm(k/k_\pm)^{4/3}\right], \tag{22.66b}$$

where

$$k_\pm = \left[\frac{\epsilon_{z\mp}^2}{\epsilon_{z\pm}\nu^3}\right]^{1/4} \tag{22.67}$$

are the dissipation wavenumber for $z^\pm$.

[1] MHD turbulence with very small and very large Pm will be described in Section 22.5.

In the limiting case when $E_{z^+}(k) = E_{z^-}(k)$ or $H_c(k) = 0$, we obtain

$$E_{z^\pm}(k) = K\left(\epsilon_{\text{tot}}\right)^{2/3} k^{-5/3} \exp\left[-\frac{3}{2}K(k/k_\nu)^{4/3}\right], \tag{22.68}$$

where $K_+ = K_- = K$ and

$$k_\pm = \left[\frac{\epsilon_{\text{tot}}}{\nu^3}\right]^{1/4} = k_\nu = k_\eta. \tag{22.69}$$

Note that k_ν and k_η are the dissipation wavenumbers for the velocity and magnetic fields respectively. Since $H_c(k) = 0$, we obtain

$$E_u(k) \sim E_b(k) \sim \left(\epsilon_{\text{tot}}\right)^{2/3} k^{-5/3} \exp\left[-\frac{3}{2}K(k/k_\nu)^{4/3}\right]. \tag{22.70}$$

22.4.2 Scaling of magnetic helicity

In this subsection, we will derive the spectrum and flux of magnetic helicity, which is an inviscid invariant of MHD turbulence. For a steady state, using Eq. (21.88), we deduce that in the inertial range, where $D_{H_M}(k) \approx 0$,

$$\Pi_{H_M}(k) = \epsilon_{H_M}, \tag{22.71}$$

where ϵ_{H_M} is the dissipation rate of magnetic helicity. Now following the derivation of kinetic helicity spectrum (see Section 8.3), we may be tempted to argue that

$$H_M(k) = K_{H_M}\Pi_{H_M}(\Pi_u)^{-1/3}k^{-5/3}, \tag{22.72}$$

where K_{H_M} is a nondimensional constant.

The solar wind observations (Matthaeus and Goldstein, 1982; Brandenburg et al., 2011; Iovieno et al., 2015) however indicate that $|H_M(k)| \sim k^{-8/3}$. This spectrum can be derived using the following arguments. In the solar wind, $H_M(k)/(kE_b(k))$ is quite small; hence, we expect some Fourier modes to yield positive $H_M(k)$, while others to yield negative $H_M(k)$. If we assume that

$$\frac{kH_M(k)}{E_b(k)} \approx r_{H_M} = \text{ a small constant } c(k), \tag{22.73}$$

then

$$H_M(k) = c(k)\frac{1}{k}E_b(k) \sim c(k)\Pi^{2/3}k^{-8/3}. \tag{22.74}$$

Note that $|c(k)| \ll 1$. Clearly, these arguments need verification from numerical simulations and solar wind observations.

22.5 MHD Turbulence for Small and Large Prandtl Numbers

The magnetic Prandtl number, Pm, of many physical systems are either too small or too large. For example, liquid metals and Earth's outer core have $\mathrm{Pm} \sim 10^{-6}$, but interstellar medium has $\mathrm{Pm} \sim 10^{12}$ (with significant uncertainties).

We derive the magnetic diffusion wavenumber, k_η, by balancing ϵ_b with the dissipation rate of the magnetic energy, that is,

$$\int_0^{k_\eta} \eta k^2 E_b(k) dk = \epsilon_b. \tag{22.75}$$

Using

$$E_b(k) \sim (\epsilon_{\mathrm{tot}})^{2/3} k^{-5/3} \exp\left[-\frac{3}{2}K(k/k_\eta)^{4/3}\right] \tag{22.76}$$

as an approximation, we estimate the magnetic dissipation wavenumber k_η as

$$k_\eta = \left(\frac{\epsilon_b^3}{\eta^3 \epsilon_{\mathrm{tot}}^2}\right)^{1/4}. \tag{22.77}$$

Similarly, matching ϵ_u with the viscous dissipation yields the viscous dissipation wavenumber as[2]

$$k_\nu = \left(\frac{\epsilon_u^3}{\nu^3 \epsilon_{\mathrm{tot}}^2}\right)^{1/4}. \tag{22.78}$$

Hence,

$$\frac{k_\eta}{k_\nu} = \left(\frac{\nu}{\eta}\right)^{3/4} \left(\frac{\epsilon_b}{\epsilon_u}\right)^{3/4} = \left(\mathrm{Pm}\frac{\epsilon_b}{\epsilon_u}\right)^{3/4}. \tag{22.79}$$

Clearly, for systems with very small or very large Pm, k_ν and k_η are very different. This feature leads to complex magnetic and kinetic energy spectra for MHD turbulence with small and large Pm's.

In the next subsection we describe the kinetic and magnetic energy spectra of MHD turbulence with small magnetic Prandtl number.

[2]Note that the viscous dissipation wavenumber for MHD turbulence differs from the corresponding wavenumber for hydrodynamic turbulence, which is Kolmogorov's wavenumber $k_d = (\epsilon_u/\nu^3)^{1/4}$.

22.5.1 Energy spectra of small Pm MHD

For MHD turbulence with small Pm, $\epsilon_b \gg \epsilon_u$ because of the dominance of Joule dissipation over viscous dissipation (Verma and Kumar, 2016). Yet, Pm dominates in Eq. (22.79) leading to $k_\eta \ll k_\nu$. Therefore, for $k < k_\eta$, where all the nonlinear terms are significant, both $E_u(k)$ and $E_b(k)$ exhibit $k^{-5/3}$ spectra.

However, for $k_\eta < k < k_\nu$, the Lorentz force is weak, thus making $\mathcal{F}_u(k) \to 0$. Therefore,

$$\Pi_u(k) = \text{const.}; \quad E_u(k) \sim \Pi^{2/3} k^{-5/3}. \tag{22.80}$$

The magnetic energy spectrum $E_b(k)$ however is expected to steepen due to strong Joule dissipation. Using the following scaling arguments (similar to that in Section 16.9), Odier et al. (1998) showed that $E_b(k) \sim k^{-11/3}$. Matching the nonlinear term $(\mathbf{b}\cdot\nabla)\mathbf{u}$ with the magnetic diffusion term yields[3]

$$k b_{k_\eta} u_k \sim \eta k^2 b_k. \tag{22.82}$$

Hence, using $E_u(k) \sim k^{-5/3}$, we obtain

$$E_b(k) = \frac{b_k^2}{k} \sim \left(\frac{b_{k_\eta}}{\eta k}\right)^2 \frac{u_k^2}{k} \sim \left(\frac{b_{k_\eta}}{\eta}\right)^2 \epsilon^{2/3} k^{-11/3}. \tag{22.83}$$

Numerical results of Verma and Kumar (2016) however showed steeper spectrum than $k^{-11/3}$ which they compensated by incorporating an exponential factor to take into account the dissipation effects:

$$E_b(k) \sim \left(\frac{B_{k_\eta}}{\eta}\right)^2 \epsilon^{2/3} k^{-11/3} \exp(-k/k_\eta). \tag{22.84}$$

Since the magnetic energy flux is dissipated by the Joule dissipation, Eqs. (21.75b, 21.77b) yield

$$\frac{d}{dk}\Pi_{b>}^{\text{all}} = -2\eta k^2 E_b(k), \tag{22.85}$$

where $\Pi_{b>}^{\text{all}} = \Pi_{b>}^{b<} + \Pi_{b>}^{u<} + \Pi_{b>}^{u>}$ is the total energy flux coming to the magnetic field at small scales. Using Eq. (22.84), we obtain

$$\Pi_{b>}^{\text{all}}(k) \sim 2\eta k^3 E_b(k) \sim k^{-2/3} \exp(-k/k_\eta). \tag{22.86}$$

[3]In the language of energy transfers, the energy transfer from u_k to b_k with the mediation of b_{k_η} is balanced by the joule dissipation:

$$k b_{k_\eta} u_k k_k \sim \eta k^2 b_k^2. \tag{22.81}$$

Note that this energy transfer is nonlocal for $k \gg k_\eta$.

For $k > k_\nu$, we expect $E_u(k)$ to follow Pao's spectrum, which is described in Section 5.5.1.

Verma and Kumar (2016) used the shell model to simulate MHD turbulence with $\mathrm{Pm} = 10^{-3}$ and 10^{-9}. The energy spectra $E_u(k)$ and $E_b(k)$ computed in these simulations are exhibited in Fig. 22.3. These figures exhibit good agreement between the aforementioned analytical models and numerical results. Here, we point out another interesting feature of the $E_u(k)$ plot. We observe a prominent dip near $k \approx k_\eta$, which may be due to nonlocal energy transfer from u_{k_η}. This issue needs further exploration.

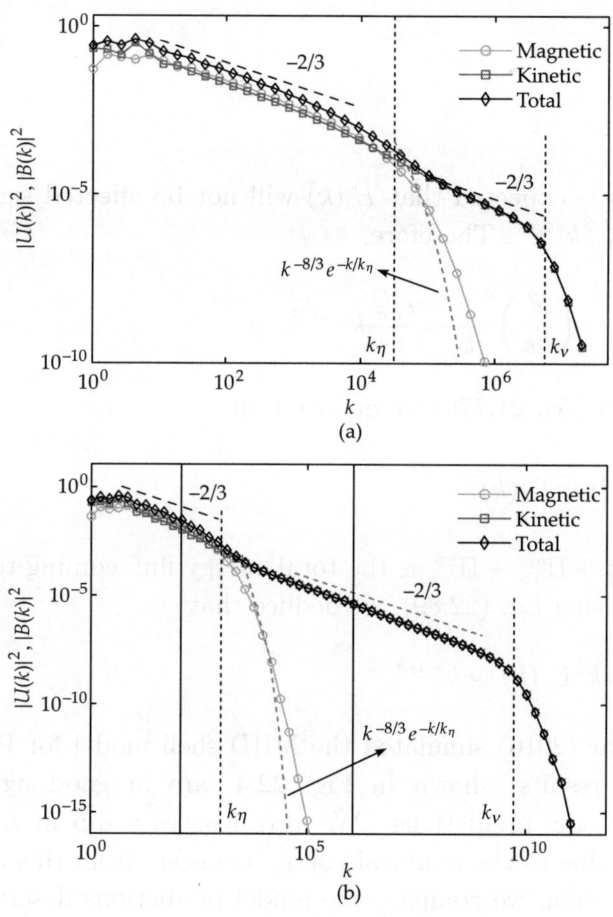

Figure 22.3 For a shell model simulation, the kinetic energy, magnetic energy, and total energy spectra with (a) $\mathrm{Pm} = 10^{-3}$ and (b) $\mathrm{Pm} = 10^{-9}$. The dashed vertical lines represent dissipation wavenumbers k_ν and k_η. Note that for the shell model, $E_u(k) = u_k^2/k$ and $E_b(k) = b_k^2/k$. From Verma and Kumar (2016). Reprinted with permission from Taylor and Francis.

22.5.2 Energy spectra of large Pm MHD

For large Pm, $\nu \gg \eta$; hence, viscous dissipation dominates Joule heating leading to $\epsilon_u \gg \epsilon_b$ (Verma and Kumar, 2016). Yet due to the larger role of Pm than the dissipation rates, Eq. (22.79) yields $k_\nu \ll k_\eta$. In the wavenumber region $k < k_\nu$, all the nonlinear terms are significant; hence, we obtain $E_u(k) \sim E_b(k) \sim k^{-5/3}$.

In the wavenumber band $k_\nu < k < k_\eta$, the momentum equation can be approximated as

$$\cancel{\frac{\partial \mathbf{u}}{\partial t} + \mathbf{u} \cdot \nabla \mathbf{u}} = -\nabla p + \mathbf{b} \cdot \nabla \mathbf{b} + \nu \nabla^2 \mathbf{u}, \tag{22.87}$$

Therefore,

$$k b_k^2 \sim \nu k^2 u_k. \tag{22.88}$$

Since $u_k \ll b_k$, it is expected that $E_b(k)$ will not be affected significantly. Hence, $E_b(k) = b_k^2/k \sim \epsilon_{\text{tot}}^{2/3} k^{-5/3}$. Therefore,

$$E_u(k) \sim \frac{u_k^2}{k} \sim \left(\frac{b_k^2}{\nu k}\right)^2 \frac{1}{k} \sim \frac{\epsilon_{\text{tot}}^{4/3}}{\nu^2} k^{-13/3}. \tag{22.89}$$

Following Eqs. (21.75a, 21.77a) we deduce that

$$\frac{d}{dk}\Pi_{u>}^{\text{all}} = -2\nu k^2 E_u(k), \tag{22.90}$$

where $\Pi_{u>}^{\text{all}} = \Pi_{u>}^{u<} + \Pi_{u>}^{b<} + \Pi_{u>}^{b>}$ is the total energy flux coming to the velocity field at small scales. Using Eq. (22.89), we deduce that

$$\Pi_{u>}^{\text{all}}(k) \sim 2\nu k^3 E_u(k) \sim k^{-4/3} \tag{22.91}$$

Verma and Kumar (2016) simulated the MHD shell model for Pm $= 10^3$ and 10^9. Their numerical results, shown in Fig. 22.4, are in good agreement with the aforementioned model predictions. We also observe a dip in $E_b(k)$ near $k \approx k_\nu$, which is possibly due to the nonlocal energy transfers from this region.

In the next section, we compare the model predictions described earlier in this chapter with the solar wind observations.

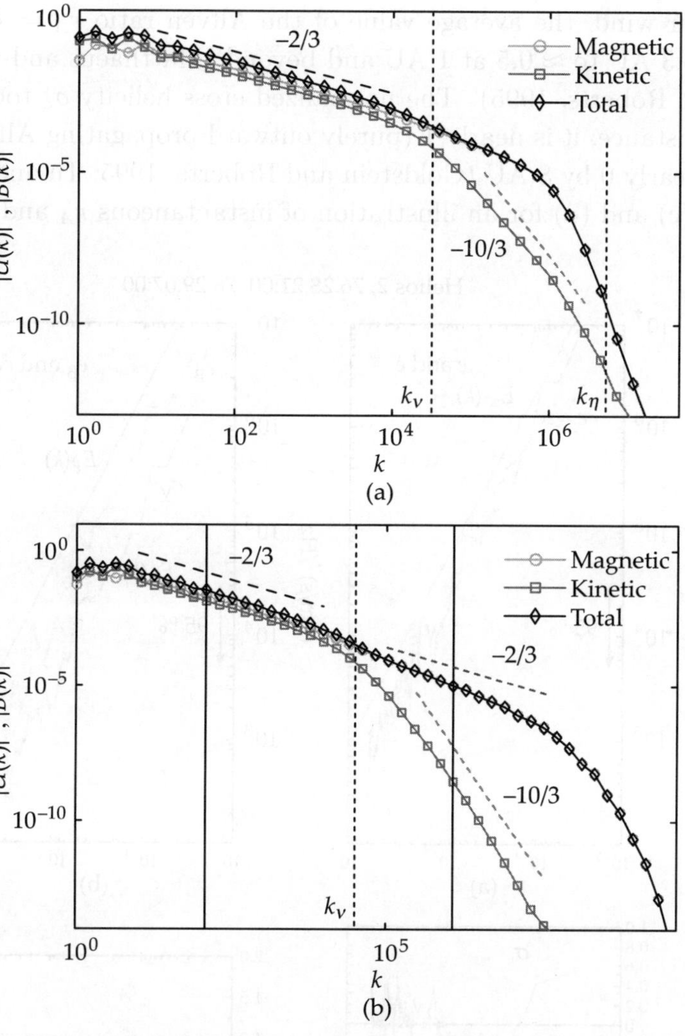

Figure 22.4 For a shell model simulation, the kinetic energy, magnetic energy, and total energy spectra with (a) $\mathrm{Pm} = 10^3$ and (b) $\mathrm{Pm} = 10^9$. From Verma and Kumar (2016). Reprinted with permission from Taylor and Francis.

22.6 Validation Using Solar Wind

Solar wind is a magnetofluid emanating from the Sun, and it flows in all directions. It is a rarified plasma with an average density of 3 to 6 protons per cubic centimeters. Solar wind travels with a speed ranging from 300 km/s to 800 km/s. See Goldstein and Roberts (1995) and Tu and Marsch (1995) for details. Here we list only some of the main results relevant to our discussion.

In the solar wind, the average value of the Alfvén ratio $r_A = E_u/E_b$ decreases from ≈ 5 at 0.3 AU to ≈ 0.5 at 1 AU and beyond (Matthaeus and Goldstein, 1982; Goldstein and Roberts, 1995). The normalized cross helicity σ_c too decreases with heliocentric distance; it is nearly 1 (purely outward propagating Alfvén waves) near 0.3 AU and nearly 0 by 8 AU (Goldstein and Roberts, 1995; Tu and Marsch, 1995). See Fig. 22.5(c) and (d) for an illustration of instantaneous r_A and σ_c.

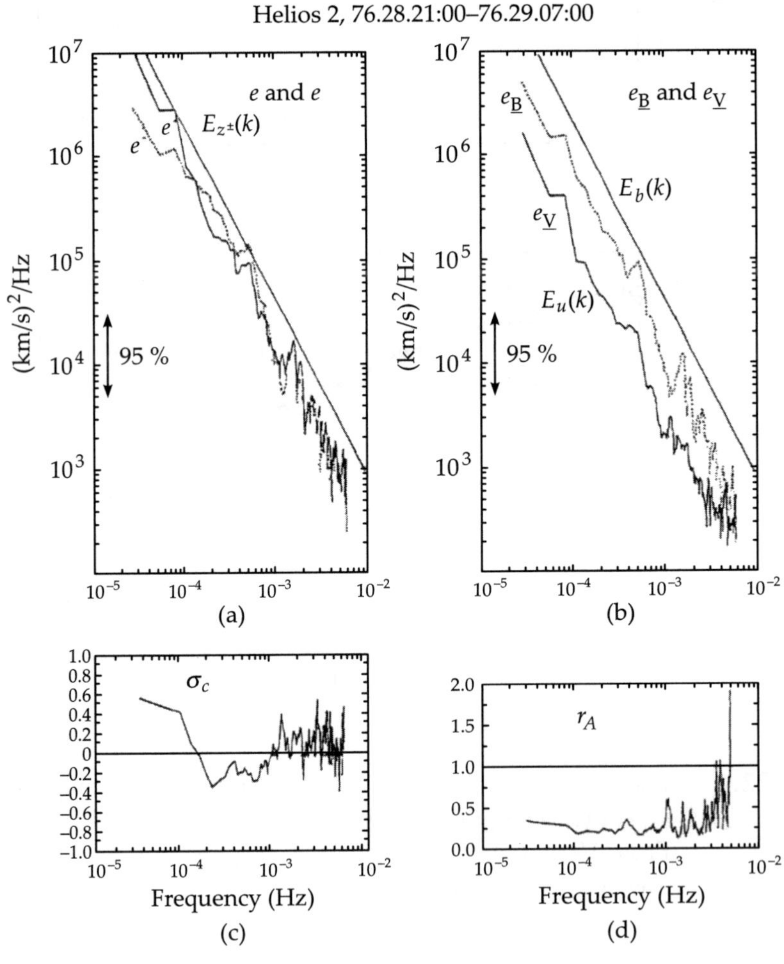

Figure 22.5 (a,b) Energy spectra $E_{z^\pm}(k)$, $E_u(k)$, and $E_b(k)$ of solar wind. (c) The normalized cross helicity $\sigma_c(k)$. (d) Alfvén ratio $r_A(k)$. The frequency f is to be converted to wavenumber k using Taylor's frozen-in turbulence hypothesis. From Tu and Marsch (1995). Reprinted with permission from AGU.

Using the solar wind data measured by spacecrafts, researchers have computed the energy spectra $E_u(k)$, $E_b(k)$, and $E_{z^\pm}(k)$ and compared them with theoretical predictions (see Goldstein and Roberts (1995), Tu and Marsch (1995), and references therein). The observed spectra[4] are in general agreement with Kolmogorov-like phenomenology of MHD turbulence. For example, refer to the energy spectra of Fig. 22.5.

The magnetic helicity spectrum of the solar wind takes both positive and negative signs, and $|H_M(k)| \sim k^{-8/3}$, as discussed in Section 22.4.2 (Iovieno et al., 2015). There is an enormous work on the cross helicity of the solar wind, but they are beyond the scope of this book. Refer to Goldstein and Roberts (1995) and Podesta (2011) for details.

It is difficult to measure the energy fluxes in the solar wind due to lack of three-dimensional data. Therefore, several researchers have measured the energy fluxes indirectly. Some of these attempts are listed here:

1. Verma et al. (1995) computed the total energy flux by substituting the observed energy spectra into Eqs. (22.9, 22.12). They reported that for the observed solar wind temperature profile, the Kolmogorov-like phenomenology yields better predictions than the Kraichnan–Iroshnikov model. Their estimate for the energy flux is 10^3 m^3/s.
2. Sorriso-Valvo et al. (2007) computed the dissipation rates using the structure function [see Eqs. (22.51)]. Their estimate for the total dissipation rate is around 200 m^2/s, which is in general agreement with Verma et al. (1995)'s estimate. Also, note that the structure functions of Sorriso-Valvo et al. (2007) are in general agreement with the predictions of Kolmogorov-like phenomenology of MHD turbulence.

The viscosity and magnetic diffusivity of solar wind are complex tensors due to the presence of mean magnetic field (Parker field) that makes the flow anisotropic. Various uncertainties in the solar wind observations and lack of sound theoretical framework make accurate estimation of ν and η impossible at present. Verma (1996) estimated these quantities using turbulence models, in particular by using Eqs. (22.78, 22.77). He assumed $\epsilon_u = \epsilon_b$, and computed them by substituting the observed energy spectra in Eqs. (22.12). He found that $\epsilon_u \approx \epsilon_b \approx 10^3$ m^2/s. In addition, he estimated $k_\nu \approx k_\eta \approx 10^{-2}$ km^{-1} based on the observed solar wind spectra that show transition from the inertial range to the dissipation range near

[4]Spacecrafts measure the velocity and magnetic fields at a point. Since the spacecraft speeds are much smaller than that of the solar wind, these measurements can be translated to scanning the solar wind along a line. This conversion is by Taylor's frozen-in turbulence hypothesis. The energy spectra $E_\pm(f)$ of Fig. 22.5 are translated to $E_\pm(k)$ using this idea.

this wavenumber. Substitution of these inputs in Eqs. (22.78, 22.77) yields

$$\nu \approx \eta \approx 5 \times 10^7 \text{ m}^2/\text{s}. \tag{22.92}$$

The energy spectrum in the dissipation range is of current interest. The spectral indices reported for the dissipative energy spectrum are -4, -3, $-7/3$, $-11/3$, etc. We believe that the generalized Pao's model of Eq. (22.68) may also be a suitable candidate for describing the inertial–dissipative range spectrum; this issue needs further exploration.

22.7 Validation Using Numerical Simulations

Numerical simulations are often used to test the predictions of MHD turbulence. Here we list key numerical results of MHD turbulence.

The competing spectral indices of MHD turbulence, $-3/2$, $-5/3$, -2, are quite close to each other. Hence, contrasting them using numerical simulations is quite difficult. Therefore, Verma et al. (1996) set out to test the validity of the MHD turbulence models using the properties of energy fluxes. According to Eq. (22.9), in the Kraichnan–Iroshnikov model, $\Pi^+ \approx \Pi^-$ irrespective of the ratio E_{z^+}/E_{z^-}. However, $\Pi^+ \neq \Pi^-$ in the Kolmogorov-like model [see Eq. (22.14)]. Verma et al. (1996) exploited this difference between the two phenomenologies and showed that the numerical results for simulations with $E_{z^+} \gg E_{z^-}$ match with the predictions of the Kolmogorov-like turbulence model (Eq. (22.14)). This was the first such numerical demonstration of the Kolmogorov-like scaling for MHD turbulence.

Later, Müller and Biskamp (2000) performed numerical simulation of MHD turbulence on the 512^3 grid and reported the spectral index to be closer to $-5/3$ than $-3/2$, hence showing consistency with the Kolmogorov-like turbulence model. Beresnyak (2011) made a similar claim, but Mason et al. (2006) argued in favor of a $-3/2$ scaling. We exhibit these results in Fig. 22.6. The plots reveal that neither $k^{-5/3}$ nor $k^{-3/2}$ spectra are conclusive. More refined simulations may resolve some of these differences.

Dar et al. (2001) performed numerical simulations of 2D MHD turbulence and reported various energy fluxes during a quasi-steady state. Later, Debliquy et al. (2005) repeated these computations for a 3D decaying MHD turbulence. In Fig. 22.7(a) we plot the fluxes computed by Debliquy et al. (2005). The maximum values of these fluxes near $k \approx 10$ are exhibited in Fig. 22.7(b). Note that all the fluxes except $\Pi^{u<}_{b<}$ are positive. $\Pi^{b<}_{b>}$ is positive indicating forward magnetic energy cascade.

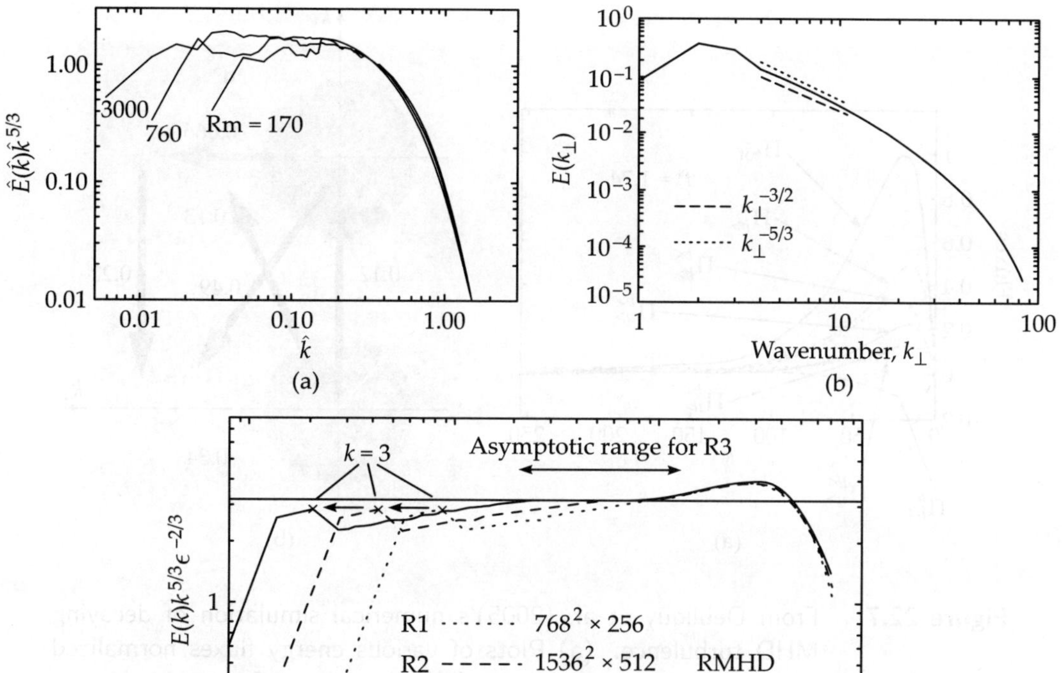

Figure 22.6 Plots of energy spectrum of MHD turbulence computed using numerical simulations: (a) From Müller and Biskamp (2000), (b) From Mason et al. (2006), (c) From Beresnyak (2011). (a,c) support $-5/3$ exponent, while (b) supports $-3/2$ exponent. Reproduced with permission from APS.

Based on the numerical results of decaying MHD turbulence, Debliquy et al. (2005) and Verma (2004) observed that

$$\frac{\Pi^{u<}_{b<}}{\Pi} \approx 0.57(r_A - 0.4). \tag{22.93}$$

Hence, the kinetic energy is transferred to the magnetic energy for $r_A > 0.4$, and vice versa. This phenomena resembles the solar wind evolution in which r_A starts from ≈ 5 near 0.3 AU, approaches $\approx 1/2$ near 1 AU, and maintains this value subsequently.

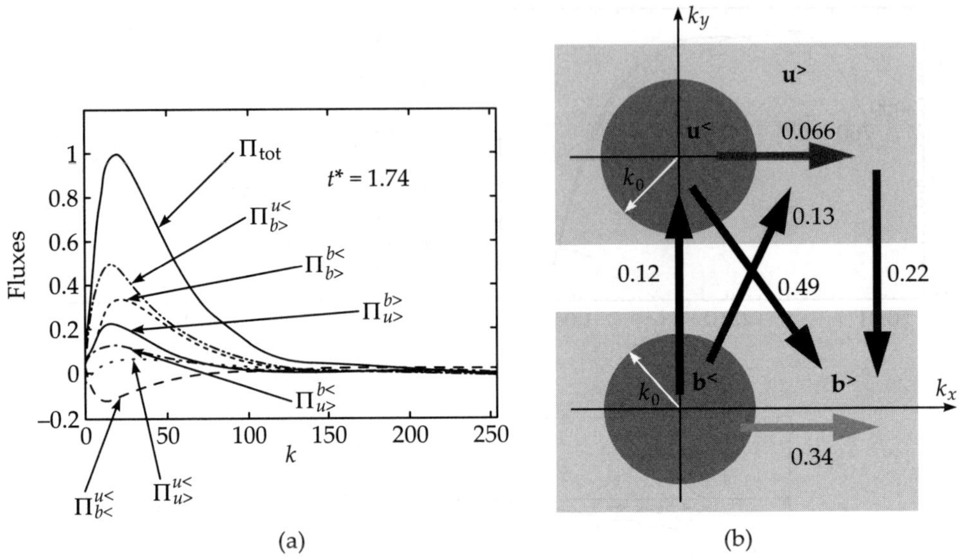

Figure 22.7 From Debliquy et al. (2005)'s numerical simulation of decaying MHD turbulence: (a) Plots of various energy fluxes normalized with $\Pi_{\rm max}$ when $r_A \approx 0.40$. (b) Max values of these fluxes near $k \approx 10$. (a) is adapted from a figure of Debliquy et al. (2005).

Debliquy et al. (2005) also computed the shell-to-shell energy transfers—$T^{u,m}_{u,n}$, $T^{b,m}_{b,n}$, $T^{u,m}_{b,n}$ for the wavenumber shells in the inertial range. These transfers are exhibited in Fig. 22.8. We summarize these observations in a schematic diagram shown in Fig. 22.8(d). From the figures we deduce that $U2U$ and $B2B$ shell-to-shell transfers are forward and local. The shell-to-shell transfers $T^{u,m}_{b,n}$ too are forward and local. Note however that for $r_A \approx 0.4$, $T^{u,m}_{b,m} < 0$ indicating energy transfers from the magnetic field to the velocity field for the same shell. Alexakis et al. (2005b), Mininni et al. (2005), and Alexakis et al. (2007) too computed the shell-to-shell energy transfers for MHD turbulence and observed similar results.

In the next section, we describe how a mean magnetic field affects MHD turbulence.

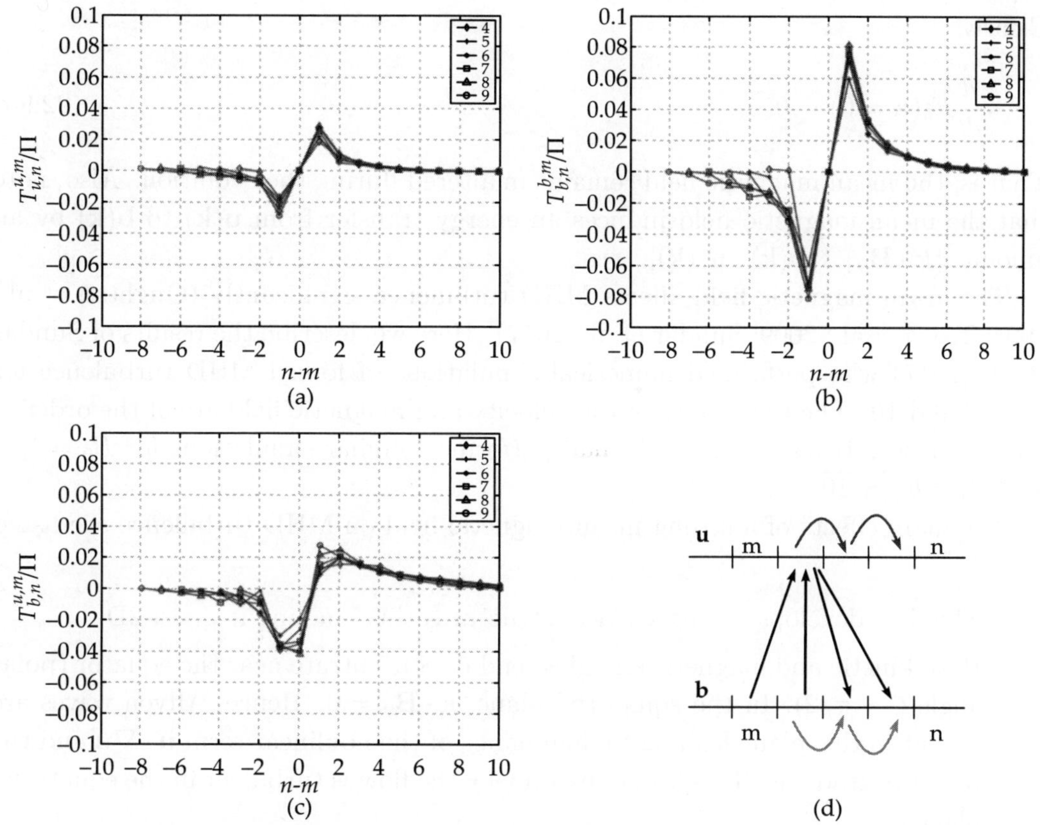

Figure 22.8 From Debliquy et al. (2005)'s numerical simulation of decaying MHD turbulence: During the stage when $r_A \approx 0.40$, (a, b, c) Shell-to-shell energy transfers. (d) A schematic diagram of the shell-to-shell energy transfers. (a), (b), and (c) from Verma (2004), which are reprinted with permission from Elsevier.

22.8 MHD Turbulence in the Presence of a Mean Magnetic Field

We start with Eqs. (20.48). Substitution of $\mathbf{B} = \mathbf{B_0} + \mathbf{b}$ in these equations yields the following set of equations:

$$\begin{aligned} \frac{d}{dt}E_u(\mathbf{k}) &= \sum_{\mathbf{p}} \Im\left[\{\mathbf{k}\cdot\mathbf{u}(\mathbf{q})\}\{\mathbf{u}(\mathbf{p})\cdot\mathbf{u}^*(\mathbf{k})\} - \{\mathbf{k}\cdot\mathbf{b}(\mathbf{q})\}\{\mathbf{b}(\mathbf{p})\cdot\mathbf{u}^*(\mathbf{k})\}\right] \\ &\quad -2\nu k^2 E_u(\mathbf{k}) - \{\mathbf{k}\cdot\mathbf{B_0}\}\Im\left[\mathbf{b}(\mathbf{k})\cdot\mathbf{u}^*(\mathbf{k})\right], \end{aligned} \tag{22.94a}$$

$$\begin{aligned} \frac{d}{dt}E_b(\mathbf{k}) &= \sum_{\mathbf{p}} \Im\left[\{\mathbf{k}\cdot\mathbf{u}(\mathbf{q})\}\{\mathbf{b}(\mathbf{p})\cdot\mathbf{b}^*(\mathbf{k})\} - \{\mathbf{k}\cdot\mathbf{b}(\mathbf{q})\}\{\mathbf{u}(\mathbf{p})\cdot\mathbf{b}^*(\mathbf{k})\}\right] \\ &\quad -2\eta k^2 E_b(\mathbf{k}) + \{\mathbf{k}\cdot\mathbf{B_0}\}\Im\left[\mathbf{b}(\mathbf{k})\cdot\mathbf{u}^*(\mathbf{k})\right]. \end{aligned} \tag{22.94b}$$

Clearly,

$$\frac{d}{dt}E_b(\mathbf{k}=0) = 0. \tag{22.95}$$

That is, the mean magnetic field remains unaltered during the evolution. Also, note that the mean magnetic field induces an energy transfer from $\mathbf{u}(\mathbf{k})$ to $\mathbf{b}(\mathbf{k})$ by an amount $\{\mathbf{k}\cdot\mathbf{B_0}\}\Im\left[\mathbf{b}(\mathbf{k})\cdot\mathbf{u}^*(\mathbf{k})\right]$.

The mean magnetic field affects MHD turbulence significantly (Oughton et al., 1994; Teaca et al., 2009; Sundar et al., 2017). Here we describe the results of Sundar et al. (2017) who performed numerical simulations of forced MHD turbulence for $B_0 = 2$ and 10. The rms values of the velocity and magnetic field are of the order of unity. Earlier, Teaca et al. (2009) had performed similar simulations for $B_0 = b_{\mathrm{rms}}$ and $B_0 = b_{\mathrm{rms}}\sqrt{10}$.

The main effects of a strong mean magnetic field on MHD turbulence are listed here:

1. The flow develops strong vortical structures thus making it quasi-2D.
2. Both kinetic and magnetic energies tend to concentrate near the equator (polar angle $\zeta = \pi/2$). In the equatorial plane, $\mathbf{k}\cdot\mathbf{B}_0 = 0$. Hence, Alfvén waves are absent in this plane leading to dominance of the nonlinear term $\mathbf{u}\cdot\nabla\mathbf{u}$ and the pressure gradient. This is one reason why the flow is turbulent in the equatorial plane.
3. The quasi-2D nature of the flow leads to a strong inverse cascade of kinetic energy. This cascade is responsible for the large-scale vortical structures observed in such flows. Note that $E_{u,\perp}/2E_{u,\parallel} > 1$ at large scales.
4. For large wavenumbers or at small length scales, $E_{u,\perp}/2E_{u,\parallel} < 1$. This is because the parallel component of the velocity field acts like a passive scalar, and it cascades in the forward direction (from small k to large k).

In the next chapter we will show that the quasi-static MHD turbulence exhibits similar behavior as described above.

Based on these results, we make the following remarks that are relevant for the present MHD turbulence models:

1. A large mean magnetic field suppresses the Alfvén effect in the equatorial plane ($\omega = \mathbf{k}\cdot\mathbf{B}_0 = 0$). Therefore, the velocity fluctuations in this plane is driven primarily by $(\mathbf{u}\cdot\nabla)\mathbf{u}$. This result is somewhat contrary to the existing weak turbulence theory of MHD turbulence.
2. A strong mean magnetic field makes the flow strongly anisotropic leading to an inverse cascade of total energy. Note that most of the MHD turbulence models

assume forward cascade of energy; hence, they would need a revision for strong B_0. This feature is relevant for the debate on suitability of the present MHD turbulence models.

3. In the solar wind, Parker field is an order of magnitude stronger than the magnetic field fluctuations (Matthaeus and Goldstein, 1982). Hence, we expect the solar wind to be quasi-two-dimensional exhibiting inverse cascade of energy. However, based on the computations of structure functions and temperature evolution, solar wind appears to exhibit a forward energy cascade. This issue needs further investigation.

In summary, the physics of MHD turbulence is still not fully understood. Some of the key topics are—energy spectra and fluxes, hierarchical interactions of Alfvén waves, effects of mean magnetic field and helicity, etc. We hope that future analytical, experimental, observational, and numerical works will help resolve some of these issues. We remark that MHD turbulence models are critical for modeling many astrophysical objects—stars, planets, galaxies, interstellar medium, dynamo, etc.

Further Reading

The present chapter provides a brief review of MHD turbulence models. For more details, refer to Biskamp (2003), Verma (2004), and original papers cited in this chapter. Also refer to Alexakis et al. (2005b), Mininni et al. (2005), and Alexakis et al. (2007) for shell-to-shell energy transfer computations using numerical simulations. For decay laws of MHD turbulence with and without helicity, refer to Brandenburg and Kahniashvili (2017).

Chapter 23

Dynamo: Magnetic Field Generation in MHD

Self-generation or self-induction of magnetic field in magnetohydrodynamics is called *dynamo mechanism*. It involves growth of initial seed (small magnitude) magnetic field (Moffatt, 1978; Krause and Rädler, 1980; Ruzmaikin et al., 1988; Cardin and Cugliandolo, 2011). Such processes occur in astrophysical objects such as stars, planets, and galaxies. Some recent laboratory experiments (Gailitis et al., 2008; Stieglitz and Müller, 2001; Monchaux et al., 2007) too have been able to produce magnetic field via dynamo mechanism.

Realistic systems exhibiting dynamo processes are quite complex. For example, planetary and stellar dynamos typically involve rotating magnetofluids with a temperature gradient. Dynamo processes are studied from different perspectives—analytical, numerical, experimental—and they are covered in many books and papers. There are many important results, such as α-dynamo, anti-dynamo theorems, experimental dynamos (Karlsruhe, Van-Karman sodium, etc.), dynamo transition, numerical dynamos, planetary and stellar dynamos, galactic dynamos. In this chapter we will discuss only the basic aspects of dynamo, primarily from energy transfer perspectives. For a more detailed discussion, the reader is referred to Moffatt (1978); Krause and Rädler (1980); Ruzmaikin et al. (1988); Verma (2004); Cardin and Cugliandolo (2011); Brandenburg and Subramanian (2005).

23.1 Definitions

In this section, we define the dynamo action. We solve MHD equations (20.10) for a given geometry, force field, and set of parameters. If a magnetic field (constant or time dependent) is sustained asymptotically ($t \to \infty$), then the system is said to exhibit dynamo action. Otherwise, the dynamo process is said to be absent. Generation of a transient magnetic field is not considered to be a dynamo action.

Typically, external forces act as control parameters. For example, in convective dynamos, the temperature difference across the two surfaces acts as a control parameter. A crossover from a no-dynamo state to a dynamo state is called *dynamo transition* (Verma and Yadav, 2013), and the corresponding transition parameters are called *critical parameters.*

A dynamo process depends on the Reynolds number, magnetic Prandtl number Pm, force fields, etc. In nature, the magnetic Prandtl number of magnetofluids is either too small or too large—liquid metals and Earth's core have small Pm ($\sim 10^{-6}$), and the interstellar medium has typically very large Pm ($\sim 10^{12}$). In this chapter we will describe properties of large Pm and small Pm dynamos.

When the magnetic field is very small, it does not have a significant influence on the evolution of the velocity field. But, the velocity field always affects the evolution of the magnetic field. In the beginning of a dynamo process, a small seed field starts to grow. Such a process is called *kinematic dynamo.* When the strength of the magnetic field becomes comparable to the velocity field, it starts to influence the evolution of the velocity field. This is called *dynamic dynamo.* In this chapter we will illustrate these dynamos using several examples.

In the next section, we state some of the anti-dynamo theorems.

23.2 Anti-dynamo Theorems

Many flow configurations do not exhibit dynamo action. Primary examples of such systems are 2D and 2D3C MHD, and axisymmetric flows (Moffatt, 1978; Cardin and Cugliandolo, 2011). We will describe these results in this section.

Theorem 1: *Dynamo action is impossible in 2D MHD.*

Proof: In 2D MHD, the vector potential evolves according to the following equation [Eq. (20.17)]:

$$\frac{\partial A}{\partial t} + \mathbf{u} \cdot \nabla A = \eta \nabla^2 A. \tag{23.1}$$

Note that $\mathbf{A} = A\hat{z}$. From Eq. (23.1) the equation for A^2 can be derived as

$$\frac{\partial}{\partial t}\frac{1}{2}A^2 + \nabla \cdot (\frac{1}{2}A^2\mathbf{u}) = -\eta(\nabla A)^2. \tag{23.2}$$

When we integrate this equation over the whole volume, the convective term vanishes for a periodic or vanishing boundary condition; hence, $\int A^2 d\mathbf{x}$ decays due to the dissipative term. Thus, $A \to 0$, and consequently, $\mathbf{B} = \nabla \times \mathbf{A} \to 0$. Therefore, dynamo action is impossible in 2D MHD. □

In terms of energy transfers, the advection term of Eq. (23.2) exchanges $E_A = A^2/2$ among themselves; hence, it cannot enhance or deplete E_A. In the absence of any source term for E_A, the dissipation term simply destroys E_A. Note however that in 3D, according to Eq. (20.15), the three-dimensional vector potential $\mathbf{A}$ is stretched by the velocity field leading to an enhancement of the magnetic field. This process is analogous to vortex stretching in 3D hydrodynamics that enhances enstrophy.

Theorem 2: *Dynamo action is impossible in 2D3C MHD.*

Proof: We assume that the velocity and magnetic fields are functions only of x, y. Following Jones (2008), 2D3C velocity and magnetic fields can be written as

$$\mathbf{B} = B_z\hat{z} + \mathbf{B}_\perp = B_z\hat{z} + \nabla \times (A\hat{z}), \tag{23.3a}$$

$$\mathbf{u} = u_z\hat{z} + \mathbf{u}_\perp, \tag{23.3b}$$

where $\perp$ represents the horizontal fields. The evolution equations for the components of $\mathbf{B}$ are

$$\frac{\partial A}{\partial t} + (\mathbf{u}_\perp \cdot \nabla)A = \eta\nabla^2 A, \tag{23.4a}$$

$$\frac{\partial B_z}{\partial t} + (\mathbf{u}_\perp \cdot \nabla)B_z = \eta\nabla^2 B_z + (\mathbf{B}_\perp \cdot \nabla)u_z. \tag{23.4b}$$

Following the same arguments as in Theorem 1, we can show that $A \to 0$ asymptotically. Hence, $\mathbf{B}_\perp \to 0$. Therefore, without the source term, B_z too vanishes asymptotically. Hence, there is no dynamo action for this field configuration. □

Theorem 3: *Axisymmetric velocity and magnetic fields vanishing at infinity cannot generate self-induced magnetic field.*

The proof for this anti-dynamo theorem is somewhat more complex. We refer the reader to Cowling (1976) and Jones (2008). There are other anti-dynamo theorems as well, but they are beyond the scope of this book.

23.3 Energetics of a Dynamo

In Chapter 21 we discussed various energy transfers in MHD. We showed that in the equation for magnetic energy E_B, the term $[(\mathbf{u} \cdot \nabla)\mathbf{B}] \cdot \mathbf{B}$ represents exchange of the magnetic energy among the magnetic modes, while the term $[(\mathbf{B} \cdot \nabla)\mathbf{u}] \cdot \mathbf{B}$ yields energy transfers from the velocity field to the magnetic field. The latter term is responsible for the growth of the magnetic field.

For the growth of the magnetic field, it is essential that the growth rate due to the $U2B$ energy transfer exceeds the Joule dissipation. Using dimensional analysis, we translate this condition to the following estimate:

$$(\mathbf{B} \cdot \nabla \mathbf{u}) \cdot \mathbf{B} \gtrsim \eta \mathbf{B} \cdot (\nabla^2 \mathbf{B}), \tag{23.5}$$

or

$$\mathrm{Rm} = \frac{UL}{\eta} \gtrsim 1. \tag{23.6}$$

Hence, the lower bound on the magnetic Reynolds number for the dynamo transition is of the order of unity. For more strict bounds on Rm, refer to Moffatt (1978); Ruzmaikin et al. (1988); Jones (2008).

Verma et al. (2013b) estimated magnetic Reynolds number for protostars and showed $\mathrm{Rm} \sim 10^9$. Using the aforementioned bound on Rm for the dynamo action, they argued that magnetic field would be generated in protostars. Tzeferacos et al. (2018) generated a laser-induced plasma whose $\mathrm{Rm} \sim 600$; based on the aforementioned bound, they argued for the presence of dynamo action in their experiment.

In the next section, we present several examples of kinematic dynamos.

23.4 Kinematic Dynamos

There are a large number of kinematic dynamo models. Here we present only two of them as examples.

23.4.1 Six-mode model—Verma et al. (2008)

Verma et al. (2008) considered the following velocity and magnetic fields containing six Fourier modes:

$$\begin{aligned}\mathbf{u} &= (C_u/\sqrt{2})(\hat{x}\sin x\cos z - \hat{z}\cos x\sin z) + (D_u/\sqrt{2})(\hat{y}\sin y\cos z - \hat{z}\cos y\sin z)\\ &\quad +(4A_u/\sqrt{6})(-\hat{x}\sin x\cos y\cos 2z - \hat{y}\cos x\sin y\cos 2z + \hat{z}\cos x\cos y\sin 2z).\end{aligned} \tag{23.7a}$$

$$\begin{aligned}\mathbf{b} &= (C_b/\sqrt{2})(\hat{x}\sin x\cos z - \hat{z}\cos x\sin z) + (D_b/\sqrt{2})(\hat{y}\sin y\cos z - \hat{z}\cos y\sin z)\\ &\quad +(4A_b/\sqrt{6})(-\hat{x}\sin x\cos y\cos 2z - \hat{y}\cos x\sin y\cos 2z + \hat{z}\cos x\cos y\sin 2z),\end{aligned} \tag{23.7b}$$

where A_u, C_u, D_u, A_b, C_b, and D_b are real numbers. We consider the box size to be π^3. The velocity and magnetic fields are nondimensionalized using

$$u \to u(\nu/d); \quad b \to b(\nu/d), \tag{23.8}$$

where d is the system size.

For a given $\mathbf{u}$, the evolution equations for the magnetic modes are (Verma et al., 2008)

$$\dot{A}_b = -\frac{6}{\text{Pm}}A_b, \tag{23.9a}$$

$$\dot{C}_b = \frac{1}{\sqrt{6}}[D_u A_b - D_b A_u] - \frac{2}{\text{Pm}}C_b, \tag{23.9b}$$

$$\dot{D}_b = \frac{1}{\sqrt{6}}[C_u A_b - C_b A_u] - \frac{2}{\text{Pm}}D_b. \tag{23.9c}$$

Clearly, $A_b \to 0$ asymptotically. Hence,

$$\frac{d}{dt}\begin{pmatrix} C_b \\ D_b \end{pmatrix} = \begin{pmatrix} -2/\text{Pm} & -A_u/\sqrt{6} \\ -A_u/\sqrt{6} & -2/\text{Pm} \end{pmatrix}\begin{pmatrix} C_b \\ D_b \end{pmatrix}. \tag{23.10}$$

The eigenvalues of the above matrix are

$$\lambda_\pm = \pm\frac{A_u}{\sqrt{6}} - \frac{2}{\text{Pm}}. \tag{23.11}$$

If we assume $A_u > 0$, then λ_- is always negative, but λ_+ can become positive if

$$\frac{A_u}{\sqrt{6}} > \frac{2}{\text{Pm}}. \tag{23.12}$$

Under this condition, C_b and D_b would grow with the growth rate of λ_+. Thus, the six-mode model of Verma et al. (2008) would exhibit dynamo action if

$$A_u > 2\sqrt{6}/\text{Pm}. \tag{23.13}$$

Another way to deduce this result is the following. In order to focus on growth rate due to the $U2B$ energy transfers only, we take $\mathrm{Pm} \to \infty$ limit. In addition, $A_b \to 0$ as $t \to \infty$. Now, taking a time derivative of Eq. (23.9b) yields

$$\ddot{C}_b = \frac{1}{6} A_u^2 C_b. \tag{23.14}$$

Thus, C_b would grow with a growth rate of $A_u/\sqrt{6}$ due to the nonlinear energy transfers. In the presence of magnetic diffusivity, this growth rate must exceed the diffusion rate, which is $2/\mathrm{Pm}$. Thus, we arrive at the same condition as Eq. (23.13).

The aforementioned condition yields the following lower bound on the magnetic Reynolds number for dynamo action:

$$\mathrm{Rm} = \frac{Ud}{\eta} > A_u \mathrm{Pm} > 2\sqrt{6}. \tag{23.15}$$

In the next subsection we describe kinematic Roberts dynamo.

23.4.2 Roberts dynamo

Roberts (1972) [also see Jones (2008)] considered the following 2D3C velocity field to excite dynamo action:

$$\mathbf{u} = \hat{x} 2 \sin y + \hat{y} 2 \sin x + \hat{z} 2 (\cos x - \cos y). \tag{23.16}$$

This is a helical field with the Fourier components (in Craya–Herring and helical basis) listed in Table 23.1. For the aforementioned flow, the following solution for the magnetic field was attempted:

$$\mathbf{B} = \mathbf{b}(x, y) \exp(pt + ikz). \tag{23.17}$$

Dynamo transition occurs when $\Re(p) > 0$. Roberts (1972) showed an existence of dynamo transition for a set of k's and Rm's. Note that the magnetic field is three-dimensional that makes the dynamo possible (recall anti-dynamo theorem for 2D3C flows).

Table 23.1 Fourier decomposition of the velocity field of the Roberts flow in Craya–Herring and helical basis. These modes are maximally helical.

Mode	(u_1, u_2)	(u_+, u_-)	$E(\mathbf{k})$	$H_K(\mathbf{k})$
$\mathbf{q} = (1, 0, 0)$	$(i, -1)$	$(-\sqrt{2}, 0)$	1	1
$\mathbf{p} = (0, 1, 0)$	$(-i, 1)$	$(\sqrt{2}, 0)$	1	1

The analysis of Roberts dynamo in real space is quite complex. In the following simplified analysis, we will consider a set of interacting triads and explore if dynamo solution exists for such systems. Here, we present two simple spectral models based on Roberts flow.

23.4.3 A 2D3C helical dynamo model?

We consider the velocity field of Eq. (23.16). Note that the wavenumbers $\mathbf{p} = (0, 1, 0)$ and $\mathbf{q} = (1, 0, 0)$ of the flow couples nonlinearly with $\mathbf{k}' = (-1, -1, 0)$. We consider helical magnetic modes at these wavenumbers—$\mathbf{b}_{s_p}(\mathbf{p})$, $\mathbf{b}_{s_q}(\mathbf{q})$, and $\mathbf{b}_{s_{k'}}(\mathbf{k}')$—and explore if they grow or not. We also choose $\mathbf{u}_{s_{k'}}(\mathbf{k}') = 0$.

In Section 20.6 we derived the equations for the helical magnetic modes in a triad as

$$\dot{b}_{s_{k'}}(\mathbf{k}') = k' s_{k'} \sum_{s_p, s_q} g[u^*_{s_q}(\mathbf{q}) b^*_{s_p}(\mathbf{p}) - u^*_{s_p}(\mathbf{p}) b^*_{s_q}(\mathbf{q})] - \frac{1}{\mathrm{Pm}} k'^2 b_{s_{k'}}(\mathbf{k}'). \tag{23.18}$$

Hence, the dynamical equations for the helical modes discussed here are

$$\dot{b}_{s_{k'}}(\mathbf{k}') = -2g s_{k'} (b^*_{s_p}(\mathbf{p}) + b^*_{s_q}(\mathbf{q})) - \frac{2}{\mathrm{Pm}} b_{s_{k'}}(\mathbf{k}'), \tag{23.19a}$$

$$\dot{b}_{s_p}(\mathbf{p}) = \sqrt{2} g s_p b^*_{s_{k'}}(\mathbf{k}') - \frac{1}{\mathrm{Pm}} b_{s_p}(\mathbf{p}), \tag{23.19b}$$

$$\dot{b}_{s_q}(\mathbf{p}) = \sqrt{2} g s_q b^*_{s_{k'}}(\mathbf{k}') - \frac{1}{\mathrm{Pm}} b_{s_q}(\mathbf{q}), \tag{23.19c}$$

where

$$g = \frac{1}{2\sqrt{2}} s_{k'} s_p s_q \{s_{k'} \sin\alpha + s_p \sin\beta + s_q \sin\gamma\}. \tag{23.20}$$

The variables in these equations are complex. Hence, we will get six real equations whose stability matrix is to be analyzed. In the following discussion, we take a shorter and approximate approach by taking the limit $\mathrm{Pm} \to \infty$.

Taking the limit $\mathrm{Pm} \to \infty$ allows us to decipher the $U2B$ energy transfer. We take another time derivative of Eq. (23.19a) and then substitute Eqs. (23.19b, 23.19c) in the resulting equation. These operations yield

$$\ddot{b}_{s_{k'}}(\mathbf{k}') = -2g^2 \sqrt{2} s_{k'} (s_p + s_q) b_{s_{k'}}(\mathbf{k}'). \tag{23.21}$$

Hence, the growth rate of the dynamo due to nonlinear energy transfer is

$$\gamma_b = g\sqrt{-2g^2 \sqrt{2} s_{k'} (s_p + s_q)}. \tag{23.22}$$

For the magnetic field to grow, this growth rate must be positive, which is possible only if

$$s_{k'}(s_p + s_q) < 0, \tag{23.23}$$

that is, when $s_{k'}$ and $(s_p + s_q)$ have opposite signs. Otherwise, the magnetic field will decay.

According to Theorem 2 of Section 23.2, the above field configuration should not exhibit dynamo action, but a truncated model does show a possibility of dynamo action. This is related to the difference between global stability and the stability of a truncated system. Hence, care needs to be exercised. We expect a decay of magnetic energy if we solve the full set of equations, or when we include more interacting modes.

23.4.4 A tetrahedron helical dynamo model—Stepanov and Plunian (2018)

Stepanov and Plunian (2018) considered another dynamo based on Roberts flow. In the following discussion, we describe a special case of Stepanov and Plunian (2018)'s dynamo. The wavenumbers of the model are shown in Fig. 23.1. The velocity field is nonzero only for wavenumbers $\mathbf{q}_1 = (1, 0, 0)$ and $\mathbf{q}_2 = (0, 1, 0)$, while the magnetic field is nonzero for the wavenumbers $\mathbf{p} = (0, 0, 1)$, $\mathbf{k}'_1 = (-1, 0, -1)$, $\mathbf{k}'_2 = (0, -1, -1)$. As in the previous example, we assume the modes to be purely helical.

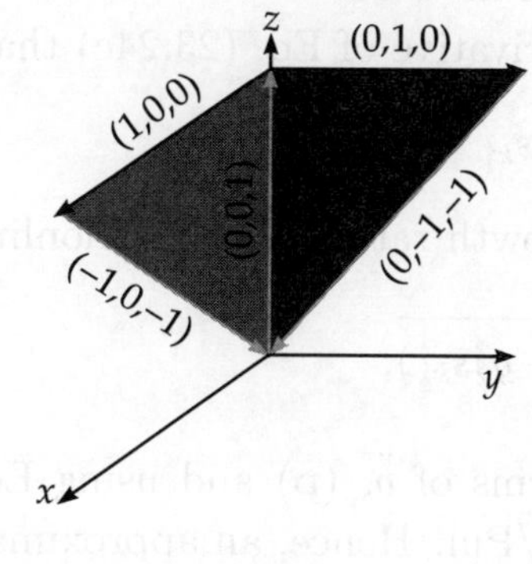

Figure 23.1 The wavenumbers of the dynamo model of Stepanov and Plunian (2018) form two triads: $\{(0, 0, 1), (1, 0, 0), (-1, 0, -1)\}$ and $\{(0, 0, 1), (0, 1, 0), (0, -1, -1)\}$. The wavenumbers (1,0,0) and (0,1,0) contain velocity modes, while (0,0,1), $(0, -1, -1)$ and $(-1, 0, -1)$ contain magnetic modes.

Following the same procedure as in the previous example, we deduce the following dynamical equations for the three helical magnetic modes:

$$\dot{b}_{s_{k'_1}}(\mathbf{k}'_1) = -2g_1 s_{k'_1} b^*_{s_p}(\mathbf{p}) - \frac{2}{\mathrm{Pm}} b_{s_{k'_1}}(\mathbf{k}'_1), \tag{23.24a}$$

$$\dot{b}_{s_{k'_2}}(\mathbf{k}'_2) = 2g_2 s_{k'_2} b^*_{s_p}(\mathbf{p}) - \frac{2}{\mathrm{Pm}} b_{s_{k'_2}}(\mathbf{k}'_2), \tag{23.24b}$$

$$\dot{b}_{s_p}(\mathbf{p}) = \sqrt{2} s_p [g_1 b^*_{s_{k'_1}}(\mathbf{k}'_1) - g_2 b^*_{s_{k'_2}}(\mathbf{k}'_2)] - \frac{1}{\mathrm{Pm}} b_{s_p}(\mathbf{p}), \tag{23.24c}$$

where g_1, g_2 are given by

$$g_1 = \frac{1}{2\sqrt{2}} s_{k'_1} s_p \{ s_{k'_1} \sin\alpha + s_p \sin\beta + \sin\gamma \}, \tag{23.25a}$$

$$g_2 = \frac{1}{2\sqrt{2}} s_{k'_2} s_p \{ s_{k'_2} \sin\alpha + s_p \sin\beta + \sin\gamma \}. \tag{23.25b}$$

Note that $s_{q_1} = s_{q_2} = 1$.

Equations (23.24) can be written in matrix form as

$$\dot{\mathbf{v}}(t) = A\mathbf{v}^*, \tag{23.26}$$

which has to be converted to six real equations with real and imaginary parts of $\mathbf{v}$. Then, the condition for the dynamo action can be determined from the eigenvalues of the matrix. This analysis is somewhat complex; hence, we look for an approximate solution by taking the limit $\mathrm{Pm} \to \infty$.

We take another time derivative of Eq. (23.24c) that yields

$$\ddot{b}_{s_p}(\mathbf{p}) = -2\sqrt{2} s_p (g_1^2 s_{k'_1} + g_2^2 s_{k'_2}) b_{s_p}(\mathbf{p}) \tag{23.27}$$

resulting in the following growth rate due to the nonlinear energy transfer:

$$\gamma_b = \sqrt{-2\sqrt{2} s_p (g_1^2 s_{k'_1} + g_2^2 s_{k'_2})}. \tag{23.28}$$

By ignoring some of the terms of $\ddot{b}_{s_p}(\mathbf{p})$ and using Eq. (23.24c), we can estimate the decay rate of $b_{s_p}(\mathbf{p})$ as $1/\mathrm{Pm}$. Hence, an approximate condition for the dynamo action is

$$\gamma_b > 1/\mathrm{Pm}. \tag{23.29}$$

From the form of γ_b, it is evident that the dynamo action is possible when

$$s_p (g_1^2 s_{k'_1} + g_2^2 s_{k'_2}) < 0, \tag{23.30}$$

or, when the helicity of $(0,0,1)$ mode is opposite to that of $\mathbf{k}_1$ and $\mathbf{k}_2$.

Example 23.1: Work out the mode-to-mode energy transfers for the aforementioned dynamo model of Stepanov and Plunian (2018).

Solution: We derive the following mode-to-mode energy transfers from the velocity field to the magnetic field using Eq. (21.102) of Section 21.9, which is

$$\begin{aligned} S^{bu}(\mathbf{k}'|\mathbf{p}|\mathbf{q}) &= \Im\left[\{\mathbf{k}'\cdot\mathbf{b}(\mathbf{q})\}\{\mathbf{u}(\mathbf{p})\cdot\mathbf{b}(\mathbf{k}')\}\right] \\ &= \sum_{s_p,s_{k'}} S^{bu}_{s_{k'}s_p}(\mathbf{k}'|\mathbf{p}|\mathbf{q}). \end{aligned}$$

Substitution of the mode amplitudes in this formula yields

$$\begin{aligned} S^{bu}(\mathbf{p}|\mathbf{q}_1|\mathbf{k}_1) &= -\frac{1}{2}\Im\{b_{s_p}b_{s_{k'_1}}\} \\ S^{bu}(\mathbf{p}|\mathbf{q}_2|\mathbf{k}_2) &= \frac{1}{2}\Im\{b_{s_p}b_{s_{k'_2}}\} \\ S^{bu}(\mathbf{k}_1|\mathbf{q}_1|\mathbf{p}) &= -\frac{1}{\sqrt{2}}(1+\frac{s_{k'_1}}{\sqrt{2}})\Im\{b_{s_p}b_{s_{k'_1}}\} \\ S^{bu}(\mathbf{k}_2|\mathbf{q}_2|\mathbf{p}) &= \frac{1}{\sqrt{2}}(1+\frac{s_{k'_2}}{\sqrt{2}})\Im\{b_{s_p}b_{s_{k'_2}}\}. \end{aligned}$$

The total energy transferred from the velocity field to the magnetic field is a sum of all the aforementioned terms.

Example 23.2: Consider the aforementioned dynamo model of Stepanov and Plunian (2018). Solve the model in Craya–Herring basis, instead of helical basis.

Solution: We label the wavenumbers as $\mathbf{q}_1=(1,0,0)$, $\mathbf{q}_2=(0,1,0)$, $\mathbf{p}=(0,0,1)$, $\mathbf{k}'_1=(-1,0,-1)$, and $\mathbf{k}'_2=(0,-1,-1)$, and take the velocity field of Eq. (23.16). From Table 23.1, we have $u_1(\mathbf{q}_1)=i$, $u_2(\mathbf{q}_1)=-1$, $u_1(\mathbf{q}_2)=-i$, $u_2(\mathbf{q}_2)=1$. Using the equations of Section 20.5 we derive the equations of motion for the b_1 and b_2 components of the magnetic modes as

$$\begin{aligned} \dot{b_1}(\mathbf{k}'_1) &= \sqrt{2}b_1^*(\mathbf{p})-\frac{2}{\mathrm{Pm}}b_1(\mathbf{k}'_1), \\ \dot{b_1}(\mathbf{k}'_2) &= -\sqrt{2}b_1^*(\mathbf{p})-\frac{2}{\mathrm{Pm}}b_1(\mathbf{k}'_2), \\ \dot{b_1}(\mathbf{p}) &= \frac{1}{\sqrt{2}}(b_1^*(\mathbf{k}_2)-b_1^*(\mathbf{k}_1))-\frac{1}{\mathrm{Pm}}b_1(\mathbf{p}), \\ \dot{b_2}(\mathbf{k}'_1) &= -b_2^*(\mathbf{p})+ib_1^*(\mathbf{p})-\frac{2}{\mathrm{Pm}}b_1(\mathbf{k}'_1), \end{aligned}$$

$$\begin{aligned}\dot{b_2}(\mathbf{k}_2') &= b_2^*(\mathbf{p}) - ib_1^*(\mathbf{p}) - \frac{2}{\mathrm{Pm}} b_1(\mathbf{k}_1'),\\ \dot{b_2}(\mathbf{p}) &= b_2^*(\mathbf{k}_1') - b_2^*(\mathbf{k}_2') - \frac{i}{\sqrt{2}}(b_1^*(\mathbf{k}_1') - b_1^*(\mathbf{k}_2')) - \frac{1}{\mathrm{Pm}} b_2(\mathbf{p}).\end{aligned}$$

Now we take the limit $\mathrm{Pm} \to \infty$. Taking another time derivative of the equations for $b_1(\mathbf{p})$ and $b_2(\mathbf{p})$ yields

$$\begin{aligned}\ddot{b_1}(\mathbf{p}) &= -2b_1(\mathbf{p}),\\ \ddot{b_2}(\mathbf{p}) &= -2b_2(\mathbf{p}).\end{aligned}$$

Hence, the nonlinear interactions induce pure oscillations on $b_1(\mathbf{p})$ and $b_2(\mathbf{p})$, and consequently on other modes as well. Therefore, the magnetic diffusion will kill the magnetic field. Hence, for a general magnetic configuration, this model does not exhibit dynamo action.

The said conclusion appears to contradict the result described in the present section according to which the magnetic field grows if the condition of Eq. (23.30) is satisfied. Note however that the aforementioned condition is derived for pure helical modes. The configuration of this example consists of a combination of several helical modes. To illustrate:

$$\dot{b}_+(\mathbf{k}_1) = \frac{1}{\sqrt{2}}(\dot{b}_2(\mathbf{k}_1) + i\dot{b}_1(\mathbf{k}_1)) = -\frac{b_2(\mathbf{p})}{\sqrt{2}} + ib_1(\mathbf{p})(1 + \frac{1}{\sqrt{2}}).$$

From Eq. (23.18), we deduce that

$$\begin{aligned}\dot{b}_+(\mathbf{k}_1) &= \sqrt{2}(-\sqrt{2}g_+ b_+^*(\mathbf{p}) - \sqrt{2}g_- b_-^*(\mathbf{p}))\\ &= -\frac{2}{2\sqrt{2}}\{(1+\sqrt{2})b_+^*(\mathbf{p}) - b_-^*(\mathbf{p})\}\\ &= -\frac{b_2(\mathbf{p})}{\sqrt{2}} + ib_1(\mathbf{p})(1 + \frac{1}{\sqrt{2}}).\end{aligned}$$

Hence, a general $\mathbf{b}$ field consists of both positive and negative helicity modes. Thus, the condition for dynamo action for a general field configuration and for purely helical modes are very different.

There are many more issues in kinematic dynamo, but they are beyond the scope of this book. Now we move on to discuss dynamic dynamo models.

23.5 Dynamic Dynamos

In dynamic models, both **u** and **b** fields affect each other. An external force is employed to the velocity field to compensate for the Joule dissipation. Without an external force, both velocity and magnetic fields would decay away.

23.5.1 Six-mode model—Verma et al. (2008) revisited

Let us reconsider the six-mode model of Verma et al. (2008), and make a dynamic model for the same. Verma et al. (2008) forced the three velocity modes with amplitudes $0, f/\sqrt{2}$, and $f/\sqrt{2}$ respectively. They derived the following evolution equations for the velocity and magnetic modes of Eqs. (23.7a, 23.7b):

$$\dot{A}_u = -\frac{2}{\sqrt{6}}[C_u D_u - C_b D_b] - 6, \tag{23.32a}$$

$$\dot{C}_u = \frac{1}{\sqrt{6}}[D_u A_u - D_b A_b] - 2 + \frac{f}{\sqrt{2}}, \tag{23.32b}$$

$$\dot{D}_b = \frac{1}{\sqrt{6}}[C_u A_u - C_b A_b] - 2 + \frac{f}{\sqrt{2}}, \tag{23.32c}$$

$$\dot{A}_b = -\frac{6}{\mathrm{Pm}} A_b, \tag{23.32d}$$

$$\dot{C}_b = \frac{1}{\sqrt{6}}[D_u A_b - D_b A_u] - \frac{2}{\mathrm{Pm}} C_b, \tag{23.32e}$$

$$\dot{D}_b = \frac{1}{\sqrt{6}}[C_u A_b - C_b A_u] - \frac{2}{\mathrm{Pm}} D_b. \tag{23.32f}$$

Verma et al. (2008) reported that for this system, dynamo transition occurs for

$$\mathrm{Pm} > 1; \quad \mathrm{Rm} > \mathrm{Rm}_c = 6\sqrt{2\mathrm{Pm}}. \tag{23.33}$$

Note that the transition parameters for the kinematic and dynamic models are quite different. Interestingly, this dynamic model yields a constant solution for the magnetic field. For details on this model, refer to Verma et al. (2008).

In the next section we describe properties of dynamo transition.

23.6 Dynamo Transition and Bifurcation Analysis

MHD systems exhibit very interesting behavior near the onset of dynamo transition. It is an extensive area of research. Here we list only some of the salient properties of dynamo transition.

1. Dynamo transition in dynamic models is an example of nonlinear instability. Without the nonlinear terms, the induction equation is a diffusion equation,

$$\partial_t \mathbf{b} = \eta \nabla^2 \mathbf{b}, \tag{23.34}$$

which yields a decaying solution. It is the nonlinear terms that generate dynamo. Note however that kinematic dynamos generate magnetic field via linear instability.

2. The nature of dynamo transition appear to depend significantly on the magnetic Prandtl number. Yadav et al. (2010, 2012) show that in Taylor–Green dynamo, dynamos with large Prandtl numbers yield supercritical pitchfork bifurcation in which the magnetic field increases with the forcing as

$$B \propto \sqrt{f - f_c}, \tag{23.35}$$

where f_c is the critical forcing. On the other hand, Taylor–Green dynamos with small Prandtl numbers exhibit subcritical pitchfork bifurcation. Morin and Dormy (2009) obtained similar behavior in spherical dynamo simulations. These features of dynamo transition have been captured reasonably well by several low-dimensional models (Verma and Yadav, 2013).

3. Beyond the dynamo transition, but not too far away from it, MHD systems exhibits rich behavior. They exhibit all sorts of dynamical states—fixed point, periodicity, quasi-periodicity, chaos, etc. Yadav et al. (2010, 2012) reported such behavior in Taylor–Green dynamo. Von-Karman Sodium (VKS) dynamo (an experimental dynamo) also shows such states (Berhanu et al., 2010). Yadav et al. (2010, 2012) obtained two chaotic states that transition to chaos via period doubling and quasi-periodic routes.

4. A dynamo is typically a higher-dimensional dynamical system. On many occasions, it exhibits multiple coexisting states that are accessed by different initial conditions (for a given set of parameters). In fact, Verma and Yadav (2013) showed that in the subcritical transition regime, for a given parameter set, some initial conditions yield a no-dynamo solution, but some others yield a dynamo solution. Some researchers believe that such mechanism may be responsible for the abrupt shutdown of Martian dynamo.

5. Interestingly, $\mathbf{B}(0, 0, 1)$ appears to be a generic magnetic mode during the transition. It occurs in Roberts dynamo (Roberts, 1972) as well as in Taylor–Green dynamo (Yadav et al., 2010). This issue needs further investigation.

6. Near a dynamo transition, the magnetic energy and kinetic energy are not necessarily equipartitioned. In fact, their ratio can differ significantly from unity (Yadav et al., 2012; Roberts and Glatzmaier, 2000). This behavior is contrary to the behavior in the turbulent regime where equipartition is expected among E_u and E_b. This deviation from equipartition is due to the existence of multiple complex attractors near the dynamo transition.

In the next section we describe energy transfers in turbulent dynamos.

23.7 Energy Transfers in Turbulent Dynamos

A magnetofluid becomes turbulent far beyond the dynamo transition. It is usually assumed that the turbulent systems are ergodic, that is, they cover all of the available phase space.[1] As a result, MHD turbulence exhibits generic properties—similar energy spectrum, an approximate equipartition of kinetic energy and magnetic energy, etc. Here we list some of the generic properties of turbulent dynamos; these properties have been deduced using turbulent models and numerical results.

In a generic dynamo simulation, we start with a small seed magnetic field. For suitable parameters, the magnetic energy grows exponentially in the initial stages, after which it saturates. The initial growth is governed by kinematic processes (because the seed field is weak), but the back-reaction from the magnetic field become significant during saturation. Though the magnetic energy growth is exponential in time, the rate of growth depends on the initial condition and Prandtl number. See Fig. 23.2 for an illustration in which two different initial conditions—narrow wavenumber band (NB) at $k = [2, 4]$, and broad band (BB) at $k = [2, 384]$—yield very different growth rates for the magnetic field.

The growth rate of the magnetic energy is also scale dependent. We start with an equation for the growth of the magnetic energy:

$$\begin{aligned}\partial_t E_b(k) \quad &\approx \quad \sum_{\mathbf{p}} -\Im\left[\{\mathbf{k}\cdot\mathbf{b}(\mathbf{q})\}\{\mathbf{u}(\mathbf{p})\cdot\mathbf{b}^*(\mathbf{k})\}\right] \\ &\quad + \sum_{\mathbf{p}} \Im\left[\{\mathbf{k}\cdot\mathbf{u}(\mathbf{q})\}\{\mathbf{b}(\mathbf{p})\cdot\mathbf{b}^*(\mathbf{k})\}\right] - 2\eta E_b(k). \end{aligned} \tag{23.36}$$

The terms in the RHS represent $U2B$ and $B2B$ transfers, and magnetic diffusion. Assuming local interactions ($k \approx p \approx q$), we obtain

$$\partial_t E_b(k) \approx k u_k E_b(k) - 2\eta k^2 E_b(k). \tag{23.37}$$

[1] The phase space structure of a chaotic system is typically quite complex, and is far from ergodic. Hence, the often-quoted assumption of ergodicity for a turbulent flow is questionable.

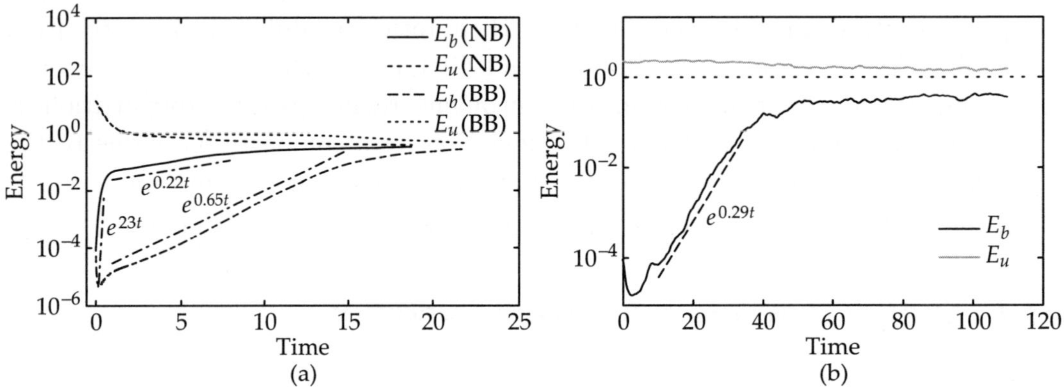

Figure 23.2 Plots of $E_u(t)$ and $E_b(t)$ vs. t: (a) For $\mathrm{Pm} = 20$ with broad band (BB) and narrow band (NB) initial condition. In NB, E_b is uniformly distributed in $k = [2, 4]$ band, but in BB, it is distributed over $k = [2, 384]$. The growth rates are very different for the two cases. (b) For $\mathrm{Pm} = 0.2$. From Kumar et al. (2014b, 2015). Reprinted with permission from Institute of Physics, and Taylor and Francis respectively.

Hence, the growth rate of $E_b(k)$ is

$$\gamma_b(k) \approx k u_k - \eta k^2 \approx \epsilon^{1/3} k^{2/3} - \eta k^2, \tag{23.38}$$

where ϵ is the total dissipation rate. The two factors in the growth rate balance each other at

$$k_\eta = \left(\frac{\epsilon}{\eta^3} \right)^{1/4}, \tag{23.39}$$

which is the diffusion wavenumber of the magnetic field. Hence, for $k \ll k_\eta$, the magnetic energy grows at a rate of

$$\gamma_b(k) \sim k^{2/3}, \tag{23.40}$$

but in the dissipative range ($k > k_\eta$), E_b decays exponentially.

Considering the complex energy exchange mechanisms in MHD such as variable energy fluxes of Section 21.6, we can claim that these scaling arguments are approximate. Yet, they explain some of the growth processes in the kinematic regime, which may be their domain of applicability.

It is useful to note that the viscous dissipation wavenumber, also called Kolmogorov wavenumber, is

$$k_\nu = \left(\frac{\epsilon}{\nu^3}\right)^{1/4}. \tag{23.41}$$

Hence,

$$\frac{k_\eta}{k_\nu} = \mathrm{Pm}^{3/4}. \tag{23.42}$$

A more sophisticated arguments of Section 22.5 indicate that

$$\frac{k_\eta}{k_\nu} = \left(\mathrm{Pm}\frac{\epsilon_b}{\epsilon_u}\right)^{3/4}, \tag{23.43}$$

where ϵ_u, ϵ_b are the respective dissipation rates for the velocity and magnetic fields. For the present discussion, we will ignore the ϵ_u/ϵ_b factor. Clearly, $k_\eta \ll k_\nu$ for small Prandtl numbers, and $k_\eta \gg k_\nu$ for large Prandtl numbers. This separation has a major impact on the energy transfers mechanism, which will be described in the following section.

Dynamos with moderate Prandtl numbers (~ 1), small Prandtl numbers, and large Prandtl numbers show different patterns of energy transfers. The behavior of dynamos with moderate Prandtl numbers are quite similar to those discussed in Chapter 21. The properties of small Pm and large Pm dynamos will be described now.

23.7.1 Small Pm dynamos

The Prandtl number of liquid metals is quite small due to their large electrical conductivity. For example, sodium and Earth's outer core have $\mathrm{Pm} \sim 10^{-6}$. Using the arguments described in Section 22.5.1, we deduce that k_η is small.

Since η is large for small Pm dynamos, using Eq. (23.38), we deduce that the growth rate $\gamma_b(k)$ is positive only for small k's or at large length scales. Hence, during the kinematic regime, the maximum energy transfer to the magnetic field occurs at large scales. Using the formula for $U2B$ energy transfers, the growth rate of $E_b(k)$ is

$$\partial_t E_b(k) \approx \sum_{\mathbf{p}} -\Im\left[\{\mathbf{k}\cdot\mathbf{b}(\mathbf{q})\}\{\mathbf{u}(\mathbf{p})\cdot\mathbf{b}^*(\mathbf{k})\}\right], \tag{23.44}$$

where $\mathbf{q} = \mathbf{k} - \mathbf{p}$. Under the assumption of local interaction ($k \approx p \approx q \approx 1/L$), we obtain

$$\partial_t E_b \approx \frac{U}{L} E_b, \tag{23.45}$$

where U, L are the large-scale velocity and length respectively. Therefore, we can deduce that (Verma, 2002; 2004)

$$E_b(t) = E_b(0) \exp(tU/L). \tag{23.46}$$

Thus, the time scale for the growth of the magnetic energy is L/U, which is the eddy turnover time of large-scale flows.

Note that the aforementioned formula is applicable in the kinematic regime only, much before the saturation of the dynamo. This growth rate is consistent with many observations and numerical simulations. Also, using these arguments we can estimate the saturation time scale as L/U, or one eddy turnover time. For galaxies, taking $L \sim 10^{18}$ km and $U \sim 10^3$ km/s yields the saturation time scale as approximately 10^7 years, which appears to be reasonable. Hence, such mathematical models are useful for assessing general features of dynamo action in astrophysical objects.

Kumar et al. (2015) studied the energy transfers in a dynamo with Pm $= 0.2$ with vanishing kinetic and magnetic helicities. Their conclusions on these transfers are illustrated in Fig. 23.3. The energy fluxes and shell-to-shell energy transfers for this case are quite similar to those for Pm $= 1$. See Kumar et al. (2015) for details.

In the next section we describe the energy transfers in large Pm dynamos.

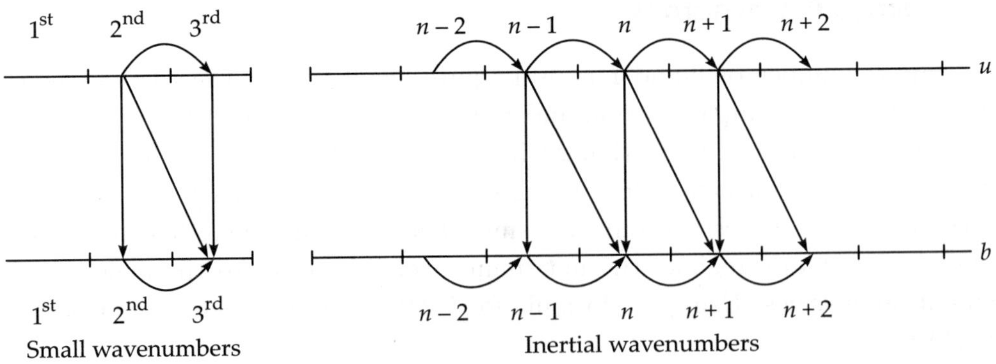

Figure 23.3 For small-Pm dynamo (with Pm $= 0.2$), a schematic diagram of shell-to-shell energy transfers among the u-shells and the b-shells. From Kumar et al. (2015). Reprinted with permission from Taylor and Francis.

23.7.2 Large Pm dynamos

Interstellar magnetofluids have large magnetic Prandtl number ($\gtrsim 10^{11}$). As argued in Section 22.5.2, large Pm dynamos have $k_\eta \gg k_\nu$. The properties of dynamos for large Pm dynamos are different from those of small Pm dynamos.

Large Pm dynamos have small η and large k_η. Hence, the nonlinear term dominates the weak magnetic diffusion. Therefore, in the kinematic regime, the magnetic energy growth is proportional to $k^{2/3}$ for $k \ll k_\eta$. Consequently, the magnetic energy at large wavenumbers grows at a fast rate. This feature is observed in many numerical simulations [see for example, Kumar et al. (2014b, 2015)]. It is however important to keep in mind that the highest growth rate and the highest energy level are two different things. The wavenumber shell with largest growth rate may not have the largest energy.

The aforementioned growth law is expected only during the kinematic regime. At a later stage, when the magnetic energy has also grown at smaller wavenumbers, the velocity field transfers energy to the magnetic field at large scales, as discussed in the previous subsection.

Using numerical simulation of nonhelical dynamo with Pm = 20, Kumar et al. (2014b, 2015) constructed a scenario of the energy transfers for large-Pm dynamo. These transfers are summarized in Fig. 23.4, according to which the magnetic energy growth is caused by nonlocal energy transfers from large-scale (or forcing-scale) velocity field to small-scale magnetic field. With time, the peak of these energy transfers moves towards lower wavenumbers. As a result, the integral length of the magnetic field increases with time.[2]

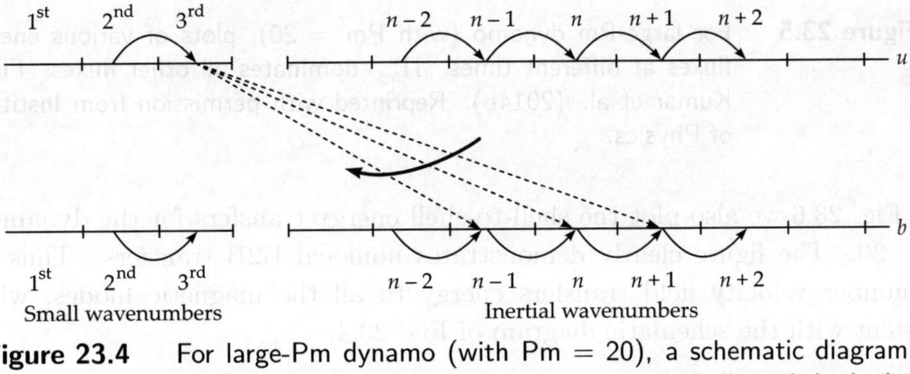

Figure 23.4 For large-Pm dynamo (with Pm = 20), a schematic diagram of shell-to-shell energy transfers among the u-shells and the b-shells. The wavenumber range with maximal $U2B$ transfer shifts leftward with time (see red thick arrow). From Kumar et al. (2015). Reprinted with permission from Taylor and Francis.

[2]The formula of Eq. (23.38) may need to be modified due to the nonlocal interactions.

For this dynamo, in Fig. 23.5 we illustrate various energy fluxes at various times. As illustrated in the figure, $\Pi^{u<}_{b>}$, which is positive, is responsible for the growth of the magnetic field. Note that the peak of $\Pi^{u<}_{b>}$ shifts leftward with time, as in the schematic diagram of Fig. 23.4. Also, note that $\Pi^{b<}_{b>} > 0$. Hence, the growth of the large-scale magnetic field is not due to an inverse cascade of magnetic energy.

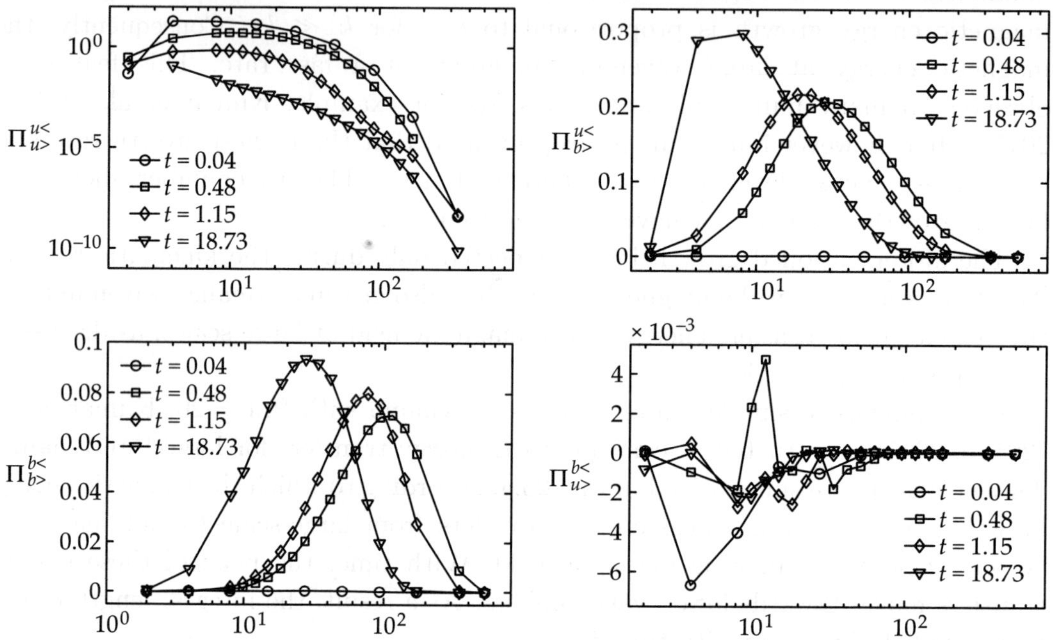

Figure 23.5 For large-Pm dynamo (with Pm = 20), plots of various energy fluxes at different times. $\Pi^{u<}_{b>}$ dominates all other fluxes. From Kumar et al. (2014b). Reprinted with permission from Institute of Physics.

In Fig. 23.6 we also plot the shell-to-shell energy transfers for the dynamo with Pm = 20. The figure clearly demonstrates nonlocal U2B transfers. Thus, small wavenumber velocity field transfers energy to all the magnetic modes, which is consistent with the schematic diagram of Fig. 23.4.

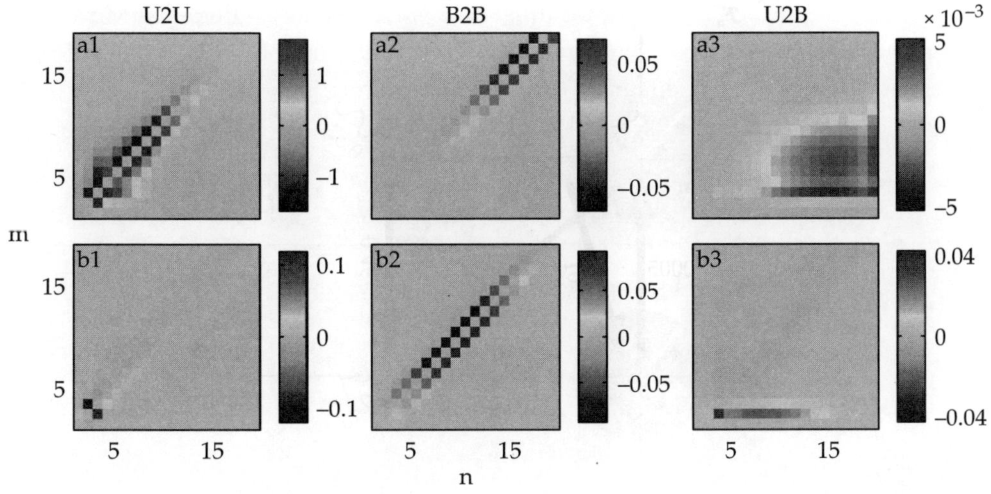

Figure 23.6 For large-Pm dynamo (with Pm = 20), plots of shell-to-shell energy transfers ($U2U$, $B2B$, $U2B$) at $t = 0.48$ (a1,a2,a3) and $t = 18.73$ (b1,b2,b3). $U2U$ and $B2B$ transfers are local and forward, but $U2B$ transfers are nonlocal from the forcing band. From Kumar et al. (2015). Reprinted with permission from Taylor and Francis (see Color Plates).

23.7.3 Large-scale dynamo with forcing at intermediate scale

The dynamos discussed in the last two sections were forced at large length scales. In the present section we consider a scenario when the velocity field is forced at an intermediate scale. Kumar and Verma (2017) analyzed one such dynamo in which the forcing was employed at scale 1/10 the box size, i.e., $k = (10, 12)$, and the kinetic and magnetic helicities were negligible. In addition, $\mathrm{Pm} = 1$.

In Fig. 23.7 we exhibit various energy fluxes and dissipation rates toward the later stages of the dynamo process when the system has reached a steady state. We compute various fluxes for wavenumber sphere of radius $k_0 = 8$ (central sphere of Fig. 23.7). The velocity field is forced at $k = [10, 12]$, which is represented as outer wavenumber shell in the figure. In the figure, $\Pi^{u>}_{b<}(k_0) \approx 0.01$ is the most dominant energy input to the magnetic sphere. Also, the $b<$ sphere loses relatively small amounts of energy to $u<$ and $b>$. The Joule dissipation (~ 0.01) in the magnetic sphere, represented by wavy line, approximately balances the energy input to the sphere. Hence, the large-scale magnetic field is sustained by the $\Pi^{u>}_{b<}$ energy flux. Note that this energy flux is only 1% of the total energy injection rate, but it is sufficient for a steady state because the Joule dissipation is quite weak at this scale ($D_\eta \propto k^2$).

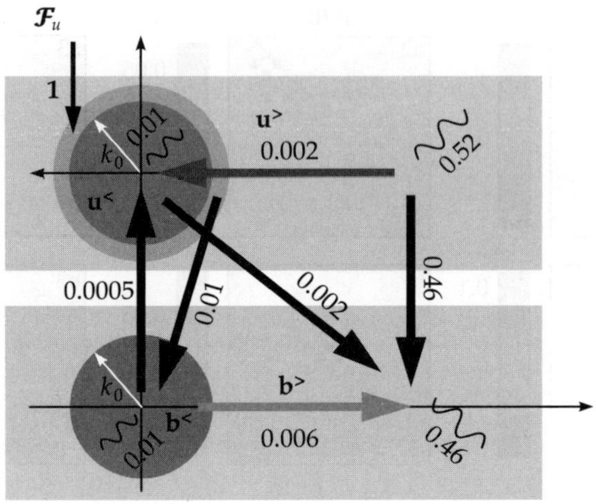

Figure 23.7 Various energy fluxes for the dynamo which is forced at $k = [10, 12]$. These fluxes are of $t = 127$ eddy turnover time. Adapted from a figure of Kumar and Verma (2017).

The energy transfers for this dynamo is schematically illustrated in Fig. 23.8. The large-scale magnetic energy grows due to $\Pi^{u>}_{b<}$ energy flux. In addition, it also receives energy from small-scale magnetic field at early stages. These two transfers arise due to the energetic velocity modes near $k = 10$. Note that at later stages, $\Pi^{b<}_{b>} > 0$, but is quite weak compared to $\Pi^{u>}_{b<}$. These results demonstrate that the magnetic field at large-scale can grow when the velocity field at an intermediate scale is forced. This is a large-scale dynamo.

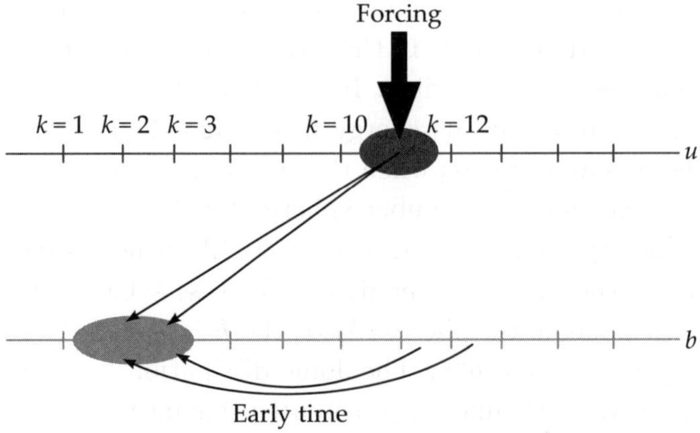

Figure 23.8 For a dynamo with forcing at k = [10, 12] and Pm = 1, a schematic diagram of shell-to-shell energy transfers among the u-shells and the b-shells. From Kumar and Verma (2017). Reprinted with permission from AIP.

In Fig. 23.9, we illustrate the kinetic and magnetic energy spectra for this dynamo. We observe that for $k < k_f$, $E_u(k) \sim k^2$, which is the prediction of the absolute equilibrium theory (Dallas et al., 2015). In Chapter 5 we presented a numerical result from hydrodynamic turbulence that yields k^2 energy spectrum for $k < k_f$ (see Fig. 5.6). The magnetic energy spectrum $E_b(k)$ decreases with k, clearly demonstrating that the magnetic energy is maximal at $k = 1$, which is the largest scale of the system.

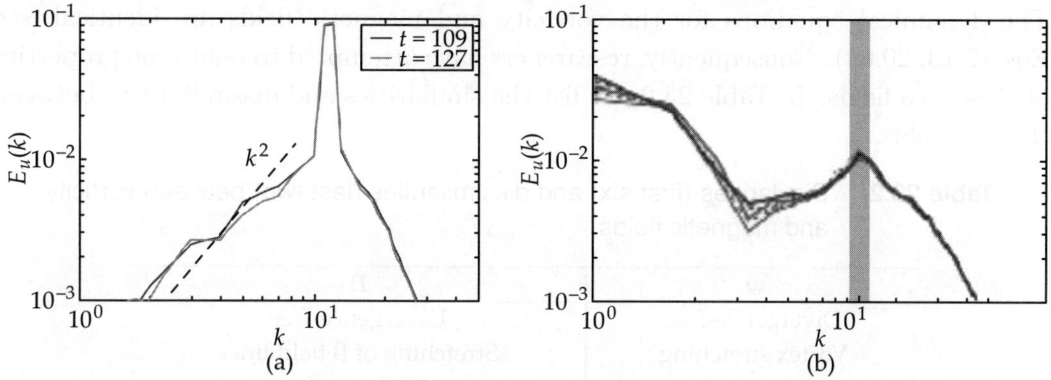

Figure 23.9 For a dynamo with forcing at $k \in [10, 12]$ and $\mathrm{Pm} = 1$, the kinetic and magnetic energy spectra. For $k < k_f$, $E_u(k) \sim k^2$ consistent with the predictions of absolute equilibrium theory. From Kumar and Verma (2017). Reprinted with permission from AIP.

In the next section we will describe the effects of helicity in a dynamo.

23.8 Role of Helicities in Dynamos

It is generally believed that both kinetic and magnetic helicities help the dynamo process (Cardin and Cugliandolo, 2011). This has been demonstrated by Pouquet et al. (1976) using eddy damped quasi normal Markovian (EDQNM) approximation, and by Brandenburg (2001) using numerical simulations. Using field-theoretic computations, Verma (2004) showed that the magnetic energy flux in the presence of kinematic and magnetic helicities is

$$\Pi^{b<}_{(b>,u>)\mathrm{helical}} = -ar_M^2 + br_M r_K, \tag{23.47}$$

where $r_K = H_K(k)/(kE_u(k))$ and $r_M = kH_M(k)/E_b(k)$. Clearly, the magnetic energy flux due to the helicity would become negative if $r_M r_K < 0$, that is, when the magnetic helicity and kinetic helicity are of opposite sign. This condition will help the dynamo process.

Alexakis et al. (2005b), Mininni et al. (2005), and Alexakis et al. (2007) have analyzed the shell-to-shell energy transfers for an Arnold–Beltrami–Childress (ABC) dynamo, which is a helical dynamo. The energy transfers reported by them have similarities with those presented in earlier sections.

23.9 Analogy between the Vorticity and Magnetic Fields

The dynamical equations for the vorticity and magnetic fields are identical [see Eqs. (2.13, 20.6a). Consequently, researchers have attempted to relate the properties of these two fields. In Table 23.2, we list the similarities and dissimilarities between the two fields.

Table 23.2 Similarities (first six) and dissimilarities (last two) between vorticity and magnetic fields.

$\boldsymbol{\omega}$	B
Divergence-free	Divergence-free
Vortex stretching	Stretching of B field lines
$U2\omega$ transfer	$U2B$ transfer
Enhancement of enstrophy	Growth of magnetic field
$\omega 2\omega$ transfer (forward)	$B2B$ transfer (forward for nonhelical)
No vortex stretching in 2D	No dynamo in 2D
$\boldsymbol{\omega} = \nabla \times \mathbf{u}$	B is independent field
No $\omega 2U$ transfer	$B2U$ transfer is defined

These similarities help us understand the two systems better. However, the two systems are different, and caution is required when we relate them.

In the last section of this chapter we describe how the magnetic field reduces turbulent drag in magnetofluids.

23.10 Turbulent Drag Reduction in MHD

In this section, we compare the turbulent drag and kinetic energy flux in MHD turbulence with the corresponding quantities of hydrodynamic turbulence. We can ask the following question: For the same Re, what is the relative magnitudes of the nonlinear term $(\mathbf{u} \cdot \nabla)\mathbf{u}$ for the two systems? In general, the magnitude of $(\mathbf{u} \cdot \nabla)\mathbf{u}$ is not computed in numerical simulations; hence, we compare the related quantities—kinetic energy flux $\Pi^{u<}_{u>}$.

As discussed in the present and previous chapters, an introduction of magnetic field induces strong energy transfers among the velocity and magnetic fields, mainly

$\Pi^{u<}_{B<}$, $\Pi^{u<}_{B>}$, and $\Pi^{u>}_{B>}$ (see Fig. 23.5). Since the injected energy is divided along many channels, $\Pi^{u<}_{u>}$ for MHD is reduced compared to its hydrodynamic counterpart. That is,

$$[\Pi^{u<}_{u>}]_{\text{MHD}} < [\Pi^{u<}_{u>}]_{\text{Hydro}} \,. \tag{23.48}$$

Since $\Pi^{u<}_{u>} \sim [(\mathbf{u}\cdot\nabla)\mathbf{u}]\cdot\mathbf{u}$, we can claim that the turbulent drag $(\mathbf{u}\cdot\nabla)\mathbf{u}$ is proportional to this flux. In other words, for the same large-scale velocity of the two systems (U), the fluxes $\Pi^{u<}_{u>}$ (or the nonlinearity $(\mathbf{u}\cdot\nabla)\mathbf{u}$, or turbulent drag) of MHD is lower than that for hydrodynamics. These observations indicate a turbulent drag reduction in MHD turbulence.

The numerical results of MHD turbulence are in agreement with this prediction. In Figs. 22.7 and 23.5, the kinetic energy fluxes $\Pi^{u<}_{u>}$ are much smaller than the total energy flux. When we compare MHD turbulence with corresponding hydrodynamic turbulence, the total energy flux would be $\Pi^{u<}_{u>}$ for hydrodynamic turbulence. Hence, these simulations demonstrate that MHD turbulence exhibits turbulent drag reduction. In Chapter 27 we will show that a similar mechanism induces turbulent drag reduction in polymeric flows.

With this, we close our discussion on dynamos.

Further Reading

The present chapter provides only a brief review of dynamos mechanism from energy transfer perspectives. For a more detailed discussion on dynamos, refer to books and reviews by Moffatt (1978); Krause and Rädler (1980); Ruzmaikin et al. (1988); Verma (2004); Cardin and Cugliandolo (2011); Brandenburg et al. (2005). The reader could also refer to the papers cited in the chapter.

Exercises

1. Derive the equations of the six-mode kinematic and dynamic models discussed in this chapter.
2. Consider the models of Section 23.4. Write down the real and imaginary parts of the equations, and construct the stability matrices. Analyze the parameters for dynamo action.
3. Consider the 2D3C helical dynamo of Section 23.4.3. Analyze the corresponding equations for the vector potential and contrast it with those appearing in the no-dynamo theorem.

Chapter 24

Phenomenology of Quasi-static MHD Turbulence

In earlier chapters on MHD it was implicitly assumed that both velocity and magnetic fields are turbulent; that is, Reynolds number, Re, and magnetic Reynolds number, Rm, are large. In the present chapter we focus on a different class of MHD flows for which $\mathrm{Rm} \to 0$. Liquid metals have very small Prandtl numbers ($\sim 10^{-6}$) or large magnetic diffusivity. Hence, for liquid metal flows with moderate Re, $\mathrm{Rm} = \mathrm{RePm} \to 0$. Such flows are referred to as quasi-static (QS) MHD.

In this chapter we describe the energy spectrum and flux of QS MHD turbulence. For simplification, we avoid the effects of the walls on the flow. In the next section, we start with the governing equations of QS MHD.

24.1 Governing Equations

The MHD equations with a mean magnetic field $\mathbf{B}_0$ are as follows.

$$\frac{\partial \mathbf{u}}{\partial t} - \mathbf{B}_0 \cdot \nabla \mathbf{b} + \mathbf{u} \cdot \nabla \mathbf{u} - \mathbf{b} \cdot \nabla \mathbf{b} = -\nabla p + \nu \nabla^2 \mathbf{u}, \tag{24.1a}$$

$$\frac{\partial \mathbf{b}}{\partial t} - \mathbf{B}_0 \cdot \nabla \mathbf{u} + \mathbf{u} \cdot \nabla \mathbf{b} - \mathbf{b} \cdot \nabla \mathbf{u} = \eta \nabla^2 \mathbf{B}. \tag{24.1b}$$

In these equations, $\mathbf{B}_0$ and $\mathbf{b}$ are in the units of velocity. We nondimensionalize these equations using system's velocity scale U_0 and length scale L:

$$\mathbf{r} \to L\mathbf{r}; \quad \mathbf{u} \to U_0\mathbf{u}; \quad \mathbf{b} \to U_0\mathbf{b}; \quad p \to (U_0^2)p. \tag{24.2}$$

As a result, the MHD equations become

$$\frac{\partial \mathbf{u}}{\partial t} - \frac{1}{U_0}\mathbf{B}_0 \cdot \nabla\mathbf{b} + \mathbf{u}\cdot\nabla\mathbf{u} - \mathbf{b}\cdot\nabla\mathbf{b} = -\nabla p + \frac{1}{\mathrm{Re}}\nabla^2\mathbf{u}, \tag{24.3a}$$

$$\mathrm{Rm}\left[\frac{\partial \mathbf{b}}{\partial t} + \mathbf{u}\cdot\nabla\mathbf{b} - \mathbf{b}\cdot\nabla\mathbf{u}\right] = \frac{\mathbf{B}_0 L}{\eta}\cdot\nabla\mathbf{u} + \nabla^2\mathbf{b}. \tag{24.3b}$$

Note that in these equations, $\mathbf{b}$ is dimensionless, but B_0 still has the units of velocity.

Under the limit $\mathrm{Rm} \to 0$ and $\eta \to \infty$, we obtain

$$\frac{\partial \mathbf{u}}{\partial t} + \mathbf{u}\cdot\nabla\mathbf{u} = -\nabla p + \frac{1}{U_0}\mathbf{B}_0\cdot\nabla\mathbf{b} + \frac{1}{\mathrm{Re}}\nabla^2\mathbf{u}, \tag{24.4a}$$

$$\nabla^2\mathbf{b} = -\frac{\mathbf{B}_0 L}{\eta}\cdot\nabla\mathbf{u}. \tag{24.4b}$$

Equation (24.4b) is Poisson's equation that can be solved given the source term and the boundary condition. Thus, in QS MHD, $\mathbf{b}$ can be determined from $\mathbf{u}$ as

$$\mathbf{b} = -\nabla^{-2}[\frac{L}{\eta}\mathbf{B}_0\cdot\nabla\mathbf{u}]. \tag{24.5}$$

Substitution of this $\mathbf{b}$ in Eq. (24.4a) yields the Lorentz force as

$$\mathbf{F}_u = -\frac{B_0^2 L}{U_0\eta}(\hat{\mathbf{B}}_0\cdot\nabla)\nabla^{-2}[\hat{\mathbf{B}}_0\cdot\nabla\mathbf{u}] = -N(\hat{\mathbf{B}}_0\cdot\nabla)\nabla^{-2}[\hat{\mathbf{B}}_0\cdot\nabla\mathbf{u}], \tag{24.6}$$

where $\hat{\mathbf{B}}_0$ is the unit vector along $\mathbf{B}_0$, and

$$N = \frac{B_0^2 L}{U_0\eta} \tag{24.7}$$

is called *interaction parameter.*[1] Note that $b \ll u$ under the QS MHD approximation.

Now we construct the equations for QS MHD in Fourier space. First, we assume that $\mathbf{B}_0$ is along the z direction, or $\mathbf{B}_0 = B_0\hat{z}$. Under this notation, in Fourier space,

---

[1] In SI units,

$$N = \frac{\sigma B_0^2 L}{\rho U_0},$$

where σ and ρ are respectively the electrical conductivity and density of the liquid metal.

Eqs. (24.5, 24.6) become

$$\mathbf{b}(\mathbf{k}) = i\frac{B_0 L}{\eta}\frac{k_z}{k^2}\mathbf{u}(\mathbf{k}), \tag{24.8}$$

$$\mathbf{F}_u(\mathbf{k}) = -\frac{B_0^2 L}{U_0\eta}\frac{k_z^2}{k^2}\mathbf{u}(\mathbf{k}) = -[N\cos^2\zeta]\mathbf{u}(\mathbf{k}), \tag{24.9}$$

where ζ is the angle between the mean magnetic field and the wavenumber $\mathbf{k}$. Substitution of these terms in the Navier–Stokes equations yields

$$\frac{d}{dt}\mathbf{u}(\mathbf{k}) + i\sum_{\mathbf{p}}\{\mathbf{k}\cdot\mathbf{u}(\mathbf{q})\}\mathbf{u}(\mathbf{p}) = -i\mathbf{k}p(\mathbf{k}) - [N\cos^2\zeta]\mathbf{u}(\mathbf{k}) - \nu k^2\mathbf{u}(\mathbf{k}), \tag{24.10a}$$

$$\mathbf{k}\cdot\mathbf{u}(\mathbf{k}) = 0. \tag{24.10b}$$

The equation for the modal kinetic energy $E_u(\mathbf{k}) = |\mathbf{u}(\mathbf{k})|^2/2$ is

$$\frac{d}{dt}E_u(\mathbf{k}) = T_u(\mathbf{k}) + \mathcal{F}_u(\mathbf{k}) - D_u(\mathbf{k}), \tag{24.11}$$

where $T_u(\mathbf{k})$ is the total energy transfer to $\mathbf{u}(\mathbf{k})$ via nonlinear interactions, and

$$\mathcal{F}_u(\mathbf{k}) = \Re[\mathbf{F}_u(\mathbf{k})\cdot\mathbf{u}^*(\mathbf{k})] = -[2N\cos^2\zeta]E_u(\mathbf{k}) = -D_J(\mathbf{k}), \tag{24.12}$$

where D_J is the Joule dissipation. The quantity

$$D_u(\mathbf{k}) = 2\nu k^2 E_u(\mathbf{k}) \tag{24.13}$$

is the viscous dissipation rate.

When we compare the above energy equation with those of MHD (see Chapter 21), we observe that

$$T_u(\mathbf{k}) = \sum_{\mathbf{p}} S^{uu}(\mathbf{k}|\mathbf{p}|\mathbf{q}) \tag{24.14}$$

yields kinetic energy flux, while the term $\mathcal{F}_u(\mathbf{k})$ originates from S^{ub}. Note that $\mathbf{B}_0$ is the mediator in this transaction, hence $\mathbf{q} = 0$ and $\mathbf{p} = \mathbf{k}$. Therefore, the normalized energy transfer term is

$$\begin{aligned} T_{ub}(\mathbf{k}) &= -\frac{1}{U_0}\Im\left[\{\mathbf{k}\cdot\mathbf{B}_0\}\{\mathbf{b}(\mathbf{k})\cdot\mathbf{u}^*(\mathbf{k})\}\right] \\ &= -[2N\cos^2\zeta]E_u(\mathbf{k}). \end{aligned} \tag{24.15}$$

Thus, $\mathbf{u}(\mathbf{k})$ transfers energy directly to $\mathbf{b}(\mathbf{k})$ because $\mathbf{q} = 0$. The transferred energy is dissipated immediately via Joule heating. In addition, under quasi-static approximation,

$$T_{bb}(\mathbf{k}) = \Im[\{\mathbf{k}\cdot\mathbf{u}(0)\}\{\mathbf{b}(\mathbf{k})\cdot\mathbf{b}^*(\mathbf{k})\}] \to 0, \tag{24.16a}$$

$$T_{bu}(\mathbf{k}) = \Im[\{\mathbf{k}\cdot\mathbf{B}_0\}\{\mathbf{u}(\mathbf{k})\cdot\mathbf{b}^*(\mathbf{k})\}] = -T_{ub}(\mathbf{k}), \tag{24.16b}$$

where $\mathbf{u}(0)$ is mean velocity of the flow, and $\mathbf{b}(\mathbf{k}) \to 0$.

Note that both viscous and Joule terms are dissipative. Hence, in the absence of an external force, the kinetic energy of the flow will decay and vanish. In this chapter we assume an additional large scale forcing for $\mathbf{u}$ that make the QS MHD flow steady. In the following sections we will describe the steady-state properties of QS MHD turbulence.

24.2 Distribution and Spectrum of Kinetic Energy

In this section we will describe the distribution of kinetic energy in Fourier space. Since the Lorentz force and the Joule dissipation are proportional to $\cos^2\zeta$, they vanish in the equatorial plane ($\zeta = \pi/2$, or $k_z = 0$). As a result, the nonlinear term $\mathbf{u}\cdot\nabla\mathbf{u}$ dominates in this plane. Hence, the dynamics in the $k_z = 0$ plane is similar to that of 2D hydrodynamic turbulence. Inverse energy cascade in this plane and near it leads to strong vortical flow structures (Reddy and Verma, 2014; Verma, 2017).

For large N, the mean magnetic field makes the flow anisotropic, yet respecting azimuthal symmetry. Hence, statistically, the modal kinetic energy is independent of the azimuthal angle, and it is a function of k and the polar angle ζ (see Chapter 11). We sum the modal energies in a ring whose inner and outer radii are k and $k + \Delta k$ respectively, and sector angles are ζ and $\zeta + \Delta\zeta$ (see Fig. 11.1). The ring spectrum is defined as

$$E(k,\zeta) = \frac{1}{\Delta k \Delta\zeta}\sum_{\mathbf{k}'} E(\mathbf{k}'). \tag{24.17}$$

Let us characterize the ring spectrum of the liquid metal flows under a steady state. We start with Eq. (4.40), sum the equation over the modes in the ring, and then divide by $\Delta k \Delta\zeta$ that yields

$$\frac{d}{dt}E_u(k,\zeta) = T_u(k,\zeta) + \mathcal{F}_u(k,\zeta) - D_u(k,\zeta) + \mathcal{F}_{\text{ext}}(k,\zeta), \tag{24.18}$$

where $T_u(k,\zeta)$ is the rate of nonlinear energy transfer to the ring; $D_u(k,\zeta)$ is the ring dissipation spectrum; and $\mathcal{F}_u(k,\zeta), \mathcal{F}_{\text{ext}}(k,\zeta)$ are respectively the ring spectra of the Lorentz force and the external force. Note that $-\mathcal{F}_u$ is the Joule dissipation.

For a steady state, in the inertial–dissipation range where $\mathcal{F}_{\text{ext}}(k,\zeta) = 0$, we obtain

$$T_u(k,\zeta) + \mathcal{F}_u(k,\zeta) - D_u(k,\zeta) = 0. \tag{24.19}$$

We also assume that in the inertial range,

$$E(k,\zeta) = E(k)\frac{g(\zeta)}{\pi}. \tag{24.20}$$

By definition,

$$\int_0^\pi d\zeta E(k,\zeta) = \int_0^\pi d\zeta E(k)\frac{g(\zeta)}{\pi} = E(k). \tag{24.21}$$

Therefore,

$$\int_0^\pi d\zeta \frac{g(\zeta)}{\pi} = 1. \tag{24.22}$$

Under the above assumption, and on integration of Eq. (24.19) over ζ,

$$T_u(k) + \mathcal{F}_u(k) - D_u(k) = 0. \tag{24.23}$$

Using $T_u(k) = -d\Pi_u(k)/dk$, we convert this equation to

$$\frac{d}{dk}\Pi_u(k) = -(2\nu k^2 + 2Nc_2)E_u(k), \tag{24.24}$$

where

$$\int_0^\pi d\zeta \cos^2\zeta \frac{g(\zeta)}{\pi} = c_2. \tag{24.25}$$

Equation (24.24) has two unknowns—$E_u(k)$ and $\Pi_u(k)$. Hence, following Pao (1968), we make an additional assumption that $E_u(k)/\Pi_u(k)$ is independent of ν and N, and that it depends only on the total dissipation rate ϵ_u and local k (also see Section 5.5.1). For large Re and for $N \lessapprox 1$, we expect the spectrum to be Kolmogorov-like. Hence,

$$\frac{E_u(k)}{\Pi_u(k)} = K_{\text{Ko}}\epsilon_u^{-1/3}k^{-5/3}, \tag{24.26}$$

substitution of which in Eq. (24.24) yields

$$\frac{d}{dk}\Pi_u(k) = -(2\nu k^2 + 2Nc_2)K_{\mathrm{Ko}}\epsilon_u^{-1/3}k^{-5/3}\Pi_u(k). \tag{24.27}$$

This equation has the following solution:

$$\begin{aligned} \log\left(\frac{\Pi_u(k)}{\Pi_u(k_0)}\right) &= -\frac{3}{2}K_{\mathrm{Ko}}[(k/k_d)^{4/3} - (k_0/k_d)^{4/3}] \\ &\quad -3c_2K_{\mathrm{Ko}}\left[(k/k_{d2})^{-2/3} - (k_0/k_{d2})^{-2/3}\right], \end{aligned} \tag{24.28}$$

where

$$k_d = \left(\frac{\epsilon_u}{\nu^3}\right)^{1/4}, \tag{24.29a}$$

$$k_{d2} = \left(\frac{N^3}{\epsilon_u}\right)^{1/2}. \tag{24.29b}$$

We take $\Pi_u(k_0) = \epsilon_u$ and obtain $E_u(k)$ using Eq. (24.26). Note however that this model is expected to work well for $N \lesssim 1$.

We (with Anas) attempted to fit the aforementioned model with the numerical data of Reddy and Verma (2014) and Verma and Reddy (2015). For $N = 0.5$, with $K_{\mathrm{Ko}} = 2.5$ and $c_2 = 0.3$, we observed very good agreement between the model predictions and numerical results. See Fig. 24.1 for an illustration. We remark that the model of Verma and Reddy (2015) is somewhat similar to the model discussed here and its predictions fit with the inertial range spectrum quite well; but this model is deficient in capturing the dissipation–range energy spectrum.

The above model is expected to work for $N \lessapprox 1$ for which $E_u(k)$ is nearly Kolmogorov-like. However, this prescription does not work for $N \gg 1$. Using the numerical data of Reddy and Verma (2014), Verma and Reddy (2015) showed that $E_u(k)$ for large N has an exponential behavior (Fig. 24.2). Hence, in Eq. (24.24), we assume that both $E_u(k)$ and $\Pi_u(k)$ are proportional to $\exp(-k/\bar{k}_d)$. This assumption yields the following solution (Verma and Reddy, 2015):

$$E_u(k) = A\exp(-k/\bar{k}_d), \tag{24.30a}$$

$$\Pi_u(k) = A\left[2\nu\bar{k}_d(k^2 + 2k\bar{k}_d + 2\bar{k}_d^2) + 2Nc_2\bar{k}_d\right]\exp(-k/\bar{k}_d), \tag{24.30b}$$

where A is the amplitude of the spectrum, while $\bar{k}_d$ is an adjustable parameter. Verma and Reddy (2015) showed that this model fits with the numerical data of $N = 130$ and 220 with $\bar{k}_d \approx 1/0.18$.

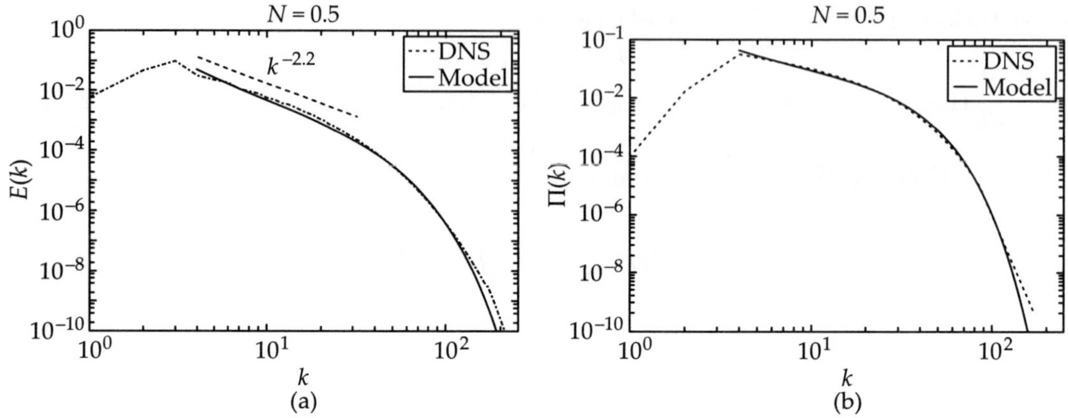

Figure 24.1 For $N = 0.5$, model predictions (solid lines) of (a) energy spectrum and (b) energy flux match with the numerical results (dashed lines) of Reddy and Verma (2014) and Verma and Reddy (2015).

Both these models exhibit steepening of energy spectrum from Kolmogorov's $k^{-5/3}$ spectrum. In addition, the energy flux too decreases with wavenumber in the inertial range itself. This feature is due to the Joule dissipation that saps the energy flux at all scales [see Eq. (24.24)]. Here, the framework of variable energy flux is useful for explaining this behavior of QS MHD turbulence.[2]

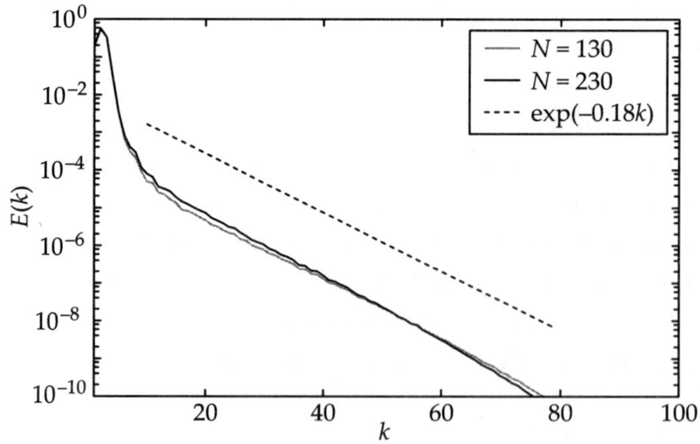

Figure 24.2 For very large N, the kinetic energy spectrum of QS MHD exhibits exponential spectrum. From Reddy and Verma (2014). Reprinted with permission from AIP.

[2]Several researchers had attributed the steepening of the energy spectrum to two-dimensionalization of the flow. The aforementioned theory provides a contrary viewpoint.

The one-dimensional energy spectrum is a sum of modal energies of all the modes of the shell, and hence it does not capture the angular dependence, which is a crucial aspect of anisotropic flows. To overcome this deficiency, Reddy and Verma (2014) computed the ring spectra for various Ns; they showed that the anisotropy in QS MHD increases with N. The ring spectra exhibited in Fig. 24.3(a, b, c) for $N = 0, 18, 130$ illustrates this feature. Naturally, the ring spectrum for $N = 0$ is isotropic. However, for finite N, maximum energy resides in the equatorial plane, that is, $E(k, \zeta)$ peaks at $\zeta = \pi/2$. The figure also contains the density plot of the ring spectrum of the Joule dissipation:

$$D_J(k, \zeta) = [2N \cos^2 \zeta] E_u(k, \zeta). \tag{24.31}$$

The spectrum of $D_J(k, \zeta)$ is somewhat similar to $E_u(k, \zeta)$ with a subtle difference; $D_J(k, \zeta) = 0$ for $\zeta = \pi/2$, but it peaks just above the equator for large N (Reddy and Verma, 2014).

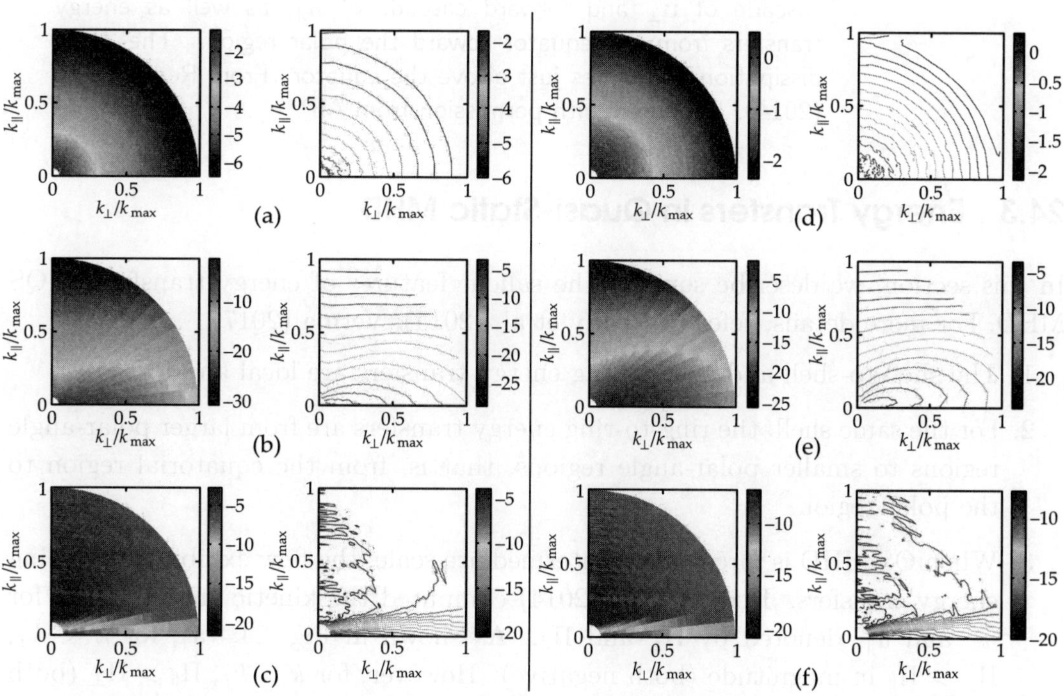

Figure 24.3 Density and contour plots of logarithm of ring energy spectra for (a) $N = 0$, (b) $N = 18$, (c) $N = 130$. (d, c, f) The corresponding ring spectra for the Joule dissipation rate. Adapted from the figures of Reddy (2015) (see Color Plates).

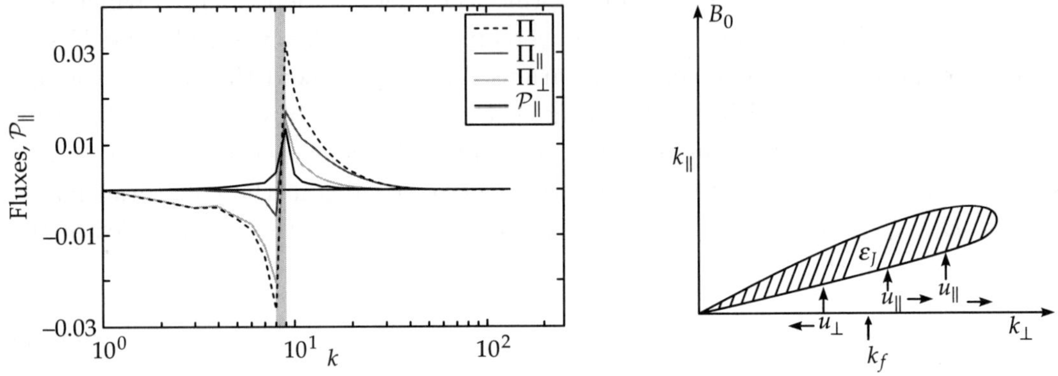

(a) (b)

Figure 24.4 (a) Plots of energy flux $\Pi_\perp(k)$ and $\Pi_\parallel(k)$, and the energy transfer from $\mathbf{u}_\perp$ to $u_\parallel$ via pressure, $\mathcal{P}_\parallel$. (b) Schematic diagram of the energy transfers in QS MHD with large N indicating inverse cascade of $\mathbf{u}_\perp$ and forward cascade of $u_\parallel$, as well as energy transfers from the equator toward the polar region. The Joule dissipation dominates just above the equator. From Reddy et al. (2014). Reprinted with permission from AIP.

24.3 Energy Transfers in Quasi-Static MHD

In this section, we describe some of the salient features of energy transfers in QS MHD. For more details, refer to Reddy et al. (2014); Verma (2017).

1. The shell-to-shell and ring-to-ring energy transfers are local and forward.
2. For the same shell, the ring-to-ring energy transfers are from larger polar-angle regions to smaller polar-angle regions, that is, from the equatorial region to the polar region.
3. When QS MHD is forced at an intermediate scale, the flow exhibits interesting energy transfers. Reddy et al. (2014) computed the kinetic energy fluxes for $u_\perp$ and $u_\parallel$, denoted by $\Pi_\perp$ and $\Pi_\parallel$. As shown in Fig. 24.4(a), for $k < k_f$, $\Pi_\perp > \Pi_\parallel$ in magnitude (both negative). However, for $k > k_f$, $\Pi_\parallel > \Pi_\perp$ (both positive). Thus, $\mathbf{u}_\perp$ flows from intermediate scales to large scales as in 2D hydrodynamic turbulence (see Chapter 7), while $u_\parallel$ cascades from intermediate scales to small scales as in passive scalar turbulence (see Chapter 14). For large N, a combination of $u_\perp$ and $u_\parallel$ makes the flow quasi-2D or 2D3C (two-dimensional flows with three components).

4. Figure 24.4(a) illustrates the energy exchange between $\mathbf{u}_\perp$ and $u_\parallel$ via pressure. We observe that $\mathcal{P}_\parallel > 0$ implying energy transfers from $\mathbf{u}_\perp$ to $u_\parallel$.

We summarize the above energy transfers in Fig. 24.4(b).

Thus, QS MHD turbulence exhibits interesting turbulent behavior: decrease of energy flux in the inertial range, steepening of the energy spectrum, anisotropic turbulence, exchange of energy among the parallel and perpendicular components of the velocity, etc. We remark that many of these features are also observed in MHD turbulence with a strong magnetic field (Teaca et al., 2009; Sundar et al., 2017). This is because the mean magnetic field plays a similar role in MHD and QS MHD.

With this, we end our discussion on QS MHD turbulence.

Further Reading

The present chapter provides a brief introduction to QS MHD turbulence. For more details, refer to Moreau (1990), Knaepen and Moreau (2008), Zikanov et al. (2014), and Verma (2017).

Chapter 25

Electron Magnetohydrodynamics

In magnetohydrodynamics (MHD), the linear momentum and electric current are carried by ions and electrons respectively. However, at very short time scales, the ions are immobile, and hence, electrons become the carriers of both linear momentum and electric current. Such a system is referred to as *electron magnetohydrodynamics*, or EMHD. In this short chapter we will describe the properties of EMHD turbulence, specially from energy transfer perspectives.

25.1 Governing Equations

We denote the mass and charge magnitude of the electron by m_e and e respectively. Throughout this chapter, the particle density and the velocity of the electron fluid are represented by n and $\mathbf{u}$ respectively, and the magnetic and electric fields by $\mathbf{B}$ and $\mathbf{E}$ respectively. In terms of these variables, the equations for the velocity field of the electron fluid in CGS system are

$$m_e n \left[\frac{\partial \mathbf{u}}{\partial t} + \mathbf{u} \cdot \nabla \mathbf{u}\right] = -\nabla p - ne\left[\mathbf{E} + \frac{1}{c}\mathbf{u} \times \mathbf{B}\right] - \mu m_e n \mathbf{u}, \tag{25.1}$$

where μ is the frictional coefficient that has dimension of $1/[T]$. We assume perfectly conducting electron fluid, hence $\mathbf{E} + (1/c)\mathbf{u} \times \mathbf{B} = 0$. Therefore, these equations and the equations for the magnetic field are as follows:

$$
\begin{aligned}
m_e \frac{\partial \mathbf{u}}{\partial t} &= m_e(\mathbf{u} \times \boldsymbol{\omega}) - \nabla(p/n + m_e u^2/2) - \mu m_e \mathbf{u}, &\text{(25.2a)}\\
\frac{\partial \mathbf{B}}{\partial t} &= \nabla \times (\mathbf{u} \times \mathbf{B}), &\text{(25.2b)}\\
\nabla \times \mathbf{B} &= \frac{4\pi}{c}\mathbf{J} = -\frac{4\pi n e}{c}\mathbf{u}, &\text{(25.2c)}\\
\nabla \cdot \mathbf{u} &= \nabla \cdot \mathbf{B} = 0, &\text{(25.2d)}
\end{aligned}
$$

where $\mathbf{J} = -ne\mathbf{u}$ is the current density. In this derivation, we employ Eq. (2.12). Note that in EMHD, B is in CGS units, not in Alfvénic units (in which B has dimension of velocity).

Equation (25.2b) yields the following equation for the vector potential $\mathbf{A}$:

$$
\frac{\partial \mathbf{A}}{\partial t} = \mathbf{u} \times \mathbf{B} + \nabla \Psi, \tag{25.3}
$$

where Ψ is a scalar. The generalized momentum is given by

$$
\mathbf{P} = m_e \mathbf{u} - \frac{e}{c}\mathbf{A}. \tag{25.4}
$$

We can sum Eq. (25.2a) and $(-e/c)\times$Eq. (25.3) to derive the following equation for $\mathbf{P}$:

$$
\frac{\partial \mathbf{P}}{\partial t} = \mathbf{u} \times (\nabla \times \mathbf{P}) - \nabla\left[p/n + m_e u^2/2 + (e/c)\Psi\right] - \mu m_e \mathbf{u}. \tag{25.5}
$$

Taking a curl of this equation yields

$$
\frac{\partial \mathbf{Q}}{\partial t} = \nabla \times (\mathbf{u} \times \mathbf{Q}) + \mu d_e^2 \nabla^2 \mathbf{B}, \tag{25.6}
$$

where

$$
\mathbf{Q} = -(c/e)\nabla \times \mathbf{P} = \mathbf{B} - \frac{mc}{e}\boldsymbol{\omega} = \mathbf{B} - d_e^2 \nabla^2 \mathbf{B}, \tag{25.7}
$$

and

$$
d_e^2 = \frac{c^2 m_e}{4\pi n e^2} = \frac{c^2}{\omega_{\mathrm{pe}}^2}, \tag{25.8}
$$

with ω_{pe} as the plasma frequency of electrons.

Therefore, the final equations for EMHD are Eq. (25.6), or

$$\frac{\partial}{\partial t}\mathbf{Q} + (\mathbf{u}\cdot\nabla)\mathbf{Q} = (\mathbf{Q}\cdot\nabla)\mathbf{u} + \mu d_e^2\nabla^2\mathbf{B}. \tag{25.9}$$

This is the only equation we need to solve in EMHD. Note that for given $\mathbf{B}$, we can compute $\mathbf{u}$ using Eq. (25.2c). The structure of this equation is analogous to the induction equation of MHD [see Eq. (20.7)].

Biskamp et al. (1999) employed the following quadratic quantity:

$$E = \int (B^2 + d_e^2 J^2)d\mathbf{r} = \int (\mathbf{Q}\cdot\mathbf{B})d\mathbf{r}. \tag{25.10}$$

Note that this quantity is not conserved.[1]

In the next section, we present EMHD equations in Fourier space.

25.2 Fourier Space Description

Fourier space description of Eq. (25.9) provides useful insights into turbulence properties. Following similar steps as in Section 20.3, we write Eq. (25.9) in Fourier space as

$$\frac{d}{dt}\mathbf{Q}(\mathbf{k}) + i\sum_{\mathbf{p}}\{\mathbf{k}\cdot\mathbf{u}(\mathbf{q})\}\mathbf{Q}(\mathbf{p}) = i\sum_{\mathbf{p}}\{\mathbf{k}\cdot\mathbf{Q}(\mathbf{q})\}\mathbf{u}(\mathbf{p}) - \mu(kd_e)^2\mathbf{B}(\mathbf{k}), \tag{25.11}$$

where

$$\mathbf{Q}(\mathbf{k}) = [1 + (kd_e)^2]\mathbf{B}(\mathbf{k}). \tag{25.12}$$

We define the modal energy of the system as

$$E(\mathbf{k}) = \frac{1}{2}\mathbf{Q}(\mathbf{k})\cdot\mathbf{B}^*(\mathbf{k}) = \frac{1}{2}[1 + (kd_e)^2]|\mathbf{B}(\mathbf{k})|^2. \tag{25.13}$$

The governing equation for $E(\mathbf{k})$ is

$$\begin{aligned}\frac{d}{dt}E(\mathbf{k}) &= \frac{1}{2}\left[\mathbf{B}^*(\mathbf{k})\cdot\dot{\mathbf{Q}}(\mathbf{k}) + \mathbf{Q}(\mathbf{k})\cdot\dot{\mathbf{B}}^*(\mathbf{k})\right] \\ &= \sum_{\mathbf{p}}[S^{QQ}(\mathbf{k}|\mathbf{p}|\mathbf{q}) + S^{Qu}(\mathbf{k}|\mathbf{p}|\mathbf{q})],\end{aligned} \tag{25.14}$$

[1] Some papers incorrectly argue that E of Eq. (25.10) is a conserved quantity.

where $\mathbf{q} = \mathbf{k} - \mathbf{p}$, and $S^{QQ}(\mathbf{k}|\mathbf{p}|\mathbf{q})$ is the mode-to-mode energy transfer via QQ channel from wavenumber $\mathbf{p}$ to wavenumber $\mathbf{k}$ with wavenumber $\mathbf{q}$ acting as a mediator. $S^{Qu}(\mathbf{k}|\mathbf{p}|\mathbf{q})$ is the corresponding mode-to-mode energy transfer via Qu channel. Using these mode-to-mode transfers we can define the energy fluxes for EMHD turbulence (similar to MHD turbulence).

In the next section we describe the turbulence phenomenology of EMHD.

25.3 Phenomenology of EMHD Turbulence

EMHD becomes turbulent when the nonlinear terms dominate the magnetic diffusion term. In literature, the turbulence phenomenology is presented for two different regimes: $kd_e \ll 1$ and $kd_e \gg 1$ (Biskamp et al., 1999; Shaikh, 2009). We present these scaling arguments in this section.

25.3.1 $kd_e \ll 1$

In this regime, $\mathbf{Q} \to \mathbf{B}$, and hence Eq. (25.9) is reduced to

$$\frac{\partial}{\partial t}\mathbf{B} = \nabla \times (\mathbf{u} \times \mathbf{B}) + \mu d_e^2 \nabla^2 \mathbf{B}, \tag{25.15}$$

which is the same as the equation for the magnetic field in MHD. By making an analogy with the magnetic field, we deduce that the nonlinear terms $(\mathbf{u} \cdot \nabla)\mathbf{B}$ and $(\mathbf{B} \cdot \nabla)\mathbf{u}$ induce the $B2B$ and $U2B$ transfers respectively, which lead to the energy cascades $\Pi^{b<}_{b>}$, $\Pi^{u<}_{b<}$, and $\Pi^{u<}_{b>}$ (see Section 21.6). As described in Section 21.6,

$$\Pi^{b<}_{b>} + \Pi^{u<}_{b<} + \Pi^{u<}_{b>} = \text{const} = \epsilon_b, \tag{25.16}$$

where ϵ_b is the total dissipation rate. Note that $U2B$ transfers can yield growth of the magnetic field, or dynamo action. This feature could be useful for understanding the dynamics of EMHD systems, such as neutron star and laser induced plasmas (Tzeferacos et al., 2018; Mondal et al., 2012).

By dimensional analysis of the energy equation and under the assumption of local interactions, we obtain

$$k u_k B_k^2 \sim \epsilon_b. \tag{25.17}$$

From Eq. (25.2c), we obtain $u_k \sim k B_k$, substitution of which in Eq. (25.17) yields

$$B_k \sim k^{-2/3} \epsilon_b^{1/3}. \tag{25.18}$$

Hence, the energy spectrum in the $kd_e \ll 1$ regime is

$$E_Q(k) \approx E_B(k) = \frac{B_k^2}{k} = \epsilon_b^{2/3} k^{-7/3}. \tag{25.19}$$

25.3.2 $kd_e \gg 1$

In this limit,

$$\mathbf{Q} \to -\frac{mc}{e}\boldsymbol{\omega}; \quad E \to j^2 \sim u^2. \tag{25.20}$$

Hence, the present limit corresponds to a weakly magnetized plasma. Therefore, we expect the dynamics to be governed by Eq. (25.2a). For small μ, the flow becomes turbulent, and it will follow Kolmogorov's spectrum, as described in Chapter 5, that is,

$$E(k) = \epsilon^{2/3} k^{-5/3}. \tag{25.21}$$

In the next section we show that this discussion can be simplified in the language of energy transfers.

25.4 Simplified Version

We present EMHD equations and energy transfers in a simpler way.

25.4.1 Governing equations and conservation laws

Assuming perfectly conducting fluid, the equations for EMHD are

$$m_e n \left[\frac{\partial \mathbf{u}}{\partial t} + \mathbf{u}\cdot\nabla\mathbf{u}\right] = -\nabla p - \mu m_e n \mathbf{u}, \tag{25.22a}$$

$$\frac{\partial \mathbf{B}}{\partial t} = \nabla \times (\mathbf{u} \times \mathbf{B}), \tag{25.22b}$$

$$\nabla \times \mathbf{B} = \frac{4\pi}{c}\mathbf{J} = -\frac{4\pi n e}{c}\mathbf{u}, \tag{25.22c}$$

$$\nabla \cdot \mathbf{u} = \nabla \cdot \mathbf{B} = 0. \tag{25.22d}$$

Equation (25.22a) are the Navier–Stokes equations whose viscous term is proportional to $\mathbf{u}$. Note that Eq. (25.22a) does not get feedback from the magnetic field. The equation for the magnetic field is same as the induction equation of

MHD except that it is diffusionless (the magnetic diffusivity $\eta = 0$). Since the magnetic field does not affect the velocity field, the above equations[2] are similar to those of kinematic dynamo with several subtle differences—(a) the new equations are diffusionless, (b) the new viscous dissipation is active at all scales. Note that the velocity field is time-dependent and it evolves simultaneously with the magnetic field.

From the above equations we can conclude that $\int d\mathbf{r}u^2/2$ is conserved when $\mu = 0$. But neither $\int d\mathbf{r}B^2/2$ nor $\int d\mathbf{r}(u^2/2 + B^2/2)$ are conserved. Note however that in 2D, $\int d\mathbf{r}A^2/2$ is conserved because $\eta = 0$. Consequently, dynamo action is possible in 2D EMHD. Thus, the conservation laws of EMHD are very different from those of MHD.

25.4.2 Energy transfers in EMHD

Making a connection with the energy transfers of MHD turbulence (Chapter 21), we conclude that

$$\begin{aligned} \frac{d}{dt}E_b(\mathbf{k}) &= \sum_{\mathbf{p}} S^{bb}(\mathbf{k}|\mathbf{p}|\mathbf{q}) + \sum_{\mathbf{p}} S^{bu}(\mathbf{k}|\mathbf{p}|\mathbf{q}) \\ &= T_{bb}(\mathbf{k}) + T_{bu}(\mathbf{k}), \end{aligned} \tag{25.23}$$

where $T_{bb}(\mathbf{k})$ is the energy transfer to $E_b(\mathbf{k})$ from other magnetic modes, while $T_{bu}(\mathbf{k})$ is the energy transfers to $E_b(\mathbf{k})$ from the velocity modes. Note that $T_{bu}(\mathbf{k})$ is responsible for growth or decay of the magnetic energy, as in dynamo action. Interestingly, without diffusion, the magnetic energy growth can saturate only via nonlinear interactions.

The above arguments indicate that the magnetic field interacts with the velocity field. However, there is no such coupling in Eq. (25.22a). We expect that such interactions is incorporated at some level.

We assume that the velocity field is forced at large scales. With this, Eq. (25.22a) yields the following equation for the variable kinetic energy flux under a steady state:

$$\frac{d}{dk}\Pi_u(k) = -2\mu E_u(k), \tag{25.24}$$

where $\Pi_u(k)$ and $E_u(k)$ are the kinetic energy flux and spectrum respectively. In the inertial range where the dissipation is weak, we expect

$$\Pi_u(k) = \text{const} = \epsilon_u, \tag{25.25a}$$

$$E_u(k) = K\epsilon_u^{2/3}k^{-5/3}. \tag{25.25b}$$

[2] In Eqs. (25.22), the magnetic field is a passive vector.

Note however that the dissipation term $2\mu E_u(k)$ is active at all scales, and it may affect the above scaling in a significant way.

The equation for the magnetic energy appears more tricky. The flux of magnetic energy is not constant due to the energy injection by the $T_{bu}(\mathbf{k})$ term. Yet, following Eq. (25.16), we can use the total magnetic energy dissipation rate ϵ_b as the parameter for dimensional analysis.[3] Therefore, following the arguments of Section 25.3.1, we deduce that

$$E_B(k) = \epsilon_b^{2/3} k^{-7/3}. \tag{25.26}$$

These discussions reveal many ambiguities in the phenomenology of EMHD turbulence, which we hope will be removed soon.

With this, we end our brief discussion on EMHD.

Further Reading

EMHD is a relatively recent field with not many rigorous results. Refer to Biskamp et al. (1996, 1999), Shaikh (2009), Cho (2016), and Das et al. (2011) for more details.

[3]Magnetic energy dissipation requires resistivity, which is absent in Eq. (25.22b). Yet, we postulate small magnetic diffusivity at small scales.

Part IV
MISCELLANEOUS FLOWS

Chapter 26

Rotating Turbulence

Many natural and industrial flows involve rotation. All planets, stars, and galaxies are rotating; hence, a good understanding of rotating flows is very important. In this chapter we will describe energy transfers in such flows.

Rotating flows are described by the Navier–Stokes equation with centrifugal and Coriolis forces. The centrifugal force is absorbed in the pressure gradient term, and hence the dynamics is dictated primarily by the Coriolis force (Kundu et al., 2015). In this chapter we will show that dynamics of rotating turbulence is quite complex. Several proposed phenomenologies of rotating turbulence invoke analogy with magnetohydrodynamics (MHD) turbulence. This is the reason why rotating turbulence appears after the chapters on MHD turbulence.

We start our discussion with the basic equations of rotating turbulence.

26.1 Governing Equations

In real space, the equations for rotating flows are

$$\frac{\partial \mathbf{u}}{\partial t} + (\mathbf{u} \cdot \nabla)\mathbf{u} = -\nabla p - 2\boldsymbol{\Omega} \times \mathbf{u} + \nu \nabla^2 \mathbf{u}, \tag{26.1a}$$

$$\nabla \cdot \mathbf{u} = 0, \tag{26.1b}$$

where $\mathbf{u}, p$ are the velocity and pressure fields respectively, $\boldsymbol{\Omega} = \omega \hat{z}$ is the rotation velocity, and ν is the kinematic viscosity. The flow is assumed to be incompressible with density chosen to be unity. The external force

$$\mathbf{F}_u = -2\boldsymbol{\Omega} \times \mathbf{u} \tag{26.2}$$

represents the *Coriolis force*.[1] The ratio of the Coriolis force and the nonlinear term is called the *Rossby number*:

$$\mathrm{Ro} = \frac{\Omega}{UL}, \tag{26.4}$$

where U, L are the large-scale velocity and length respectively. Since the Coriolis force does no work (note that $(\boldsymbol{\Omega} \times \mathbf{u}) \cdot \mathbf{u} = 0$), the equation for the kinetic energy is same as that for hydrodynamic flows [Eq. (2.27)]:

$$\frac{\partial}{\partial t}\frac{u^2}{2} + \nabla \cdot \left[\frac{u^2}{2}\mathbf{u}\right] = -\nabla \cdot (p\mathbf{u} - \nu \mathbf{u} \times \boldsymbol{\omega}) - \nu\omega^2. \tag{26.5}$$

Hence, the kinetic energy of an inviscid rotating flow is conserved under periodic or vanishing boundary conditions.

In Fourier space, the corresponding equations are

$$\frac{d}{dt}\mathbf{u}(\mathbf{k}) + \mathbf{N}_u(\mathbf{k}) = -i\mathbf{k}p(\mathbf{k}) - 2\boldsymbol{\Omega} \times \mathbf{u}(\mathbf{k}) - \nu k^2 \mathbf{u}(\mathbf{k}), \tag{26.6a}$$

$$\mathbf{k} \cdot \mathbf{u}(\mathbf{k}) = 0, \tag{26.6b}$$

where

$$\mathbf{N}_u(\mathbf{k}) = i\sum_{\mathbf{p}} [\mathbf{k} \cdot \mathbf{u}(\mathbf{k}-\mathbf{p})]\mathbf{u}(\mathbf{p}). \tag{26.7}$$

Since the nonlinear term is same as that for hydrodynamic flows [Eq. (3.16)], the formulas for the kinetic energy flux and shell-to-shell energy transfers are same as those for hydrodynamic turbulence. See Chapter 4 for the same.

In Fourier space, the kinetic energy injection rate in rotating turbulence is

$$\begin{aligned}
\mathcal{F}_u(\mathbf{k}) &= \frac{1}{2}\mathbf{F}_\mathbf{u}(\mathbf{k}) \cdot \mathbf{u}^*(\mathbf{k}) + c.c. \\
&= \{-\boldsymbol{\Omega} \times \mathbf{u}(\mathbf{k})\} \cdot \mathbf{u}^*(\mathbf{k}) + c.c. \\
&= -\boldsymbol{\Omega} \cdot [\mathbf{u}(\mathbf{k}) \times \mathbf{u}^*(\mathbf{k})] + c.c. = 0,
\end{aligned} \tag{26.8}$$

where $c.c.$ stands for complex conjugate. Similarly, the enstrophy injection rate is

[1]The centrifugral force can be written as

$$\mathbf{F}_c = -\boldsymbol{\Omega} \times (\boldsymbol{\Omega} \times \mathbf{r}) = -\boldsymbol{\Omega}(\boldsymbol{\Omega} \cdot \mathbf{r}) + \Omega^2 \mathbf{r} = -\frac{1}{2}\nabla\left[(\boldsymbol{\Omega} \cdot \mathbf{r})^2 - \Omega^2 r^2\right], \tag{26.3}$$

which is absorbed in the pressure gradient term.

$$\begin{aligned} 2\mathcal{F}_\omega(\mathbf{k}) &= i\mathbf{k} \times \mathbf{F_u}(\mathbf{k}) \cdot \boldsymbol{\omega}^*(\mathbf{k}) + c.c. \\ &= i\mathbf{F_u}(\mathbf{k}) \cdot [\boldsymbol{\omega}^*(\mathbf{k}) \times \mathbf{k}] + c.c. \\ &= \mathbf{F_u}(\mathbf{k}) \cdot \mathbf{u}^*(\mathbf{k}) k^2 + c.c. = 0. \end{aligned} \tag{26.9}$$

The kinetic helicity injection rate is

$$\begin{aligned} \mathcal{F}_{H_K}(\mathbf{k}) &= \Re[\boldsymbol{\omega}^*(\mathbf{k}) \cdot \mathbf{F}_u(\mathbf{k})] \\ &= -2\Re[(-i\mathbf{k} \times \mathbf{u}^*(\mathbf{k})) \cdot (\boldsymbol{\Omega} \times \mathbf{u}(\mathbf{k})] \\ &= 2\Re[i\mathbf{k} \cdot \{\mathbf{u}^*(\mathbf{k}) \times (\boldsymbol{\Omega} \times \mathbf{u}(\mathbf{k}))\}) \\ &= 2\Re[i(\boldsymbol{\Omega} \cdot \mathbf{k})|\mathbf{u}(\mathbf{k})|^2] = 0. \end{aligned} \tag{26.10}$$

Thus, the Coriolis force does not inject any kinetic energy, enstrophy, or kinetic helicity.

In the next section, we describe the properties of linear rotating hydrodynamics.

26.2 Properties of Linear Rotating Hydrodynamics

The linearized version of Eqs. (26.1) admits columnar structures called *Taylor's column* and *intertial waves.* We will describe these phenomena in the present section.

26.2.1 Taylor–Proudman theorem

Taylor–Proudman theorem predicts Taylor's columns using the following arguments. The linearized and inviscid version of Eqs. (26.1) is

$$\frac{\partial \mathbf{u}}{\partial t} = -\nabla p - 2\boldsymbol{\Omega} \times \mathbf{u}. \tag{26.11}$$

For slow or steady motion, $\partial \mathbf{u}/\partial t \approx 0$. Under this approximation, taking a curl of Eq. (26.11) yields

$$\nabla \times (\boldsymbol{\Omega} \times \mathbf{u}) = 0. \tag{26.12}$$

Under the assumption of constant $\boldsymbol{\Omega}$ and irrotational flow ($\nabla \times \mathbf{u} = 0$), Eq. (26.12) becomes

$$\boldsymbol{\Omega} \cdot \nabla \mathbf{u} = \Omega \frac{\partial}{\partial z} \mathbf{u} = 0; \tag{26.13}$$

that is, the velocity field is invariant along $\hat{z}$. In other words, such rotating flows have strong vortical structures that do not vary along $\hat{z}$. This is the *Taylor–Proudman theorem* (Chandrasekhar, 2013).

Taylor's columns are time-independent solution of Eq. (26.11). The time-dependent rotating hydrodynamics admits wave solution, which will be described in the next section.

26.2.2 Inertial waves in rotating flows

In Fourier space, the linearized and inviscid version of Eqs. (26.6) yields (Chandrasekhar, 2013)

$$\frac{d}{dt}\mathbf{u}(\mathbf{k}) = -i\mathbf{k}p(\mathbf{k}) - 2\boldsymbol{\Omega} \times \mathbf{u}(\mathbf{k}), \tag{26.14a}$$

$$\mathbf{k} \cdot \mathbf{u}(\mathbf{k}) = 0. \tag{26.14b}$$

In Craya–Herring basis, the Coriolis force is (see Figs. 9.1 and 9.2)

$$\begin{aligned} \mathbf{F}(\mathbf{k}) &= -2\boldsymbol{\Omega} \times \mathbf{u}(\mathbf{k}) \\ &= -2\Omega\{\hat{e}_3 \cos\zeta - \hat{e}_2 \sin\zeta\} \times \{u_1\hat{e}_1 + u_2\hat{e}_2\} \\ &= 2\Omega\{\hat{e}_1 u_2 \cos\zeta - \hat{e}_2 u_1 \cos\zeta - \hat{e}_3 u_1 \sin\zeta\}, \end{aligned} \tag{26.15}$$

where ζ is the angle between $\mathbf{k}$ and $\hat{z}$. Hence, the dynamical equations along $\hat{e}_1(\mathbf{k}), \hat{e}_2(\mathbf{k})$ are

$$\dot{u}_1(\mathbf{k}) = \omega u_2(\mathbf{k}), \tag{26.16a}$$

$$\dot{u}_2(\mathbf{k}) = -\omega u_1(\mathbf{k}), \tag{26.16b}$$

where

$$\omega = 2\Omega \cos\zeta. \tag{26.17}$$

Equations (26.16) have oscillatory solutions, which are inertial waves. These solutions are represented more conveniently in helical basis as (see Section 9.5):

$$\begin{aligned} \dot{u}_+(\mathbf{k}) &= \frac{1}{2}(\dot{u}_2(\mathbf{k}) + i\dot{u}_1(\mathbf{k})) \\ &= \frac{1}{2}[-\omega u_1(\mathbf{k}) + i\omega u_2(\mathbf{k})] = i\omega u_+(\mathbf{k}), \end{aligned} \tag{26.18}$$

and

$$\dot{u}_-(\mathbf{k}) = -i\omega u_-(\mathbf{k}). \tag{26.19}$$

These equations admit the following solution:

$$u_+(\mathbf{k}, t) = u_+(\mathbf{k}, 0)\exp(i\omega t), \tag{26.20a}$$

$$u_-(\mathbf{k},t) = u_-(\mathbf{k},0)\exp(-i\omega t), \tag{26.20b}$$

with Eq. (26.17) as the dispersion relation. Note that $\mathbf{u}_+(\mathbf{k})$ and $\mathbf{u}_-(\mathbf{k})$ modes have maximal helicity with $H_K(k)/(kE(k)) = 1$ and -1 respectively (Chandrasekhar, 2013).

Since $u_+(\mathbf{r},t) \sim u_+(\mathbf{k},t)\exp(i\mathbf{k}\cdot\mathbf{r})$, we deduce that $\mathbf{u}_+$ moves along $-\mathbf{k}$, while $\mathbf{u}_-$ moves along $\mathbf{k}$. The phase velocity of the wave modes $\mathbf{u}_+$ and $\mathbf{u}_-$ are

$$\mathbf{c}_\pm = \mp\hat{k}\frac{2\Omega}{k}\cos\zeta. \tag{26.21}$$

These waves are depicted in Fig. 26.1.

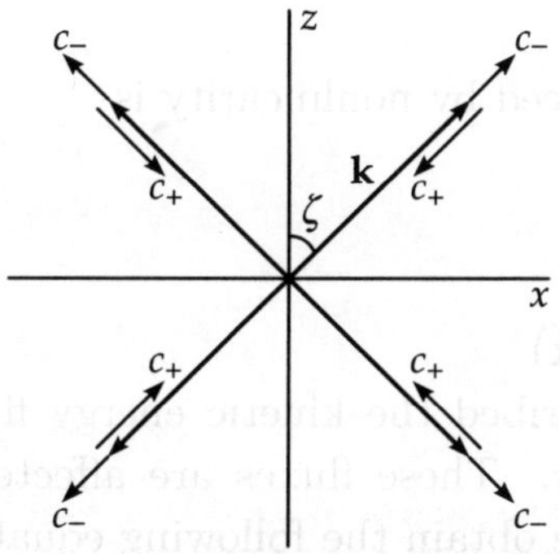

Figure 26.1 Wave $\mathbf{u}_-$ that travels along $\mathbf{k}$ has phase velocity $\mathbf{c}_-$, while $\mathbf{u}_+$ traveling along $-\mathbf{k}$ has phase velocity $\mathbf{c}_+$.

Takng a projection of Eq. (26.14a) along $\hat{e}_3(\mathbf{k})$ yields

$$-ikp(\mathbf{k}) - 2\Omega u_1(\mathbf{k})\sin\zeta = 0, \tag{26.22}$$

using which we determine the pressure $p(\mathbf{k})$ as

$$p(\mathbf{k}) = \frac{i}{k}2\Omega u_1(\mathbf{k})\sin\zeta. \tag{26.23}$$

In the next section we describe the nonlinear aspects of rotating flows.

26.3 Nonlinear Regime in Rotating Flows

Rotation induces anisotropy that makes the flow quite complex. Though Coriolis force does not inject any kinetic energy or kinetic helicity, it induces an exchange of energy between the parallel and perpendicular components of the velocity field via pressure (see Section 11.4). We describe these transfers here.

The pressure gets contributions from the nonlinear term $(\mathbf{u} \cdot \nabla)\mathbf{u}$ and the Coriolis force. The pressure due to the Coriolis force induces an energy exchange between $u_\perp$ and $u_\parallel$. We derive this energy transfer by invoking Eq. (11.19d):

$$\begin{aligned} \mathcal{P}_\parallel(\mathbf{k}) &= k_\parallel \Im[p(\mathbf{k}) u_\parallel^*(\mathbf{k})] \\ &= -2k_\parallel \Im[i\frac{\Omega}{k} \sin\zeta u_1(\mathbf{k}) u_2^*(\mathbf{k}) \sin\zeta] \\ &= -\frac{2\Omega k_\parallel}{k} \sin^2\zeta \Re[u_1(\mathbf{k}) u_2^*(\mathbf{k})], \end{aligned} \tag{26.24}$$

and

$$\mathcal{P}_\perp(\mathbf{k}) = -\mathcal{P}_\parallel(\mathbf{k}). \tag{26.25}$$

Note that the pressure induced by nonlinearity is

$$p(\mathbf{k}) = \frac{i}{k^2}\mathbf{k} \cdot \mathbf{N}_u(\mathbf{k}) \tag{26.26}$$

that too contributes to $\mathcal{P}_\parallel(\mathbf{k})$.

In Section 11.4, we described the kinetic energy fluxes $\Pi_\parallel$ and $\Pi_\perp$ for $u_\parallel$ and $u_\perp$ components respectively. These fluxes are affected by $\mathcal{P}_\parallel(\mathbf{k})$. Following the discussion of Section 4.5, we obtain the following equations:

$$\frac{d}{dk}\Pi_\parallel(k) = \mathcal{P}_\parallel(k) - \nu k^2 E_\parallel(k), \tag{26.27a}$$

$$\frac{d}{dk}\Pi_\perp(k) = -\mathcal{P}_\parallel(k) - \nu k^2 E_\perp(k). \tag{26.27b}$$

Hence, $\mathcal{P}_\parallel(k)$ induces variations in $\Pi_\parallel(k)$ and $\Pi_\perp(k)$ with k. We will describe these effects in the subsequent discussion.

In the next section we will describe the leading turbulence phenomenologies of rotating turbulence.

26.4 Phenomenology of Rotating Turbulence

26.4.1 Zeman's phenomenology

Making an analogy with stably stratified turbulence, Zeman (1994) argued that the Bologiano–Obukhov (BO) spectrum is applicable to rotating turbulence. In this framework, they substitute $N \to \Omega$ and $\epsilon_\rho \to \epsilon_u$ in Eq. (15.39a) that leads to

$$E_u(k) = \epsilon_u^{2/5}\Omega^{4/5}k^{-11/5} \quad \text{for } k < k_\Omega \tag{26.28a}$$

$$E_u(k) = \epsilon_u^{2/3}k^{-5/3} \quad \text{for } k > k_\Omega \tag{26.28b}$$

where k_Ω, called the *Zeman wavenumber*, is obtained by matching the aforementioned two spectra, which is

$$k_\Omega = \sqrt{\frac{\Omega^3}{\epsilon_u}}. \tag{26.29}$$

The above arguments however have several flaws. In the BO phenomenology the kinetic energy flux decreases with k due to the conversion of kinetic energy to potential energy (see Section 15.5.1). But there is no potential energy in rotating turbulence. In addition, in BO phenomenology, buoyancy is equated to the nonlinear term [see Eq. (15.37)]. Similar matching of the Coriolis force with the nonlinear term does not yield the above scaling. Hence, a simple extension of the Bolgiano–Obukhov scaling to rotating turbulence is not appropriate.

26.4.2 Zhou's phenomenology

Rotating flow has two kinds of inertial waves that travel along opposite directions (see Section 26.2.2). Zhou (1995) made an analogy between inertial waves and Alfvén waves of MHD, and argued that the time scale for the interactions between the the inertial waves in rotating turbulence is

$$\tau \approx 1/\omega \sim 1/\Omega. \tag{26.30}$$

Now, using Eq. (22.7), Zhou (1995) argued that the kinetic energy flux is

$$\Pi_u \approx \Pi_{u_+} \approx \Pi_{u_-} \approx \frac{1}{\Omega}[E_{u_+}(k)E_{u_-}(k)k^4], \tag{26.31}$$

where $E_{u_\pm}$ are the kinetic energies of $u_\pm$ fluctuations. For $E_{u_+}(k) \approx E_{u_-}(k) \approx E_u(k)$, we obtain

$$E_u(k) \approx (\Pi_u \Omega)^{1/2} k^{-2}. \tag{26.32}$$

Thus, Zhou (1995)'s phenomenology yields $E_u(k) \sim k^{-2}$.

26.4.3 Smith and Waleffe's phenomenology

A simplified version of Smith and Waleffe (1999)'s phenomenology is as follows. They equated the Coriolis force with the nonlinear advection term that leads to

$$k u_k^2 \sim u_k \Omega. \tag{26.33}$$

Hence,

$$E_u(k) = \frac{u_k^2}{k} = \Omega^2 k^{-3}. \tag{26.34}$$

26.4.4 Kuznetsov–Zakharov–Kolmogorov spectrum

Galtier (2003) employed weak wave turbulence theory to rotating turbulence and derived the following Kuznetsov–Zakharov–Kolmogorov spectrum for the rotating turbulence:

$$E_u(k_\perp, k_\|) \sim k_\perp^{-5/2} k_\|^{-1/2}. \tag{26.35}$$

A derivation for this energy spectrum is as follows. Verma (2004) derived the following formula for MHD turbulence under the weak turbulence limit:

$$\Pi \sim k_\| k_\perp^6 E_u^2(\mathbf{k})/B_0, \tag{26.36}$$

where B_0 is the Alfvén speed, and $E_u(\mathbf{k})$ is the modal energy. Note that the model energy is related to the one-dimensional spectrum $E_u(k)$ as

$$E_u(\mathbf{k}) = \frac{E_u(k)}{k_\perp k_\|}. \tag{26.37}$$

We can translate the formula of Eq. (26.36) to rotating turbulence by employing

$$B_0 \to \Omega k_\perp^{-1}, \tag{26.38}$$

substitution of which in Eq. (26.36) yields

$$E_u(k_\perp, k_\|) \sim (\Pi_u \Omega)^{1/2} k_\perp^{-5/2} k_\|^{1/2} \tag{26.39}$$

that has the same $k_\perp$ dependence as Eq. (26.35), although with a sign difference for $k_\|$. For $k_\| \approx k_\perp$, we recover Zhou (1995)'s formula [Eq. (26.32)].

The aforementioned phenomenologies predict divergent spectral exponents. In the following subsection, we present arguments based on energy fluxes that can provide an alternative explanation for the variations in the spectral exponents depending on the strength of Ω.

26.4.5 Inferences from the energy transfers in rotating turbulence

Using Eq. (26.27) we deduce that in the inertial range where $D_u(k) = 0$, the energy flux would be affected by $\mathcal{P}_\parallel(k)$. Unfortunately, these effects have not been studied in detail, except in some works that will be presented in the next section. Here we make certain conjectures on the energy spectrum and flux based on $\mathcal{P}_\parallel(k)$.

For low Reynolds number, the nonlinear term can be ignored, and the sign of $\mathcal{P}_\parallel(k)$ will determine the variations in $\Pi_\parallel$ and $\Pi_\perp$. However, for large Reynolds numbers, a strong nonlinearity makes $u_\perp$ stronger than $u_\parallel$. Thus, the flow becomes quasi-two-dimensional that leads to an inverse cascade of kinetic energy as in two-dimensional turbulence (see Chapter 7). Thus, in rotating turbulence, anisotropy induced by rotation exhibits a major complexity depending on the strength of Ω.

In the next section, we will describe some of the leading experimental and numerical results of rotating turbulence, and relate them to the aforementioned phenomenologies.

26.5 Experimental and Numerical Results on Rotating Turbulence

There have been a large number of experimental and numerical works on rotating turbulence. Here, we present only a small fraction of these results.

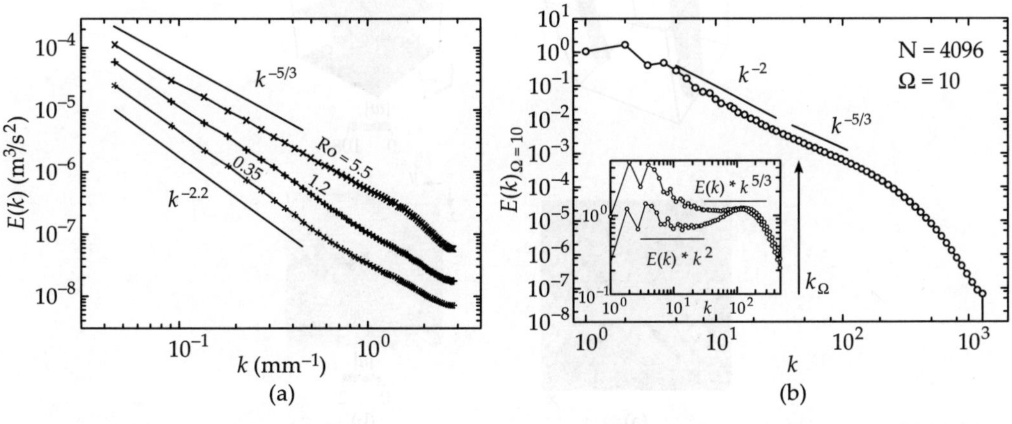

Figure 26.2 (a): $E_u(k)$ for the rotating turbulence of Morize et al. (2005)'s experiment. The spectral exponent varies from $-5/3$ to -2.2 as Ro decreases from 5.5 to 0.35. (b): For a numerical simulation for $\mathrm{Ro} = 0.25$, Biferale et al. (2016) reported dual spectrum for $E_u(k)$ with exponents -2 and $-5/3$. Figures (a) from Morize et al. (2005). Reprinted with permission from AIP. Figure (b) from Biferale (2016). Reprinted under Creative Commons Attribution 3.0 License.

Morize et al. (2005) performed experiments on rotating turbulence with slow and moderate rotation and showed that $E_u(k) \sim k^{-5/3}$ for Ro from 100 to 1, but $E_u(k)$ steepens gradually from $k^{-5/3}$ to $k^{-2.2}$ as the Rossby number decreases from 1 to 0.1 (see Figure 26.2(a)). Mininni et al. (2009) studied scale interactions and scaling laws in rotating flows with moderate Ro. Biferale et al. (2016) simulated rotating turbulence for Ro $= 0.25$ and reported $E_u(k)$ with dual spectrum, a combination of k^{-2} and $k^{-5/3}$. See Figure 26.2(b) for an illustration. It will be interesting to relate these spectra to energy flux variations, as described in Section 26.4.5.

Sharma et al. (2018b) simulated rapidly rotating decaying turbulence with the asymptotic Ro ranging from 10^{-3} to 10^{-2}. They reported a strong inverse cascade of kinetic energy leading to strong columnar structures as shown in Figure 26.3(a). In fact, these structures have dominant contributions from the Fourier modes $\mathbf{u}(1, 0, 1)$ and $\mathbf{u}(0, 1, 1)$, where the indices represent k_x, k_y, k_z respectively. Such structures lead to a weakening of energy at intermediate and small scales. It is important to note that these structures arise due to a strong inverse cascade of energy (a nonlinear effect). Hence, the structures of Fig. 26.3 are not Taylor columns that arise in linear rotating systems.

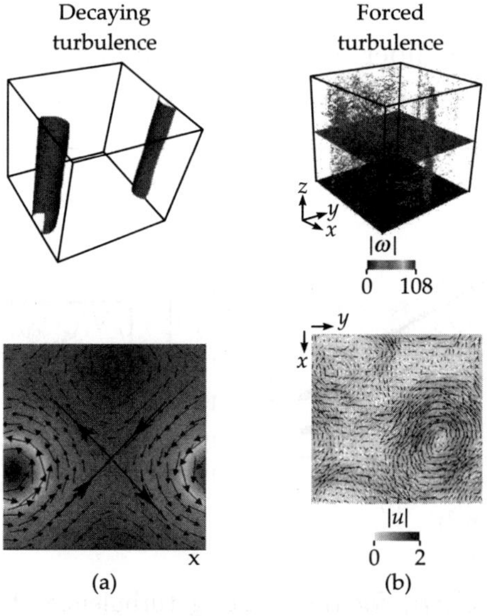

Figure 26.3 Strong columnar structures appear (a) in decaying rotating turbulence simulations of Sharma et al. (2018b) during the asymptotic regime, and (b) in forced rotating turbulence simulations ($k_f = 80$–82) of Sharma et al. (2018a). The structures in forced turbulence are more diffused compared to those of decaying turbulence. Adopted from the two figures of Sharma et al. (2018b) and Sharma et al. (2018a).

Sharma et al. (2018b) observed that the energy flux is very weak in rapidly rotating decaying turbulence though the Reynolds number is as high as 3000. As in 2D hydrodynamics, the enstrophy flux $\Pi_\omega(k)$ dominates $\Pi_u(k)$ for large k (see Fig. 26.4(c, d)). Here the large-scale structures provide effective forcing at their respective scale, hence $k_f \approx 1/L$. It is believed that the energy buildup at large scales is stronger in rotating turbulence than that in 2D turbulence.

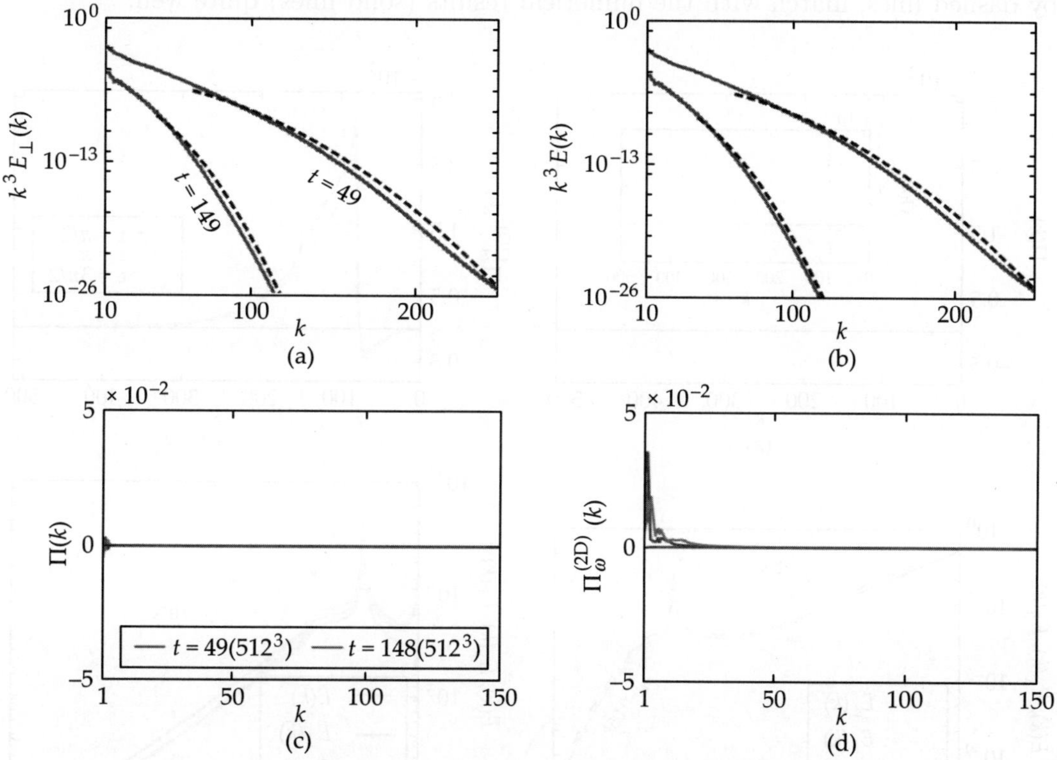

Figure 26.4 For decaying rotating turbulence simulations of Sharma et al. (2018b) at $t = 49$ and 148: (a) the normalized energy spectrum $k^3 E_\perp(k)$ for a horizontal cross-section; (b) the energy spectrum $k^3 E(k)$; (c) the energy flux $\Pi_u(k)$; (d) the enstrophy flux $\Pi_\omega(k)$. The predictions of Eqs. (26.40), plotted as dashed lines, match with the numerical results (solid lines) quite well. $E_\perp(k)$ and $\Pi_\omega(k)$ are computed for a horizontal cross-section.

Note however that $\Pi_\omega(k)$ is much weaker than its 2D hydrodynamic counterpart. This is due to the weakening of fluctuations at $k > 10$ due to the strong inverse cascade of energy. Sharma et al. (2018b) modeled the energy spectrum and flux using Pao's model for 2D turbulence in the enstrophy cascade regime (see Section 7.3):

$$\Pi_\omega(k) = \epsilon_\omega \exp\left(-K'_{2D}(k/k_{d2D})^2\right), \tag{26.40a}$$

$$E_\omega(k) = K'_{2D}\epsilon_\omega^{2/3}k^{-1}\exp\left(-K'_{2D}(k/k_{d2D})^2\right), \quad (26.40b)$$
$$E_u(k) = K'_{2D}\epsilon_\omega^{2/3}k^{-3}\exp\left(-K'_{2D}(k/k_{d2D})^2\right). \quad (26.40c)$$

The enstrophy flux and $E_\perp(k)$ are computed for a horizontal cross-section of the flow. See Fig. 26.4(a, b) for an illustration of the energy spectrum at $t = 49$ and $t = 149$ during the evolution of turbulence. The predictions of Eqs. (26.40), represented by dashed lines, match with the numerical results (solid lines) quite well.

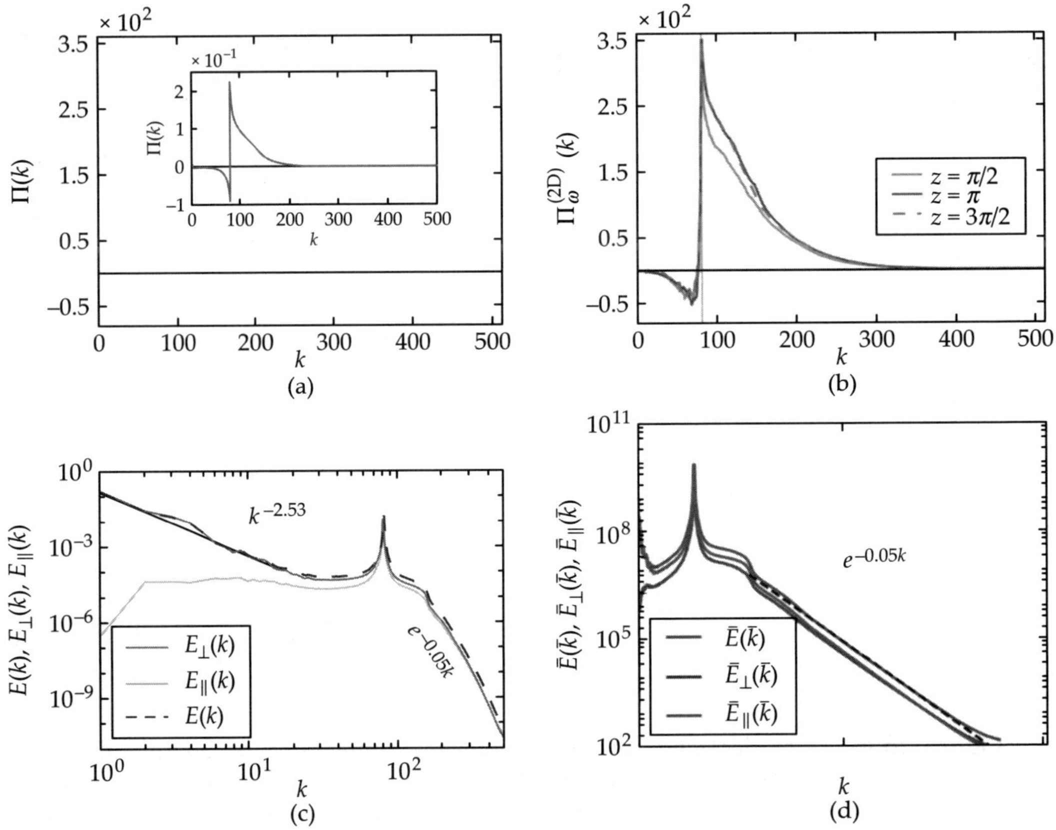

Figure 26.5 For rapidly rotating forced turbulence simulations of Sharma et al. (2018a) with $k_f = (80\text{–}82)$: (a) the energy flux $\Pi_u(k)$; (b) the enstrophy flux $\Pi_\omega(k)$. (c) energy spectrum $E(k), E_\perp(k), E_\parallel(k)$ in log-log scale; (d) $E(k), E_\perp(k), E_\parallel(k)$ in semilog-y scale. For $k < k_f$, $E_u(k) \sim k^{-5/2}$, but for $k > k_f$, the predictions of Eqs. (5.40) match with the numerical results quite well.

Later, Sharma et al. (2018a) simulated rapidly rotating forced turbulence with isotropic forcing employed at the intermediate range ($k_f = (80\text{–}82)$). They showed that the vertical structures are diffused compared to the decaying case. See Fig. 26.3(b) for an illustration. For $k < k_f$, $E_u(k) \sim k^{-5/2}$ consistent with the Kuznetsov–Zakharov–Kolmogorov spectrum (Galtier, 2003). However, at the intermediate and small scales, the flow is nearly three-dimensional with very small amount of energy. Consequently, the flux and spectrum are similar to those for laminar flows [of the form $\exp(-ak)$ as in Eqs. (5.40). See Fig. 26.5 for an illustration of the energy spectrum, and the energy and enstrophy fluxes.

Sharma et al. (2019) computed $\Pi_\perp, \Pi_\parallel$, and $\mathcal{P}_\parallel$ for the aforementioned rapidly rotating decaying and forced turbulence. These quantities are plotted in Fig. 26.6. The figure exhibits strong variations in these quantities. In addition, $\mathcal{P}_\parallel > 0$ implying energy transfers from $u_\perp$ to $u_\parallel$. However $\mathcal{P}_\parallel$ is weak.

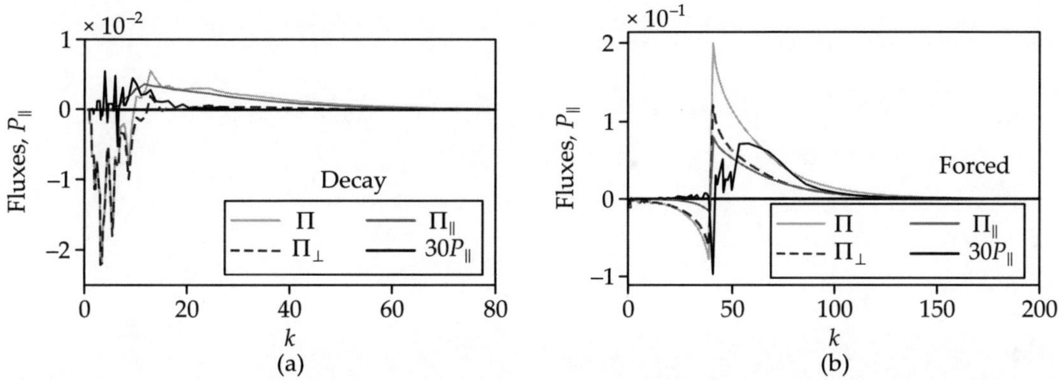

Figure 26.6 For the decaying and forced rotating turbulence simulations of Sharma et al. (2018b) and Sharma et al. (2018a), plots of Π, $\Pi_\perp, \Pi_\parallel$, and $\mathcal{P}_\parallel$. Note that $\mathcal{P}_\parallel > 0$. Adopted from a figure of Sharma et al. (2019).

In summary, these simulations indicate that for slow and moderate rotation, the energy spectrum is a dual spectrum with two power laws (see Fig. 26.2(b)). However, for the rapidly rotating turbulence, the spectrum for $k > k_f$ is exponential due to a strong inverse cascade of energy. The aforementioned results illustrate the usefulness of energy flux and energy transfers for describing rotating turbulence.

With these remarks, we end our discussion on rotating turbulence.

Further Reading

The literature on rotating turbulence is extensive. For an introduction to this topic, we recommend books by Davidson (2013) and Greenspan (1968). For the phenomenological theories, refer to Zeman (1994), Zhou (1995), Smith and Waleffe (1999), Galtier (2003), and the reference therein. For further details on experimental investigations of rotating turbulence, refer to the papers cited in this chapter as well as in Moisy et al. (2010). Refer to Sharma et al. (2018a,b) for more details on numerical results presented in this chapter.

Chapter 27

Flow with a Tensor

On many occasions we encounter fluid flows that advect tensorial fields, for example, polymers, elastic material, etc. A polymer, which will be discussed in this chapter is modeled as a second rank tensor. Following the categorization of scalar and vector fields, we classify a tensor as *passive* or *active* depending on whether it affects the velocity field or not.

In this short chapter, we will provide a cursory view of the energy spectra and fluxes of a tensor flow. Some of the predictions discussed in this chapter are new, and they need to be tested in future. For a more detailed discussion on earlier works on tensor flows, specially those with polymers, refer to de Gennes (1979) and Benzi and Ching (2018).

27.1 Governing Equations

Consider a tensor $\mathcal{C}$ being advected by an incompressible flow. The equations for the two fields—$\mathcal{C}$ and velocity field $\mathbf{u}$—are

$$\frac{\partial \mathbf{u}}{\partial t} + (\mathbf{u}\cdot\nabla)\mathbf{u} = -\nabla(p/\rho) + \nu\nabla^2\mathbf{u} + \mathbf{F}_u, \tag{27.1a}$$

$$\frac{\partial \mathcal{C}_{ij}}{\partial t} + (\mathbf{u}\cdot\nabla)\mathcal{C}_{ij} = \eta\nabla^2\mathcal{C}_{ij} + F_{\mathcal{C},ij}, \tag{27.1b}$$

$$\nabla\cdot\mathbf{u} = 0, \tag{27.1c}$$

where p is the pressure field, ρ is density which is assumed to be unity, ν is the kinematic viscosity, η is the diffusion coefficient of the tensor, and $\mathbf{F}_u, F_{\mathcal{C}}$ are the force fields to $\mathbf{u}, \mathcal{C}$ respectively. In this equation, $\mathcal{C}_{ij}$ represents a component of the tensor field; here, the indices i, j take values $1, 2, 3$ or x, y, z. In Eq. (27.1b), the ratio of the nonlinear term and the diffusion term is termed as Reynolds number based on $\mathcal{C}$:

$$\mathrm{Re}_{\mathcal{C}} = \frac{(\mathbf{u} \cdot \nabla)\mathcal{C}}{\eta \nabla^2 \mathcal{C}} = \frac{UL}{\eta}. \tag{27.2}$$

For the tensor field, we define the energy density as

$$E_{\mathcal{C}}(\mathbf{r}) = \frac{1}{2}\mathcal{C}^2 = \frac{1}{2}\mathcal{C}_{ij}\mathcal{C}_{ji}, \tag{27.3}$$

whose evolution equation is

$$\frac{\partial}{\partial t}\frac{\mathcal{C}^2}{2} + \nabla \cdot \left(\frac{1}{2}\mathcal{C}^2 \mathbf{u}\right) = F_{\mathcal{C},ij}\mathcal{C}_{ji} + \eta \mathcal{C}_{ji}(\nabla^2 \mathcal{C}_{ij}). \tag{27.4}$$

In the above equations, we follow Einstein's notation in which the repeated indices are summed. Clearly, in the absence of $F_{\mathcal{C}}$ and η, for periodic or vanishing boundary conditions,

$$\int \frac{1}{2}\mathcal{C}^2 d\mathbf{r} = \text{const.} \tag{27.5}$$

This is a statement of *conservation of tensor energy* in a nondiffusive and force-free tensor flow.

In Fourier space, the dynamical equations are

$$\frac{d}{dt}\mathbf{u}(\mathbf{k}) + \mathbf{N}_u(\mathbf{k}) = -i\mathbf{k}p(\mathbf{k}) + \mathbf{F}_u(\mathbf{k}) - \nu k^2 \mathbf{u}(\mathbf{k}), \tag{27.6a}$$

$$\frac{d}{dt}\mathcal{C}_{ij}(\mathbf{k}) + N_{\mathcal{C},ij}(\mathbf{k}) = F_{\mathcal{C},ij}(\mathbf{k}) - \eta k^2 \mathcal{C}_{ij}(\mathbf{k}), \tag{27.6b}$$

$$\mathbf{k} \cdot \mathbf{u}(\mathbf{k}) = 0, \tag{27.6c}$$

where the nonlinear terms are

$$\mathbf{N}_u(\mathbf{k}) = i\sum_{\mathbf{p}} \{\mathbf{k} \cdot \mathbf{u}(\mathbf{k}-\mathbf{p})\}\mathbf{u}(\mathbf{p}), \tag{27.7a}$$

$$N_{\mathcal{C},ij}(\mathbf{k}) = i\sum_{\mathbf{p}} \{\mathbf{k} \cdot \mathbf{u}(\mathbf{k}-\mathbf{p})\}\mathcal{C}_{ij}(\mathbf{p}). \tag{27.7b}$$

The equations for the modal kinetic energy and modal tensor energy, $\mathcal{C}_{ij}(\mathbf{k})\mathcal{C}_{ji}^*(\mathbf{k})/2$, are as follows:

$$\frac{d}{dt}E_u(\mathbf{k}) = \sum_{\mathbf{p}} \Im\left[\{\mathbf{k}\cdot\mathbf{u}(\mathbf{q})\}\{\mathbf{u}(\mathbf{p})\cdot\mathbf{u}^*(\mathbf{k})\}\right] + \Re[\mathbf{F}_u(\mathbf{k})\cdot\mathbf{u}^*(\mathbf{k})] - 2\nu k^2 E_u(\mathbf{k}), \tag{27.8a}$$

$$\frac{d}{dt}E_{\mathcal{C}}(\mathbf{k}) = \sum_{\mathbf{p}} \Im\left[\{\mathbf{k}\cdot\mathbf{u}(\mathbf{q})\}\{\mathcal{C}_{ij}(\mathbf{p})\mathcal{C}^*_{ji}(\mathbf{k})\}\right] + \Re[F_{\mathcal{C},ij}(\mathbf{k})\mathcal{C}^*_{ji}(\mathbf{k})] - 2\eta k^2 E_{\mathcal{C}}(\mathbf{k}). \tag{27.8b}$$

In the next section we describe the properties of energy transfers in a tensor flow.

27.2 Mode-to-mode Tensor Energy Transfer and Tensor Energy Flux

The energy exchange among the velocity modes is facilitated by the nonlinear term $(\mathbf{u}\cdot\nabla)\mathbf{u}$. Hence, the formulas for the mode-to-mode kinetic energy transfer and kinetic energy flux in a tensor flow would be the same as those derived in Chapter 4. We do not repeat them here. Rather, we focus on the energy transfers for the tensor field.

The energy transfers among the tensor Fourier modes occur via the nonlinear term $(\mathbf{u}\cdot\nabla)\mathcal{C}_{ij}$. For the triad $(\mathbf{k}',\mathbf{p},\mathbf{q})$ that satisfies $\mathbf{k}'+\mathbf{p}+\mathbf{q} = 0$, using Eq. (27.8b), we derive that

$$\begin{aligned}\frac{d}{dt}E_{\mathcal{C}}(\mathbf{k}') &= S^{\mathcal{CC}}(\mathbf{k}'|\mathbf{p},\mathbf{q}) \\ &= -\Im\left[\{\mathbf{k}'\cdot\mathbf{u}(\mathbf{q})\}\{\mathcal{C}_{ij}(\mathbf{p})\mathcal{C}_{ij}(\mathbf{k}')\}\right] + \Im\left[\{\mathbf{k}'\cdot\mathbf{u}(\mathbf{p})\}\{\mathcal{C}_{ij}(\mathbf{q})\mathcal{C}_{ij}(\mathbf{k}')\}\right],\end{aligned} \tag{27.9}$$

where $S^{\mathcal{CC}}(\mathbf{k}'|\mathbf{p},\mathbf{q})$ is the combined tensor energy transfer to the mode $\mathcal{C}(\mathbf{k}')$ from $\mathcal{C}(\mathbf{p})$ and $\mathcal{C}(\mathbf{q})$. It is easy to show that

$$S^{\mathcal{CC}}(\mathbf{k}'|\mathbf{p},\mathbf{q}) + S^{\mathcal{CC}}(\mathbf{p}|\mathbf{q},\mathbf{k}') + S^{\mathcal{CC}}(\mathbf{q}|\mathbf{k}',\mathbf{p}) = 0, \tag{27.10}$$

which is a statement on the *detailed conservation of tensor energy in a triad.*

We can derive the mode-to-mode tensor energy transfer following the same arguments as in Section 6.1.1. We skip the details and just state the result. The mode-to-mode tensor energy transfer from tensor mode $\mathcal{C}(\mathbf{p})$ to tensor mode $\mathcal{C}(\mathbf{k}')$ with the mediation of velocity mode $\mathbf{u}(\mathbf{q})$ is

$$S^{\mathcal{CC}}(\mathbf{k}'|\mathbf{p}|\mathbf{q}) = -\Im\left[\{\mathbf{k}'\cdot\mathbf{u}(\mathbf{q})\}\{\mathcal{C}_{ij}(\mathbf{p})\mathcal{C}_{ij}(\mathbf{k}')\}\right]. \tag{27.11}$$

In Eq. (27.8b), the nonlinear term $[(\mathbf{u}\cdot\nabla)\mathcal{C}]\cdot\mathcal{C}$ represents advection of the tensor field $\mathcal{C}$ by the velocity field $\mathbf{u}$. Therefore, $\mathbf{u}$ mediates the tensor energy transfer between $\mathcal{C}$ fields (see Section 4.1.1 for a detailed argument). In Fourier space, this translates to the tensor energy transfer from mode $\mathcal{C}(\mathbf{p})$ to mode $\mathcal{C}(\mathbf{k}')$ with mode $\mathbf{u}(\mathbf{q})$ acting as a mediator. In Fig. 27.1 we illustrate this transfer.

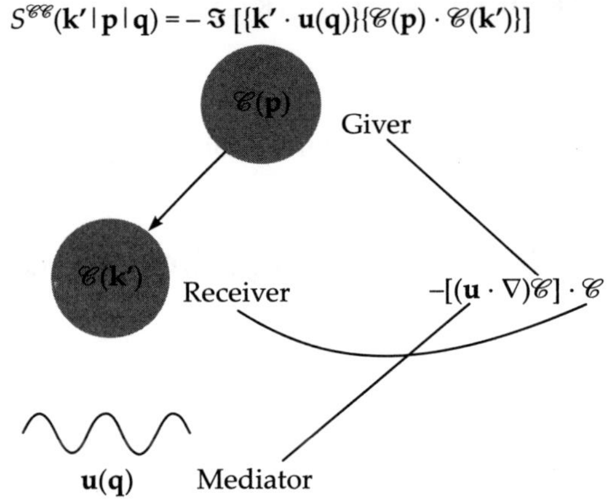

Figure 27.1 A schematic diagram exhibiting mode-to-mode tensor energy transfer from mode $\mathcal{C}(\mathbf{p})$ to mode $\mathcal{C}(\mathbf{k}')$ with mode $\mathbf{u}(\mathbf{q})$ acting as a mediator. In this energy transfer, mode $\mathbf{u}(\mathbf{q})$ essentially advects tensor modes $\mathcal{C}(\mathbf{p})$ and $\mathcal{C}(\mathbf{k}')$ who exchange energy among themselves.

Using the aforementioned mode-to-mode energy transfer, we deduce that the tensor energy flux for a wavenumber sphere of radius k_0 is

$$\Pi_C(k_0) = \sum_{|\mathbf{k}'|>k_0} \sum_{|\mathbf{p}|\le k_0} S^{CC}(\mathbf{k}'|\mathbf{p}|\mathbf{q}). \tag{27.12}$$

Refer to Chapter 4 for a definition of energy flux.

Following the arguments of Section 4.4, we deduce the evolution equation for the tensor energy spectrum $E_C(k)$ as

$$\frac{\partial}{\partial t}E_C(k,t) = -\frac{\partial}{\partial k}\Pi_C(k,t) + \mathcal{F}_C(k,t) - D_C(k,t), \tag{27.13}$$

where $\mathcal{F}_C(k)$ is the tensor energy supply rate by F_C, and $D_C(k)$ is the diffusion rate of tensor energy in shell k, that is,

$$\mathcal{F}_C(k)dk = \sum_{k<k'\le k+dk} \Re[F_{C,ij}(\mathbf{k}')C^*_{ji}(\mathbf{k}')], \tag{27.14}$$

$$D_C(k)dk = \sum_{k<k'\le k+dk} 2\eta k^2 E_C(\mathbf{k}'). \tag{27.15}$$

Under a steady state, $\partial E_{\mathcal{C}}(k)/\partial t \approx 0$, we obtain

$$\frac{d}{dk}\Pi_{\mathcal{C}}(k) = \mathcal{F}_{\mathcal{C}}(k) - D_{\mathcal{C}}(k). \tag{27.16}$$

In the inertial range, $D_{\mathcal{C}}(k)$ is negligible. Hence, in the inertial range, $\Pi_{\mathcal{C}}(k)$ depends on the properties of $\mathcal{F}_{\mathcal{C}}(k)$:

$$\frac{d}{dk}\Pi_{\mathcal{C}}(k) = \mathcal{F}_{\mathcal{C}}(k). \tag{27.17}$$

As argued for the velocity, scalar, and vector fields (see Section 4.5),

1. $\Pi_{\mathcal{C}}(k) = \text{const.}$ when $\mathcal{F}_{\mathcal{C}}(k) = 0$.
2. $\Pi_{\mathcal{C}}(k)$ increases with k when $\mathcal{F}_{\mathcal{C}}(k) > 0$.
3. $\Pi_{\mathcal{C}}(k)$ decreases with k when $\mathcal{F}_{\mathcal{C}}(k) < 0$.

In the next section, we will derive the spectrum and flux of a passive tensor field.

27.3 Energy Spectrum and Flux in a Passive Tensor

A tensor field is called passive when $\mathbf{F}_u$ is independent of the tensor field. For such a configuration, we derive the turbulent spectra and fluxes for both velocity and tensor fields. Since the velocity field is unaffected by the tensor field, its spectrum and flux follow Kolmogorov's phenomenology if $\mathbf{F}_u$ acts at large scales.

Now for the tensor field, if we assume that $\mathbf{F}_{\mathcal{C}}$ is active only at large scales, then in the inertial range, Eq. (27.17) yields

$$\Pi_{\mathcal{C}}(k) = \text{const.} \tag{27.18}$$

We derive the energy spectrum for the tensor field following similar arguments as in Chapters 14 and 19. Scaling arguments indicate that the inertial $E_{\mathcal{C}}(k)$ would depend on Π_u, $\Pi_{\mathcal{C}}$, and k, whose dimensions are

$$[\Pi_u] = [L^2/T^3]; \quad [\Pi_{\mathcal{C}}] = [\mathcal{C}^2/T]; \quad [E_{\mathcal{C}}(k)] = [\mathcal{C}^2 L]. \tag{27.19}$$

Now we postulate that

$$E_{\mathcal{C}}(k) = (\Pi_{\mathcal{C}})^{\alpha}(\Pi_u)^{\beta}k^{\gamma}. \tag{27.20}$$

Matching the dimensions of $[\mathcal{C}]$, $[L]$, and $[T]$ yields

$$\alpha = 1; \quad \beta = -1/3; \quad \gamma = -5/3. \tag{27.21}$$

Therefore,

$$E_{\mathcal{C}}(k) = K\Pi_{\mathcal{C}}(\Pi_u)^{-1/3}k^{-5/3}, \tag{27.22}$$

where K is a nondimensional constant. Thus, we derive the spectrum and flux of the tensor energy; these formulas are similar to those of a passive scalar and a passive vector.

In the next section we describe the properties of a turbulent flow of dilute polymer solution in which the polymer acts as an active tensor.

27.4 Flow with an Active Tensor Field: FENE-p Model

When a small concentration of polymer is dispersed in a turbulent flow, the flow exhibits many interesting phenomena, one among them being turbulent drag reduction. In such flows, the polymer is often described by the finitely extensible nonlinear elastic-Peterlin (FENE-P) model. In this section we briefly describe the equations and turbulence properties of such flows.

27.4.1 Governing equations

For the FENE-p model, the equations for the flow velocity and the polymer-conformation tensor $\mathcal{C}$ are given by the following simplified equations (Fouxon and Lebedev, 2003; Perlekar et al., 2006):

$$\frac{\partial \mathbf{u}}{\partial t} + (\mathbf{u}\cdot\nabla)\mathbf{u} = -c_1\nabla(p/\rho) + \nu\nabla^2\mathbf{u} + c_1\nabla\cdot(f\mathcal{C}), \tag{27.23a}$$

$$\frac{\partial \mathcal{C}}{\partial t} + (\mathbf{u}\cdot\nabla)\mathcal{C} = (\nabla\mathbf{u})^T\cdot\mathcal{C} + \mathcal{C}\cdot(\nabla\mathbf{u}) + \eta\nabla^2\mathcal{C}, \tag{27.23b}$$

$$\nabla\cdot\mathbf{u} = 0, \tag{27.23c}$$

where c_1 is a constant, and f is a function of $\mathcal{C}$. When we compare these equations with Eqs. (27.1), we deduce that the following forces are associated with $\mathbf{u}$ and $\mathcal{C}$:

$$F_{u,i} = c_1\partial_j(f\mathcal{C}_{ji}), \tag{27.24}$$

$$F_{\mathcal{C},ij} = [(\nabla\mathbf{u})^T\cdot\mathcal{C} + \mathcal{C}\cdot(\nabla\mathbf{u})]_{ij} \sim \mathcal{C}_{il}\partial_l u_j, \tag{27.25}$$

and the energy feed to kinetic energy and tensor energy in Fourier space are respectively:

$$\mathcal{F}_u(\mathbf{k}) = -c_1\sum_{\mathbf{p}}\Im\left[k_j f(\mathbf{q})\mathcal{C}_{ij}(\mathbf{p})\mathbf{u}_i^*(\mathbf{k})\right], \tag{27.26}$$

$$\mathcal{F}_C(\mathbf{k}) \quad = \quad -\sum_{\mathbf{p}} \Im \left[C_{il}(\mathbf{q}) p_l u_j(\mathbf{p}) C_{ij}^*(\mathbf{k}) \right], \tag{27.27}$$

where $\mathbf{q} = \mathbf{k} - \mathbf{p}$. Both the injection rates are convolutions due to their nonlinearity. From the above form of energy injection, we deduce that the polymer in the FENE-p model is an active tensor.

In a turbulent flow of dilute polymer solution, we expect that the velocity field stretches the polymers. This mechanism is similar to the stretching of the magnetic field by the velocity field in MHD or dynamo ($\mathbf{u} \to \mathbf{b}$), and vortex stretching by the velocity field ($\mathbf{u} \to \boldsymbol{\omega}$) (Tabor and de Gennes, 1986; Fouxon and Lebedev, 2003; Thais et al., 2013). For such energy transfers associated with polymers stretching,

$$\mathcal{F}_u(\mathbf{k}) < 0; \quad \mathcal{F}_C > 0. \tag{27.28}$$

These relations imply energy transfers from the velocity field to the tensor field.

After this, we describe energy spectra of the aforementioned system.

27.4.2 Energy spectra and fluxes in the FENE-p model

Depending on the relative strengths of the nonlinear terms, a tensor flow can be divided into four categories, which will be described in the following. This classification is similar to those adopted for scalar and vector flows.

Based on Eq. (27.28) we argue that $E_u(k)$ would be steeper and $E_C(k)$ shallower than the corresponding spectra for the passive tensor (see Section 27.3).

$\mathrm{Re} \gg 1; \mathrm{Re}_C \gg 1$

This case corresponds to a turbulent flow of dilute polymer solution. As argued in Section 4.5 case (3), negative $\mathcal{F}_u(\mathbf{k})$ leads to a reduction of the flux $\Pi_u(k)$ with k. Hence, we expect that $E_u(k)$ will be steeper than the $k^{-5/3}$ spectrum. However, positive $\mathcal{F}_C$ will make Π_C increase with k, thus making it shallower than the $k^{-5/3}$ spectrum of a passive tensor. Determination of the spectral exponents however would require exact forms of $\mathcal{F}_u$ and $\mathcal{F}_C$.

$\mathrm{Re} \ll 1; \mathrm{Re}_C \gg 1$

This case corresponds to gels that move very slowly. We derive the energy spectra for this case following arguments similar to that in Section 16.9.4. Since $\mathrm{Re} \ll 1$, Eq. (27.23a) yields (assuming $f \sim 1$)

$$\nu k^2 u_k \approx k C_k. \tag{27.29}$$

Since $\mathrm{Re}_C \gg 1$, we expect a cascade of tensor energy. An assumption of a constant tensor energy flux yields

$$\Pi_C \approx k u_k C_k^2. \tag{27.30}$$

Equations (27.29, 27.30) yield

$$u_k \approx \nu^{-2/3} \Pi_C^{1/3} k^{-1}; \quad C_k = \nu^{1/3} \Pi_C^{1/3}. \tag{27.31}$$

Therefore,

$$E_u(k) \approx \nu^{-4/3} \Pi_C^{2/3} k^{-3}, \tag{27.32a}$$

$$E_C(k) \approx \nu^{2/3} \Pi_C^{2/3} k^{-1}. \tag{27.32b}$$

The aforementioned spectra get modified when the tensor energy flux varies with k, or when we include the effects of f of Eq. (27.23a). Note that these predictions are in general agreement with some of the experimental results (Majumdar and Sood, 2011).

Due to uncertainties in the scaling of $\mathbf{F}_u$ and F_C, it is difficult to model the other two cases: $\mathrm{Re} \gg 1; \mathrm{Re}_C \ll 1$, and $\mathrm{Re} \ll 1; \mathrm{Re}_C \ll 1$. For the former, $E_C(k)$ will be steeper than the $k^{-5/3}$ spectrum, but for the latter, both $E_u(k)$ and $E_C(k)$ would be much steeper than the $k^{-5/3}$ spectrum, possibly of an exponential form as in laminar hydrodynamics (see Section 5.6).

In the next section, we will invoke Eq. (27.28) to describe polymer-induced turbulent drag reduction.

27.5 Turbulent Drag Reduction in Polymeric Flows

Many experiments and numerical simulations report turbulent drag reduction in a flow of dilute polymer solution. This phenomena remains an unsolved problem in turbulence. Researchers attribute this drag reduction to boundary layers, nonlinearity, back-reaction by polymers to the flow, etc. In this section we will show that the energy transfers play an important role in turbulent drag reduction. An analogy with MHD turbulence provides interesting clues to this intriguing problem.

In MHD turbulence and dynamo, the magnetic field typically gains energy from the velocity field. These transfers are responsible for the growth of the magnetic energy in the early stages of the dynamo, and for saturation at later stages. As argued in the previous section, in polymeric turbulent flows, we expect energy transfers from the velocity field to the configuration tensor (Tabor and de Gennes,

1986; Fouxon and Lebedev, 2003; Thais et al., 2013). These energy transfers lead to stretching of the polymers.

In analogy with the discussion of Section 23.10, we argue that the injected energy in the polymeric fluid at large scales is divided among the kinetic energy and tensor energy. Though tensor energy flux is not as clearly defined as the energy fluxes of MHD turbulence, we can safely argue that the kinetic energy flux $\Pi^{u<}_{u>}$ in polymeric turbulent flow would be smaller than the corresponding flux of hydrodynamic turbulence, as in MHD turbulence (see Section 23.10). Since the flux $\Pi^{u<}_{u>}$ arises due to the term $[(\mathbf{u}\cdot\nabla)\mathbf{u}]\cdot\mathbf{u}$, and turbulent drag is proportional to $(\mathbf{u}\cdot\nabla)\mathbf{u}$, we deduce that

$$[(\mathbf{u}\cdot\nabla)\mathbf{u}]_{\text{Tensor}} < [(\mathbf{u}\cdot\nabla)\mathbf{u}]_{\text{Hydro}}\,. \tag{27.33}$$

Or, the turbulent drag is reduced in a polymeric flow compared to its hydrodynamic counterpart. Thus, we argue for turbulent drag reduction in polymeric fluids from energy transfer perspectives.

With this, we conclude our discussion on tensor flows. We provide certain predictions for the energy spectra and fluxes that need to be tested using numerical simulations and experiments.

Further Reading

There is extensive literature on polymeric flows. Here we provide only some of the references. de Gennes (1979) provides an introduction to polymers. Benzi and Ching (2018) and Sreenivasan and White (2000) have written reviews on turbulent drag reduction in polymeric flows.

Chapter 28

Shell Models of Turbulence

Analytical and numerical works on turbulence with all the Fourier modes are quite involved. Therefore, researchers often employ shell models that contain much fewer variables. There are a large number of research papers (e.g., Biferale (2003); Plunian et al. (2012)) and books (e.g., Ditlevsen (2010)) that cover a variety of shell models and their properties, for example, energy spectrum and flux, multiscaling exponents, etc. In this chapter we will focus on the energy transfers in a shell model (Plunian et al., 2012; Kumar and Verma, 2015; Verma and Kumar, 2016).

We start with a shell model for hydrodynamic turbulence.

28.1 Shell Model for Hydrodynamic Turbulence

Researchers have constructed many shell models for hydrodynamic turbulence. In this section, we will describe one such shell model, and then we will discuss the energy transfers in the model.

28.1.1 Shell model

In a shell model, the wavenumber space is divided into concentric shells. In Fig. 28.1 we illustrate one such shell whose inner and outer radii are k_n and k_{n+1} respectively. The velocity field for the shell is denoted by a single complex variable U_n. Considering that turbulence exhibits power law energy spectrum, the radii $\{k_n\}$ are binned logarithmically, that is,

$$k_n = k_0 \lambda^n, \tag{28.1}$$

where $\lambda > 1$ is a constant (see Appendix A).

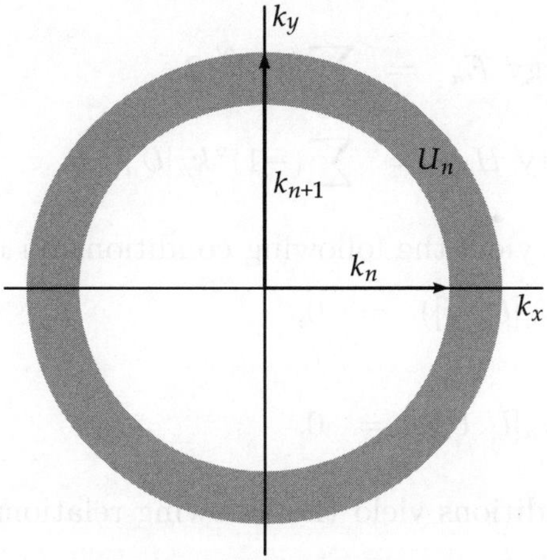

Figure 28.1 For a shell model, a schematic diagram of a shell whose inner and outer radii are k_n and k_{n+1} respectively. A single variable U_n represents the velocity field of this shell.

As described in Chapter 5, the turbulence interactions among the shells are local. Hence, shell k_n interacts with shells k_{n-1} and k_{n+1}. In the shell model, the condition on the wavenumbers, $\mathbf{k} = \mathbf{p} + \mathbf{q}$, corresponds to

$$k_{n+1} = k_n + k_{n-1}. \tag{28.2}$$

Substitution of $k_0 \lambda^n$ in this equation leads to

$$\lambda^2 = \lambda + 1, \tag{28.3}$$

hence $\lambda = (\sqrt{5} + 1)/2$, which is the golden mean.

Using the above properties, a Sabra-based shell model (L'vov et al., 1999) is written as

$$\frac{dU_n}{dt} = N_n[U, U] - \nu k_n^2 U_n + F_n, \tag{28.4}$$

where

$$N_n[U, U] = -i(a_1 k_n U_{n+1}^* U_{n+2} + a_2 k_{n-1} U_{n+1} U_{n-1}^* - a_3 k_{n-2} U_{n-1} U_{n-2}). \tag{28.5}$$

Here, a_1, a_2, a_3 are constants. In this equation, the complex conjugates are chosen to mimic the reality condition $[U(-k_n) = U^*(k_n)]$. For example, for the first term in the right-hand-side, $k_n = k_{n+2} - k_{n+1}$; hence, the corresponding product is $U_{n+2}U^*_{n+1}$.

When $\nu = 0$ and $F_u = 0$, the conserved quantities of the shell model are as follows.

$$\text{Total kinetic energy } E_u = \sum_n |U_n|^2/2, \tag{28.6a}$$

$$\text{Total kinetic helicity } H_K = \sum_n (-1)^n k_n |U_n|^2. \tag{28.6b}$$

These conservation laws yield the following conditions on the nonlinear term:

$$\sum_n \Re\left(U_n^* N_n[U,U]\right) = 0, \tag{28.7a}$$

$$\sum_n \Re\left((-1)^n k_n U_n^* N_n[U,U]\right) = 0. \tag{28.7b}$$

The aforementioned conditions yield the following relations for the constants a_i's:

$$a_1 + a_2 + a_3 = 0, \tag{28.8a}$$

$$a_1 - \lambda a_2 + \lambda^2 a_3 = 0. \tag{28.8b}$$

These two equations do not provide a unique solution for a_i's. Therefore, we write a_1 and a_3 in terms of a_2 as

$$a_1 = -(1+\lambda)a_2; \qquad a_3 = \frac{1+\lambda}{\lambda}a_2. \tag{28.9}$$

In the next section we will describe energy transfers and energy flux in the above shell model.

28.1.2 Energy transfers in the shell model

For the derivation of shell-to-shell energy transfers in the shell model, we focus on a set of three consecutive shells (k_{n-1}, k_n, k_{n+1}). This set is analogous to the triad $(\mathbf{k}, \mathbf{p}, \mathbf{q})$ with $\mathbf{k} = \mathbf{p} + \mathbf{q}$ in Fourier space. Using Eqs. (28.4, 28.5) with $\nu = 0$ and $F_u = 0$, we derive the following energy equations for the above three shells:

$$\frac{d}{dt}\frac{1}{2}|U_{n-1}|^2 = S^{uu}(n-1|n, n+1) = a_1 k_{n-1} \Im(U^*_{n-1}U^*_n U_{n+1}) \tag{28.10a}$$

$$\frac{d}{dt}\frac{1}{2}|U_n|^2 = S^{uu}(n|n-1, n+1) = a_2 k_{n-1} \Im(U^*_{n-1}U^*_n U_{n+1}) \tag{28.10b}$$

$$\frac{d}{dt}\frac{1}{2}|U_{n+1}|^2 = S^{uu}(n+1|n, n-1) = a_3 k_{n-1} \Im(U^*_{n-1}U^*_n U_{n+1}), \tag{28.10c}$$

where $\Im$ denotes the imaginary part of the argument, and $S^{uu}(X|Y,Z)$ represents the combined energy transfer from shells Y and Z to shell X. Using the condition Eq. (28.8a), we derive

$$\frac{1}{2}[|U_{n-1}|^2 + |U_n|^2 + |U_{n+1}|^2] = \text{const.}, \tag{28.11}$$

which is a statement on the *detailed kinetic energy conservation* in a triadic unit of shells. Naturally, when all the triads are included, the total kinetic energy is also conserved (see Section 4.1).

Now, let us derive shell-to-shell transfer $S^{uu}(X|Y|Z)$, which is the energy transfer from shell Y to shell X with the mediation of shell Z. By definition,

$$S^{uu}(X|Y|Z) + S^{uu}(X|Z|Y) = S^{uu}(X|Y,Z), \tag{28.12a}$$

$$S^{uu}(X|Y|Z) + S^{uu}(Y|X|Z) = 0. \tag{28.12b}$$

See Fig. 28.2 for an illustration.

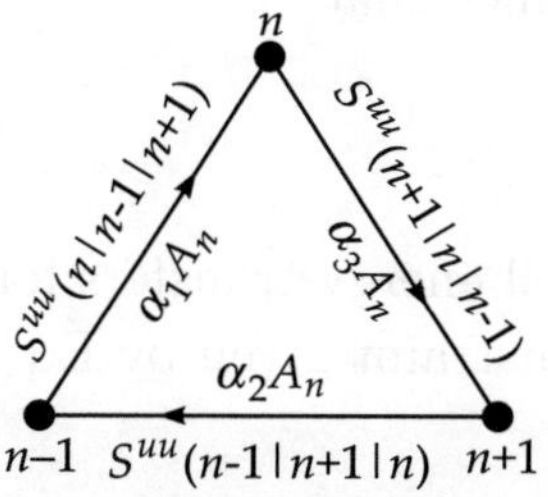

Figure 28.2 Schematic diagram of the shell-to-shell energy transfers in a triad $(n-1, n, n+1)$ of the shell model.

As discussed in Section 4.1, these set of equations do not have a unique solution. In addition, in the absence of the condition $\mathbf{k} \cdot \mathbf{u}(\mathbf{k}) = 0$, we do not have a nice formula like Eq. (4.7) for the shell model. We postulate that

$$S^{UU}(n|n-1|n+1) = \alpha_1 A_n, \tag{28.13a}$$

$$S^{UU}(n+1|n|n-1) = \alpha_3 A_n, \tag{28.13b}$$

$$S^{UU}(n-1|n+1|n) = \alpha_2 A_n, \tag{28.13c}$$

where

$$A_n = k_{n-1} \Im(U_{n-1}^* U_n^* U_{n+1}). \tag{28.14}$$

Using Eqs. (28.12), we obtain

$$\alpha_1 - \alpha_3 = a_2, \tag{28.15a}$$

$$\alpha_2 - \alpha_1 = a_1, \tag{28.15b}$$

$$\alpha_3 - \alpha_2 = a_3. \tag{28.15c}$$

We take α_1 as a free parameter in terms of which

$$\alpha_3 = \alpha_1 - a_2, \tag{28.16a}$$

$$\alpha_2 = \alpha_1 + a_1. \tag{28.16b}$$

In 3D hydrodynamic turbulence, the energy transfer is forward, that is, from a small wavenumber shell to a large wavenumber shell. Therefore, given $A_n > 0$, we demand that

$$\alpha_1 > 0; \quad \alpha_2 < 0; \quad \alpha_3 > 0. \tag{28.17}$$

Using Eqs. (28.9, 28.16) we deduce that

$$a_2 < \alpha_1 < (1+\lambda)a_2. \tag{28.18}$$

Thus, we derive the shell-to-shell energy transfers for the shell model. Note that α_1 is a free parameter with the constraint given by Eq. (28.18).

We choose

$$\alpha_1 = (1+\frac{\lambda}{2})a_2 \tag{28.19}$$

with $a_2 > 0$. Using Eq. (28.9) we derive that

$$\alpha_2 = -\frac{\lambda}{2}a_2, \tag{28.20a}$$

$$\alpha_3 = \frac{\lambda}{2}a_2. \tag{28.20b}$$

Also, the above solution is not unique; we can add a *circulating transfer* that traverses $(n-1) \to n \to (n+1) \to (n-1)$ (Dar et al., 2001; Verma, 2004). This transfer however does not alter the energy flux, which is a measurable quantity.

The energy flux in fluid turbulence is defined as the energy leaving a wavenumber sphere of radius K due to nonlinear interactions. For a shell model,

$\Pi^{U<}_{U>}(K)$ is the net energy transfer from all the shells within a sphere of radius K to the shells outside the sphere:

$$\Pi^{U<}_{U>}(K) = \sum_{m\le K}\sum_{n>K}\sum_{p} S^{UU}(n|m|p). \tag{28.21}$$

Due to local interactions among the shells,

$$\begin{aligned}\Pi^{U<}_{U>}(K) &= \sum_{n>K}\sum_{m\le K}\sum_{p} S^{UU}(n|m|p)\\ &= S^{UU}(K+1|K|K-1) + S^{UU}(K+1|K|K+2)\\ &\quad + S^{UU}(K+1|K-1|K) + S^{UU}(K+2|K|K+1)\\ &= \alpha_3 k_{K-1}\Im(U^*_{K-1}U^*_K U_{K+1}) + \alpha_1 k_K \Im(U^*_K U^*_{K+1}U_{K+2})\\ &\quad -\alpha_2 k_{K-1}\Im(U^*_{K-1}U^*_K U_{K+1}) - \alpha_2 k_K \Im(U^*_K U^*_{K+1} U_{K+2}).\end{aligned} \tag{28.22}$$

Similarly, the shell-to-shell kinetic energy transfer from shell K to $K+1$, and from shell K to $K+2$ are

$$\begin{aligned}T^{U,K}_{U,K+1} &= S^{UU}(K+1|K|K-1) + S^{UU}(K+1|K|K+2)\\ &= \alpha_3 k_{K-1}\Im(U^*_{K-1}U^*_K U_{K+1}) + \alpha_1 k_K \Im(U^*_K U^*_{K+1} U_{K+2}),\end{aligned} \tag{28.23a}$$

$$T^{U,K}_{U,K+2} = S^{UU}(K+2|K|K+1) = -\alpha_2 k_K \Im(U^*_K U^*_{K+1} U_{K+2}). \tag{28.23b}$$

Using the numerical simulations of the shell model, we observe that in these equations, all the terms of the form $\Im(U^*U^*U)$ are positive. Hence, all the terms Eqs. (28.22, 28.23) are positive. Thus, our formulas for the energy flux and shell-to-shell energy transfers are consistent with the forward energy transfer observed in 3D hydrodynamics.

Researchers have performed numerical simulations of the shell models of 3D hydrodynamics. Interestingly, they observed Kolmogorov's $k^{-5/3}$ energy spectrum and constant energy flux for these models. In addition, numerically computed intermittency exponents of the shell models and direct numerical simulations are quite close to each other. Here we omit discussions on the numerical results of shell models due to lack of space. We refer the reader to Ditlevsen (2010), Biferale (2003), Plunian et al. (2012), and references therein for further details on this topic.

28.2 Shell Model for Scalar, Vector, and Tensor Flows

As discussed in Chapters 13, 18, and 27, the velocity field often advects vector, scalar, or tensor fields. The advection term is of the form $\mathbf{u} \cdot \nabla\theta$, where θ represents a scalar, vector, or tensor field. In addition, the force fields could be complex functions of $\mathbf{u}$ and θ.

We construct the following shell model for such flows:

$$\frac{dU_n}{dt} = N_n[U,U] - \nu k_n^2 U_n + F_n, \tag{28.24a}$$

$$\frac{d\theta_n}{dt} = M_n[U,\theta] - \kappa k_n^2 \theta_n + G_n, \tag{28.24b}$$

where F_n, G_n are the force fields for U_n and θ_n respectively, ν is the kinematic viscosity, and κ is the diffusivity of θ. The nonlinear term $N_n[U,U]$ is same as Eq. (28.5), while $M_n[U,\theta]$ corresponds to the advection term $\mathbf{u} \cdot \nabla\theta$ that yields conservation of the total scalar energy for diffusionless and force-free flows:

$$E_\theta = \sum_n \frac{1}{2}|\theta_n|^2. \tag{28.25}$$

This condition implies that

$$\sum_n \Re\left(\theta_n^* M_n[U,\theta]\right) = 0. \tag{28.26}$$

The following form of $M_n[U,\theta]$ satisfies the aforementioned condition:

$$\begin{aligned} M_n[U,\theta] = & -i[k_n(d_1 U_{n+1}^* \theta_{n+2} + d_3 \theta_{n+1}^* U_{n+2}) \\ & + k_{n-1}(d_2 U_{n-1}^* \theta_{n+1} - d_3 \theta_{n-1}^* U_{n+1}) \\ & + k_{n-2}(d_1 U_{n-1} \theta_{n-2} + d_2 \theta_{n-1} U_{n-2})], \end{aligned} \tag{28.27}$$

where d_1, d_2, and d_3 are arbitrary constants.

Now we derive mode-to-mode transfers for the θ field in a triadic interactions among the shells (k_{n-1}, k_n, k_{n+1}). For this triad, the combined $\theta^2/2$ transfers are

$$S^{\theta\theta}(n-1|n,n+1) = d_1 k_{n-1} \Im(U_n^* \theta_{n-1}^* \theta_{n+1}) + d_3 k_{n-1} \Im(U_{n+1} \theta_{n-1}^* \theta_n^*), \tag{28.28a}$$

$$S^{\theta\theta}(n|n-1,n+1) = d_2 k_{n-1} \Im(U_{n-1}^* \theta_n^* \theta_{n+1}) - d_3 k_{n-1} \Im(U_{n+1} \theta_{n-1}^* \theta_n^*), \tag{28.28b}$$

$$S^{\theta\theta}(n+1|n,n-1) = -d_1 k_{n-1} \Im(U_n^* \theta_{n-1}^* \theta_{n+1}) - d_2 k_{n-1} \Im(U_{n-1}^* \theta_n^* \theta_{n+1}). \tag{28.28c}$$

Clearly,

$$S^{\theta\theta}(n-1|n,n+1) + S^{\theta\theta}(n|n-1,n+1) + S^{\theta\theta}(n+1|n,n-1) = 0, \tag{28.29}$$

which is a statement on the detailed conservation law of $\theta^2/2$ in a triad.

Now we seek a formula for the shell-to-shell $\theta_n^2/2$ transfer from θ_Y to θ_X with the mediation of U_Z. Such transfers, denoted by $S^{\theta\theta}(X|Y|Z)$, satisfy the following conditions:

$$S^{\theta\theta}(X|Y|Z) + S^{\theta\theta}(X|Z|Y) = S^{\theta\theta}(X|Y,Z), \tag{28.30a}$$

$$S^{\theta\theta}(X|Y|Z) + S^{\theta\theta}(Y|X|Z) = 0. \tag{28.30b}$$

As argued in the earlier section, Eqs. (28.30) do not have a unique solution. Yet, the following convenient formulas satisfy the aforementioned equations:

$$S^{\theta\theta}(n-1|n+1|n) = d_1 k_{n-1} \Im(U_n^* \theta_{n-1}^* \theta_{n+1}), \tag{28.31a}$$

$$S^{\theta\theta}(n-1|n|n+1) = d_3 k_{n-1} \Im(U_{n+1} \theta_{n-1}^* \theta_n^*), \tag{28.31b}$$

$$S^{\theta\theta}(n|n+1|n-1) = d_2 k_{n-1} \Im(U_{n-1}^* \theta_n^* \theta_{n+1}), \tag{28.31c}$$

$$S^{\theta\theta}(n|n-1|n+1) = -d_3 k_{n-1} \Im(U_{n+1} \theta_{n-1}^* \theta_n^*), \tag{28.31d}$$

$$S^{\theta\theta}(n+1|n-1|n) = -d_1 k_{n-1} \Im(U_n^* \theta_{n-1}^* \theta_{n+1}), \tag{28.31e}$$

$$S^{\theta\theta}(n+1|n|n-1) = -d_2 k_{n-1} \Im(U_{n-1}^* \theta_n^* \theta_{n+1}). \tag{28.31f}$$

Note that the velocity field acts as a mediator in all the above scalar/vector/tensor energy transfers. If $\Im(U_{n-1}^* \theta_n^* \theta_{n+1}) > 0$, then the forward shell-to-shell transfer implies that $d_2 < 0$. Similar conditions are derived for other constants.

Kumar and Verma (2015) used a generalized version of the shell model of Eq. (28.24) to simulate buoyancy-driven turbulence—stably stratified turbulence and turbulent thermal convection. In their model, F_n is the buoyancy, while G_n is proportional to U_n. The results of the shell model and direct numerical simulations are in general agreement with each other. Refer to the original paper, and Verma (2018) for more details.

Verma and Kumar (2016) simulated magnetohydrodynamic (MHD) turbulence with another generalized shell model with F_n, G_n representing $\mathbf{B} \cdot \nabla \mathbf{B}$ and $\mathbf{B} \cdot \nabla \mathbf{U}$ respectively. They showed a preferential energy transfer from the velocity field to the magnetic field that leads to generation of large-scale magnetic field. A limitation of the present MHD model is lack of nonlocal interactions that is important for small-scale dynamo (Plunian et al., 2012; Kumar et al., 2014b). Refer to Verma and Kumar (2016) for details.

Kalelkar et al. (2005), and Ray and Vincenzi (2016) constructed a shell model for a turbulent flow with dilute polymers. The overall structure of their shell model for polymeric flow is similar to Verma and Kumar (2016)'s model for MHD turbulence with some differences. It will be interesting to study the energy transfers in the shell model for polymeric flows. We do not detail any of these works, but make a remark that the connection between the shell models of MHD and polymeric fluid leads us to believe that the kinetic energy may indeed be transferred to the elastic energy of the polymers [see Eq. (27.28)]. These results have direct implications on the turbulence drag reduction in polymeric flows, as described in Section 27.5.

With this, we end our discussion on the shell model.

Further Reading

Ditlevsen (2010)'s book described various shell models and their properties. Also refer to review articles by Biferale (2003) and Plunian et al. (2012); the latter article focuses on the energy transfers. The energy transfer formalism discussed in this chapter are covered in Kumar and Verma (2015) and Verma and Kumar (2016).

Chapter 29

Burgers Turbulence

In this chapter we describe the energy spectrum and flux of Burgers turbulence. Burgers equation describes fully compressible flows with vanishing sound speed. As we show in the following sections, removal of the incompressibility condition introduces new complexities.

29.1 Governing Equations

Burgers equation is

$$\partial_t u + u\partial_x u = \nu\partial_{xx} u + F_u, \tag{29.1}$$

where u is the velocity field along the x-axis, F_u is the external force, and ν is the kinematic viscosity. This equation is a limiting case of compressible Navier–Stokes equations with vanishing pressure gradient. This condition corresponds to zero sound speed for the system.

Force-free Burgers equation with $\nu \to 0$ limit has an exact solution (Kida, 1979; Verma, 2000):

$$u(x) = \frac{1}{t}(x - \eta_i) \quad \text{for } \xi_i < x \leq \xi_{i+1}, \tag{29.2}$$

where ξ_i is the position of the ith shock where the velocity jumps by μ_i; η_i, which lies between ξ_i and ξ_{i+1}, is the ith zero of $u(x)$. See Fig. 29.1 for an illustration. Also, note that $u(x)$ has a slope of $1/t$, and that the kinetic energy is dissipated at the shocks. For finite viscosity, the shocks become smoother (Kida, 1979). In the presence of F_u, the flow becomes steady with the dissipation rate and energy injection rate balancing each other.

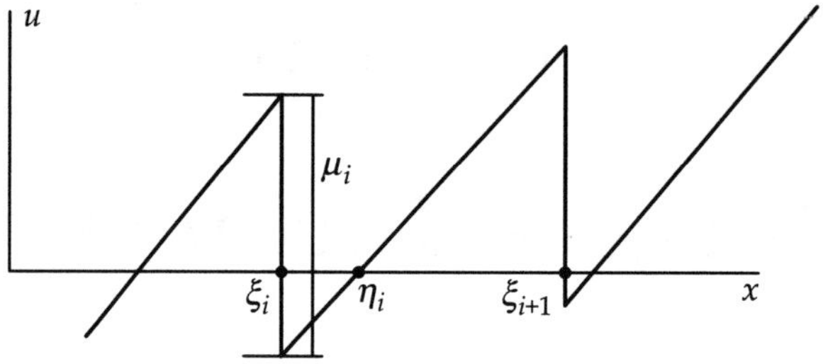

Figure 29.1 A schematic diagram of the velocity field of the force-free inviscid Burgers equation. ξ_i is the position of the ith shock, where $\Delta u = \mu_i$, and $u(\eta_i) = 0$.

The equation for the kinetic energy density, $u^2/2$, is

$$\partial_t u^2/2 + \partial_x u^3/3 = \nu u \partial_{xx} u + u F_u. \tag{29.3}$$

For $F_u = 0$, $\nu = 0$, and periodic or vanishing boundary condition, using Eqs. (29.1, 29.3) we deduce the following conserved quantities for the Burgers equation:

$$\text{Total linear momentum} = \int_0^L u dx, \tag{29.4a}$$

$$\text{Total kinetic energy } E_u = \int_0^L \frac{1}{2} u^2 dx, \tag{29.4b}$$

where L is the system size. It is easy to show that $\int_0^L u^n dx$, where n is an integer, is conserved. Note that the conservation laws of Eq. (29.4) follow from the conservation of $\int_0^L u^n dx$.

In Fourier space, Eq. (29.1) transforms to

$$\frac{d}{dt} u(k) = -i \frac{k}{2} \sum_p u(q) u(p) - \nu k^2 u(k) + F_u(k), \tag{29.5}$$

with $q = k - p$. We multiply Eq. (29.5) with $u^*(k)$, and then add the resulting equation with its complex conjugate. This operation yields the following equation for the modal kinetic energy, $E_u(k) = |u(k)|^2/2$:

$$\begin{aligned}\frac{d}{dt}E_u(k) &= \frac{k}{2}\sum_p \Im[u(q)u(p)u^*(k)] - 2\nu k^2 E_u(k) + \Re[F_u(k)u^*(k)] \\ &= T_u(k) - D_u(k) + \mathcal{F}_u(k),\end{aligned} \tag{29.6}$$

where $T_u(k)$ is the nonlinear energy transfer to $u(k)$, $D_u(k)$ is the dissipation rate, and $\mathcal{F}_u(k)$ is the energy injection rate by the external force.

In the next section we will discuss the energy transfer and flux of Burgers turbulence.

29.2 Energy Transfers in Burgers Turbulence

As discussed in Chapter 4, we focus on the energy equations for triads (k', p, q) and $(-k', -p, -q)$ with $k' + p + q = 0$. Note that $k' = -k$. We also set $\nu = 0$ and $F_u = 0$. Under these conditions, the equations for the modal energies are

$$\frac{d}{dt}E_u(k) = S(k'|p,q) = -k'\Im[u(q)u(p)u(k')], \tag{29.7a}$$

$$\frac{d}{dt}E_u(p) = S(p|q,k') = -p\Im[u(q)u(p)u(k')], \tag{29.7b}$$

$$\frac{d}{dt}E_u(q) = S(q|k',p) = -q\Im[u(q)u(p)u(k')]. \tag{29.7c}$$

Clearly, when we sum Eqs. (29.7), using $k' + p + q = 0$, we obtain

$$E_u(k) + E_u(p) + E_u(q)] = \text{const.} \tag{29.8}$$

This is a statement of *detailed conservation of energy in a triad.*

In the aforementioned equations, $S^{uu}(X|Y,Z)$ is the *combined kinetic energy transfer* to $u(X)$ from $u(Y)$ and $u(Z)$. Here, $X+Y+Z = 0$. An important question is whether we can derive mode-to-mode kinetic energy transfer, $S^{uu}(X|Y|Z)$, which is the energy transfer from $u(Y)$ to $u(X)$ with the mediation of $u(Z)$. As in Chapter 4, the nature of transactions yield the following conditions for $S^{uu}(X|Y|Z)$:

$$S^{uu}(X|Y|Z) + S^{uu}(X|Z|Y) = S(X|Y,Z), \tag{29.9a}$$

$$S^{uu}(X|Y|Z) + S^{uu}(Y|X|Z) = 0. \tag{29.9b}$$

As argued in Section 4.1, these equations do not have a unique solution. For incompressible flows, we could obtain a convenient formula, Eq. (4.7), that satisfies Eqs. (29.9). Unfortunately, no such formula exists for Burgers equation.[1]

[1] For example, $S^{uu}(X|Y|Z) = -Y\Im[u(X)u(Y)u(Z)]$ satisfies Eq. (29.9a), but not Eq. (29.9b).

Though Burgers equation does not have a convenient formula for the mode-to-mode kinetic energy transfer, we can derive a formula for its energy flux. As in Section 4.4, we define energy flux $\Pi_u(k_0)$ as the net nonlinear energy transfer from all the modes in the band $|k| \leq k_0$ to the modes outside the band. Following arguments similar to those in Section 4.4,

$$\Pi_u(k_0) = -\int_{-k_0}^{k_0} T_u(k')dk'. \tag{29.10}$$

In addition,

$$\frac{\partial}{\partial t}E_u(k,t) = -\frac{\partial}{\partial k}\Pi_u(k,t) + \mathcal{F}_u(k,t) - D_u(k,t), \tag{29.11}$$

whose physical interpretation is described in Section 4.4. Also see Fig. 4.9.

In the next section, we describe the phenomenology of Burgers turbulence.

29.3 Phenomenology of Burgers Turbulence

We assume a steady Burgers turbulence with F_u active only at large scales. Under these conditions, following Eq. (29.11), we obtain the following equation for the inertial–dissipation range:

$$\frac{d}{dk}\Pi_u(k) = -D_u(k). \tag{29.12}$$

In the inertial range where $D_u(k) \to 0$, we obtain

$$\Pi_u(k) = \text{const.} = \Pi_u. \tag{29.13}$$

Using dimensional analysis, Girimaji and Zhou (1995) and Verma (2000) derived that

$$\Pi_u = \frac{\mu^3}{L}, \tag{29.14}$$

where $\mu = \langle \mu_i \rangle$ is the average velocity difference across the shocks. For a steady turbulence, μ is maintained at a constant value. However, μ decreases with time for a decaying turbulence.

The derivation of energy spectrum $E_u(k)$ is somewhat tricky. $E_u(k)$ depends on three parameters: Π_u, μ, and k. Therefore, we cannot derive it using dimensional analysis because dimension matching yields only two equations.[2] In fact, $E_u(k)$ is derived using the structure function.

[2]Based on constancy of $\Pi_u(k)$ and the derivation of Section 5.1, we may be tempted to claim $k^{-5/3}$ spectrum for Burgers turbulence. But this is not correct because $E_u(k)$ depends on μ as well.

Using the velocity profile of Eq. (29.2), we derive the qth order structure function for Burgers turbulence as (Kida, 1979; Verma, 2000)

$$\begin{aligned} S_q(l) &= \frac{1}{L}\int_0^L |u(x+l) - u(x)|^q dx \\ &= \frac{1}{L}\sum_i \left[\int_{\xi_i}^{\xi_{i+1}-r} \left(\frac{l}{t}\right)^q dx + \int_{\xi_{i+1}-r}^{\xi_{i+1}} \left(\mu_{i+1} - \frac{l}{t}\right)^q dx \right] \\ &= \left(\frac{l}{t}\right)^q + \frac{l}{L}\sum_i \left(\mu_{i+1} - \frac{l}{t}\right)^q. \end{aligned} \quad (29.15)$$

Therefore, for small l,

$$S_2(l) \approx \mu^2 \frac{l}{L}. \quad (29.16)$$

Hence,

$$\langle u(x+l)u(x)\rangle = \langle u^2\rangle - \frac{1}{2}S_2(l) \approx \langle u^2\rangle - \mu^2\frac{l}{2L}, \quad (29.17)$$

whose Fourier transform yields

$$E_u(k) = \frac{\mu^2}{L}k^{-2} = \frac{\Pi_u}{\mu}k^{-2} \quad (29.18)$$

apart from a dimensionless proportionality constant. Thus, the energy spectrum for Burgers turbulence differs from $k^{-5/3}$ spectrum for incompressible hydrodynamic turbulence. Note that the above derivation extends to the Burgers equation with small ν because their shocks are quite sharp.

The aforementioned formulas for $E_u(k)$ and $\Pi_u(k)$ are applicable to the inertial range. But these formulas could be extended to the inertial–dissipation range by extending Pao's hypothesis for hydrodynamic turbulence to Burgers turbulence. We argue that in the inertial–dissipation range, $E_u(k)/\Pi_u(k)$ is independent of ν and L, and it is a function of k and μ. Hence, using Eq. (29.18), we deduce that

$$\frac{E_u(k)}{\Pi_u(k)} = \frac{k^{-2}}{\mu}. \quad (29.19)$$

Substitution of this equation in Eq. (29.12) yields

$$\frac{d}{dk}\Pi_u(k) = -2\frac{\nu}{\mu}\Pi_u(k) \quad (29.20)$$

whose solution is

$$\Pi_u(k) = \Pi_u \exp\left(-\frac{2\nu}{\mu}k\right). \quad (29.21)$$

Substitution of Eq. (29.21) in Eq. (29.19) yields

$$E_u(k) = \frac{\Pi_u}{\mu} k^{-2} \exp\left(-\frac{2\nu}{\mu} k\right). \tag{29.22}$$

The aforementioned forms of energy spectrum and flux needs to be tested using numerical simulations.

There is a large body of numerical works on Burgers turbulence that describe energy spectrum and structure function, for example,Verma (2000), Girimaji and Zhou (1995), and references therein. However, we do not detail them here due to lack of space.

With this, we end our discussion on Burgers turbulence.

Further Reading

Kida (1979) describes asymptotic properties of decaying Burgers turbulence. For a detailed discussion on the topics of this chapter, refer to Verma (2000), and Girimaji and Zhou (1995).

Chapter 30

Compressible Turbulence

All the past chapters except Chapter 29 dealt with incompressible flows. Note however that most astrophysical flows are compressible. In addition, the flows around high speed jets and vehicles are also compressible. Hence, compressible turbulence is widely studied by scientists and engineers. In the present chapter we will briefly describe the equations and energy transfers in compressible turbulence.

We start with the governing equations of compressible flows.

30.1 Governing Equations

The material density ϱ varies in compressible turbulence, and its evolution equation under diffusionless limit is

$$\frac{\partial \varrho}{\partial t} + \nabla \cdot (\varrho \mathbf{u}) = 0. \tag{30.1}$$

The equation for the linear momentum density $\varrho \mathbf{u}$ is

$$\frac{\partial}{\partial t}[\varrho u_i] + \partial_j[\varrho u_j u_i] = -\partial_i \sigma + \mu[\nabla^2 u_i + \frac{1}{3}\partial_i(\nabla \cdot \mathbf{u})], \tag{30.2}$$

where σ is the pressure field, and μ is the dynamic viscosity. Using Eq. (30.1) we can rewrite Eq. (30.2) as

$$\varrho \left[\frac{\partial}{\partial t}\mathbf{u} + (\mathbf{u} \cdot \nabla)\mathbf{u}\right] = -\nabla \sigma + \mu[\nabla^2 \mathbf{u} + \frac{1}{3}\nabla(\nabla \cdot \mathbf{u})]. \tag{30.3}$$

We focus on the nonlinear interactions in the compressible flow, hence we set $\mu = 0$.

By taking a dot product of Eq. (30.2) with $\mathbf{u}$, we derive the following equation for the kinetic energy density:

$$\frac{\partial}{\partial t}\left[\frac{1}{2}\varrho u^2\right] + \nabla\cdot\left[\frac{1}{2}\varrho u^2\mathbf{u}\right] = -(\mathbf{u}\cdot\nabla)\sigma. \tag{30.4}$$

Note that the volume integral of the second term of Eq. (30.4) vanishes for a periodic or vanishing boundary condition. However, the integral of the third term of the equation is nonzero for a compressible flow. Recall that the corresponding integral for an incompressible flow vanishes because of the constraint $\nabla\cdot\mathbf{u} = 0$. Consequently, the total kinetic energy

$$\int d\mathbf{r}\frac{1}{2}\varrho u^2 \tag{30.5}$$

is not conserved for a compressible flow.

According to the first law of thermodynamics,

$$de = dQ - \sigma dv, \tag{30.6}$$

where de is the change in internal energy per unit mass, dQ is the elemental heat source in the form of heat conduction and viscous dissipation, and dv is the change in specific volume. We assume adiabatic process, that is, $dQ = 0$. In addition, using $v = 1/\rho$, we obtain

$$de = \frac{\sigma}{\varrho^2}d\varrho. \tag{30.7}$$

Note that in an incompressible flow,

$$d\varrho = 0 \implies de = 0. \tag{30.8}$$

Thus, the internal energy is constant for an incompressible flow.

After some more steps, we derive (Kundu et al., 2015)

$$\frac{\partial}{\partial t}[\varrho e] + \nabla\cdot[\varrho e\mathbf{u}] = -\sigma\nabla\cdot\mathbf{u}. \tag{30.9}$$

Clearly, the total internal energy

$$\int d\mathbf{r}\,\varrho e \tag{30.10}$$

is not conserved due to the last term of Eq. (30.9) whose volume integral is nonzero; this term represents the rate of increase in internal energy density due to the work done by pressure. Note that $\nabla \cdot \mathbf{u}$ represents the rate of change of volume [see Eq. (2.35)].

The total energy density is

$$E = e + \frac{1}{2}u^2, \tag{30.11}$$

whose evolution equation is obtained by adding Eqs. (30.4, 30.9):

$$\frac{\partial}{\partial t}[\varrho E] + \nabla \cdot [\varrho E \mathbf{u}] = -\nabla \cdot [\sigma \mathbf{u}]. \tag{30.12}$$

In this equation, the volume integral of $-\nabla \cdot [\sigma \mathbf{u}]$ vanishes for periodic and vanishing boundary condition. Hence, the total energy

$$\int d\mathbf{r}[e + \frac{1}{2}u^2] \tag{30.13}$$

is conserved for a compressible flow. The pressure induces an exchange between kinetic energy and internal energy keeping the total energy fixed.

For incompressible flows, $e =$ const., and σ is *dynamic pressure* that is determined using

$$-\nabla^2 \sigma = \varrho \nabla \cdot [(\mathbf{u} \cdot \nabla)\mathbf{u}], \tag{30.14}$$

which is independent of thermodynamic variables, for example, temperature, density. For compressible flows, σ is the thermodynamic pressure. However, dynamic pressure too exists in compressible flows, and it is determined using the incompressible part of the nonlinear term. We will describe these computations in Section 30.5.

A compressible flow has five unknown variables—three components of $\mathbf{u}$, σ, and ϱ. To solve for them, in addition to the four equations [Eq. (30.1, 30.3)], we invoke the equation of state:

$$\sigma = \sigma(\rho). \tag{30.15}$$

For polytropic processes,

$$\sigma = C\rho^n, \tag{30.16}$$

where C, n are constants. For an adiabatic process,

$$n = \gamma = \frac{C_p}{C_v}, \tag{30.17}$$

where C_p, C_v are specific heats at constant pressure and volume respectively. Note that in some literature, internal energy e and its corresponding evolution equation, Eq. (30.9), are added for compressible flows. In comparison, incompressible hydrodynamics has four unknowns—u_x, u_y, u_z, and σ—and four equations, Eqs. (2.1a, 2.1b). Note that Eq. (2.1b) is a constraint equation that helps solve for σ using Eq. (30.14).

In the next three sections we will illustrate the limiting cases of compressible hydrodynamics.

30.2 Linear Compressible Flow; Sound Waves

For linearization, we divide the density ϱ into mean density ρ_0 and density fluctuation ρ:

$$\varrho = \rho_0 + \rho, \tag{30.18}$$

and assume that $\rho \ll \rho_0$. We also assume that the velocity fluctuations are small; hence, the nonlinear term can be ignored.

Under this approximation, the linearized version of Eq. (30.1) is

$$\frac{\partial \rho}{\partial t} = -\rho_0 \nabla \cdot \mathbf{u}. \tag{30.19}$$

Fourier transform of this equation yields

$$\frac{d}{dt}\rho(\mathbf{k}) = -i\rho_0[\mathbf{k} \cdot \mathbf{u}(\mathbf{k})]. \tag{30.20}$$

We assume that $\mathbf{k} = k\hat{x}$, and

$$\mathbf{u}(\mathbf{k}) = u_x(\mathbf{k})\hat{x} + u_y(\mathbf{k})\hat{y}. \tag{30.21}$$

The linearized momentum equations are

$$\rho_0 \frac{d}{dt} u_x(\mathbf{k}) = -ik\sigma(\mathbf{k}), \tag{30.22a}$$

$$\rho_0 \frac{d}{dt} u_y(\mathbf{k}) = 0. \tag{30.22b}$$

Using Eqs. (30.20, 30.22a) and $\sigma(\mathbf{k})/\rho(\mathbf{k}) = c_s^2$, where c_s is the speed of sound, we derive that

$$\ddot{\rho}(\mathbf{k}) = -k^2 c_s^2 \rho(\mathbf{k}), \tag{30.23}$$

which is the equation of the sound wave. Thus, the longitudinal velocity mode $u_x(\mathbf{k})$ helps propagate sound waves with the assistance of density and thermodynamic pressure. Equation (30.22b) yields $u_y = \text{const}$. Hence, the incompressible component of the velocity field is not affected by the thermodynamic pressure.

In the next section we will describe a nearly incompressible flow.

30.3 Nearly Incompressible Flow

Under nearly incompressible limit, in addition to the division of density as in Eq. (30.18), the total velocity field is decomposed into incompressible and compressible components:

$$\mathbf{u} = \mathbf{u}^{\text{inc}} + \mathbf{u}^{\text{comp}}. \tag{30.24}$$

Under nearly incompressible approximation, we assume that

$$\rho \ll \rho_0; \quad |\mathbf{u}^{\text{comp}}| \ll |\mathbf{u}^{\text{inc}}|; \tag{30.25a}$$

$$\nabla \cdot \mathbf{u} \approx \nabla \cdot \mathbf{u}^{\text{inc}} = 0. \tag{30.25b}$$

Under this approximation, to leading order, Eqs. (30.1, 30.3) reduce to

$$\frac{\partial}{\partial t}\mathbf{u}^{\text{inc}} + (\mathbf{u}^{\text{inc}} \cdot \nabla)\mathbf{u}^{\text{inc}} = -\frac{1}{\rho}\nabla\sigma + \nu\nabla^2\mathbf{u}^{\text{inc}}, \tag{30.26a}$$

$$\frac{\partial}{\partial t}\rho + (\mathbf{u}^{\text{inc}} \cdot \nabla)\rho = 0. \tag{30.26b}$$

An expansion of Eq. (30.3) to the next order yields

$$\frac{\partial}{\partial t}\mathbf{u}^{\text{comp}} + (\mathbf{u}^{\text{inc}} \cdot \nabla)\mathbf{u}^{\text{comp}} = \nu\nabla^2\mathbf{u}^{\text{comp}}. \tag{30.27}$$

These equations clearly indicate that ρ is a passive scalar, and $\mathbf{u}^{\text{comp}}$ is a passive vector, both being advected by the incompressible velocity field $\mathbf{u}^{\text{inc}}$. The pressure field, which is the dynamic pressure, affects $\mathbf{u}^{\text{inc}}$ only. Using these equations, we deduce that all the three fields follow $k^{-5/3}$ spectrum—$\mathbf{u}^{\text{inc}}$ on the basis of Kolmogorov's theory (see Section 5.1), while ρ and $\mathbf{u}^{\text{comp}}$ follow the phenomenology of passive scalar and passive vector respectively (see Chapters 14 and 19).

Zank and Matthaeus (1991) derived a generalized expansion of the Navier–Stokes equations using singular expansion technique. Our arguments given here are a simplified version of Zank and Matthaeus (1991)'s derivation. We also remark that the density spectrum of interstellar medium follows $k^{-5/3}$ spectrum, and it is explained using the aforementioned scaling (Zank and Matthaeus, 1993).

In the next section we will connect the equations of compressible flow with Burgers equation.

30.4 Fully Compressible Turbulence: Burgers Turbulence Revisited

Fully compressible flow with $\mathbf{u}^{\mathrm{inc}} = 0$ is the opposite of the incompressible limit. In this limit,

$$\mathbf{u} = \mathbf{u}^{\mathrm{comp}}. \tag{30.28}$$

Therefore, the equations of the flow are

$$\frac{\partial}{\partial t}\mathbf{u} + \mathbf{u}\cdot\nabla\mathbf{u} = -\frac{1}{\varrho}\nabla\sigma + \nu[\nabla^2\mathbf{u} + \frac{1}{3}\nabla(\nabla\cdot\mathbf{u})], \tag{30.29a}$$

$$\frac{\partial}{\partial t}\varrho + \nabla\cdot(\varrho\mathbf{u}) = 0, \tag{30.29b}$$

$$\sigma = C\varrho^{\gamma}, \tag{30.29c}$$

where C and γ are constants. Note that σ is the thermodynamic pressure; for an adiabatic process, $\gamma = C_p/C_v$.

Limiting the aforementioned set of equations to 1D and setting $\sigma = 0$ (zero sound speed) yield the Burgers equation. The equations for this case with $\mathbf{u} = u\hat{x}$ are

$$\partial_t u + u\partial_x u = \nu\partial_{xx}u, \tag{30.30a}$$

$$\partial_t\varrho + \partial_x(\rho u) = 0. \tag{30.30b}$$

In Chapter 29 we described the properties of Eq. (30.30a); hence, they will not be repeated here. Solving Eq. (30.30b) is an interesting exercise, but it is beyond the scope of this book.

In the next section we will discuss the properties of compressible turbulence using Craya–Herring decomposition.

30.5 Equation of Motion of a Compressible Flow in Craya–Herring Basis

In this section we describe the equations of a compressible flow in Fourier space. For simplification, we further assume that the density fluctuations are small, that is, $\rho \approx \rho_0$. Moreover, to disentangle various components of nonlinearity, we focus on a single triad $\mathbf{k}', \mathbf{p}, \mathbf{q}$, and set $\mu = 0$.

We expand the velocity field in Craya–Herring basis as (see Figs. 9.1 and 9.2):

$$\begin{aligned} \mathbf{u}(\mathbf{X}) &= u_1(\mathbf{X})\hat{e}_1(\mathbf{X}) + u_2(\mathbf{X})\hat{e}_2(\mathbf{X}) + u_3(\mathbf{X})\hat{e}_3(\mathbf{X}) \\ &= \mathbf{u}_\perp(\mathbf{X}) + u_3(\mathbf{X})\hat{e}_3(\mathbf{X}), \end{aligned} \tag{30.31}$$

where $\mathbf{X}$ is the wavenumber. Note that for the said triad,

$$\hat{e}_3(\mathbf{p})\cdot\hat{e}_3(\mathbf{q}) = -\cos\alpha; \quad \hat{e}_3(\mathbf{q})\cdot\hat{e}_3(\mathbf{k}') = -\cos\beta; \quad \hat{e}_3(\mathbf{k}')\cdot\hat{e}_3(\mathbf{p}) = -\cos\gamma. \tag{30.32}$$

Using

$$\dot{u}_i = -u_j\partial_j u_i - \frac{1}{\rho_0}\partial_i\sigma, \tag{30.33}$$

we obtain the following equations for the Fourier mode $\mathbf{u}(\mathbf{k}')$:

$$\begin{aligned} \frac{d}{dt}\mathbf{u}(\mathbf{k}') &= -i[(-\mathbf{p})\cdot\mathbf{u}(-\mathbf{q})]\mathbf{u}(-\mathbf{p}) + \mathbf{p}\leftrightarrow\mathbf{q} - \frac{i}{\rho_0}\mathbf{k}\sigma(\mathbf{k}) \\ &= -i[(-\mathbf{p})\cdot\{\mathbf{u}_\perp^*(\mathbf{q}) + u_3^*(\mathbf{q})\hat{e}_3(\mathbf{q})\}][\mathbf{u}_\perp^*(\mathbf{p}) + u_3^*(\mathbf{p})\hat{e}_3(\mathbf{p})] \\ &\quad +(\mathbf{p}\leftrightarrow\mathbf{q}) - \frac{i}{\rho_0}\mathbf{k}'\sigma(\mathbf{k}') \\ &= -\mathbf{N}_\perp^{\mathrm{inc}}(\mathbf{k}') - \mathbf{N}_\perp^{\mathrm{int}}(\mathbf{k}') - \left[N_3^{\mathrm{inc}}(\mathbf{k}') + N_3^{\mathrm{int}}(\mathbf{k}') + N_3^{\mathrm{comp}}(\mathbf{k}')\right]\hat{e}_3(\mathbf{k}') \\ &\quad -\frac{i}{\rho_0}\mathbf{k}'\sigma(\mathbf{k}'), \end{aligned} \tag{30.34}$$

where $\perp = (1,2)$ stands for the components perpendicular to $\mathbf{k}'$, while the index 3 represents the component along $\mathbf{k}'$. In these equations, the incompressible components are

$$\begin{aligned} N_1^{\mathrm{inc}}(\mathbf{k}') &= i[\mathbf{k}'\cdot\mathbf{u}_\perp^*(\mathbf{q})][\hat{e}_1(\mathbf{p})\cdot\hat{e}_1(\mathbf{k}')]u_1^*(\mathbf{p}) + \mathbf{p}\leftrightarrow\mathbf{q} \\ &= -i[\mathbf{k}'\cdot\mathbf{u}_\perp^*(\mathbf{q})]\cos\gamma u_1^*(\mathbf{p}) + \mathbf{p}\leftrightarrow\mathbf{q}, \end{aligned} \tag{30.35a}$$

$$N_2^{\mathrm{inc}}(\mathbf{k}') = i[\mathbf{k}'\cdot\mathbf{u}_\perp^*(\mathbf{q})][\hat{e}_2(\mathbf{p})\cdot\hat{e}_2(\mathbf{k}')]u_2^*(\mathbf{p}) + \mathbf{p}\leftrightarrow\mathbf{q}$$

$$= i[\mathbf{k}' \cdot \mathbf{u}_\perp^*(\mathbf{q})]u_2^*(\mathbf{p}) + \mathbf{p} \leftrightarrow \mathbf{q}, \tag{30.35b}$$

$$\begin{aligned} N_3^{\text{inc}}(\mathbf{k}') &= i[\mathbf{k}' \cdot \mathbf{u}_\perp^*(\mathbf{q})][\hat{e}_1(\mathbf{p}) \cdot \hat{e}_3(\mathbf{k}')]u_1^*(\mathbf{p}) + \mathbf{p} \leftrightarrow \mathbf{q} \\ &= -i[\mathbf{k}' \cdot \mathbf{u}_\perp^*(\mathbf{q})](\sin\gamma)u_1^*(\mathbf{p}) + \mathbf{p} \leftrightarrow \mathbf{q}; \end{aligned} \tag{30.35c}$$

the compressible component is

$$\begin{aligned} N_3^{\text{comp}}(\mathbf{k}') &= i[-\mathbf{p} \cdot \hat{e}_3(\mathbf{q})][\hat{e}_3(\mathbf{p}) \cdot \hat{e}_3(\mathbf{k}')]u_3^*(\mathbf{q})u_3^*(\mathbf{p}) + \mathbf{p} \leftrightarrow \mathbf{q} \\ &= -ip(\cos\alpha\cos\gamma)u_3^*(\mathbf{q})u_3^*(\mathbf{p}) + \mathbf{p} \leftrightarrow \mathbf{q}; \end{aligned} \tag{30.36a}$$

while the mixed terms involving interactions among the $\mathbf{u}_\perp$ and u_3 components are

$$\begin{aligned} N_1^{\text{int}}(\mathbf{k}') &= i[-\mathbf{p} \cdot \hat{e}_3(\mathbf{q})][\hat{e}_1(\mathbf{p}) \cdot \hat{e}_1(\mathbf{k}')]u_3^*(\mathbf{q})u_1^*(\mathbf{p}) \\ &= -ip(\cos\alpha\cos\gamma)u_3^*(\mathbf{q})u_1^*(\mathbf{p}) + \mathbf{p} \leftrightarrow \mathbf{q}, \end{aligned} \tag{30.37a}$$

$$\begin{aligned} N_2^{\text{int}}(\mathbf{k}') &= i[-\mathbf{p} \cdot \hat{e}_3(\mathbf{q})][\hat{e}_2(\mathbf{p}) \cdot \hat{e}_2(\mathbf{k}')]u_3(\mathbf{q})u_2^*(\mathbf{p}) + \mathbf{p} \leftrightarrow \mathbf{q} \\ &= ip(\cos\alpha)u_3(\mathbf{q})u_2^*(\mathbf{p}) + \mathbf{p} \leftrightarrow \mathbf{q}, \end{aligned} \tag{30.37b}$$

$$\begin{aligned} N_3^{\text{int}}(\mathbf{k}') &= i[\mathbf{k}' \cdot \mathbf{u}_\perp^*(\mathbf{q})][\hat{e}_3(\mathbf{p}) \cdot \hat{e}_3(\mathbf{k}')]u_3^*(\mathbf{p}) + \mathbf{p} \leftrightarrow \mathbf{q} \\ &= -i[\mathbf{k}' \cdot \mathbf{u}_\perp^*(\mathbf{q})](\cos\gamma)u_3^*(\mathbf{p}) + \mathbf{p} \leftrightarrow \mathbf{q}. \end{aligned} \tag{30.37c}$$

The equations of motion along $\hat{e}_1(\mathbf{k})$ and $\hat{e}_2(\mathbf{k})$ are

$$\dot{u}_1(\mathbf{k}') = -N_1^{\text{inc}}(\mathbf{k}') - N_1^{\text{int}}(\mathbf{k}'), \tag{30.38a}$$

$$\dot{u}_2(\mathbf{k}') = -N_2^{\text{inc}}(\mathbf{k}') - N_2^{\text{int}}(\mathbf{k}'). \tag{30.38b}$$

For consistency with the incompressible limit ($u_3 = 0$), we postulate that

$$N_3^{\text{inc}}(\mathbf{k}') = -\frac{i}{\rho_0}k'\sigma_{\text{dyn}}(\mathbf{k}') \tag{30.39}$$

and

$$\dot{u}_3(\mathbf{k}') = -N_3^{\text{int}}(\mathbf{k}') - N_3^{\text{comp}}(\mathbf{k}') - \frac{i}{\rho_0}k'\sigma_{\text{thermo}}(\mathbf{k}'). \tag{30.40a}$$

Here, σ_{thermo} and σ_{dyn} represent thermodynamic and dynamic pressure fields respectively. Note that

$$\dot{u}_3(\mathbf{k}') \neq -N_3(\mathbf{k}') - \frac{i}{\rho_0}k'\sigma_{\text{thermo}}(\mathbf{k}') \tag{30.41}$$

with

$$N_3 = N_3^{\text{comp}} + N_3^{\text{inc}} + N_3^{\text{int}}. \tag{30.42}$$

The evolution of the density field is given by

$$\dot{\rho} + \rho_0 \nabla \cdot \mathbf{u} = 0, \tag{30.43}$$

whose Fourier space counterpart is

$$\dot{\rho}(\mathbf{k}') = -i\rho_0 k u_3(\mathbf{k}'). \tag{30.44}$$

We can derive the following limiting cases from the aforementioned set of equations:

1. Incompressible flows: $N_3^{\text{comp}} = 0$, $N_i^{\text{int}} = 0$, and $\sigma_{\text{thermo}} = 0$. Therefore, the equations are

$$\dot{\mathbf{u}}_\perp^{\text{inc}}(\mathbf{k}') = -\mathbf{N}_\perp^{\text{inc}}(\mathbf{k}'); \quad N_3^{\text{inc}}(\mathbf{k}') = -\frac{i}{\rho_0} k' \sigma_{\text{dyn}}(\mathbf{k}'). \tag{30.45}$$

2. Sound waves: All the nonlinear terms are zero. Hence,

$$\dot{u}_3(\mathbf{k}') = -\frac{i}{\rho_0} k' \sigma_{\text{thermo}}(\mathbf{k}'). \tag{30.46}$$

3. Nearly incompressible flow: $N_3^{\text{comp}} = 0$, $N_{1,2}^{\text{int}} = 0$, and $\sigma_{\text{thermo}} = 0$ yielding

$$\dot{\mathbf{u}}_\perp^{\text{inc}}(\mathbf{k}') = -\mathbf{N}_\perp^{\text{inc}}(\mathbf{k}'); \quad \dot{u}_3(\mathbf{k}') = -N_3^{\text{int}}(\mathbf{k}'); \tag{30.47}$$

$$N_3^{\text{inc}}(\mathbf{k}') = \frac{i}{\rho_0} k' \sigma_{\text{dyn}}(\mathbf{k}'). \tag{30.48}$$

4. Burgers equation: All nonlinear terms except N_3^{comp} vanish. In addition, $\sigma = 0$. Due to 1D nature, $\alpha = \pi$ and $\gamma = 0$. Therefore,

$$\dot{u}_3(k') = -N_3^{\text{comp}}(k') = ik' u_3^*(p) u_3^*(q), \tag{30.49}$$

which is same as Eq. (29.5) for an interacting triad.

In the Navier–Stokes equations for incompressible flows, $\sigma = \sigma_{\text{dyn}}$, and for compressible flows, $\sigma = \sigma_{\text{thermo}}$. Note however that dynamic pressure exists in compressible flows, and that it is computed internally using Eq. (30.39); σ_{dyn} does not appear explicitly in the momentum equation. Thus, care is required while computing pressure in compressible flows.

We can also derive equations for the kinetic energy using Eqs. (30.34). In these equations,

1. $\mathbf{N}_{\text{inc}}$ induces energy transfers among the incompressible modes, as in Chapter 9.
2. $\mathbf{N}_{\text{comp}}$ induces energy transfers among the compressible modes.
3. $\mathbf{N}_{\text{int}}$ facilitates energy transfers across incompressible and compressible components.
4. σ_{thermo} induces transfers between the compressible component of the kinetic energy and internal energy.
5. The incompressible components satisfy the locality property in Fourier space. However, the interactions of compressible flows are local in real space, and nonlocal in Fourier space.
6. Neither kinetic energy nor internal energy are conserved for triadic interactions. In addition, the total energy for a triad too is not conserved, that is,

$$E_u(\mathbf{k}') + E_u(\mathbf{p}) + E_u(\mathbf{q}) + e(\mathbf{k}') + e(\mathbf{p}) + e(\mathbf{q}) \neq \text{const.} \tag{30.50}$$

   This property arises due to the nonlocal interactions among the Fourier modes.
7. For compressible flows, we do not have simple formulas for the mode-to-mode energy transfers [like Eqs. (4.7)]. This is due to the nonlocal interactions in Fourier space.

In the next section we will discuss energy transfers in compressible flows.

30.6 Energy Transfers in Compressible Flows

For compressible flows, the kinetic energy density $\rho u^2/2$ is a product of three fields. Hence, it is not possible to define a corresponding modal kinetic energy, which is a quadratic quantity (see Parceval's theorem in Chapter 3). To express kinetic energy as a product of two variables, we define

$$\mathbf{v} = \rho \mathbf{u}. \tag{30.51}$$

In terms of $\mathbf{v}$ and $\mathbf{u}$, the modal kinetic energy is

$$E_u(\mathbf{k}) = \frac{1}{2}\Re[\mathbf{v}(\mathbf{k}) \cdot \mathbf{u}^*(\mathbf{k})]. \tag{30.52}$$

In the next subsection we describe the equations for the modal kinetic energy and modal internal energy.

30.6.1 Equations for modal kinetic and internal energies

Using the momentum equation, we derive

$$\partial_t \mathbf{v}(\mathbf{k}) = -i \sum_{\mathbf{p}} \{\mathbf{k} \cdot \mathbf{u}(\mathbf{q})\} \mathbf{v}(\mathbf{p}) - i\mathbf{k}\sigma(\mathbf{k}) \tag{30.53a}$$

$$\partial_t \mathbf{u}(\mathbf{k}) = -i \sum_{\mathbf{p}} \{\mathbf{k} \cdot \mathbf{u}(\mathbf{q})\} \mathbf{u}(\mathbf{p}) - i\mathbf{k}\tilde{\sigma}(\mathbf{k}), \tag{30.53b}$$

where $\mathbf{q} = \mathbf{k} - \mathbf{p}$ and

$$\nabla\tilde{\sigma} = \frac{\nabla\sigma}{\rho}; \quad \tilde{\sigma}(\mathbf{k}) = FT[\frac{\nabla\sigma}{\rho}](\mathbf{k}), \tag{30.54}$$

where FT stands for the Fourier transform. Equation (30.53a)$\times\mathbf{u}^*(\mathbf{k})$ + Equation (30.53b)$\times\mathbf{v}^*(\mathbf{k})$, and subsequent addition of the resulting equations with its complex conjugate yields

$$\partial_t E_u(\mathbf{k}) = T_u(\mathbf{k}) + \mathcal{F}_\sigma(\mathbf{k}), \tag{30.55}$$

where

$$T_u(\mathbf{k}) = \sum_{\mathbf{p}} \frac{1}{2} \Im[\{\mathbf{k} \cdot \mathbf{u}(\mathbf{q})\}\{\mathbf{u}(\mathbf{p}) \cdot \mathbf{v}^*(\mathbf{k}) + \mathbf{v}(\mathbf{p}) \cdot \mathbf{u}^*(\mathbf{k})\}] \tag{30.56}$$

is the nonlinear energy transfer to $E_u(\mathbf{k})$. In this equation, $\mathbf{u}(\mathbf{q})$ advects $\mathbf{u} \cdot \mathbf{v}$. Moreover,

$$\mathcal{F}_\sigma(\mathbf{k}) = \Im\left[\{\mathbf{k} \cdot \mathbf{u}^*(\mathbf{k})\}\sigma(\mathbf{k}) + \{\mathbf{k} \cdot \mathbf{v}^*(\mathbf{k})\}\tilde{\sigma}(\mathbf{k})\right] \tag{30.57}$$

is the kinetic energy gain from the internal energy via pressure. Note that $\mathcal{F}_\sigma(\mathbf{k})$ vanishes in the incompressible limit.

We define internal energy per unit volume as

$$\tilde{e} = \rho e. \tag{30.58}$$

Using Eq. (30.12), we derive the spectral equation for $\tilde{e}$ as

$$\begin{aligned} \partial_t \tilde{e}(\mathbf{k}) &= -i \sum_{\mathbf{p}} \{\mathbf{k} \cdot \mathbf{u}(\mathbf{q})\} \tilde{e}(\mathbf{p}) - i \sum_{\mathbf{p}} \{\mathbf{p} \cdot \mathbf{u}(\mathbf{p})\} \sigma(\mathbf{q}) \\ &= T_e(\mathbf{k}) + \mathcal{F}_{e,\sigma}(\mathbf{k}), \end{aligned} \tag{30.59}$$

where $T_e(\mathbf{k})$ is the internal energy transfer to $\tilde{e}(\mathbf{k})$, and $\mathcal{F}_{e,\sigma}(\mathbf{k})$ is the gain in internal energy via pressure. Note that

$$\sum_{\mathbf{k}} [\mathcal{F}_\sigma(\mathbf{k}) + \mathcal{F}_{e,\sigma}(\mathbf{k})] = 0 \tag{30.60}$$

because $\int d\mathbf{r}\nabla \cdot (\sigma\mathbf{u}) = 0$. Kinetic and internal energies are exchanged via thermodynamic pressure while keeping the total energy constant.

Note that $(\mathbf{u} \cdot \nabla)\tilde{e}$ represents the local transfer of internal energy in real space, that is, energy transfer from position $\mathbf{r}$ to $\mathbf{r} + d\mathbf{r}$. This is because the energy transfers mediated by the thermodynamic pressure is transmitted at the speed of sound (finite), and hence they are local in real space. These transfers however are nonlocal in Fourier space. Contrast these against the energy transfers among the incompressible components of the velocity field that are transmitted at an infinite speed; these transfers are nonlocal in real space, and local in Fourier space.

In the next subsection we explore whether we can derive formulas for the mode-to-mode energy transfers for compressible flows.

30.6.2 Triadic interactions in a compressible flow?

We focus on triads $(\mathbf{k}', \mathbf{p}, \mathbf{q})$ and $(-\mathbf{k}', -\mathbf{p}, -\mathbf{q})$ satisfying $\mathbf{k}' + \mathbf{p} + \mathbf{q} = 0$. Using Eq. (30.56) and using $\mathbf{k} \cdot \mathbf{u}_\perp(\mathbf{k}) = 0$, we show that

$$\begin{aligned} T_u(\mathbf{k}') + T_u(\mathbf{p}) + T_u(\mathbf{q}) &= \Im[qu_3(\mathbf{q})\{\mathbf{u}(\mathbf{p}) \cdot \mathbf{v}(\mathbf{k}') + \mathbf{v}(\mathbf{p}) \cdot \mathbf{u}(\mathbf{k}')\}] \\ &\quad + \Im[pu_3(\mathbf{p})\{\mathbf{u}(\mathbf{q}) \cdot \mathbf{v}(\mathbf{k}') + \mathbf{v}(\mathbf{q}) \cdot \mathbf{u}(\mathbf{k}')\}] \\ &\quad + \Im[k'u_3(\mathbf{k}')\{\mathbf{u}(\mathbf{q}) \cdot \mathbf{v}(\mathbf{p}) + \mathbf{v}(\mathbf{q}) \cdot \mathbf{u}(\mathbf{p})\}]. \end{aligned} \tag{30.61}$$

Clearly, detailed energy conservation does not hold for triadic interactions in compressible turbulence. Note however that

$$\sum_{\mathbf{k}} T_u(\mathbf{k}) = 0 \tag{30.62}$$

because $\int d\mathbf{r}\nabla \cdot [(u^2/2)\mathbf{u}] = 0$.

For compressible flows, there is no simple formula for the mode-to-mode kinetic energy transfer. This is essentially because of the lack of incompressibility condition that makes it very difficult to satisfy $S^{uu}(\mathbf{k}'|\mathbf{p}|\mathbf{q}) + S^{uu}(\mathbf{p}|\mathbf{k}'|\mathbf{q}) = 0$. Still we can define energy flux, as we will show in the next subsection.

Also note that

$$T_e(\mathbf{k}') + T_e(\mathbf{p}) + T_e(\mathbf{q}) \neq 0. \tag{30.63}$$

In addition, the total energy is not conserved for triadic interactions, that is,

$$T_u(\mathbf{k}') + T_u(\mathbf{p}) + T_u(\mathbf{q}) + T_e(\mathbf{k}') + T_e(\mathbf{p}) + T_e(\mathbf{q}) \neq 0. \tag{30.64}$$

This feature is due to the nonlocal interactions of compressible flows in Fourier space.

In the next subsection we describe energy fluxes for compressible turbulence.

30.6.3 Energy fluxes in compressible turbulence

Following the discussion of the previous section and Section 4.4, we define the fluxes of kinetic and internal energies for a wavenumber sphere of radius k_0 as [see Eq. (4.41)]

$$\Pi_u(k_0) = -\sum_{k \le k_0} T_u(\mathbf{k}), \tag{30.65a}$$

$$\Pi_e(k_0) = -\sum_{k \le k_0} T_e(\mathbf{k}), \tag{30.65b}$$

with $T_u(\mathbf{k})$ and $T_e(\mathbf{k})$ defined using Eqs. (30.56, 30.59) respectively. Under steady state, in the inertial range,

$$\frac{d}{dk}\Pi_u(k) = \mathcal{F}_\sigma(k), \tag{30.66a}$$

$$\frac{d}{dk}\Pi_e(k) = \mathcal{F}_{e,\sigma}(k). \tag{30.66b}$$

Thus, in the inertial range, $\Pi_u(k)$ and $\Pi_e(k)$ are not constant due to the exchange among kinetic energy and internal energy via pressure. These computations are quite complex. Here we present approximate formulas.

We can separate fluxes into incompressible and compressible parts. For the same, we start with Eq. (30.34), and construct an approximate equation for the energy evolution, using which we derive $T_u(\mathbf{k})$ and then divide it into various components:

$$\begin{aligned} T_u(\mathbf{k}) &= T_u^{\mathrm{inc}}(\mathbf{k}) + T_u^{\mathrm{comp}}(\mathbf{k}) + T_u^{\mathrm{inc,int}}(\mathbf{k}) + T_u^{\mathrm{comp,int}}(\mathbf{k}) \\ &= -\Re[\mathbf{N}_\perp^{\mathrm{inc}}(\mathbf{k}) \cdot \mathbf{u}_\perp^*(\mathbf{k})] - \Re[N_3^{\mathrm{comp}}(\mathbf{k}) u_3^*(\mathbf{k})] \\ &\quad -\Re[\mathbf{N}_\perp^{\mathrm{inc,int}}(\mathbf{k}) \cdot \mathbf{u}_\perp^*(\mathbf{k})] - \Re[N_3^{\mathrm{comp,int}}(\mathbf{k}) u_3^*(\mathbf{k})], \end{aligned} \tag{30.67}$$

where $T_u^{\mathrm{inc}}(\mathbf{k}), T_u^{\mathrm{comp}}(\mathbf{k})$ represent interactions among incompressible and compressible components respectively, while $T_u^{\mathrm{inc,int}}(\mathbf{k})$ and $T_u^{\mathrm{comp,int}}(\mathbf{k})$ arise due to cross interactions between $\mathbf{u}_\perp$ and the compressible components. Again, following the definition of Eq. (4.41), we define

$$\Pi_u^{\mathrm{inc}}(k_0) = -\sum_{k \le k_0} T_u^{\mathrm{inc}}(\mathbf{k}), \tag{30.68a}$$

$$\Pi_u^{\mathrm{comp}}(k_0) = -\sum_{k \le k_0} T_e^{\mathrm{comp}}(\mathbf{k}). \tag{30.68b}$$

Under a steady state, the fluxes $\Pi_u^{\mathrm{inc}}(k)$ and $\Pi_u^{\mathrm{inc}}(k)$ would not be constant due to the cross-interaction terms, and hence,

$$\frac{d}{dk}\Pi_u^{\mathrm{inc}}(k) = T_u^{\mathrm{inc,int}}(k) + \mathcal{F}_\sigma(k), \tag{30.69a}$$

$$\frac{d}{dk}\Pi_u^{\mathrm{comp}}(k) = T_u^{\mathrm{comp,int}}(k) + \mathcal{F}_{e,\sigma}(k). \tag{30.69b}$$

Note that the above equations assume $\rho \approx \rho_0$; the general set of equations are too complex and involved to be presented here. The aforementioned fluxes need to be studied carefully using numerical simulations.

With this, we end our discussion on equations and energy transfers in compressible flows.

Further Reading

For introduction to the equations and formalism of compressible flows, refer to textbooks by Kundu et al. (2015) and Choudhuri (1998). The energy transfer formalism for compressible flows is not well developed. The reader is referred to recent papers by Ristorcelli (2018) and Wang et al. (2018) on numerical studies of spectra of compressible and incompressible components. Zank and Matthaeus (1991) discuss nearly incompressible flows. Also refer to Eyink and Aluie (2009) for locality studies of compressible flows, and Galtier and Banerjee (2011) for exact relations of energy transfers in real space.

The discussion in the present chapter provides an alternate perspective in terms of disentangling contributions of compressible and incompressible components, as well as those of dynamic and thermodynamic pressures.

Chapter 31

Miscellaneous Applications of Energy Transfers

So far in the book we described scalar, vector, and compressible flows, as well as energy transfers in such flows. In the present chapter we will describe three additional flows—a flow with Ekman friction, gyrokinetic turbulence, and a flow in spherical geometry. These examples further illustrate the power of energy transfer formalism.

In the next section we describe a flow with Ekman friction.

31.1 Variable Enstrophy Flux in 2D Turbulence with Ekman Friction

In a thin layer of fluid, the bottom surface offers frictional resistance to the flow. A generic form for such friction is $-\alpha\mathbf{u}$, where α is a constant, and $\mathbf{u}$ is the velocity field. Consequently, the governing equations for an incompressible 2D flow with Ekman friction is

$$\frac{\partial \mathbf{u}}{\partial t} + \mathbf{u}\cdot\nabla\mathbf{u} = -\nabla p - \alpha\mathbf{u} + \nu\nabla^2\mathbf{u} + \mathbf{F}_\mathbf{u}, \tag{31.1a}$$

$$\nabla\cdot\mathbf{u} = 0. \tag{31.1b}$$

Hence, the equation for the vorticity field is

$$\frac{\partial \omega}{\partial t} + \mathbf{u}\cdot\nabla\omega = (-\alpha + \nu\nabla^2)\omega + F_\omega, \tag{31.2}$$

where $\omega = [\nabla \times \mathbf{u}]_z$ and $F_\omega = [\nabla \times \mathbf{F_u}]_z$.

Let us assume that the flow is forced at $k_f \approx 1/L$, where L is the system size. In such a 2D flow, for $\alpha = 0$, we expect a constant enstrophy cascade and $E_\omega(k) \sim k^{-1}$ in the inertial range (see Section 7.2). An addition of Ekman friction leads to a stronger enstrophy dissipation; hence, there is steepening of enstrophy flux and spectrum compared to pure 2D hydrodynamics. In the following discussion we will derive the extent of steepening as a function of α using the variable enstrophy flux formalism (Verma, 2012). The present derivation, which is a variant of Verma (2012)'s work, invokes Pao (1968)'s model.

Following the discussion of Section 7.3, we deduce that in the inertial–dissipation range where $F_\omega(k) = 0$, under a steady state,

$$\frac{d}{dk}\Pi_\omega(k) = -(\alpha + \nu k^2)E_\omega(k). \tag{31.3}$$

To solve this equation, we extend Pao (1968)'s model for 3D hydrodynamics to the present system and propose that $E_\omega(k)/\Pi_\omega(k)$ is independent of ν, and that it is a function of the enstrophy injection rate, ϵ_ω, and k. That is,

$$\frac{E_\omega(k)}{\Pi_\omega(k)} = K'_{2D}\epsilon_\omega^{-1/3}k^{-1}. \tag{31.4}$$

Substitution of $E_\omega(k)$ of Eq. (31.4) in Eq. (31.3) yields

$$\frac{d}{dk}\Pi_\omega(k) = -K'_{2D}(2\nu k + \alpha/k)\epsilon_\omega^{-1/3}\Pi_\omega(k), \tag{31.5}$$

whose solution is

$$\Pi_\omega(k) = \Pi_\omega(k_0)\left(\frac{k}{k_0}\right)^{-\alpha K'_{2D}\epsilon_\omega^{-1/3}} \exp\left(-\frac{K'_{2D}}{k_{d2D}^2}(k^2 - k_0^2)\right), \tag{31.6}$$

where

$$k_{d2D} = \frac{\epsilon_\omega^{1/6}}{\sqrt{\nu}}, \tag{31.7}$$

and $\Pi_\omega(k_0)$ is the reference value of the enstrophy flux at $k = k_0$. Note that $\Pi_\omega(k_f) \approx \epsilon_\omega$, where k_f is the forcing wavenumber. Substitution of the above enstrophy flux in Eq. (31.4) yields the following enstrophy spectrum:

$$E_\omega(k) = K'_{2D}\epsilon_\omega^{2/3}k^{-1}\left(\frac{k}{k_0}\right)^{-\alpha K'_{2D}\epsilon_\omega^{-1/3}} \exp\left(-\frac{K'_{2D}}{k_{d2D}^2}(k^2 - k_0^2)\right). \tag{31.8}$$

Thus, compared to the $\alpha = 0$ case, the enstrophy flux and spectrum are steeper by a factor of $k^{-\alpha K'_{2D}\epsilon_\omega^{-1/3}}$. This result is similar to that predicted by Verma (2012).

In the next section we describe energy transfers in gyrokinetic plasma turbulence.

31.2 Energy Transfers in Gyrokinetic Plasma Turbulence

Typically, plasma turbulence is more complex than hydrodynamic turbulence. Interestingly, Teaca et al. (2014) showed that the mode-to-mode energy transfer formalism could be extended to the gyrokinetic plasma turbulence (GKT), one of the important plasmas prevalent in fusion reactors. Since this turbulence is quite complex, here, we only state the result, and refer the reader to the original paper for details. The derivation is similar to that in Section 4.1.

In GKT, g is the perturbed distribution function from the mean, and ϕ is the gyro-averaged self-consistent electrostatic field. Teaca et al. (2014) performed Fourier transform for g and ϕ perpendicular to the direction of the mean magnetic field that yields $h(\mathbf{k}_\perp, z)$ and $g(\mathbf{k}_\perp, z)$, where $\mathbf{k}_\perp = (k_x, k_y)$. The formula for the energy transfers were derived for the $(\mathbf{k}_\perp, z)$ space.

Teaca et al. (2014) showed that the nonlinear term in GKT is

$$N(\mathbf{k}_\perp, z) = \sum_{\mathbf{p}_\perp} [\mathbf{q}_\perp \times \mathbf{p}_\perp]_z \phi(\mathbf{q}_\perp, z) h(\mathbf{p}_\perp, z), \tag{31.9}$$

where $\mathbf{q}_\perp = \mathbf{k}_\perp - \mathbf{p}_\perp$, and the modal energy is of the following form (apart from a proportionality constant):

$$E(\mathbf{k}_\perp) = \Re[h^*(\mathbf{k}_\perp) g(\mathbf{k}_\perp)]. \tag{31.10}$$

The nonlinear energy transfer term is

$$T(\mathbf{k}_\perp) = \Re[h^*(\mathbf{k}_\perp) N(\mathbf{k}_\perp)] = \sum_{\mathbf{p}_\perp, \mathbf{q}_\perp} S(\mathbf{k}_\perp | \mathbf{p}_\perp, \mathbf{q}_\perp), \tag{31.11}$$

where $S(\mathbf{k}_\perp | \mathbf{p}_\perp, \mathbf{q}_\perp)$ is the combined energy transfer from $\mathbf{p}_\perp$ and $\mathbf{q}_\perp$ to $\mathbf{k}_\perp$:

$$S(\mathbf{k}_\perp | \mathbf{p}_\perp, \mathbf{q}_\perp) = [\mathbf{q}_\perp \times \mathbf{p}_\perp]_z [\phi(\mathbf{q}_\perp) h(\mathbf{p}_\perp) - \phi(\mathbf{p}_\perp) h(\mathbf{q}_\perp)] h^*(\mathbf{k}_\perp). \tag{31.12}$$

As in Section 4.1, we consider a triad $(\mathbf{k'}_\perp, \mathbf{p}_\perp, \mathbf{q}_\perp)$ satisfying the property $\mathbf{k'}_\perp + \mathbf{p}_\perp + \mathbf{q}_\perp = 0$. Note that $\mathbf{k'}_\perp = -\mathbf{k}_\perp$. For such a triad, using Eq. (31.12), we can show that

$$S(\mathbf{k}'_\perp|\mathbf{p}_\perp, \mathbf{q}_\perp) + S(\mathbf{p}_\perp|\mathbf{q}_\perp, \mathbf{k}'_\perp) + S(\mathbf{q}_\perp|\mathbf{k}'_\perp, \mathbf{p}_\perp) = 0; \tag{31.13}$$

this is a statement of the *detailed energy conservation in a triad.*

Using the structure of the nonlinear term, Teaca et al. (2014) derived the following formula for the mode-to-mode energy transfer from $h(\mathbf{p}_\perp)$ to $h(\mathbf{k}_\perp)$ with the mediation of $\phi(\mathbf{q}_\perp)$ as

$$S(\mathbf{k}'_\perp|\mathbf{p}_\perp|\mathbf{q}_\perp) = [\mathbf{q}_\perp \times \mathbf{p}_\perp]_z \phi(\mathbf{q}_\perp) h(\mathbf{p}_\perp) h(\mathbf{k}'_\perp). \tag{31.14}$$

Note that this function satisfies the following properties required for the mode-to-mode energy transfers:

$$S(\mathbf{k}'_\perp|\mathbf{p}_\perp, \mathbf{q}_\perp) = S(\mathbf{k}'_\perp|\mathbf{p}_\perp|\mathbf{q}_\perp) + S(\mathbf{k}'_\perp|\mathbf{q}_\perp|\mathbf{p}_\perp), \tag{31.15}$$

$$S(\mathbf{k}'_\perp|\mathbf{p}_\perp|\mathbf{q}_\perp) = -S(\mathbf{p}_\perp|\mathbf{k}'_\perp|\mathbf{q}_\perp). \tag{31.16}$$

Hence, $S(\mathbf{k}'_\perp|\mathbf{p}_\perp|\mathbf{q}_\perp)$ is indeed the desired formula for mode-to-mode energy transfer.

Using the above mode-to-mode energy transfer formula, one can construct energy flux that is very useful for modeling such flows. For further details, refer to Teaca et al. (2014) and subsequent papers on this topic.

31.3 Energy Transfers in Spherical Geometry

Many natural systems like planets and stars are spherical. Hence, the spherical coordinate system is best suited for describing fluid flows in these systems. Unfortunately, the flow equations in the spherical coordinate system are quite complicated. Therefore, we illustrate the main ingredients of nonlinear interactions in spherical geometry using a simplified equation.

We assume that a scalar field $f(\zeta, \phi)$ is on the surface of a sphere, and its evolution is described by

$$\frac{\partial}{\partial t} f = f^2 + \nabla^2 f. \tag{31.17}$$

In the spherical coordinate system, f is expanded using spherical harmonics Y_l^m as

$$f(\zeta, \phi) = \sum_{l,m} f_l^m Y_l^m(\zeta, \phi), \tag{31.18}$$

where f_l^m is the coefficient of the expansion. It is important to note that $-l \le m \le l$. See Fig. 31.1 for an illustration of (l, m) pairs of spherical harmonics.

Substitution of f of Eq. (31.18) in Eq. (31.17) yields

$$\sum_{l,m} \dot{f}_l^m Y_l^m(\zeta,\phi) = \sum_{l_1,m_1,l_2,m_2} f_{l_1}^{m_1} f_{l_2}^{m_2} Y_{l_1}^{m_1}(\zeta,\phi) Y_{l_2}^{m_2}(\zeta,\phi) + \sum_{l,m} f_l^m \nabla^2 Y_l^m(\zeta,\phi). \quad (31.19)$$

Now we employ the following identities of $Y_l^m(\zeta,\phi)$ (Sakurai, 1994):

$$\int d\Omega Y_l^m(\zeta,\phi) \left[Y_{l'}^{m'}(\zeta,\phi)\right]^* = \delta_{ll'}\delta_{mm'}, \quad (31.20a)$$

$$\nabla^2 Y_l^m(\zeta,\phi) = -l(l+1) Y_l^m(\zeta,\phi), \quad (31.20b)$$

$$\int d\Omega Y_{l_1}^{m_1}(\zeta,\phi) Y_{l_2}^{m_2}(\zeta,\phi) \left[Y_{l'}^m(\zeta,\phi)\right]^* = g(l_1,m_1,l_2,m_2,l,m), \quad (31.20c)$$

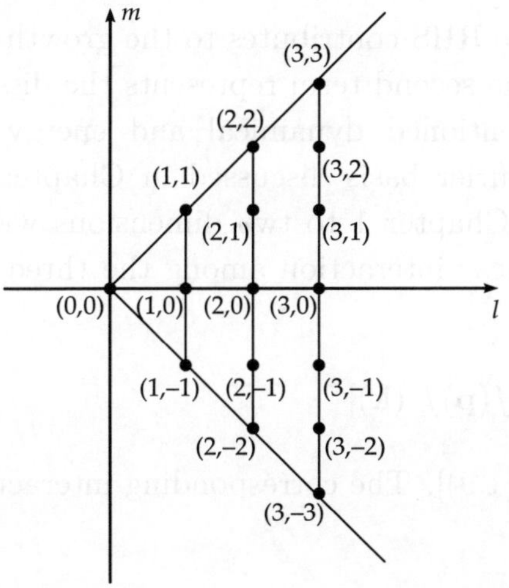

Figure 31.1 For a spherical system, an illustration of coordinates (l, m). Here $-l \le m \le l$. To compute energy E_L $(l \le L)$, we sum over all the modes of the triangle up to $l = L$.

where $d\Omega$ is the solid angle, and

$$g(l_1,m_1,l_2,m_2,l,m) = \sqrt{\frac{(2l_1+1)(2l_2+1)(2l_2+1)}{4\pi}} \times \begin{pmatrix} l_1 & l_2 & l \\ 0 & 0 & 0 \end{pmatrix} \begin{pmatrix} l_1 & l_2 & l \\ m_1 & m_2 & -m \end{pmatrix} \quad (31.21)$$

with the bracketed terms as the Clebsch–Gordan coefficients. Substitution of the above identities in Eq. (31.19) yields

$$\dot{f}_l^m = \sum_{l_1,m_1,l_2,m_2} g(l_1,m_1,l_2,m_2,l,m) f_{l_1}^{m_1} f_{l_2}^{m_2} - l(l+1) f_l^m. \tag{31.22}$$

The modal energy in spherical coordinates is defined as

$$E_l^m = \frac{1}{2}|f_l^m|^2, \tag{31.23}$$

whose evolution equation is

$$\begin{aligned} \dot{E}_l^m &= \sum_{l_1,m_1,l_2,m_2} g(l_1,m_1,l_2,m_2,l,m) \Re[f_{l_1}^{m_1} f_{l_2}^{m_2} f_l^{*m}] - 2l(l+1)E_l^m \\ &= T_l^m - 2l(l+1)E_l^m. \end{aligned} \tag{31.24}$$

Here, the first term in the RHS contributes to the growth rate of E_l^m via nonlinear energy transfers, while the second term represents the dissipation of E_l^m.

Contrast the aforementioned dynamical and energy equations [Eqs. (31.22, 31.24)] with those in Fourier basis discussed in Chapter 1. For comparison, we extend the equations of Chapter 1 to two dimensions with $k \to \mathbf{k} = (k_x, k_y)$. In Fourier space, the nonlinear interaction among the three interacting modes of the triad is

$$T(\mathbf{k},\mathbf{p},\mathbf{q}) = \Re[f(\mathbf{q})f(\mathbf{p})f^*(\mathbf{k})] \tag{31.25}$$

with $\mathbf{q} = \mathbf{k} - \mathbf{p}$ [see Eq. (1.9)]. The corresponding interaction term in the spherical coordinate system is

$$T(l_1,m_1,l_2,m_2,l,m) = g(l_1,m_1,l_2,m_2,l,m) \Re[f_{l_1}^{m_1} f_{l_2}^{m_2} f_l^{*m}], \tag{31.26}$$

where g is a complex function involving Clebsch–Gordan coefficients, and

$$m_1 + m_2 = m; \quad |l_1 - l_2| \le l \le l_1 + l_2. \tag{31.27}$$

These conditions are similar to the selection rules[1] for l, m in quantum mechanics (Sakurai, 1994). Thus, the nonlinear term in the spherical basis is much more complex than the corresponding term in Fourier basis. In Table 31.1, we make correspondence between the Fourier and spherical representations.

[1]The second selection rule is similar to that for sin–cos basis functions. Note that interactions of $\sin(px)$ and $\sin(qx)$ yield harmonic functions with wavenumbers $|p-q|$ and $p+q$.

Table 31.1 Contrasting 2D flows in Cartesian and spherical coordinate systems. In the fourth row, $\bigoplus$ represents nonlinear interaction.

Property	Cartesian system	Spherical system
Basis function	$\exp(i(k_x x + k_y y))$	$Y_l^m(\zeta, \phi)$
Real space coord.	x, y	ζ, ϕ
Fourier space wavenumbers	$\mathbf{k} = (k_x, k_y)$	(l, m)
Interacting wavenumbers	$(\mathbf{p}, \mathbf{q}, \mathbf{k})$ with $\mathbf{p} + \mathbf{q} = \mathbf{k}$	$(l_1, m_1) \bigoplus (l_2, m_2) = (l, m)$ with $m_1 + m_2 = m$ and $\lvert l_1 - l_2 \rvert \le l \le l_1 + l_2$
Modal energy	$\lvert f(k_x, k_y)\rvert^2/2$	$\lvert f_l^m \rvert^2/2$
$-\nabla^2$	k^2	$l(l+1)$
Triad interaction	$\Re[f(\mathbf{q}) f(\mathbf{p}) f^*(\mathbf{k})]$	$\Re[f_{l_1}^{m_1} f_{l_2}^{m_2} f_l^{*m}]$ $\times g(l_1, m_1, l_2, m_2, l, m)$

Due to the above complexities, it is quite difficult to derive combined energy transfer and mode-to-mode energy transfer in spherical geometry. Yet, we can derive a formula for the energy flux as follows. We define the energy E_L by summing the modes of the triangle of Fig. 31.1 up to $l = L$, that is,

$$E_L = \sum_{l \le L} E_l^m. \tag{31.28}$$

This is equivalent to the wavenumber sphere of radius L discussed in Chapter 4. Note that this sum includes m-sums with $-L \le m \le L$. We derive an equation for the temporal evolution of E_L by summing Eq. (31.24) for $l \le L$ that yields

$$\begin{aligned} \frac{d}{dt} E_L &= \sum_{l \le L} T_l^m - \sum_{l \le L} 2l(l+1) E_l^m \\ &= -\Pi(L) - \sum_{l \le L} 2l(l+1) E_l^m. \end{aligned} \tag{31.29}$$

Here, the first term in the RHS represents the energy gained by E_L due to the nonlinear interactions, or $-\Pi(L)$, where $\Pi(L)$ is the energy flux leaving the sphere of radius L. The second term is the dissipation rate of E_L. This is how we can define the energy flux in spherical geometry. The definition of $\Pi(L)$ presented here is similar to Eq. (4.41) of Chapter 4.

The equations for the velocity field are even more complex, and we skip them altogether. We refer the reader to Amit and Olson (2010) and references therein for some details on hydrodynamics in spherical geometry,

Further Reading

For the three topics covered in this chapter we refer the reader to Verma (2012), Teaca et al. (2014), and Amit and Olson (2010) respectively.

Chapter 32

Conclusions

In this book, we describe nonlinear energy transfers in incompressible and compressible hydrodynamics; scalar flows including stably stratified and thermal convection; vector flows including magnetofluids, electron magnetohydrodynamics (MHD), tensor flows, etc. After introducing such flows, we describe the formalism of mode-to-mode energy, enstrophy, kinetic helicity, and magnetic helicity transfers in respective flows.

Using the formulas for the mode-to-mode transfers, we can easily derive fluxes and shell-to-shell transfers for the corresponding quantities (e.g., kinetic energy, kinetic helicity, etc.). Complex flows like thermal convection and MHD have various kinds of energy fluxes that are covered in this book. For anisotropic flows, the useful transfer functions are ring-to-ring transfers, and energy transfer from $u_\perp$ to $u_\parallel$ via pressure. In the book we also describe various phenomenologies of energy flux and spectrum, and compare the phenomenological predictions with numerical and experimental results.

This book also includes discussions on energy transfers in Craya–Herring and helical basis. These formulations provide alternate perspectives, as well as simplify many derivations. For example, the field-theoretic computation of energy transfers, and the energy transfers in helical turbulence and dynamo are much more compact in Craya–Herring and helical basis.

The formalism of energy transfers is very powerful and it provides valuable tools for analyzing turbulent flows. Yet this formulation has limitations. The scheme of mode-to-mode energy transfer cannot be extended to compressible flows in a straightforward manner, yet energy flux can be defined for such flows. Moreover, the derivation of energy transfers for flows in spherical geometry are quite involved due to the complex nature of spherical harmonics.

In compressible flows, there are interesting energy transfers among compressible and incompressible components of the flows. In addition, there are exchanges between the internal energy and kinetic energy. Separation of the equations for the incompressible and incompressible components also contrasts the dynamic and thermodynamic pressures, the former involves incompressible components, while the latter participates with the compressible components. These ideas are covered in the chapters on compressible turbulence.

Most discussions in this book are based on Fourier space formalism, but we also cover energy transfers in real space, namely, Kolmogorov's theory of turbulence. We highlight connections between the multiscale energy transfers in real space and in Fourier space.

I believe that it may be possible to extend the energy transfer formalism described in this book to other fields like electrodynamics, optics, quantum systems, etc. The constraints $\nabla \cdot \mathbf{E} = \nabla \cdot \mathbf{B} = 0$ (where $\mathbf{E}$ and $\mathbf{B}$ are electric and magnetic fields respectively) of free-space electrodynamics is similar to the incompressible condition $\nabla \cdot \mathbf{u} = 0$; these similarities may help decipher connections between electrodynamics and hydrodynamics. Also note that energy fluxes in equilibrium field theory are zero, while those in nonequilibrium field theory are nonzero (see Chapter 4). The latter field theories are active areas of research, and the energy transfer formulation may provide interesting insights into this complex field.

A large fraction of the material covered in this book are based on published material, but some topics, for example, compressible turbulence, EMHD, field-theoretic computation of energy transfers, energy transfers in tensor flows are unpublished works. In the present book we present the formalism of energy transfers in a coherent manner and provide connections between a variety of flows. We do hope that you as a reader would have enjoyed these discussions.

Appendix A

Power Law Physics

Multiscale phenomena have many interesting features. Some of the analytical tools used to describe such phenomena are unique. We illustrate them using examples including turbulent flows.

Consider a one-dimensional rod whose density is $\rho(x)$ with $x = [0, L]$. The mass of the rod segments $[0, x]$ and $[0, x + dx]$ are respectively

$$M(x) = \int_0^x \rho(x')dx', \tag{A.1}$$

$$M(x + dx) = \int_0^{x+dx} \rho(x')dx'. \tag{A.2}$$

Hence, the elemental mass of the rod segment $[x, x + dx]$ is $\rho(x)dx$.

Contrast this with the multiscale mass distribution in our universe. Our universe has asteroids, planets, stars, galaxies, galaxy clusters, etc. at different length scales, which are denoted by r. Hence, it is meaningful to think of variations of length in the universe as

$$dr = \beta r. \tag{A.3}$$

For example, the Earth is many times smaller than our solar system, the next object in the hierarchy. Hence, the increment from one system to another is not small.

Now, let us consider a mass function $M(r)$, which is the total mass up to length scale r. It is expected that

$$M(r) = Cr^\alpha, \tag{A.4}$$

where α lies between 2 and 3 due to the fractal nature of the universe. Note that a sphere with a uniform mass density ρ has $M(r) = (4/3)\pi\rho r^3$, but in the universe, the voids at multiple scales lead to α between 2 and 3.

When we go from scale r to $r + dr = (1 + \beta)r$, the change in mass is

$$\delta M(r) = M(r + dr) - M(r) = Cr^\alpha[(1 + \beta)^\alpha - 1] \sim M(r). \tag{A.5}$$

Contrast this with the example of the rod for which dx and $\delta M(x)$ are small, and $\delta M(x) \ll M(x)$.

Similar analysis is employed to the inertial range of a turbulent flow that exhibits multiscale physics. We estimate the energy content in a scale range of dr as follows. Using $dr \sim r$, wavenumber $k \sim 1/r$, and $dk \sim k$, we obtain

$$\delta E(r) = \frac{u_r^2}{2} = \int_k^{2k} E(k)dk. \tag{A.6}$$

Using $E(k) \sim \epsilon_u^{2/3} k^{-5/3}$ from Kolmogorov's theory, we deduce that

$$\frac{u_r^2}{2} \sim \epsilon_u^{2/3} k^{-2/3} \sim \epsilon_u^{2/3} r^{2/3}. \tag{A.7}$$

Hence, the velocity magnitude at length scale r is

$$u_r \sim \epsilon_u^{1/3} r^{1/3}. \tag{A.8}$$

Also recall that for the shell-to-shell energy transfers, the radii of the wavenumber shells were chosen as $k_n = k_0 2^{ns}$, where n is the shell index, and s is a positive real number. This choice of k_n is motivated by the arguments given earlier.

The arguments given here are general, and they are applicable to all multiscale phenomena. For example, water flow in streams and rivers at different scales, energy contents in multiscale fractures of an earthquake system, oxygen content in the lungs at different scales, wealth distribution (to be discussed in the next Appendix), etc.

Further Reading

Refer to Newman (2005) and reference therein for further discussion and examples of power law physics.

Appendix B

Wealth Distribution and Cascade in an Economy

In this appendix we make a simple multiscale model for the wealth distribution in an economic system. This model is motivated by Kolmogorov's picture of turbulence.

Let us assume that the wealth is generated at the largest scale, for example, by government, large companies, etc., and it flows from large scales to small scales in a steady manner. We consider a 2D wavenumber mesh with each grid point representing a financial entity with wealth of $W(\mathbf{k})$. The number density of mesh points in a 2D shell of radius k is[1]

$$n(k) = 2\pi k. \tag{B.1}$$

We aim to solve for the wealth distribution $n(W)$, which is the number density of financial entities with wealth W. To illustrate, there are fewer financial entities at small k, and they correspond to financial giants like Google and Apple. Intermediate and large k's represent intermediate-scale and small-scale companies. The smallest unit in the financial model are consumers of the wealth.

As in the shell model of turbulence discussed in Chapter 28, we assume local interactions (in wavenumber space) among financial entities, and a constant cascade of wealth from large scales to intermediate scale and then to small scales.

[1] The real number density of financial entities can be determined from the dimension of the network of the companies. This exercise however requires extensive data analysis.

Hence, taking a cue from the shell model of turbulence, we construct the following multiscale model for finance:

$$\frac{dW_k}{dt} = ak^\alpha W_{k-1}W_{k+1} - bk^\beta W_k + Q_{k,1}, \tag{B.2}$$

where a, b, α, and β are constants, and W_k is the wealth spectrum, which is analogous to the shell spectrum in turbulence.[2] Hence,

$$W_k = 2\pi k W(\mathbf{k}). \tag{B.3}$$

In Eq. (B.2), the first term in the RHS represents nonlinear interactions among the financial entities at scales $k-1$, k, and $k+1$; while the second term represents financial losses at scale k. The third term $Q_{k,1}$ represents wealth generation at the largest scale, which is $k = 1$.

This simple model ignores nonlocal interactions, as well as other complex processes like loans, savings, banks, generation of wealth at the intermediate and small scales, etc. Further, we assume a steady state with money flowing from large scales to small scales. The wealth is finally consumed at the smallest scale of the system.

For this system, we model the flux and distribution of wealth by making an analogy with Kolmogorov's theory of turbulence. We assume no pilferage of wealth due to corruption; hence, the money flux in the intermediate scale is constant, as shown in Fig. B.1(a). In the intermediate range, we expect a power law behavior for W_k (to be derived in the following).

Since the financial losses at the intermediate scales are assumed to be negligible, we expect that at these scales,

$$\Pi_m = \frac{dW_k}{dt} \approx k^\alpha W_k^2 \approx \text{const.}, \tag{B.4}$$

where Π_m is the cascade of wealth. Inversion of this equation yields

$$W_k \sim \Pi_m^{1/2} k^{-\alpha/2}. \tag{B.5}$$

This is the wealth distribution as a function of scale (k). We illustrate the aforementioned spectrum in Fig. B.1(b).

[2] In the shell model of hydrodynamic turbulence discussed in Chapter 28, the amplitude of the velocity, U_n is a complex variable, while the kinetic energy, $|U_n|^2$, is a real quantity. In Eq. (B.2), W_n is real. It may also be possible to write an equation for a complex W_n. These issues need further investigation.

In Eq. (B.5), we replace k and W_k using Eqs. (B.1, B.3) that yields

$$n(W) \sim \Pi_m^{\frac{1}{\alpha+2}} W^{-\frac{2}{\alpha+2}}, \tag{B.6}$$

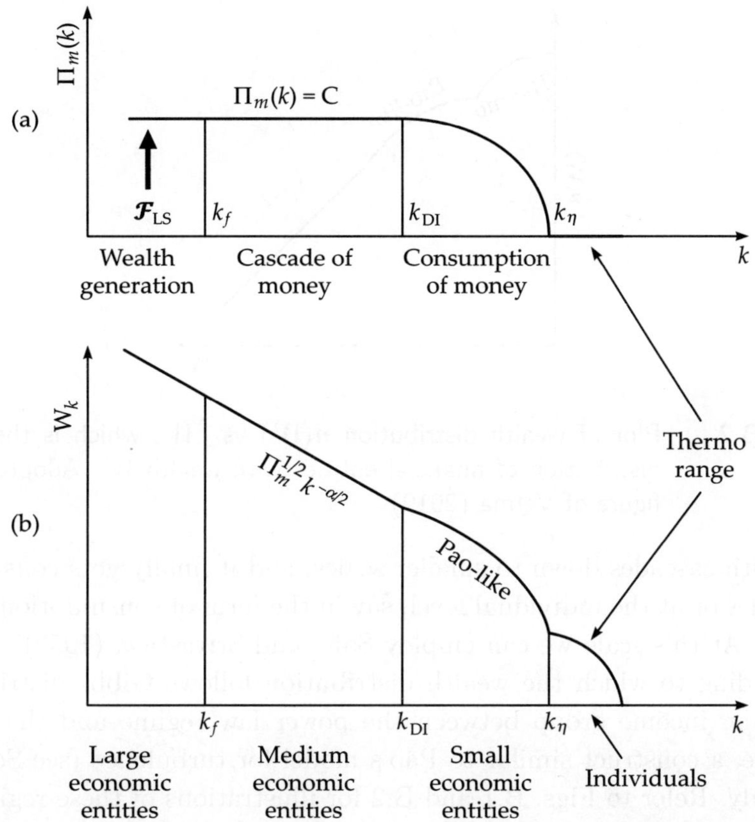

Figure B.1 For the finance model of Eq. (B.2): (a) the flux of wealth $\Pi_m(k)$, and (b) the wealth distribution of financial entities. The large and medium scale entities have constant wealth flux and power law distribution of wealth ($\Pi_m^{1/2} k^{-\alpha/2}$). Small scale entities exhibit exponential distribution (Pao's model), while the smallest ones exhibit Gibbs distribution. Adopted from a figure of Verma (2019).

where we write $W(\mathbf{k})$ as W. Thus, we derive the number of financial entities $n(W)$ with wealth W. Clearly, $\alpha = -1$ gives

$$n(W) \sim \Pi_m W^{-2}, \tag{B.7}$$

which is the Pareto's law for the wealth distribution (Pareto, 1964; Chakrabarti et al., 2013) of the large financial entities. Note however that $n(W)$ depends quite crucially on the choice of α. In Fig. B.2, we plot wealth distribution $n(W)$ vs. W. Note that in some sense, $n(W)$ vs. W plot is an inverted form of W_k vs. k plot.

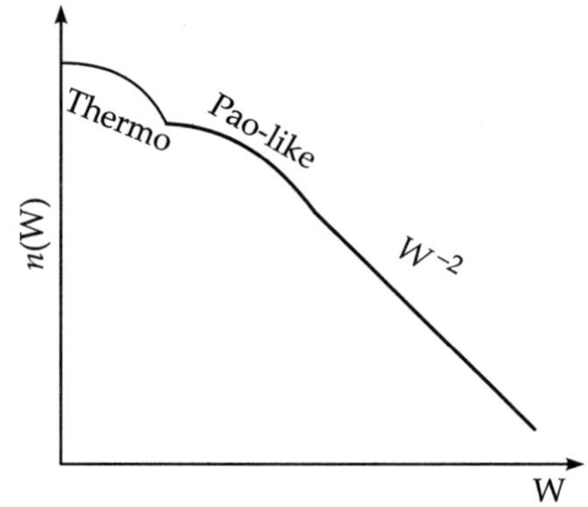

Figure B.2 Plot of wealth distribution $n(W)$ vs. W, which is the number distribution of financial entities with wealth W. Adopted from a figure of Verma (2019).

The wealth cascades down to smaller scales, and it finally gets consumed at the smallest scales or at the individual level, say in the form of consumption of food and basic needs. At this scale we can employ Saha and Srivastava (1950)'s equilibrium model according to which the wealth distribution follows Gibbs distribution. We also expect an income group between the power law regime and the equilibrium regime. Here, a construct similar to Pao's model for turbulence (see Section 5.5.1) may be handy. Refer to Figs. B.1 and B.2 for illustrations of these regimes.

The above finance model is very simple since it ignores many important ingredients such as savings, banking, pilferage of wealth, nonlocal interactions, etc. The model however has a certain novelty. It emphasizes nonequilibrium and multiscale nature of the financial system, cascade of money at different scales, etc., which are normally absent in equilibrium models (Saha and Srivastava, 1950).

Further Reading

The model discussed in this Appendix is taken from Verma (2019). Saha and Srivastava (1950) constructed one of the earliest financial model based on equilibrium theory. For more detailed discussion on this topic, refer to Chakrabarti et al. (2013).

Appendix C

Renormalization Group Analysis of Hydrodynamic Turbulence

Hydrodynamic turbulence involves multiscale physics; hence, renormalization group (RG) formulation is a useful tool for studying hydrodynamic turbulence. Some of the leading efforts on RG formulation of hydrodynamic turbulence are: Yakhot–Orszag perturbative approach (Yakhot and Orszag, 1986); and self-consistent iterative RG (i-RG) of McComb (1990) and Zhou (2010). In this Appendix we present i-RG procedure (McComb, 1990, 2014; McComb and Shanmugasundaram, 1983, 1984; Zhou et al., 1988; Zhou, 2010; Verma, 2004).

In Chapter 3 we derived the Navier–Stokes equations in Fourier space. An inclusion of pressure gradient in Eq. (3.17a) yields (McComb, 1990)

$$(-i\omega + \nu k^2)u_i(\hat{k}) = -\frac{i}{2}P_{ijm}(\mathbf{k})\int_{\hat{p}+\hat{q}=\hat{k}} d\hat{p}\,[u_j(\hat{p})u_m(\hat{q})] + f_i(\hat{k}), \tag{C.1a}$$

$$k_i u_i(\mathbf{k}) = 0, \tag{C.1b}$$

where

$$P_{ijm}(\mathbf{k}) = k_j P_{im}(\mathbf{k}) + k_m P_{ij}(\mathbf{k}), \tag{C.2}$$

$$\hat{k} = (\omega, \mathbf{k}), \hat{p} = (\omega', \mathbf{p}), \text{and } \hat{q} = (\omega'', \mathbf{q}).$$

In the RG scheme, the wavenumber range (k_N, k_0) is divided logarithmically into N shells with the nth shell as (k_n, k_{n-1}), where $k_n = h^n k_0$ $(h < 1)$ and $k_N = h^N k_0$. See Fig. C.1 for an illustration.

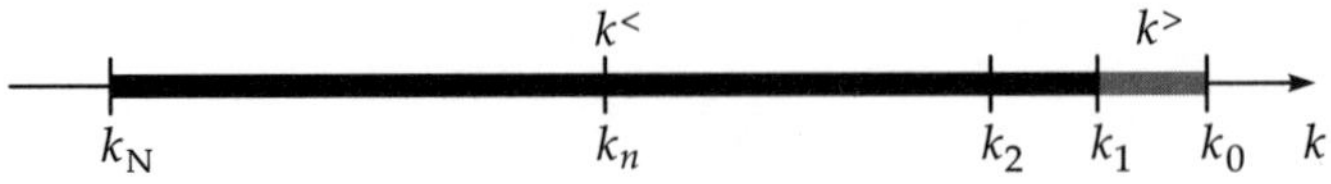

Figure C.1 Division of wavenumber range into shells. In the first iteration of RG, we average the velocity field of the shell (k_1, k_0), which is denoted by $k^>$. The remaining modes (k_N, k_1) belong to the band $k^<$. This procedure is repeated subsequently to shells (k_2, k_1), (k_3, k_2), and so on.

In the first step, the spectral space is divided into two bands: $(k_1, k_0) = k^>$ (to be eliminated) and $(k_N, k_1) = k^<$ (to be retained). The equation for a Fourier mode $\hat{k}$ in $k^<$ is

$$\begin{aligned}\left[-i\omega + \nu_{(0)}k^2\right]u_i^<(\hat{k}) &= -\frac{i}{2}P_{ijm}(\mathbf{k})\int_{\hat{p}+\hat{q}\hat{k}} d\hat{p}([u_j^<(\hat{p})u_m^<(\hat{q})] \\ &\quad +2[u_j^<(\hat{p})u_m^>(\hat{q})] + [u_j^>(\hat{p})u_m^>(\hat{q})]) + F_i^<(\hat{k}),\end{aligned} \tag{C.3}$$

where $\nu_{(0)} = \nu$. The equation for $u_i^>(\hat{k})$ modes is obtained by interchanging $<$ and $>$ in the aforementioned equations.

In the RG procedure, we compute the corrections to the viscosity, $\delta\nu_{(0)}$, due to the second and third terms in the RHS of Eq. (C.3). The steps involved in this process are described as follows.

1. The terms in the second and third brackets in the right-hand side of Eq. (C.3) are computed perturbatively. It is assumed that $\{\mathbf{u}^>(\hat{k})\}$ has a Gaussian distribution with a zero mean, while $\{\mathbf{u}^<(\hat{k})\}$ is unaffected by the averaging process. We perform an ensemble average of the system (McComb, 1990) that yields

$$\left\langle u_i^>(\hat{k})\right\rangle = 0; \quad \left\langle u_i^<(\hat{k})\right\rangle = u_i^<(\hat{k}). \tag{C.4}$$

For homogeneous and isotropic flows [Eq. (10.1)],

$$\langle u_i^>(\hat{p})u_j^>(\hat{q})\rangle = \left[P_{ij}(\mathbf{p})C(\hat{p}) - i\epsilon_{ijl}p_l\frac{H_K(\hat{p})}{p^2}\right]\delta(\hat{p}+\hat{q}), \tag{C.5}$$

where $C(\hat{p}), H_K(\hat{p})$ are related to the modal energy and modal kinetic helicity respectively (Batchelor, 1953). Using Eq. (C.5) and $P_{ijm}(\mathbf{k} = 0) = 0$, we observe that the second and third terms vanish to the zeroth order. Hence we go to the first order.

For the expansion to the first order, we employ the properties that the triple order correlations $\left\langle u_i^>(\hat{k})u_j^>(\hat{p})u_m^>(\hat{q})\right\rangle$ are zero due to the Gaussian nature of the fluctuations, and the fourth order correlations are expressed as products of second order correlations. In addition, the contributions from the triple nonlinearity $\left\langle u^<(\hat{k})u_j^<(\hat{p})u_m^<(\hat{q})\right\rangle$ are neglected (Zhou et al., 1988). Zhou et al. (1988) and Zhou (2010) performed a more refined calculation to take into account the effects of triple nonlinearity, but this topic is beyond the scope of this appendix.

2. To first order, the second bracketed term of Eq. (C.3) vanishes, but the nonvanishing third bracketed term yields corrections to $\nu_{(0)}$ (McComb, 1990, 2014). Consequently, Eq. (C.3) gets transformed to

$$\begin{aligned}&\left[-i\omega \;+\; (\nu_{(0)}(k)+\delta\nu_{(0)}(k))k^2\right]u_i^<(\hat{k}) = \\ &\qquad -\frac{i}{2}P_{ijm}(\mathbf{k})\int_{\hat{p}+\hat{q}=\hat{k}}\frac{d\mathbf{p}d\omega'}{(2\pi)^{d+1}}[u_j^<(\hat{p})u_m^<(\hat{k}-\hat{p})]+F_i^<(\hat{k})\end{aligned} \tag{C.6}$$

with

$$\delta\nu_{(0)}(\hat{k})k^2 = \frac{1}{d-1}\int_{\hat{p}+\hat{q}=\hat{k}}^{\Delta}\frac{d\mathbf{p}d\omega'}{(2\pi)^{d+1}}[B(k,p,q)G(\hat{q})C(\hat{p})], \tag{C.7}$$

where

$$\begin{aligned}B(k,p,q) &= P_{ijm}(\mathbf{k})P_{mai}(\mathbf{p})P_{ja}(\mathbf{q}) \\ &= kp[(d-3)z+2z^3+(d-1)xy]\end{aligned} \tag{C.8}$$

with d is the space dimensionality that is taken to be 3 in this appendix, x, y, z are the direction cosines of $\mathbf{k}, \mathbf{p}, \mathbf{q}$ (McComb, 1990; Verma, 2004). It is important to note that the helical component of Eq. (C.5) does not contribute to Eq. (C.6) because the tensor corresponding to the helical component is of the following form:

$$\begin{aligned}B'(k,p,q)G(\hat{q})H_K(\hat{p}) &= -\frac{i}{q^2}P_{ijm}(\mathbf{k})P_{mai}(\mathbf{p})q_l\epsilon_{jal}G(\hat{q})H_K(\hat{p}) \\ &= -\frac{i}{q^2}[k_jP_{im}(\mathbf{k})+k_mP_{ij}(\mathbf{k})][p_aP_{mi}(\mathbf{p})+p_iP_{ma}(\mathbf{p})] \\ &\quad \times q_l\epsilon_{jal}G(\hat{q})H_K(\hat{p}) \\ &= 0.\end{aligned} \tag{C.9}$$

due to the asymmetric nature of ϵ_{jal}. Thus, $\delta\nu$ is independent of the kinetic helicity H_K.

Green's function $G(\hat{q})$ is defined as

$$G(\hat{q}) = \frac{1}{-i\omega'' + \nu_{(0)}(q)q^2}. \tag{C.10}$$

In addition, we assume that the correlation function and Green's function have the same frequency dependence (McComb, 1990). Hence, the correlation function $C(\hat{p})$ is defined as

$$C(\hat{p}) = \frac{C(\mathbf{p})}{-i\omega' + \nu_{(0)}(p)p^2}, \tag{C.11}$$

where $C(\mathbf{p})$ is the modal kinetic energy.

3. We substitute this Green's function and the correlation function in Eq. (C.7), and take the limit $\omega \to 0$ (long time scale) that yields

$$\delta\nu_{(0)}(k)k^2 = \frac{1}{d-1}\int_{\mathbf{p}+\mathbf{q}=\mathbf{k}}^{\Delta} \frac{d\mathbf{p}}{(2\pi)^d} \frac{B(k,p,q)C(\mathbf{p})}{\nu_{(0)}(p)p^2 + \nu_{(0)}(q)q^2}. \tag{C.12}$$

4. We perform the integral of Eq. (C.12) over the first shell (k_1, k_0) that yields the following renormalized viscosity after the first step:

$$\nu_{(1)}(k) = \nu_{(0)}(k) + \delta\nu_{(0)}(k). \tag{C.13}$$

We keep eliminating the shells one after the other in a similar manner. After $n+1$ iterations, we obtain

$$\nu_{(n+1)}(k) = \nu_{(n)}(k) + \delta\nu_{(n)}(k), \tag{C.14a}$$

$$\delta\nu_{(n)}(k)k^2 = \frac{1}{d-1}\int_{\mathbf{p}+\mathbf{q}=\mathbf{k}}^{\Delta} \frac{d\mathbf{p}}{(2\pi)^d} \frac{B(k,p,q)C(\mathbf{p})}{\nu_{(n)}(p)p^2 + \nu_{(n)}(q)q^2}, \tag{C.14b}$$

with the integration performed over the nth shell.

5. We compute Eqs. (C.14a, C.14b) self-consistently for which we employ Kolmogorov's spectrum for energy, that is,

$$C(\mathbf{p}) = \frac{2(2\pi)^d}{S_d(d-1)} p^{-(d-1)} E(p); \quad E(p) = K_{\mathrm{Ko}} \Pi^{2/3} p^{-5/3}, \tag{C.15}$$

where S_d is the surface area of a d-dimensional sphere of unit radius, and $E(p)$ is the one-dimensional Kolmogorov's spectrum. In addition, we attempt the following form of solution for $\nu_{(n)}(k)$:

$$\nu_{(n)}(k) = \nu_{(n)}(k_n k') = (K_{\mathrm{Ko}})^{1/2} \Pi^{1/3} k_n^{-4/3} \nu_{*(n)}(k') \tag{C.16}$$

with $k = k_n k'$ and $k' < 1$. We expect $\nu_{*(n)}(k')$ to be a universal function for large n. Here, we follow the notation of Verma (2004).

Substitutions of these forms of $C(\mathbf{p})$ and $\nu_{(n)}(k)$ in Eqs. (C.14a, C.14b) yields the following equations:

$$\delta\nu_{*(n)}(k') = \frac{1}{(d-1)} \int_{\mathbf{p}'+\mathbf{q}'=\mathbf{k}'} d\mathbf{q}' \frac{2}{(d-1)S_d} \frac{E^u(q')}{q'^{d-1}} \times \left[\frac{S(k',p',q')}{\nu_{*(n)}(hp')p'^2 + \nu_{*(n)}(hq')q'^2} \right], \tag{C.17a}$$

$$\nu_{*(n+1)}(k') = h^{4/3}\nu_{*(n)}(hk') + h^{-4/3}\delta\nu_{*(n)}(k'), \tag{C.17b}$$

where the integral in this equation is performed over a region $1 \le p', q' \le 1/h$ with the constraint $\mathbf{p}' + \mathbf{q}' = \mathbf{k}'$. Note that $\mathbf{k}' = \mathbf{k}/k_n$, $\mathbf{p}' = \mathbf{p}/k_n$, $\mathbf{q}' = \mathbf{q}/k_n$. Fournier and Frisch (1978) showed that the above volume integral in d dimensions is

$$\int_{\mathbf{p}'+\mathbf{q}'=\mathbf{k}'} d\mathbf{p}' = S_{d-1} \int dp' dq' \left(\frac{p'q'}{k'}\right)^{d-2} (\sin\alpha)^{d-3}, \tag{C.18}$$

where α is the angle between vectors $\mathbf{p}'$ and $\mathbf{q}'$.

6. We solve for $\nu_{*(n)}(k')$ iteratively using Eqs. (C.17). For h in the range 0.55 to 0.75, Zhou et al. (1997) showed that the Kolmogorov constant computed using RG is approximately 1.6 independent of h. Therefore, we choose $h = 0.7$ for our computation. We start with a constant value of $\nu_{*(0)}(k')$, and compute the integral using Gaussian quadrature. We iterate the above process till it converges, that is, $\nu_{*(n+1)}(k') \approx \nu_{*(n)}(k')$. The result of our RG analysis, exhibited in Fig. C.2, shows a constancy of $\nu_*(k')$ with k'.

For large n, $\nu_{*(n)}(k')$ converges asymptotically to $\nu_* \approx 0.38$ as $k' \to 0$ (see Fig. C.2). A downward profile of $\nu_{*(n+1)}(k')$ near $k' = 1$ is attributed to the neglect of the triple nonlinearity of the unresolved modes (see item 1, and Zhou et al. (1988)).

Thus, Kolmogorov's spectrum and renormalized viscosity of Eq. (C.16) are self-consistent solutions of the RG equations. Here we also show that the renormalized viscosity is independent of kinetic helicity.

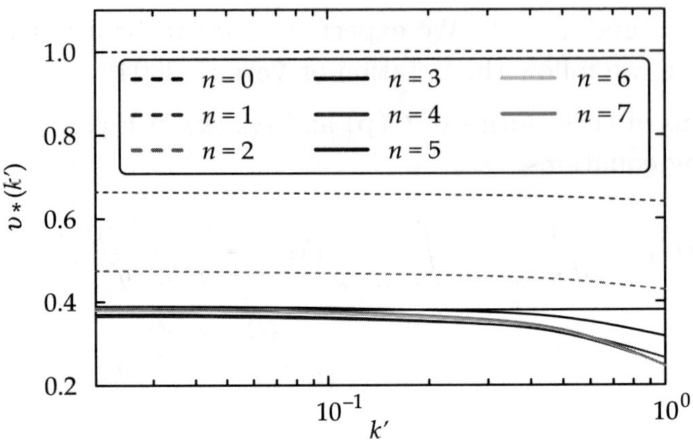

Figure C.2 Plot of $\nu_*(k')$ vs k' for various RG iterations (n). The function approaches $\nu_*(k') \approx 0.38$ (the black horizontal line) asymptotically.

Further Reading

There are a large number of books and reviews on the renormalization group treatment of turbulence. We refer the reader to the following leading articles and books: Yakhot and Orszag (1986), McComb (1990), Zhou (2010), and references therein.

Notation

Fields and Fourier transform:

$\mathbf{r}$	Real space coordinate
$\mathbf{k}$	Wavenumber in Fourier space
k	Magnitude of $\mathbf{k}$
$f(\mathbf{k})$	Fourier transforms of $f(\mathbf{r})$
$k_{\parallel}, k_z$	Component of $\mathbf{k}$ along $\hat{z}$
$k_{\perp}$	$\sqrt{k_x^2 + k_y^2}$

Miscellaneous Variables:

ζ	Polar angle in spherical coordinate system
ϕ	Azimuthal angle in spherical coordinate system
ϕ	Polar angle in 2D coordinate system (for Craya–Herring basis)
$\hat{e}_{1,2,3}$	Unit vectors of Craya–Herring basis
$u_{1,2,3}$	The components of the velocity field in Craya–Herring basis
$\hat{e}_{\pm}$	Unit vectors of helical basis
$u_{\pm}$	The components of the velocity field in helical basis
β	Sector index of a ring
$\Re(.)$	Real part of the argument
$\Im(.)$	Imaginary part of the argument

Velocity, pressure, scalar, vector, tensor, and magnetic fields:

$\mathbf{u}$	Velocity field
$u_{\parallel}$	Velocity field parallel to the anisotropy direction
$\mathbf{u}_{\perp}$	Velocity field perpendicular to the anisotropy direction
$\boldsymbol{\omega}$	Vorticity field
θ	Scalar field (temperature, density, etc.)
$\mathbf{w}$	Vector field
$\mathcal{C}$	Tensor field
$\mathbf{b}$	Magnetic field fluctuation
$\mathbf{B}$	Total magnetic field (mean $\mathbf{B}_0$ + fluctuation)
$\mathbf{z}^{\pm}$	Elsässer variables ($\mathbf{u} \pm \mathbf{b}$)
$\mathbf{A}$	Vector potential
$\mathbf{F}_u, \mathbf{F}_\omega$	Forcing field for the velocity/vorticity field
$F_\theta, \mathbf{F}_w, F_\mathcal{C}$	Forcing field for the scalar/vector/tensor field
p	Pressure field
ν, μ	Kinematic/Dynamic viscosity
κ	Diffusivity of scalar field
η	Diffusivity of vector/tensor/magnetic field
U	Large scale velocity ($= U_{\mathrm{rms}}$)
ϵ_{u,ω,H_K}	Viscous/enstrophy/kinetic helicity dissipation rate

Definitions of quadratic quantities:

$$\text{Kinetic energy (KE) density } E_u(\mathbf{r}) = \frac{1}{2}u^2$$

$$\text{KE density in Craya–Hering basis } E_{u_i}(\mathbf{r}) = \frac{1}{2}u_i^2 \quad (i = 1, 2, 3)$$

$$\text{KE density in helical basis } E_{u_\pm}(\mathbf{r}) = \frac{1}{2}u_\pm^2$$

$$\text{Kinetic helicity density } H_K(\mathbf{r}) = \frac{1}{2}\mathbf{u} \cdot \boldsymbol{\omega}$$

$$\text{Magnetic helicity density } H_M(\mathbf{r}) = \frac{1}{2}\mathbf{A}\cdot\mathbf{B}$$

$$\text{Enstrophy density } E_\omega(\mathbf{r}) = \frac{1}{2}\omega^2$$

$$\text{Magnetic energy density } E_b(\mathbf{r}) = \frac{1}{2}b^2$$

$$\text{Energy density of } \mathbf{z}^\pm : E_{z^\pm}(\mathbf{r}) = \frac{1}{2}(z^\pm)^2$$

$$\text{Cross helicity density } H_c(\mathbf{r}) = \frac{1}{2}\mathbf{u}\cdot\mathbf{B}$$

$$\text{Mean square vector potential density } E_A(\mathbf{r}) = \frac{1}{2}A^2$$

$$\text{Total of quadratic quantities (real space) } (f,g): \quad \int d\mathbf{r}\frac{1}{2}(fg)$$

$$\text{Modal energy } (f,g): \quad \frac{1}{2}\Re[f(\mathbf{k})g^*(\mathbf{k})]$$

$$\text{Total of quadratic quantities (Fourier space) } (f,g): \quad \sum_{\mathbf{k}}\frac{1}{2}\Re[f(\mathbf{k})g^*(\mathbf{k})]$$

$$\text{KE parallel/perp to the anisotropy direction} = E_{u_{\parallel,\perp}} = \int d\mathbf{r}\frac{1}{2}|u_{\parallel,\perp}|^2$$

Energy spectrum:

$E_f(\mathbf{k})$	Modal energy of f
$\mathcal{F}_f(\mathbf{k})$	Energy supply rate to f by external force
$E_f(k)$	One-dimensional energy spectrum for f
$E_{f,\parallel}(k)$	Energy spectrum of f parallel to anisotropy direction
$E_{f,\perp}(k)$	Energy spectrum of f perpendicular to anisotropy direction
$E_f(k,\beta)$	Ring spectrum of f in spherical system (β= sector index)
$E_f(k,i)$	Ring spectrum of f in cylindrical system (i = height index)

Mode-to-mode transfer:

$S^{XY}(\mathbf{k}'\|\mathbf{p}\|\mathbf{q})$	Mode-to-mode energy transfer from field $Y(\mathbf{p})$ to field $X(\mathbf{k}')$
$S^{H_K,H_M}(\mathbf{k}'\|\mathbf{p}\|\mathbf{q})$	Mode-to-mode transfer of kinetic helicity/ magnetic helicity
$S^{XY}(\mathbf{k}'\|\mathbf{p},\mathbf{q})$	Combined energy transfer from field $Y(\mathbf{p})$ and $Y(\mathbf{q})$ to field $X(\mathbf{k}')$

Energy transfers:

$\Pi_f(k)$	Spectral flux of quadratic quantity f
$T_{Y,n}^{X,m}$	Shell-to-shell transfer from field X of shell m to field Y of shell n
$T_{Y,B}^{X,A}$	Transfer of quadratic quantity from field X of region A to field Y of region B
$T_{(Y,n,\beta)}^{(X,m,\alpha)}$	Ring-to-ring transfer from field X of ring (m,α) to field Y of (n,β) (spherical)
$T_{(Y,n,h_2)}^{(X,m,h_1)}$	Ring-to-ring transfer from field X of ring (m,h_1) to field Y of (n,h_2) (cylindrical)
$\mathcal{P}_\parallel(\mathbf{k}')$	Kinetic energy gained by $u_\parallel(\mathbf{k})$ from $\mathbf{u}_\perp(\mathbf{k})$ via pressure
$\mathcal{P}_\perp(\mathbf{k}')$	Kinetic energy gained by $\mathbf{u}_\perp(\mathbf{k})$ from $u_\parallel(\mathbf{k})$ via pressure

System parameters:

λ	Taylor's microscale
Re	Reynolds number (UL/ν)
Re_λ	Reynolds number based on Taylor scale $(U\lambda/\nu)$
Re_M	Magnetic Reynolds number (UL/η)
Pe	Péclet number (UL/κ)
Pr	Prandtl number (ν/κ)
Sc	Schmidt number (ν/κ)
Ra	Rayleigh number $(\alpha g \Delta d^3/(\nu\kappa))$
Ro	Rossby number $(U/(\Omega L))$
N	Brunt Väisälä frequency (in Chapter 15)
N	Interaction parameter (in Chapter 24)

Structure functions:

l	Distance between the two points $[\mathbf{u}(\mathbf{r}), \mathbf{u}(\mathbf{r}+\mathbf{l})]$
$S_q(l)$	qth order Structure function for the velocity field
$S_q^\theta(l)$	qth order Structure function for the scalar field θ

$S_q^{z^{\pm}}(l)$	qth order Structure function for the Elsässer variables of MHD
$C(\mathbf{l})$	Real space second order correlation of the fields

Shell model:

k_n	Wavenumber of shell n
U_n	Velocity field of shell n
θ_n	Scalar/vector/tensor field of shell n

References

Adzhemyan, L. Ts, Nikolai V. Antonov, and Alexander N. Vasiliev. 1999. *Field Theoretic Renormalization Group in Fully Developed Turbulence.* CRC Press.

Ahlers, Guenter, Siegfried Grossmann, and Detlef Lohse. 2009. “Heat Transfer and Large-scale Dynamics in Turbulent Rayleigh–Bénard Convection.” *Reviews of Modern Physics* 81 (2): 503.

Alam, Shadab, Anirban Guha, and Mahendra K. Verma. 2018. “Revisiting Bolgiano-Obukhov Scaling for Stably Stratified Turbulence.” *arXiv preprint arXiv:1811.12848.*

Alexakis, Alexandros, and Luca Biferale. 2018. “Cascades and Transitions in Turbulent Flows.” *Physics Reports* 767–769: 1.

Alexakis, Alexandros, Pablo D. Mininni, and Annick Pouquet. 2005a. “Imprint of Large-scale Flows on Turbulence.” *Physical Review Letters* 95 (26): 264503.

Alexakis, Alexandros, Pablo D. Mininni, and Annick Pouquet. 2005b. “Shell-to-shell Energy Transfer in Magnetohydrodynamics. I. Steady State Turbulence.” *Physical Review E* 72 (4): 046301.

Alexakis, Alexandros, Pablo D. Mininni, and Annick Pouquet. 2007. “Turbulent Cascades, Transfer, and Scale Interactions in Magnetohydrodynamics.” *New Journal of Physics* 9 (8): 298.

Amit, Hagay, and Peter Olson. 2010. “A Dynamo Cascade Interpretation of the Geomagnetic Dipole Decrease.” *Geophysical Journal International* 181 (3): 1411–27.

Arnold, Vladimir I. 1989. *Mathematical Methods of Classical Mechanics.* 2nd edn. New York: Springer-Verlag.

Avinash, V., Mahendra K. Verma, and Amar V. Chandra. 2006. "Field-theoretic Calculation of Kinetic Helicity Flux." *Pramana* 66 (2): 447–53.

Basu, Abhik, Anirban Sain, Sujan K. Dhar, and Rahul Pandit. 1998. "Multiscaling in Models of Magnetohydrodynamic Turbulence." *Physical Review Letters* 81 (13): 2687.

Batchelor, George Keith. 1953. *The Theory of Homogeneous Turbulence.* Cambridge University Press.

Batchelor, George K. 1959. "Small-scale Variation of Convected Quantities Like Temperature in Turbulent Fluid Part 1. General Discussion and the Case of Small Conductivity." *Journal of Fluid Mechanics* 5 (1): 113–33.

Batchelor, G. K., I. D. Howells, and A. A. Townsend. 1959. "Small-scale Variation of Convected Quantities Like Temperature in Turbulent Fluid Part 2. The Case of Large Conductivity." *Journal of Fluid Mechanics* 5 (1): 134–39.

Benzi, Roberto, and Emily SC Ching. 2018. "Polymers in Fluid Flows." *Annual Review of Condensed Matter Physics* 9: 163–181.

Beresnyak, Andrey. 2011. "Spectral Slope and Kolmogorov Constant of MHD Turbulence." *Physical Review Letters* 106 (7): 075001.

Berhanu, Michaël, Gautier Verhille, Jean Boisson, Basile Gallet, Christophe Gissinger, Stéphan Fauve, Nicolas Mordant et al. 2010. "Dynamo Regimes and Transitions in the VKS Experiment." *The European Physical Journal B* 77 (4): 459–468.

Bhattacharya, Shashwat, Ambrish Pandey, Abhishek Kumar, and Mahendra K. Verma. 2018a. "Complexity of Viscous Dissipation in Turbulent Thermal convection." *Physics of Fluids* 30 (3): 031702.

Bhattacharya, Shashwat, Shubhadeep Sadhukhan, Anirban Guha, and Mahendra K. Verma. 2018b. "Structure Functions of Turbulent Thermal Convection." *arXiv*: 1903.00871.

Biferale, Luca. 2003. "Shell Models of Energy Cascade in Turbulence." *Annual Review of Fluid Mechanics* 35 (1): 441–68.

Biferale, L., S. Musacchio, and F. Toschi. 2013. "Split Energy–Helicity Cascades in Three-dimensional Homogeneous and Isotropic Turbulence." *Journal of Fluid Mechanics* 730: 309–27.

Biferale, Luca, Fabio Bonaccorso, Irene M. Mazzitelli, Michel AT van Hinsberg, Alessandra S. Lanotte, Stefano Musacchio, Prasad Perlekar, and Federico Toschi. 2016. "Coherent Structures and Extreme Events in Rotating Multiphase Turbulent Flows." *Physical Review X* 6 (4): 041036.

Biskamp, D., E. Schwarz, and James F. Drake. 1996. "Two-dimensional Electron Magnetohydrodynamic Turbulence." *Physical Review Letters* 76 (8): 1264.

Biskamp, Dieter. 2003. *Magnetohydrodynamic Turbulence.* Cambridge: Cambridge University Press.

Biskamp, Dieter, and Wolf-Christian Müller. 2000. "Scaling Properties of Three-dimensional Isotropic Magnetohydrodynamic Turbulence." *Physics of Plasmas* 7 (12): 4889–4900.

Biskamp, D., E. Schwarz, A. Zeiler, A. Celani, and J. F. Drake. 1999. "Electron Magnetohydrodynamic Turbulence." *Physics of Plasmas* 6 (3): 751–58.

Bodenschatz, Eberhard, Werner Pesch, and Guenter Ahlers. 2000. "Recent Developments in Rayleigh-Bénard Convection." *Annual Review of Fluid Mechanics* 32 (1): 709–78.

Boffetta, Guido, and Robert E. Ecke. 2012. "Two-dimensional Turbulence." *Annual Review of Fluid Mechanics* 44 (1): 427–51.

Boldyrev, Stanislav. 2006. "Spectrum of Magnetohydrodynamic Turbulence." *Phys. Rev. Lett.*, 96: 115002.

Bolgiano Jr, R. 1959. "Turbulent Spectra in a Stably Stratified Atmosphere." *Journal of Geophysical Research* 64 (12): 2226–29.

Boyd, John P. 2003. *Chebyshev and Fourier Spectral Methods.* 2nd Revised edn. New York: Dover Publications.

Brachet, Marc. 2000. "A Primer in Classical Turbulence Theory." In *Instabilities and Nonequilibrium Structures VI*, pp. 5–34. Dordrecht: Springer.

Brandenburg, Axel. 2001. "The Inverse Cascade and Nonlinear Alpha-effect in Simulations of Isotropic Helical Hydromagnetic Turbulence." *The Astrophysical Journal* 550 (2): 824.

Brandenburg, Axel, and Tina Kahniashvili. 2017. "Classes of Hydrodynamic and Magnetohydrodynamic Turbulent Decay." *Physical Review Letters* 118 (5): 055102.

Brandenburg, Axel, and Kandaswamy Subramanian. 2005. "Astrophysical Magnetic Fields and Nonlinear Dynamo Theory." *Physics Reports* 417 (1-4): 1–209.

Brandenburg, Axel, Kandaswamy Subramanian, André Balogh, and Melvyn L. Goldstein. 2011. "Scale Dependence of Magnetic Helicity in the Solar Wind." *The Astrophysical Journal* 734 (1): 9.

Canuto, Claudio, Yousuff Hussaini, Alfio Quarteroni, and Thomas Zang. 1988. *Spectral Methods in Fluid Dynamics.* Berlin Heidelberg: Springer-Verlag.

Cardin, Philippe, and L. G. Cugliandolo. 2011. *Dynamos: Lecture Notes of the Les Houches Summer School 2007.* Vol. 88. Elsevier.

Chakrabarti, Bikas K., Anirban Chakraborti, Satya R. Chakravarty, and Arnab Chatterjee. 2013. *Econophysics of Income and Wealth Distributions.* Cambridge: Cambridge University Press.

Chandra, Mani, and Mahendra K. Verma. 2011. "Dynamics and Symmetries of Flow Reversals in Turbulent Convection." *Physical Review E* 83 (6): 067303.

Chandra, Mani, and Mahendra K. Verma. 2013. "Flow Reversals in Turbulent Convection via Vortex Reconnections." *Physical Review Letters* 110 (11): 114503.

Chandrasekhar, Subrahmanyan. 2013. *Hydrodynamic and Hydromagnetic Stability.* Oxford: Oxford University Press.

Chatterjee, Anando G., Mahendra K. Verma, Abhishek Kumar, Ravi Samtaney, Bilel Hadri, and Rooh Khurram. 2018. "Scaling of a Fast Fourier Transform and a Pseudo-spectral Fluid Solver up to 196608 cores." *Journal of Parallel and Distributed Computing* 113: 77–91.

Chillà, F., and J. Schumacher. 2012. "New Perspectives in Turbulent Rayleigh–Bénard Convection." *The European Physical Journal E* 35 (7): 58.

Cho, Jungyeon. 2016. "Forward and Inverse Cascades in EMHD Turbulence." In *Journal of Physics: Conference Series*, 719 (1): 012001. IOP Publishing.

Choudhuri, Arnab Rai. 1998. *The Physics of Fluids and Plasmas: An Introduction for Astrophysicists.* Cambridge: Cambridge University Press.

Cowling, T. G. 1976. *Magnetohydrodynamics.* London: Adam Hilger.

Craya, Antoine. 1958. *Contribution à l'analyse de la turbulence associée à des vitesses moyennes.* Ph.D. thesis, Université de Granoble.

Cross, Michael, and Pierre C. Hohenberg. 1993. "Pattern Formation Outside of Equilibrium." *Reviews of Modern Physics* 65 (3): 851.

Cross, Michael, and Henry Greenside. 2009. *Pattern Formation and Dynamics in Nonequilibrium Systems.* Cambridge: Cambridge University Press.

Dallas, Vassilios, Stephan Fauve, and Alexandros Alexakis. 2015. "Statistical Equilibria of Large Scales in Dissipative Hydrodynamic Turbulence." *Physical Review Letters* 115 (20): 204501.

Dar, Gaurav, Mahendra K. Verma, and Vinayak Eswaran. 2001. “Energy Transfer in Two-dimensional Magnetohydrodynamic Turbulence: Formalism and Numerical Results.” *Physica D: Nonlinear Phenomena* 157 (3): 207–25.

Das, Amita, Sharad K. Yadav, Predhiman Kaw, and Sudip Sengupta. 2011 “Collisionless Stopping of Electron Current in an Inhomogeneous Electron Magnetohydrodynamics Plasma.” *Pramana* 77 (5): 949–57.

Davidson, Peter Alan. 2004. *Turbulence: An Introduction for Scientists and Engineers.* Oxford: Oxford University Press.

Davidson, Peter Alan. 2013. *Turbulence in Rotating, Stratified and Electrically Conducting Fluids.* Cambridge: Cambridge University Press.

Davidson, Peter Alan. 2017. *An Introduction to Magnetohydrodynamics.* Second edn. Cambridge: Cambridge University Press.

De Gennes, Pierre-Gilles. 1979. *Scaling Concepts in Polymer Physics.* Cornell University Press.

Debliquy, Olivier, Mahendra K. Verma, and Daniele Carati. 2005. “Energy Fluxes and Shell-to-shell Transfers in Three-dimensional Decaying Magnetohydrodynamic Turbulence.” *Physics of Plasmas* 12 (4): 042309.

Ditlevsen, Peter D. 2010. *Turbulence and Shell Models.* Cambridge: Cambridge University Press.

Dobrowolny, M., A. Mangeney, and P. Veltri. 1980. “Fully Developed Anisotropic Hydromagnetic Turbulence in Interplanetary Space.” *Physical Review Letters* 45 (2): 144.

Domaradzki, J. Andrzej. 1992. “Nonlocal Triad Interactions and the Dissipation Range of Isotropic Turbulence.” *Physics of Fluids A: Fluid Dynamics* 4 (9): 2037–45.

Domaradzki, J. Andrzej, and Robert S. Rogallo. 1990. “Local Energy Transfer and Nonlocal Interactions in Homogeneous, Isotropic Turbulence.” *Physics of Fluids A: Fluid Dynamics* 2 (3): 413–26.

Domaradzki, J. Andrzej, Bogdan Teaca, and Daniele Carati. 2009. “Locality Properties of the Energy Flux in Turbulence.” *Physics of Fluids* 21 (2): 025106.

Eyink, Gregory L., and Hussein Aluie. 2009. “Localness of Energy Cascade in Hydrodynamic Turbulence. I. Smooth Coarse Graining.” *Physics of Fluids* 21 (11): 115107.

Falkovich, Gregory. 1994. “Bottleneck Phenomenon in Developed Turbulence.” *Physics of Fluids* 6 (4): 1411–14.

Falkovich, Gregory, K. Gawędzki, and Massimo Vergassola. 2001. "Particles and Fields in Fluid Turbulence." *Reviews of Modern Physics* 73 (4): 913.

Ferziger, Joel H, and Milovan Peric. 2001. *Computational Methods for Fluid Dynamics*. 3 edn. Berlin Heidelberg: Springer-Verlag.

Fjϕrtoft, Ragnar. 1953. "On the Changes in the Spectral Distribution of Kinetic Energy for Two-dimensional, Nondivergent Flow." *Tellus* 5 (3): 225–30.

Fournier, J. D., and Uriel Frisch. 1978. "D-dimensional Turbulence." *Phys. Rev. A*, 17 (2): 747–62.

Fouxon, A., and V. Lebedev. 2003. "Spectra of Turbulence in Dilute Polymer Solutions." *Physics of Fluids* 15 (7): 2060–72.

Frisch, Uriel. 1995. *Turbulence: The Legacy of A. N. Kolmogorov.* Cambridge: Cambridge University Press.

Frisch, Uriel, and G. Parisi. 1985. On the Singularity Structure of Fully Developed Turbulence. In *Turbulence and Predictability in Geophysical uid Dynamics and Climate Dynamics.*

Frisch, Uriel, Pierre-Louis Sulem, and Mark Nelkin. 1978. "A Simple Dynamical Model of Intermittent Fully Developed Turbulence." *Journal of Fluid Mechanics* 87 (4): 719–36.

Gailitis, Agris, Gunter Gerbeth, Thomas Gundrum, Olgerts Lielausis, Ernests Platacis, and Frank Stefani. 2008. "History and Results of the Riga Dynamo Experiments." *C. R. Physique*, 9 (7): 721–28.

Galtier, Sébastien. 2003. "Weak Inertial-wave Turbulence Theory." *Physical Review E* 68 (1): 015301.

Galtier, Sébastien, and Supratik Banerjee. 2011. "Exact Relation for Correlation Functions in Compressible Isothermal Turbulence." *Phys. Rev. Lett.*, 96, 134501.

Galtier, Sébastien, Sergei V. Nazarenko, Alan C. Newell, and Annick G. Pouquet. 2000. "A Weak Turbulence Theory for Incompressible Magnetohydrodynamics." *J. Plasma Phys.*, 63 (Jan.): 447–88.

Getling, Alexander V. 1998. *Rayleigh-Bnard Convection: Structures and Dynamics.* Singapore: World Scientific.

Girimaji, Sharath S, and Ye Zhou. 1995. "Spectrum and Energy Transfer in Steady Burgers Turbulence." *Phys. Lett. A*, 202 (4): 279–87.

Goldreich, P., and S. Sridhar. 1995. "Toward a Theory of Interstellar Turbulence. 2: Strong Alfvenic Turbulence." *The Astrophysical Journal* 438: 763–75.

Goldstein, Melvyn L., D. Aaron Roberts, and W. H. Matthaeus. 1995. "Magnetohydrodynamic Turbulence in the Solar Wind." *Annual Review of Astronomy and Astrophysics* 33 (1): 283–325.

Gotoh, T., P. K. Yeung, P. A. Davidson, Y. Kaneda, and K. R. Sreenivasan. 2013. "Passive Scalar Transport in Turbulence: A Computational Perspective." In *Ten Chapters in Turbulence* 87–131.

Gotoh, Toshiyuki, Takeshi Watanabe, and Hideaki Miura. 2014. "Spectrum of Passive Scalar at Very High Schmidt Number in Turbulence." *Plasma and Fusion Research*, 9: 3401019.

Greenspan, H. P. 1968. *The Theory of Rotating Fluids.* Cambridge: Cambridge University Press.

Gupta, Akansha, Rohith Jayaram, Anando Chatterjee, Shubhadeep Shadhukhan, Ravi Samtaney, and Mahendra K. Verma. 2019. "Energy and Enstrophy Spectra and Fluxes for the Inertial-dissipation Range of Two Dimensional Turbulence." *ArXiv*: 1902.03572.

Herring, J. R. 1974. "Approach of Axisymmetric Turbulence to Isotropy." *Physics of Fluids* 17 (5): 859–72.

Hoyle, Rebecca. 2006. *Pattern Formation: An Introduction to Methods.* Cambridge: Cambridge University Press.

Iovieno, M, Gallana, L, Fraternale, F, Richardson, J D, Opher, M, and Tordella, D. 2015. "Cross and Magnetic Helicity in the Outer Heliosphere from Voyager 2 Observations." *Eur. J. Mech. B. Fluids*, 55 (Part 2): 394–401.

Iroshnikov, P S. 1964. "Turbulence of a Conducting Fluid in a Strong Magnetic Field." *Soviet Astronomy*, 7: 566–71.

Jones, Chris A. 2008. "Dynamo theory." In Les Houches Vol. 88, *Dynamo.* Elsevier.

Kalelkar, Chirag, Rama Govindarajan, and Rahul Pandit. 2005. "Drag Reduction by Polymer Additives in Decaying Turbulence." *Physical Review E* 72 (1): 017301.

Kessar, Mouloud, Franck Plunian, Rodion Stepanov, and Guillaume Balarac. 2015. "Non-Kolmogorov Cascade of Helicity-driven Turbulence." *Physical Review E* 92 (3): 031004.

Kida, Shigeo. 1979. "Asymptotic Properties of Burgers Turbulence." *Journal of Fluid Mechanics* 93 (2): 337–77.

Knaepen, Bernard, and René Moreau. 2008. "Magnetohydrodynamic Turbulence at Low Magnetic Reynolds Number." *Annu. Rev. Fluid Mech.* 40: 25–45.

Knobloch, E. 1992. "Onset of Zero Prandtl Number Convection." *Journal de Physique II* 2 (5): 995–99.

Kolmogorov, Andrey Nikolaevich. 1941a. "Dissipation of Energy in Locally Isotropic Turbulence." *Dokl Acad Nauk SSSR*, 32 (1): 16–18.

Kolmogorov, Andrey Nikolaevich. 1941b "On the Degeneration of Isotropic Turbulence in an Incompressible Viscous Fluid." *Dokl. Akad. Nauk SSSR*, 31: 319–23.

Kolmogorov, Andrey Nikolaevich. 1941c. "The Local Structure of Turbulence in Incompressible Viscous Fluid for Very Large Reynolds Numbers." *Dokl Acad Nauk SSSR*, 30 (4): 301–05.

Kolmogorov, Andrey Nikolaevich. 1962. "A Refinement of Previous Hypotheses Concerning the Local Structure of Turbulence in a Viscous Incompressible Fluid at High Reynolds Number." *Journal of Fluid Mechanics* 13 (1): 82–85.

Kraichnan, Robert H. 1959. "The Structure of Isotropic Turbulence at Very High Reynolds Numbers." *Journal of Fluid Mechanics* 5 (4): 497–543.

Kraichnan, Robert H. 1964. "Kolmogorov's Hypotheses and Eulerian Turbulence Theory." *Physics of Fluids* 7 (11): 1723–34.

Kraichnan, Robert H. 1965. "Inertial-range Spectrum of Hydromagnetic Turbulence." *Physics of Fluids* 8 (7): 1385–87.

Kraichnan, Robert H. 1967. "Inertial Ranges in Two-dimensional Turbulence." *Physics of Fluids* 10 (7): 1417–23.

Kraichnan, Robert H. 1968. "Small-scale Structure of a Scalar Field Convected by Turbulence." *Physics of Fluids* 11 (5): 945–53.

Krause, F, and K.-H. Rädler. 1980. *Mean-Field Magnetohydrodynamics and Dynamo Theory.* Oxforc: Pergamon Press.

Kreyszig, Erwin, Herbert Kreyszig, and Edward J. Norminton. 2011. *Advanced Engineering Mathematics.* 10th edn. New York: Wiley & Sons.

Krstulovic, G., P. D. Mininni, M. E. Brachet, and A. Pouquet. 2009. "Cascades, Thermalization, and Eddy Viscosity in Helical Galerkin Truncated Euler Flows." *Physical Review E* 79 (5): 056304.

Kumar, Abhishek, and Mahendra K. Verma. 2015. "Shell Model for Buoyancy-driven Turbulence." *Physical Review E* 91 (4): 043014.

Kumar, Abhishek, and Mahendra K. Verma. 2018. "Applicability of Taylor's Hypothesis in Thermally Driven Turbulence." *Royal Society Open Science* 5 (4): 172152.

Kumar, Abhishek, Anando G. Chatterjee, and Mahendra K. Verma. 2014a. "Energy Spectrum of Buoyancy-driven Turbulence." *Physical Review E* 90 (2): 023016.

Kumar, Rohit, and Mahendra K. Verma. 2017. "Amplification of Large-scale Magnetic Field in Nonhelical Magnetohydrodynamics." *Physics of Plasmas* 24 (9): 092301.

Kumar, Rohit, Mahendra K. Verma, and Ravi Samtaney. 2014b. "Energy Transfers and Magnetic Energy Growth in Small-scale Dynamo." *EPL (Europhysics Letters)* 104 (5): 54001.

Kumar, Rohit, Mahendra K. Verma, and Ravi Samtaney. 2015. "Energy Transfers in Dynamos with Small Magnetic Prandtl Numbers." *Journal of Turbulence* 16 (11): 1114–34.

Kundu, Pijush K, Ira M. Cohen, and David R. Dowling. 2015. *Fluid Mechanics.* 6th edn. San Diego: Academic Press.

Landau, Lev Davidovich, and Evgeny Mikhailovich Lifshitz. 1976. *Mechanics.* 3rd edn. Course of Theoretical Physics. Oxford: Elsevier.

Landau, Lev Davidovich, and Evgeny Mikhailovich Lifshitz. 1987. *Fluid Mechanics.* 2nd edn. Course of Theoretical Physics. Oxford: Elsevier.

Lesieur, Marcel. 2008. *Turbulence in Fluids.* Dordrecht: Springer-Verlag.

Leslie, D. C. 1973. *Developments in the Theory of Turbulence.* Oxford: Clarendon Press.

Lin, Bin-Shei, Chien C. Chang, and Chi-Tzung Wang. 2000. "Renormalization Group Analysis for Thermal Turbulent Transport." *Physical Review E* 63 (1): 016304.

Lindborg, Erik. 2006. "The Energy Cascade in a Strongly Stratified Fluid." *Journal of Fluid Mechanics* 550: 207–42.

Lohse, Detlef, and Ke-Qing Xia. 2010. "Small-scale Properties of Turbulent Rayleigh–Bénard Convection." *Annual Review of Fluid Mechanics* 42: 335–64.

L'vov, Victor S. 1991. "Spectra of Velocity and Temperature Fluctuations with Constant Entropy Flux of Fully Developed Free-convective Turbulence." *Physical Review Letters* 67 (6): 687.

L'vov, Victor S., and Gregory E. Falkovich. 1992. "Conservation Laws and Two-flux Spectra of Hydrodynamic Convective Turbulence." *Physica D: Nonlinear Phenomena* 57 (1–2): 85–95.

L'vov, Victor S., Evgenii Podivilov, and Itamar Procaccia. 1999. "Hamiltonian Structure of the Sabra Shell Model of Turbulence: Exact Calculation of an Anomalous Scaling Exponent." *EPL (Europhysics Letters)* 46 (5): 609.

Ma, Shang-Keng. 1985. *Statistical Mechanics.* Singapore: World Scientific.

Majumdar, Sayantan, and A. K. Sood. 2011. "Universality and Scaling Behavior of Injected Power in Elastic Turbulence in Wormlike Micellar Gel." *Physical Review E* 84 (1): 015302.

Manneville, Paul. 2014. *Dissipative Structures and Weak Turbulence.* San Diego: Academic Press.

Marsch, Eckart. 1991. Turbulence in the Solar Wind. Pages 145–156 of: Klare, G. (ed), *Reviews in Modern Astronomy.* Berlin, Heidelberg: Springer.

Mason, Joanne, Fausto Cattaneo, and Stanislav Boldyrev. 2006. "Dynamic Alignment in Driven Magnetohydrodynamic Turbulence." *Physical Review Letters* 97 (25): 255002.

Mathieu, Jean, and Julian Scott. 2000. *An Introduction to Turbulent Flow.* Cambridge University Press.

Matthaeus, William H., and Melvyn L. Goldstein. 1982. "Measurement of the Rugged Invariants of Magnetohydrodynamic Turbulence in the Solar Wind." *Journal of Geophysical Research: Space Physics* 87 (A8): 6011–28.

McComb, William D., and V. Shanmugasundaram. 1983. "Fluid Turbulence and the Renormalization Group: A Preliminary Calculation of the Eddy Viscosity." *Physical Review A* 28 (4): 2588.

McComb, William D., and V. Shanmugasundaram. 1984. "Numerical Calculation of Decaying Isotropic Turbulence Using the LET Theory." *Journal of Fluid Mechanics* 143: 95–123.

McComb, William D., W. Roberts, and A. G. Watt. 1992. "Conditional-averaging Procedure for Problems with Mode-mode Coupling." *Physical Review A* 45 (6): 3507.

McComb, William D. 1990. *The Physics of Fluid Turbulence.* Oxford University Press.

McComb, William D. 2014. *Homogeneous, Isotropic Turbulence: Phenomenology. Renormalization and Statistical Closures.* Oxford University Press.

Meneveau, Charles, and Katepalli R. Sreenivasan. 1987. "Simple Multifractal Cascade Model for Fully Developed Turbulence." *Physical review letters* 59 (13): 1424.

Mininni, Pablo, Alexandros Alexakis, and Annick Pouquet. 2005. "Shell-to-shell Energy Transfer in Magnetohydrodynamics. II. Kinematic Dynamo." *Physical Review E* 72 (4): 046302.

Mininni, Pablo D., Alexandros Alexakis, and Annick Pouquet. 2009. "Scale Interactions and Scaling Laws in Rotating Flows at Moderate Rossby Numbers and Large Reynolds Numbers." *Physics of Fluids* 21 (1): 015108.

Mishra, Pankaj Kumar, and Mahendra K. Verma. 2010. "Energy Spectra and Fluxes for Rayleigh-Bénard Convection." *Physical Review E* 81 (5): 056316.

Moffatt, Henry K. 1978. *Field Generation in Electrically Conducting Fluids.* Cambridge University Press.

Moisy, F, C. Morize, M. Rabaud, and Jöel Sommeria. 2010. "Decay Laws, Anisotropy and Cyclone–Anticyclone Ssymmetry in Decaying Rotating Turbulence." *Journal Fluid Mech.,* 666 (Oct.): 5–35.

Monchaux, Romain, Michaël Berhanu, Mickaël Bourgoin, Marc Moulin, Ph Odier, J-F. Pinton, Romain Volk et al. 2007. "Generation of a Magnetic Field by Dynamo Action in a Turbulent Flow of Liquid Sodium." *Physical Review Letters* 98 (4): 044502.

Mondal, Sudipta, V. Narayanan, Wen Jun Ding, Amit D. Lad, Biao Hao, Saima Ahmad, Wei Min Wang et al. 2012. "Direct Observation of Turbulent Magnetic Fields in Hot, Dense Laser Produced Plasmas." *Proceedings of the National Academy of Sciences* 109 (21): 8011–15.

Moreau, René J. 1990. *Magnetohydrodynamics.* Berlin: Springer.

Morin, Vincent, and Emmanuel Dormy. 2009. "The Dynamo Bifurcation in Rotating Spherical Shells." *International Journal of Modern Physics B* 23 (28n29): 5467–82.

Morize, Cyprien, Frédéric Moisy, and Marc Rabaud. 2005. "Decaying Grid-generated Turbulence in a Rotating Tank." *Physics of Fluids* 17 (9): 095105.

Müller, Wolf-Christian, and Dieter Biskamp. 2000. "Scaling Properties of Three-dimensional Magnetohydrodynamic Turbulence." *Physical Review Letters* 84 (3): 475.

Nath, Dinesh, Ambrish Pandey, Abhishek Kumar, and Mahendra K. Verma. 2016. "Near Isotropic Behavior of Turbulent Thermal Convection." *Physical Review Fluids* 1 (6): 064302.

Nazarenko, Sergey V. 2011. *Wave Turbulence.* Berlin: Springer.

Nemirovskii, Sergey K. 2013. "Quantum Turbulence: Theoretical and Numerical Problems." *Physics Reports* 524 (3): 85–202.

Newman, Mark E. J. 2005. "Power Laws, Pareto Distributions and Zipf's Law." *Contemporary Physics* 46 (5): 323–51.

Novikov, E., A. Novikov, D. Shannahoff-Khalsa, B. Schwartz, and J. Wright. 1997. "Scalesimilar Activity in the Brain." *Physical Review E* 56 (3): R2387.

Obukhov, A. M. 1959. "Effect of Archimedean Forces on the Structure of the Temperature Field in a Turbulent Flow." *Dokl. Akad. Nauk SSSR*, 125 (6): 1246–48.

Obukhov, A. M. 1962. "Some Specific Features of Atmospheric Turbulence." *Journal Geophysical Research,* 13: 77–81.

Odier, P., J-F. Pinton, and Stephan Fauve. 1998. "Advection of a Magnetic Field by a Turbulent Swirling Flow." *Physical Review E* 58 (6): 7397.

Oughton, Sean, Eric R. Priest, and William H. Matthaeus. 1994. "The Influence of a Mean Magnetic Field on Three-dimensional Magnetohydrodynamic Turbulence." *Journal of Fluid Mechanics* 280: 95–117.

Pandey, Ambrish, and Mahendra K. Verma. 2016. "Scaling of Large-scale Quantities in Rayleigh-Bénard Convection." *Physics of Fluids* 28 (9): 095105.

Pandey, Ambrish, Mahendra K. Verma, and Pankaj K. Mishra. 2014. "Scaling of Heat Flux and Energy Spectrum for Very Large Prandtl Number Convection." *Physical Review E* 89 (2): 023006.

Pao, Yih-Ho. 1965. "Structure of Turbulent Velocity and Scalar Fields at Large Wavenumbers." *Physics of Fluids* 8 (6): 1063–75.

Pao, Yih-Ho. 1968. "Transfer of Turbulent Energy and Scalar Quantities at Large Wavenumbers." *Physics of Fluids* 11 (6): 1371–72.

Pareto, Vilfredo. 1964. *Cours d'économie Politique.* Vol. 1. Librairie Droz.

Perlekar, Prasad, Dhrubaditya Mitra, and Rahul Pandit. 2006. "Manifestations of Drag Reduction by Polymer Additives in Decaying, Homogeneous, Isotropic Turbulence." *Physical Review Letters* 97 (26): 264501.

Perlekar, Prasad, Nairita Pal, and Rahul Pandit. 2017. "Two-dimensional Turbulence in Symmetric Binary-fluid Mixtures: Coarsening Arrest by the Inverse Cascade." *Scientific Reports* 7: 44589.

Plunian, Franck, Rodion Stepanov, and Peter Frick. 2013. "Shell Models of Magnetohydrodynamic Turbulence." *Physics Reports* 523 (1): 1–60.

Podesta, J. J. 2011. "On the Cross-helicity Dependence of the Energy Spectrum in Magnetohydrodynamic Turbulence." *Physics of Plasmas* 18 (1): 012907.

Politano, H, and Annick G. Pouquet. 1995. "Model of Intermittency in Magnetohydrodynamic Turbulence." *Physical Review E* 52 (1): 636–41.

Politano, H., and A. G. Pouquet. 1998. "von Kármán–Howarth Equation for Magnetohydrodynamics and its Consequences on Third-order Longitudinal Structure and Correlation Functions." *Physical Review E* 57 (1): R21.

Pope, Stephen B. 2000. *Turbulent flows.* Cambridge University Press.

Pouquet, A., U. Frisch, and J. Léorat. 1976. "Strong MHD Helical Turbulence and the Nonlinear Dynamo Effect." *Journal of Fluid Mechanics* 77 (2): 321–54.

Ramaswamy, Sriram. 2010. "The Mechanics and Statistics of Active Matter." *Annu. Rev. Condens. Matter Phys.* 1 (1): 323–45.

Ray, Samriddhi Sankar, and Dario Vincenzi. 2016. "Elastic Turbulence in a Shell Model of Polymer Solution." *EPL (Europhysics Letters)* 114 (4): 44001.

Reddy, K. Sandeep. 2015. "Anisotropic Energy Spectrum, Flux and Transfers in Quasi-static Magnetohydrodynamic Turbulence." PhD Thesis IIT, Kanpur.

Reddy, K. Sandeep, and Mahendra K. Verma. 2014. "Strong Anisotropy in Quasi-static Magnetohydrodynamic Turbulence for High Interaction Parameters." *Physics of Fluids* 26 (2): 025109.

Reddy, K. Sandeep, Raghwendra Kumar, and Mahendra K. Verma. 2014. "Anisotropic Energy Transfers in Quasi-static Magnetohydrodynamic Turbulence." *Physics of Plasmas* 21 (10): 102310.

Ristorcelli, J. R. 2018. "A Pseudo-sound Constitutive Relationship for the Dilatational Covariances in Compressible Turbulence." *J. Fluid Mech.,* 347: 37–70.

Roberts, Gareth Owen. 1972. "Dynamo Action of Fluid Motions with Two-dimensional Periodicity." *Phil. Trans. R. Soc. Lond. A* 271 (1216): 411–54.

Roberts, Paul H., and Gary A. Glatzmaier. 2000. "Geodynamo Theory and Simulations." *Reviews of Modern Physics* 72 (4): 1081.

Ruiz, Ricardo, and David R. Nelson. 1981. "Turbulence in Binary Fluid Mixtures." *Physical Review A* 23 (6): 3224.

Ruzmaikin, Alexander A, Dmitry D. Sokoloff, and A. M. Shukurov. 1988. *Magnetic Fields of Galaxies.* Dordrecht: Kluwer Academic Publishers.

Sadhukhan, Shubhadeep, Mahendra K. Verma, Rodion Stepanov, Franck Plunian, and Ravi Samtaney. 2018. Kinetic Helicity and Enstrophy Transfers in Helical Hydrodynamic Turbulence. Preprint.

Sagaut, Pierre, and Claude Cambon. 2008. *Homogeneous Turbulence Dynamics.* Cambridge: Cambridge University Press.

Saha, Meghnad, and B. N. Srivastava. 1950. *A Treatise on Heat.* 5th edn. Kolkata: Indian Press.

Sahoo, Ganapati, and Luca Biferale. 2018. "Energy Cascade and Intermittency in Helically Decomposed Navier–Stokes Equations." *Fluid Dynamics Research* 50 (1): 011420.

Sakurai, J J. 1994. *Modern Quantum Mechanics.* Revised edn. San Francisco: Addison-Wesley.

Shaikh, Dastgeer. 2009. "Whistler Wave Cascades in Solar Wind Plasma." *Monthly Notices of the Royal Astronomical Society* 395 (4): 2292–98.

Sharma, Manohar K., Mahendra K. Verma, and Sagar Chakraborty. 2018a. "On the Energy Spectrum of Rapidly Rotating Forced Turbulence." *Physics of Fluids* 30 (11): 115102.

Sharma, Manohar K., Abhishek Kumar, Mahendra K. Verma, and Sagar Chakraborty. 2018b. "Statistical Features of Rapidly Rotating Decaying Turbulence: Enstrophy and Energy Spectra and Coherent Structures." *Physics of Fluids* 30 (4): 045103.

Sharma, Manohar K, Mahendra K. Verma, and Sagar Chakraborty. 2019. Anisotropic Energy Transfer in Rapidly Rotating Forced and Decaying Turbulence. Preprint.

She, Zhen-Su, and Emmanuel Leveque. 1994. "Universal Scaling Laws in Fully Developed Turbulence." *Physical Review Letters* 72 (3): 336.

Smith, Leslie M., and Fabian Waleffe. 1999. "Transfer of Energy to Two-dimensional Large Scales in Forced, Rotating Three-dimensional Turbulence." *Physics of fluids* 11 (6): 1608–22.

Sorriso-Valvo, Luca, Raffaele Marino, Vincenzo Carbone, A. Noullez, F. Lepreti, P. Veltri, Roberto Bruno, Bruno Bavassano, and Ermanno Pietropaolo. 2007. "Observation of Inertial Energy Cascade in Interplanetary Space Plasma." *Physical Review Letters* 99 (11): 115001.

Spandan, Vamsi, Roberto Verzicco, and Detlef Lohse. 2018. "Physical Mechanisms Governing Drag Reduction in Turbulent Taylor–Couette Flow with Finite-size Deformable Bubbles." *J. Fluid Mech.,* 849: 143–13.

Spiegel, Edward A., ed. 2010. *The Theory of Turbulence: Subrahmanyan Chandrasekhar's 1954 Lectures.* Springer.

Sreenivasan, Katepalli R. 1991. "Fractals and Multifractals in Fluid Turbulence." *Annual Review of Fluid Mechanics* 23 (1): 539–604.

Sreenivasan, Katepalli R., and Christopher M. White. 2000. "The Onset of Drag Reduction by Dilute Polymer Additives, and the Maximum Drag Reduction Asymptote." *Journal of Fluid Mechanics* 409: 149–64.

Stanisic, M. M. 1984. *The Mathematical Theory of Turbulence.* New York: Springer-Verlag.

Stepanov, Rodion, and Franck Plunian. 2018. "Kinematic Dynamo in a Tetrahedron of Fourier Modes." *Fluid Dynamics Research,* 50 (5): 051409.

Stepanov, Rodion, Ephim Golbraikh, Peter Frick, and Alexander Shestakov. 2015. "Hindered Energy Cascade in Highly Helical Isotropic Turbulence." *Physical Review Letters* 115 (23): 234501.

Stieglitz, Robert, and Ulrich Müller. 2001. "Experimental Demonstration of a Homogeneous Two-scale Dynamo." *Physics of Fluids* 13 (3): 561–64.

Sundar, Sita, Mahendra K. Verma, Alexandros Alexakis, and Anando G. Chatterjee. 2017. "Dynamic Anisotropy in MHD Turbulence Induced by Mean Magnetic Field." *Physics of Plasmas* 24 (2): 022304.

Tabeling, Patrick. 2002. "Two-dimensional Turbulence: a Physicist Approach." *Physics Reports* 362 (1): 1–62.

Tabor, M., and P. G. De Gennes. 1986. "A Cascade Theory of Drag Reduction." *EPL (Europhysics Letters)* 2 (7): 519.

Taylor, Geoffrey Ingram. 1954. "The Dispersion of Matter in Turbulent Flow Through a Pipe." *Proc. R. Soc. Lond. A* 223 (1155): 446–68.

Teaca, Bogdan, Mahendra K. Verma, Bernard Knaepen, and Daniele Carati. 2009. "Energy Transfer in Anisotropic Magnetohydrodynamic Turbulence." *Physical Review E* 79 (4): 046312.

Teaca, Bogdan, Alejandro Banón Navarro, and Frank Jenko. 2014. "The Energetic Coupling of Scales in Gyrokinetic Plasma Turbulence." *Physics of Plasmas* 21 (7): 072308.

Teimurazov, A. S., R. A. Stepanov, Mahendra K. Verma, S. Barman, A. Kumar, and S. Sadhukhan. 2018. "Direct Numerical Simulation of Homogeneous Isotropic Helical Turbulence with the TARANG Code." *Journal of Applied Mechanics and Technical Physics* 59 (7): 1279–87.

Tennekes, Hendrik, and John Lumley. 1972. *A First Course in Turbulence.* MIT press.

Thais, Laurent, Thomas B. Gatski, and Gilmar Mompean. 2013. "Analysis of Polymer Drag Reduction Mechanisms from Energy Budgets." *International Journal of Heat and Fluid Flow* (43): 52–61.

Tritton, D. J. 1988. *Physical Fluid Dynamics.* Oxord: Clarendon Press.

Tu, C-Y., and Eckart Marsch. 1995. "MHD Structures, Waves and Turbulence in the Solar Wind: Observations and Theories." *Space Science Reviews* 73 (1–2): 1–210.

Tzeferacos, P., A. Rigby, A. F. A. Bott, A. Bell, R. Bingham, R. Casner, et al. 2018. "Laboratory Evidence of Dynamo Amplification of Magnetic Fields in a Turbulent Plasma." *Nature Communications,* 1–8.

Vashishtha, Sumit, Mahendra K. Verma, and Roshan Samuel. 2018. "Large-eddy Simulations of Turbulent Thermal Convection Using Renormalized Viscosity and Thermal Diffusivity." *Physical Review E* 98 (4): 043109.

Verma, Mahendra K. 1996. "Nonclassical Viscosity and Resistivity of the Solar Wind Plasma." *Journal of Geophysical Research: Space Physics* 101 (A12): 27543–27548.

Verma, Mahendra K. 1999. "Mean Magnetic Field Renormalization and Kolmogorovs Energy Spectrum in Magnetohydrodynamic Turbulence." *Physics of Plasmas* 6 (5): 1455–60.

Verma, Mahendra K. 2000. "Intermittency Exponents and Energy Spectrum of the Burgers and KPZ Equations with Correlated Noise." *Physica A* 277 (3-4): 359–88.

Verma, Mahendra K. 2001a. "Field Theoretic Calculation of Renormalized Viscosity, Renormalized Resistivity, and Energy Fluxes of Magnetohydrodynamic Turbulence." *Physical Review E* 64 (2): 026305.

Verma, Mahendra K. 2001b. "Field Theoretic Calculation of Scalar Turbulence." *International Journal of Modern Physics B* 15 (26): 3419–28.

Verma, Mahendra K. 2002. "On Generation of Magnetic Field in Astrophysical Bodies." *Current Science,* 83 (5): 620–22.

Verma, Mahendra K. 2003a. "Energy Fluxes in Helical Magnetohydrodynamics and Dynamo Action." *Pramana* 61 (4): 707–24.

Verma, Mahendra K. 2003b. "Field Theoretic Calculation of Energy Cascade Rates in Non-helical Magnetohydrodynamic Turbulence." *Pramana* 61 (3): 577–94.

Verma, Mahendra K. 2004. "Statistical Theory of Magnetohydrodynamic Turbulence: Recent Results." *Physics Reports* 401 (5-6): 229–380.

Verma, Mahendra K. 2012. "Variable Enstrophy Flux and Energy Spectrum in Two-dimensional Turbulence with Ekman Friction." *EPL (Europhysics Letters)* 98 (1): 14003.

Verma, Mahendra K. 2016. *Introduction to Mechanics.* 2nd edn. Hyderabad: Universities Press.

Verma, Mahendra K. 2017. "Anisotropy in Quasi-static Magnetohydrodynamic Turbulence." *Reports on Progress in Physics* 80 (8): 087001.

Verma, Mahendra K. 2018. *Physics of Buoyant Flows: From Instabilities to Turbulence.* Singapore: World Scientific.

Verma, Mahendra K. 2019. "Hierarchical Financial Structures with Money Cascade." In: Abergel, F, Chakrabarti, B., Chakraborti, Anirban, Deo, Nivedita, and Sharma, Kiran (eds), *New Perspectives and Challenges in Econophysics and Sociophysics.*

Verma, Mahendra K., and Diego Donzis. 2007. "Energy Transfer and Bottleneck Effect in Turbulence." *Journal of Physics A: Mathematical and Theoretical* 40 (16): 4401.

Verma, Mahendra K., and Rohit Kumar. 2016. "Dynamos at Extreme Magnetic Prandtl Numbers: Insights from Shell Models." *Journal of Turbulence* 17 (12): 1112–41.

Verma, Mahendra K., and K. Sandeep Reddy. 2015. "Modeling Quasi-static Magnetohydrodynamic Turbulence with Variable Energy Flux." *Physics of Fluids* 27 (2): 025114.

Verma, Mahendra K., and Rakesh K. Yadav. 2013. "Supercriticality to Subcriticality in Dynamo Transitions." *Physics of Plasmas* 20 (7): 072307.

Verma, Mahendra K., D. A. Roberts, and M. L. Goldstein. 1995. "Turbulent Heating and Temperature Evolution in the Solar Wind Plasma." *Journal of Geophysical Research: Space Physics* 100 (A10): 19839–50.

Verma, Mahendra K., D. A. Roberts, M. L. Goldstein, S. Ghosh, and W. T. Stribling. 1996. "A Numerical Study of the Nonlinear Cascade of Energy in Magnetohydrodynamic Turbulence." *Journal of Geophysical Research: Space Physics* 101 (A10): 21619–625.

Verma, Mahendra K., Arvind Ayyer, Olivier Debliquy, Shishir Kumar, and Amar V. Chandra. 2005. "Local shell-to-shell Energy Transfer via Nonlocal Interactions in Fluid Turbulence." *Pramana* 65 (2): 297.

Verma, Mahendra K., Thomas Lessinnes, Daniele Carati, Ioannis Sarris, Krishna Kumar, and Meenakshi Singh. 2008. "Dynamo Transition in Low-dimensional Models." *Physical Review E* 78 (3): 036409.

Verma, Mahendra K., Anando Chatterjee, K. Sandeep Reddy, Rakesh K. Yadav, Supriyo Paul, Mani Chandra, and Ravi Samtaney. 2013. "Benchmarking and Scaling Studies of Pseudospectral Code Tarang for Turbulence Simulations." *Pramana* 81 (4): 617–29.

Verma, Mahendra K., Bidya Binay Karak, and Rohit Kumar. 2013. "Dynamo in Protostars." *Pramana* 81 (6): 1037–43.

Verma, Mahendra K., Abhishek Kumar, and Ambrish Pandey. 2017. "Phenomenology of Buoyancy-driven Turbulence: Recent Results." *New Journal of Physics* 19 (2): 025012.

Verma, Mahendra K., Abhishek Kumar, Praveen Kumar, Satyajit Barman, Anando G. Chatterjee, Ravi Samtaney, and Rodion Stepanov. 2018. "Energy Spectra and Fluxes in Dissipation Range of Turbulent and Laminar Flows." *Fluid Dyn.*, 53: 728–39.

Waleffe, Fabian. 1992. "The Nature of Triad Interactions in Homogeneous Turbulence." *Physics of Fluids A: Fluid Dynamics* 4 (2): 350–63.

Wang, Jianchun, Minping Wan, Song Chen, and Shiyi Chen. 2018. "Kinetic Energy Transfer in Compressible Isotropic Turbulence." *Journal of Fluid Mechanics* 841: 581–613.

Yadav, R., M. Chandra, Mahendra K. Verma, S. Paul, and P. Wahi. 2010. "Dynamo Transition Under Taylor-Green Forcing." *EPL (Europhysics Letters)* 91 (6): 69001.

Yadav, Rakesh K., Mahendra K. Verma, and Pankaj Wahi. 2012. "Bistability and Chaos in the Taylor-Green Dynamo." *Physical Review E* 85 (3): 036301.

Yaglom, A. M. 1949. "On the Local Structure of a Temperature Field in a Turbulent Flow." In *Dokl. Akad. Nauk SSSR*, 69 (6): 743–46.

Yakhot, Victor, and Steven A. Orszag. 1986. "Renormalization Group Analysis of Turbulence. I. Basic Theory." *Journal of scientific computing* 1 (1): 3–51.

Yeung, P. K., and K. R. Sreenivasan. 2013. "Spectrum of Passive Scalars of High Molecular Diffusivity in Turbulent Mixing." *Journal of Fluid Mechanics* 716.

Yeung, P. K., S. Xu, D. A. Donzis, and K. R. Sreenivasan. 2004. "Simulations of Three-dimensional Turbulent Mixing for Schmidt Numbers of the Order 1000." *Flow, turbulence and combustion* 72 (2-4): 333–47.

Yeung, P. K., D. A. Donzis, and K. R. Sreenivasan. 2005. "High-Reynolds-number Simulation of Turbulent Mixing." *Physics of Fluids* 17 (8): 081703.

Zank, G. P., and W. H. Matthaeus. 1991. "The Equations of Nearly Incompressible Fluids. I. Hydrodynamics, Turbulence, and Waves." *Physics of Fluids A: Fluid Dynamics* 3 (1): 69–82.

Zank, G. P., and W. H. Matthaeus. 1993. "Nearly Incompressible Fluids. II: Magnetohydrodynamics, Turbulence, and Waves." *Physics of Fluids A: Fluid Dynamics* 5 (1): 257–73.

Zeman, O. 1994. "A Note on the Spectra and Decay of Rotating Homogeneous Turbulence." *Physics of Fluids* 6 (10): 3221–23.

Zhou, Ye. 1993. "Degrees of Locality of Energy Transfer in the Inertial Range." *Physics of Fluids A: Fluid Dynamics* 5 (5): 1092–94.

Zhou, Ye. 1995. "A Phenomenological Treatment of Rotating Turbulence." *Physics of Fluids* 7 (8): 2092–94.

Zhou, Ye. 2010. "Renormalization Group Theory for Fluid and Plasma Turbulence." *Physics Reports* 488 (1): 1–49.

Zhou, Ye, and George Vahala. 1993. "Renormalization-group Estimates of Transport Coefficients in the Advection of a Passive Scalar by Incompressible Turbulence." *Physical Review E* 48 (6): 4387.

Zhou, Ye, George Vahala, and Murshed Hossain. 1988. "Renormalization-group Theory for the Eddy Viscosity in Subgrid Modeling." *Physical Review A* 37 (7): 2590.

Zhou, Ye, W. David McComb, and George Vahala. 1997. "Renormalization Group (RG) in Turbulence: Historical and Comparative Perspective." NASA Technical Report ICAS–97–36

Zikanov, Oleg, Dmitry Krasnov, Thomas Boeck, André Thess, and Maurice Rossi. 2014. "Laminar-turbulent Transition in Magnetohydrodynamic Duct, Pipe, and Channel Flows." *Applied Mechanics Reviews* 66 (3): 030802.

Subject Index

Color Plates

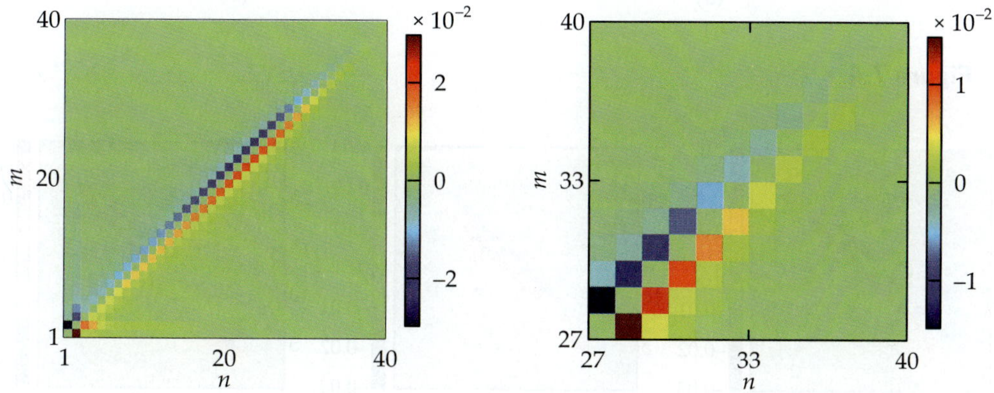

(a) (b)

Figure 5.4

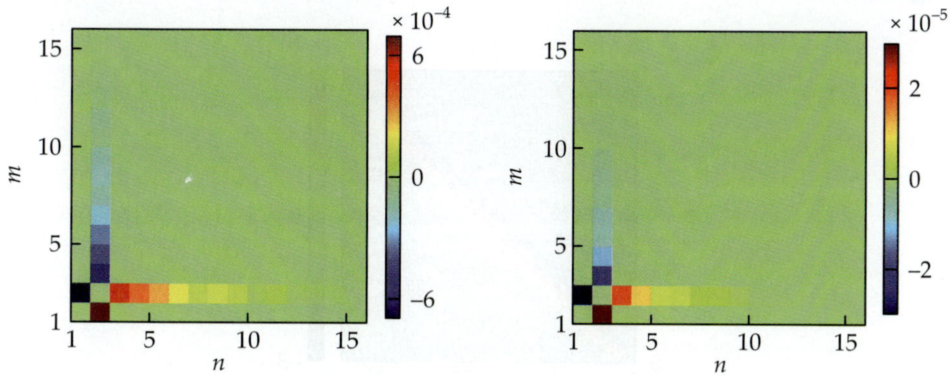

(a) (b)

Figure 5.9

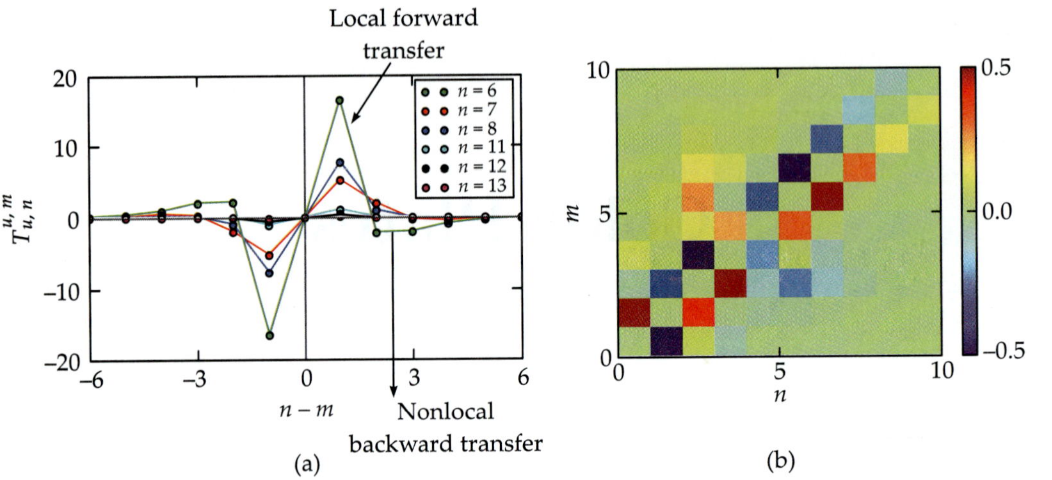

Figure 7.4

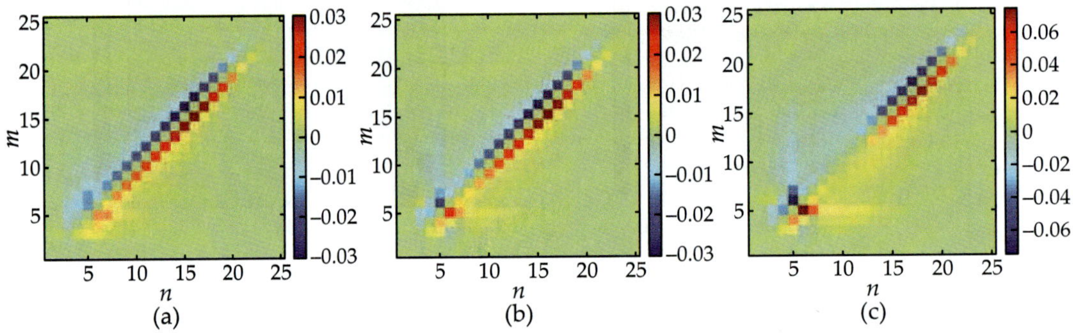

Figure 8.4

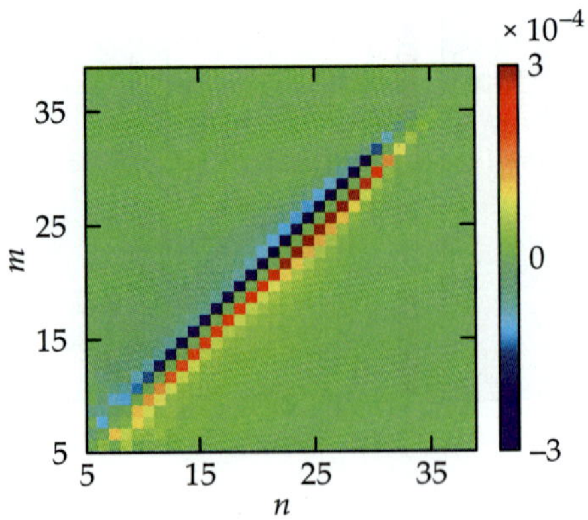

Figure 16.9